Pseudolinear
Functions
and
Optimization

Pseudolinear Functions
and
Optimization

Shashi Kant Mishra

Balendu Bhooshan Upadhyay

CRC Press
Taylor & Francis Group
Boca Raton London New York

CRC Press is an imprint of the
Taylor & Francis Group, an **informa** business

A CHAPMAN & HALL BOOK

CRC Press
Taylor & Francis Group
6000 Broken Sound Parkway NW, Suite 300
Boca Raton, FL 33487-2742

First issued in paperback 2019

ISBN-13: 978-1-4822-5573-7 (hbk)
ISBN-13: 978-0-367-37792-2 (pbk)

Library of Congress Cataloging-in-Publication Data

Mishra, Shashi Kant, 1967-
    Pseudolinear functions and optimization / Shashi Kant Mishra, Balendu Bhooshan Upadhyay.
        pages cm
    "A CRC title."
    Includes bibliographical references and index.
    ISBN 978-1-4822-5573-7 (hardcover : alk. paper) 1. Pseudoconvex domains. 2. Convex domains. 3. Calculus of variations. 4. Convex functions. I. Upadhyay, Balendu Bhooshan. II. Title.

QA331.M567 2015
515'.882--dc23   2015000793

**Visit the Taylor & Francis Web site at
http://www.taylorandfrancis.com**

**and the CRC Press Web site at
http://www.crcpress.com**

To
Our Beloved Parents
Smt. Shyama Mishra, Shri Gauri Shankar Mishra
and
Smt. Urmila Upadhyay, Shri Yadvendra D. Upadhyay

Contents

Foreword

Optimization is a discipline that plays a key role in modeling and solving problems in most areas of engineering and sciences. Optimization tools are used to understand the dynamics of information networks, financial markets, solve logistics and supply chain problems, design new drugs, solve biomedical problems, design renewable and sustainable energy systems, reduce pollution, and improve health care systems. In general, we have convex optimization models, which, at least in theory, can be solved efficiently, and nonconvex optimization models that are computationally very hard to solve. Designing efficient algorithms and heuristics for solving optimization problems requires a very good understanding of the mathematics of optimization. For example, the development of optimality conditions, the theory of convexity, and computational complexity theory have been instrumental for the development of computational optimization algorithms.

The book *Pseudolinear Functions and Optimization* by Shashi Kant Mishra and Balendu Bhooshan Upadhyay, sets the mathematical foundations for a class of optimization problems. Although convexity theory plays the most important role in optimization, several attempts have been made to extend the theory to generalized convexity. Such extensions were necessary to understand optimization models in a wide spectrum of applications.

The book *Pseudolinear Functions and Optimization* is to my knowledge the first book that is dedicated to a specific class of generalized convex functions that are called pseudolinear functions. This is an in-depth study of the mathematics of pseudolinear functions and their applications. Most of the recent results on pseudolinear functions are covered in this book.

The writing is pleasant and rigorous and the presentation of material is very clear. The book will definitely will be useful for the optimization community and will have a lasting effect.

Panos M. Pardalos
Distinguished Professor
Paul and Heidi Brown Preeminent Professor
in Industrial and Systems Engineering
University of Florida
Gainesville, USA

Preface

In 1967 Kortanek and Evans [149] studied the properties of the class of functions, which are both pseudoconvex and pseudoconcave. This class of functions were later termed as pseudolinear functions. In 1984 Chew and Choo [47] derived first and second order characterizations for pseudolinear functions. The linear and quadratic fractional functions are particular cases of pseudolinear functions. Several authors have studied pseudolinear functions and their characterizations, see Cambini and Carosi [34], Schaible and Ibaraki [246], Rapcsak [233], Komlosi [147], Kaul *et al.* [139], Lu and Zhu [171], Dinh *et al.* [67], Zhao and Tang [298], Ansari and Rezaei [4] and Mishra *et al.* [200].

Chapter 1 is introductory and contains basic definitions and concepts needed in the book.

Chapter 2, presents basic properties and characterization results on pseudolinear functions. Further, it includes semilocal pseudolinear functions, Dini differentiable pseudolinear functions, locally Lipschitz pseudolinear functions, h-pseudolinear functions, directionally differentiable pseudolinear functions, weakly pseudolinear functions and their characterizations.

Chapter 3, presents characterizations of solution sets of pseudolinear optimization problems, linear fractional optimization problems, directionally differentiable pseudolinear optimization problems, h-pseudolinear optimization problems and locally Lipschitz optimization problems.

Chapter 4, presents characterizations of solution sets in terms of Lagrange multipliers for pseudolinear optimization problems and its other generalizations given in Chapter 3.

Chapter 5, considers multiobjective pseudolinear optimization problems and multiobjective fractional pseudolinear optimization problems and presents optimality conditions and duality results for these two problems.

Chapter 6, extends the results of Chapter 5 to locally Lipschitz functions using the Clarke subdifferentials.

Chapter 7, considers static minmax pseudolinear optimization problems and static minmax fractional pseudolinear optimization problems and presents optimality conditions and duality results for these two problems.

Chapter 8, extends the results of Chapter 7 to locally Lipschitz functions using the Clarke subdifferentials.

Chapter 9, presents optimality and duality results for h-pseudolinear optimization problems.

Chapter 10, presents optimality and duality results for semi-infinite pseudolinear optimization problems.

Chapter 11, presents relationships between vector variational inequalities and vector optimization problems involving pseudolinear functions. Moreover, relationships between vector variational inequalities and vector optimization problems involving locally Lipschitz pseudolinear functions using the Clarke subdifferentials are also presented.

Chapter 12, presents an extension of pseudolinear functions are used to establish results on variational inequality problems.

Chapter 13, presents results on η-pseudolinear functions and characterizations of solution sets of η-pseudolinear optimization problems.

Chapter 14, presents pseudolinear functions on Riemannian manifolds and characterizations of solution sets of pseudolinear optimization problems on Riemannian manifolds. Moreover, η-pseudolinear functions and characterizations of solution sets of η-pseudolinear optimization problems on differentiable manifolds are also presented.

Chapter 15, presents results on pseudolinearity of quadratic fractional functions.

Chapter 16, extends the class of pseudolinear functions and η-pseudolinear functions to pseudolinear and η-pseudolinear fuzzy mappings and characterizations of solution sets of pseudolinear fuzzy optimization problems and η-pseudolinear fuzzy optimization problems.

Finally, in Chapter 17, some applications of pseudolinear optimization problems to hospital management and economics are given.

The authors are thankful to Prof. Nicolas Hadjisavvas for his help and discussion in Chapter 4. The authors are indebted to Prof. Juan Enrich Martinez-Legaz, Prof. Pierre Marechal, Prof. Dinh The Luc, Prof. Le Thi Hoai An, Prof. Sy-Ming Guu, Prof. King Keung Lai, and Prof. S.K. Neogy for their help, support, and encouragement in the course of writing this book. The authors are also thankful to Ms. Aastha Sharma from CRC Press for her patience and effort in handling the book.

Shashi Kant Mishra
Balendu Bhooshan Upadhyay

List of Figures

List of Tables

Symbol Description

$:=$	equal to by definition		derivative of f at x in the direction d
ϕ	empty set		
$\forall$	for all	$f'_-(x;d)$	left sided directional derivative of f at x in the direction d
∞	infinity		
$\langle\cdot,\cdot\rangle$	Euclidean inner product		
$\|\cdot\|$	Euclidean norm	$f^\circ(x;d)$	Clarke directional derivative of f at x in the direction d
$\exists$	there exists		
0	zero element in the vector space $\mathbb{R}^n$		
		$D^+f(x;d)$	Dini upper directional derivative of f at x in the direction d
2^X	family of all subsets of a set X		
A^T	transpose of a matrix A	$D_+f(x;d)$	Dini lower directional derivative of f at x in the direction d
$\text{lin}(A)$	linear hull of a set A		
$\text{aff}(A)$	affine hull of a set A		
$\arg\min f$	set of all minima of a function f	$f^{DH}(x;d)$	Dini-Hadamard upper directional derivative of f at x in the direction d
$\mathbb{B}_r(x)$	open ball with center at x and radius r	$f_{DH}(x;d)$	Dini-Hadamard lower directional derivative of f at x in the direction d
$\mathbb{B}_r[x]$	closed ball with center at x and radius r		
$\mathbb{B}$	open unit ball	$\partial f(x)$	subdifferential of a function f at x
$[x,y]$	closed line segment joining x and y	$\text{supp } u$	support of fuzzy set u
$]x,y[$	open line segment joining x and y	$[u]_\alpha$	$\alpha-$cut set of fuzzy set u
		$\mathfrak{I}_0$	family of fuzzy numbers
$\text{bd}(A)$	boundary of a set A	$\partial^c f(x)$	Clarke subdifferential of a function f at x
$\text{cl}(A)$	closure of a set A		
$\text{co}(A)$	convex hull of a set A	$d_C(\cdot)$	distance function of a set C
$\text{con}(A)$	conic hull of a set A		
$d(x,y)$	distance between x and y	$\text{graph}(f)$	graph of a function f
$\text{d}(f)$	domain of a function f	$\text{hyp }(f)$	hypograph of a function f
$\text{dom}(f)$	effective domain of map f		
$\text{epi}(f)$	epigraph of a function f	$\delta(\cdot|C)$	indicator function of a set C
$\nabla f(x)$	gradient of a function f at x	$X^\perp$	orthogonal complement of X
$\nabla^2 f(x)$	Hessian matrix of a function f at x	H^+	upper closed half-space
		H^{++}	upper open half-space
$f'(x;d)$	directional derivative of f at x in the direction d	H^-	lower closed half-space
$f'_+(x;d)$	right sided directional	H^{--}	lower open half-space
		$\text{int}(A)$	interior of a set A

$\mathrm{ri}(A)$	relative interior of a set A	$\mathbb{R}_+$	set of all nonnegative real numbers
$J(f)(x)$	Jacobian matrix of a function at x	$\sigma_C(.)$	support function of a set C
X^*	dual cone of the cone X	$\mathbb{R}_{++}$	set of all positive real numbers
$\Lambda(f, \alpha)$	lower level set of a function f at level α	$\mathbb{R}^n$	n-dimensional Euclidean space
$\mathbb{N}$	set of all natural numbers	$\mathbb{R}^n_+$	$\{(x_1, x_2, \ldots, x_n) \in \mathbb{R}^n : x_i \geq 0 \ \forall \ i\}$
$N_C(x)$	normal cone to a set C at x	$\mathbb{R}^n_{++}$	$\{(x_1, x_2, \ldots, x_n) \in \mathbb{R}^n : x_i > 0 \ \forall \ i\}$
0^+X	recession cone of a set X	$\Omega(f, \alpha)$	upper level set of a function f at level α
$\mathbb{R}$	set of all real numbers		
$\overline{\mathbb{R}}$	$\mathbb{R} \cup \{+\infty, -\infty\}$		

Chapter 1

Basic Concepts in Convex Analysis

1.1 Introduction

Optimization is everywhere, as nothing at all takes place in the universe, in which some rule of maximum or minimum does not appear. It is the human nature to seek for the best among the available alternatives. An optimization problem is characterized by its specific objective function that is to be maximized or minimized, depending upon the problem and, in the case of a constrained problem, a given set of constraints. Possible objective functions include expressions representing profits, costs, market share, portfolio risk, etc. Possible constraints include those that represent limited budgets or resources, nonnegativity constraints on the variables, conservation equations, etc.

The concept of convexity is of great importance in the study of optimization problems. It extends the validity of a local solution of a minimization problem to global one and the first order necessary optimality conditions become sufficient for a point to be a global minimizer. We mention the earlier work of Jensen [112], Fenchel [80, 81] and Rockafellar [238]. However, in several real-world applications, the notion of convexity does no longer suffice. In many cases, the nonconvex functions provide more accurate representation of reality. Nonconvex functions preserve one or more properties of convex functions and give rise to models which are more adaptable to the real-world situations, than convex models. This led to the introduction of several generalizations of the classical notion of convexity.

In 1949, the Italian mathematician Bruno de Finetti [62] introduced one of the fundamental generalized convex functions, known as quasiconvex function having wider applications in economics, management sciences and engineering. Mangasarian [176] introduced the notion of pseudoconvex and pseudoconcave functions as generalizations of convex and concave functions, respectively. In 1969, Kortanek and Evans [149] studied the properties of a class of functions, which are both pseudoconvex and pseudoconcave, later termed as pseudolinear functions. In case of stationary points the behavior of the class of pseudolinear functions is as good as linear functions. Several other generalizations of these functions have been introduced to find weakest conditions in order to establish sufficient optimality conditions and duality results for optimization problems.

The main aim of this chapter is to explore the properties of convex sets and convex functions. We give some basic definitions and preliminary results from algebra, geometry and topology, that will be used throughout the book to develop some important results. Taking into consideration that, a function f is concave if and only if $-f$ is convex, the proofs of the results are provided just for convex functions.

1.2   Basic Definitions and Preliminaries

Throughout the book, suppose that $\mathbb{R}^n$ be the n-dimensional Euclidean space. Let $\mathbb{R}^n_+$ and $\mathbb{R}^n_{++}$ denote the nonnegative and positive orthants of $\mathbb{R}^n$, respectively. Let $\mathbb{R}$ denote the real number system and $\overline{\mathbb{R}} = \mathbb{R} \bigcup \{\pm\infty\}$ be an extended real line. The number 0 will denote either the real number zero or the zero vector in $\mathbb{R}^n$, all components of which are zero.

If $x = (x_1, \ldots, x_n), y = (y_1, \ldots, y_n) \in \mathbb{R}^n$, then, the following convention for equalities and inequalities will be adopted:

$$x = y \Leftrightarrow x_i = y_i, \forall i = 1, \ldots, n;$$

$$x < y \Leftrightarrow x_i < y_i, \forall i = 1, \ldots, n;$$

$$x \leqq y \Leftrightarrow x_i \leqq y_i, \forall i = 1, \ldots, n;$$

$$x \leq y \Leftrightarrow x_i \leqq y_i, \forall i = 1, \ldots, n, \text{ but } x \neq y.$$

The following rules are natural extensions of the rules of arithmetic: For every $x \in \mathbb{R}$,

$$x + \infty = \infty \text{ and } x - \infty = -\infty;$$

$$x \times \infty = \infty, \text{ if } x > 0; x \times \infty = -\infty \text{ if } x < 0.$$

For every extended real number x,

$$x \times 0 = 0.$$

The expression $\infty - \infty$ is meaningless.

Let $\langle ., . \rangle$ denote the Euclidean inner product, that is, for $x, y \in \mathbb{R}^n$

$$\langle x, y \rangle = x^T y = x_1 y_1 + \ldots + x_n y_n.$$

Norm of any point $x \in \mathbb{R}^n$ is given by

$$\|x\| = \sqrt{\langle x, x \rangle}.$$

The symbol $\|.\|$ will denote the Euclidean norm, unless otherwise specified.

The following proposition illustrates that any two norms on $\mathbb{R}^n$ are equivalent:

Proposition 1.1 *If $\|.\|_1$ and $\|.\|_2$ are any two norms on $\mathbb{R}^n$, then there exist constants $c_1 \geq c_2 > 0$, such that*

$$c_1 \|x\|_1 \leq \|x\|_2 \leq c_2 \|x\|_1, \forall x \in \mathbb{R}^n.$$

The Euclidean distance between two points $x, y \in \mathbb{R}^n$, is given by

$$d(x, y) := \|x - y\| = \langle x - y, x - y \rangle^{\frac{1}{2}}.$$

The closed and open balls with center at any point $\bar{x}$ and radius ε, denoted by $\mathbb{B}_\varepsilon[\bar{x}]$ and $\mathbb{B}_\varepsilon(\bar{x})$, respectively are defined as follows:

$$\mathbb{B}_\varepsilon[\bar{x}] := \{x \in \mathbb{R}^n : \|x - \bar{x}\| \leq \varepsilon\}$$

and

$$\mathbb{B}_\varepsilon(\bar{x}) := \{x \in \mathbb{R}^n : \|x - \bar{x}\| < \varepsilon\}.$$

Moreover, the open unit ball is the set

$$\mathbb{B} := \{x \in \mathbb{R}^n : \|x\| \leq 1\}.$$

Definition 1.1 (Lower and upper bound) *Let $X \subseteq \mathbb{R}$ be a nonempty set. Then a number $\alpha \in \mathbb{R}$, is called a lower bound of X, if*

$$x \geq \alpha, \forall x \in X.$$

A number $\beta \in \mathbb{R}$, is called an upper bound of X, if

$$x \leq \beta, \forall x \in X.$$

A set $X \subseteq \mathbb{R}$ is said to be bounded if lower and upper bound of the set exists.

Definition 1.2 (Supremum and infimum of set of real numbers) *Let $X \subseteq \mathbb{R}^n$ be a nonempty set. Then a number M is called the least upper bound or the supremum of the set X, if*

(i) M is the upper bound of the set, i.e.,

$$x \leq M, \forall x \in X.$$

(ii) No number less than M can be an upper bound of X, i.e., for every $\epsilon > 0$ however small, there exists a number $y \in X$ such that

$$y > M - \epsilon;$$

A number m is called the greatest lower bound or the infimum of set X, if

(i) m is the lower bound of the set, i.e.,

$$x \geq m, \forall x \in X.$$

(ii) No number greater than m can be a lower bound of X, i.e., for every $\epsilon > 0$, however small, there exists a number $z \in X$, such that

$$z < m + \epsilon.$$

By the **completeness axiom** it is known that every nonempty set $X \subseteq \mathbb{R}$, which has a lower (upper) bound has a greatest (least) lower (upper) bound in $\mathbb{R}$.

Definition 1.3 *(ε-Neighborhood of a point) Let $\bar{x} \in \mathbb{R}^n$ and $\varepsilon > 0$ be given. Then, a subset N of $\mathbb{R}^n$ is said to be an ε-neighborhood of $\bar{x}$, if there exists an open ball $\mathbb{B}_\varepsilon(\bar{x})$, with center at $\bar{x}$ and radius ε, such that*

$$\bar{x} \in \mathbb{B}_\varepsilon(\bar{x}) \subseteq N.$$

Definition 1.4 *(Interior of a set) A point $\bar{x} \in X \subseteq \mathbb{R}^n$ is said to be an interior point of X, if there exists a neighborhood N of $\bar{x}$ contained in X, i.e.,*

$$\bar{x} \in N \subseteq X.$$

The set of all interior points of X is called the interior of X and is denoted by $\text{int}(X)$. *For example, interior of $\mathbb{R}_+^n$ is the set*

$$\mathbb{R}_{++}^n := \{x \in \mathbb{R}^n : x_i > 0, \forall i = 1, ..., n\}.$$

Definition 1.5 *(Open set) Any set $X \subseteq \mathbb{R}^n$ is said to be open, if every point of X is an interior point and conversely. In other words, $X = \text{int}(X)$. For example, every open ball $\mathbb{B}_\varepsilon(x)$ is an open set. It is easy to see that $\text{int}(X)$ is the largest open set contained in X.*

Definition 1.6 *(Closure of a set) A point $\bar{x} \in \mathbb{R}^n$ is said to be a point of closure of the set $X \subseteq \mathbb{R}^n$, denoted by $\text{cl}(X)$, if for each $\varepsilon > 0$,*

$$\mathbb{B}_\varepsilon(\bar{x}) \bigcap X \neq \emptyset.$$

In other words, $\bar{x} \in \text{cl}(X)$, if and only if every ball around $\bar{x}$ contains at least one point of X.

Definition 1.7 *(Closed set) A set $X \subseteq \mathbb{R}^n$ is said to be a closed set if every point of closure of X is in X. In other words, $X = \text{cl}(X)$. For example, every closed ball $\mathbb{B}_\varepsilon[\bar{x}]$ is a closed set.*

Definition 1.8 *(Relatively open and closed) Let X and Y be two nonempty subsets of $\mathbb{R}^n$, such that $X \subseteq Y \subseteq \mathbb{R}^n$. Then, X is said to be relatively open with respect to Y, if*

$$X = Y \bigcap \Omega,$$

where Ω is some open set in $\mathbb{R}^n$.

Furthermore, the set X is said to be relatively closed with respect to Y, if

$$X = Y \bigcap \Lambda,$$

where Λ is some closed set in $\mathbb{R}^n$.

Definition 1.9 (*Boundary of a set*) *Let $X \subseteq \mathbb{R}^n$ be a nonempty set. Any point $\bar{x} \in \mathbb{R}^n$ is said to be a point of boundary of the set X, if for each $\varepsilon > 0$, open ball $\mathbb{B}_\varepsilon(\bar{x})$ contains points of X as well as points not belonging to X.*

The set of all boundary points of X is called the boundary of X and it is denoted by $bd(X)$. Let $X \subseteq \mathbb{R}^n$ be a nonempty set equipped with the Euclidean norm $\|.\|$.

Definition 1.10 (*Angle between two vectors*) *Let x and y be two nonzero vectors in $\mathbb{R}^n$. Then the angle θ between x and y, is given by*

$$\cos\theta = \frac{\langle x, y \rangle}{\|x\| \|y\|}, \ \ 0 \le \theta \le \pi.$$

The nonzero vector x and y are said to

(i) *be orthogonal if $\langle x, y \rangle = 0$, that is, if $\theta = \frac{\pi}{2}$;*

(ii) *form an acute (strictly acute) angle if*

$$\langle x, y \rangle \ge 0, \ 0 \le \theta \le \frac{\pi}{2} \left(\langle x, y \rangle > 0, \left(0 < \theta < \frac{\pi}{2} \right) \right);$$

(iii) *form an obtuse (strictly obtuse) angle if $\langle x, y \rangle \le 0$, $\frac{\pi}{2} \le \theta \le \pi \left(\langle x, y \rangle < 0, \left(\frac{\pi}{2} < \theta < \pi \right) \right)$.*

Given a subspace X of $\mathbb{R}^n$, the orthogonal complement of X, denoted by $X^\perp$, is defined as

$$X^\perp := \{ x \in \mathbb{R}^n : \langle x, y \rangle = 0, \ \forall \ y \in \mathbb{R}^n \}.$$

It is easy to see that $X^\perp$ is another subspace of $\mathbb{R}^n$ and $dim X + dim X^\perp = n$. Furthermore, any vector x can be uniquely expressed as the sum of a vector from X and a vector from $X^\perp$.

Proposition 1.2 (*Pythagorean theorem*) *For any two orthogonal vectors x and y, in $\mathbb{R}^n$, we have*

$$\|x + y\|^2 = \|x\|^2 + \|y\|^2.$$

Proposition 1.3 (*Schwarz inequality*) *Let any two vectors $x, y \in \mathbb{R}^n$. Then,*

$$|\langle x, y \rangle| \le \|x\| \|y\|.$$

The above inequality holds as equality, if and only if $x = \lambda y$, for some $\lambda \in \mathbb{R}$.

1.3   Matrices

A matrix is a rectangular array of numbers called elements. Any matrix A with m rows and n columns is called a $m \times n$ matrix and is written as:

$$A = \begin{bmatrix} a_{11} & a_{12} & \cdots & a_{1n} \\ \vdots & \vdots & \vdots & \vdots \\ a_{m1} & a_{m2} & \cdots & a_{mn} \end{bmatrix}.$$

The ijth element of A is denoted by a_{ij}. The transpose of A, denoted by A^T is defined as $A^T = [a_{ij}]^T = [a_{ji}]$. A matrix A is said to be nonvacuous, if it contains at least one element.

The ith row of the matrix A is denoted by A_i and is given by

$$A_i = (a_{i1}, \ldots, a_{in}), i = 1, \ldots, m.$$

The jth column of the matrix A is denoted by $A_{.j}$ and is given by

$$A_{.j} = \begin{bmatrix} a_{1j} \\ \vdots \\ a_{mj} \end{bmatrix}, j = 1, \ldots, n.$$

The rank of a matrix is equal to the number of linearly independent columns of A, which is also equal to the maximum number of linearly independent rows of A. The matrix A and transpose of A, that is A^T have the same rank. The matrix A is said to have full rank, if

$$rank(A) = \min\{m, n\}.$$

In other words, for a matrix A having full rank either all the rows are linearly independent or all the columns of A are linearly independent.

Let A be a square matrix of order $n \times n$. Then A is said to be symmetric if $A^T = A$. The square matrix A is said to be diagonal, if $a_{ij} = 0$, $for\ i \neq j$. The diagonal matrix A is said to be identity, if $a_{ij} = 1$, for $i = j$, i.e. diagonal elements are equal to 1.

Definition 1.11 *Let A be a square matrix of order $n \times n$. Then A is said to be*

(i) Nonsingular, if

$$rank\ (A) = n.$$

(ii) Positive semidefinite matrix, if

$$\langle x, Ax \rangle \geq 0, \forall x \in \mathbb{R}^n.$$

(iii) Negative semidefinite, if

$$\langle x, Ax \rangle \le 0, \forall x \in \mathbb{R}^n.$$

Definition 1.12 *Let A be a square matrix of order $n \times n$. Then, A is said to be*

(i) Positive definite matrix, if

$$\langle x, Ax \rangle > 0, \forall x \in \mathbb{R}^n, x \ne 0.$$

(ii) Negative definite matrix, if

$$\langle x, Ax \rangle < 0, \forall x \in \mathbb{R}^n, x \ne 0.$$

Remark 1.1 *It is clear by the definitions, that every positive (negative) definite matrix is positive (negative) semidefinite matrix. Furthermore, the negative of a positive definite (positive semidefinite) matrix is negative definite (negative semidefinite) and vice versa. It is easy to see that each positive (negative) definite matrix is nonsingular.*

Next, we state the following result from Bertsekas *et al.* [23].

Proposition 1.4 *Let A and B be two square matrices of order $n \times n$. Then,*

(i) The matrix A is symmetric and positive definite if and only if it is invertible and its inverse is symmetric and positive definite.

(ii) If A and B are symmetric positive semidefinite matrices, then, their sum $A + B$ is positive semidefinite matrix. In addition, if A or B is positive definite, then $A + B$ is positive definite matrix.

(iii) If A is symmetric positive semidefinite matrix and X be any $m \times n$ matrix, then, the matrix XAX^T is positive semidefinite matrix. If A is positive definite matrix and X be invertible, then, XAX^T is positive definite matrix.

(iv) If A is symmetric positive definite matrix, then there exists positive real numbers α and β, such that

$$\alpha \parallel x \parallel^2 \le \langle x, Ax \rangle \le \beta \parallel x \parallel^2 .$$

(v) If A is symmetric positive definite matrix, then there exists unique symmetric positive definite matrix that yields A, when multiplied with itself. This matrix is called square root of A and is denoted by $A^{1/2}$ and its inverse is denoted by $A^{-1/2}$.

Definition 1.13 *(Mapping) Let X and Y be two sets. Then a correspondence $f : X \to Y$ which associates to each $x \in X$, a subset of Y is called a mapping. For each $x \in X$ the set $f(x)$ is called image of x. The subset of points of X, for which the image $f(x)$ is nonempty is called the domain of f, i.e., the set*

$$d(f) = \{x \in X : f(x) \neq \phi\}$$

is called domain of f. The union of image of points of $d(f)$ is called range of f, denoted by $f(d(f))$, i.e.,

$$f(d(f)) = \bigcup_{x \in d(f)} f(x).$$

Definition 1.14 *(Function) Let X and Y be two sets. A single valued mapping $f : X \to Y$ is called a function. In other words, for each $x \in X$, the image set $f(x)$ consists of a single element of Y. The domain of f is X and the range of f is*

$$f(x) = \bigcup_{x \in X} f(x).$$

Definition 1.15 *(**Affine and linear functions**) A function $f : \mathbb{R}^n \to \mathbb{R}$ is said to be affine, if it has the form*

$$f(x) = \langle a, x \rangle + b,$$

where $a, x \in \mathbb{R}^n$ and $b \in \mathbb{R}$.
Similarly, a function $f : \mathbb{R}^n \to \mathbb{R}^m$ is called affine, if it has the form

$$f(x) = \langle A, x \rangle + b,$$

where A is any $m \times n$ matrix, $x \in \mathbb{R}^n$ and $b \in \mathbb{R}^m$. If $b = 0$, then f is said to be a linear function or a linear transformation.

In optimization problems, we often encounter the objective functions, which can take values on an extended real line, referred to as an extended real-valued function.

Definition 1.16 *(**Effective domain**) Let $X \subseteq \mathbb{R}^n$ be a nonempty set and $f : X \to \overline{\mathbb{R}}$ be an extended real-valued function. The effective domain of f, denoted by $\mathrm{dom}(f)$, is defined by*

$$\mathrm{dom}(f) := \{x \in X : f(x) < \infty\}.$$

Definition 1.17 *Let $X \subseteq \mathbb{R}^n$ be a nonempty set and $f : X \to \overline{\mathbb{R}}$ be an extended real-valued function.*

(i) The graph of f is defined by

$$graph(f) := \{(x, \alpha) : x \in \mathbb{R}^n, \alpha \in \mathbb{R} : f(x) = \alpha\}.$$

(ii) *The epigraph of the function f, denoted by* epi(f), *is a subset of* $\mathbb{R}^{n+1}$ *and is defined as follows*

$$\text{epi}(f) := \{(x, \alpha) : x \in X, \alpha \in \mathbb{R} : f(x) \le \alpha\}.$$

(iii) *The hypograph of the function f, denoted by* hyp(f), *is a subset of* $\mathbb{R}^{n+1}$ *and is defined as follows*

$$\text{hyp}(f) := \{(x, \alpha) : x \in X, \alpha \in \mathbb{R} : f(x) \ge \alpha\}.$$

For the epigraph of a function f, see Figure 1.1.

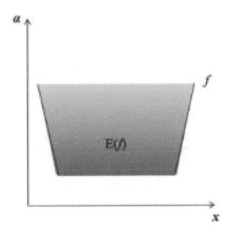

FIGURE 1.1: Epigraph of a Function f

Definition 1.18 *Let* $X \subseteq \mathbb{R}^n$ *be a nonempty set and* $f : X \to \overline{\mathbb{R}}$ *be an extended real-valued function.*

(i) *Lower level set at the level* α, *is defined as*

$$\Lambda(f, \alpha) := \{x \in \mathbb{R}^n : f(x) \le \alpha\}.$$

(ii) *Upper level set at the level* α, *is defined as*

$$\Omega(f, \alpha) := \{x \in \mathbb{R}^n : f(x) \ge \alpha\}.$$

It can be seen that

$$\text{dom}(f) := \{x \in X : \exists\, y \in \mathbb{R} \text{ such that } (x, y) \in epi(f)\}.$$

In other words, dom(f) *is the projection of epigraph of f on* $\mathbb{R}^n$.

Definition 1.19 (Proper function) *Let $f : X \to \overline{\mathbb{R}}$ be an extended real-valued function. The function f is said to be proper, if $f(x) < \infty$, for at least one $x \in X$ and $f(x) > -\infty$, for all $x \in X$. In other words, a function f is proper if* $\mathrm{dom}(f)$ *is nonempty and f is finite valued on* $\mathrm{dom}(f)$.

Definition 1.20 (Sequence) *A sequence on a set $X \subseteq \mathbb{R}^n$ is a function f from the set $\mathbb{N}$ of natural number to the set X. If $f(n) = x_n \in X$, for $n \in N$, then the sequence, f is denoted by the symbol $\{x_n\}$ or by $x_1, x_2, \ldots$.*

Definition 1.21 (Bounded sequence) *Let $\{x_n\}$ be any sequence in $\mathbb{R}^n$. Then, $\{x_n\}$ is said to be bounded above, if there exists $\bar{x} \in \mathbb{R}^n$, such that*

$$x_n \leq \bar{x}, \forall n \in \mathbb{N}.$$

The sequence $\{x_n\}$ is said to be bounded below, if there exists $\hat{x} \in \mathbb{R}^n$, such that

$$x_n \geq \hat{x}, \forall n \in \mathbb{N}.$$

A sequence $\{x_n\}$ is said to be bounded, if it is bounded above as well as bounded below.

Definition 1.22 (Monotonic sequence) *Let $\{x_n\}$ be any sequence in $\mathbb{R}^n$. Then, $\{x_n\}$ is said to be monotonically increasing (respectively, decreasing), if*

$$x_n \leq x_{n+1}, \forall n$$

(respectively, $x_n \geq x_{n+1}, \forall n$).

The sequence $\{x_n\}$ is said to be a monotonic sequence, if it is either monotonically increasing or monotonically decreasing.

Definition 1.23 (Limit point) *Let $\{x_n\}$ be any sequence in $\mathbb{R}^n$. Then, a number $\bar{x} \in \mathbb{R}^n$ is said to be the limit point (accumulation point) of the sequence $\{x_n\}$, if for given $\varepsilon > 0$, one has*

$$\| x_n - \bar{x} \| < \varepsilon,$$

for infinitely many values of n.

Definition 1.24 (Limit) *Let $\{x_n\}$ be any sequence in $\mathbb{R}^n$. Then a number $\bar{x} \in \mathbb{R}^n$, is said to be limit of the sequence, if for each $\varepsilon > 0$, there exists a positive integer m (depending on ε), such that*

$$\| x_n - \bar{x} \| < \varepsilon, \forall n \geq m.$$

Moreover, the sequence $\{x_n\}$ is said to converge to $\bar{x}$, and is denoted by $x_n \to \bar{x}$, whenever $n \to \infty$ or $\lim_{n \to \infty} x_n \to \bar{x}$.

Remark 1.2 *It is obvious, that a limit of a sequence is also limit point of the sequence, but not conversely. For example, the sequence $\{1, -1, 1, -1, \ldots, \}$ has limit points 1 and -1 but having no limit.*

Definition 1.25 (Subsequence) *Let $\{x_n\}$ be any sequence. If there exists a sequence of positive integers $n_1, n_2, \ldots$, such that $n_1 \leq n_2 \leq \ldots$, then, the sequence $\{x_{n_i}\}$ is called a subsequence of the sequence $\{x_n\}$.*

It is known that if $\bar{x}$ is a limit of a sequence $\{x_n\}$, then, it is limit of every subsequence of $\{x_n\}$.

Now, we state the following classical result on bounded sequences.

Proposition 1.5 (Bolzano-Weierstrass theorem) *Every bounded sequence in $\mathbb{R}^n$, has a convergent subsequence*

Now, we have the following important results.

Proposition 1.6 *Every bounded and monotonically nondecreasing (nonincreasing) sequence in $\mathbb{R}^n$, has a limit.*

Proposition 1.7 (Cauchy convergence criteria) *A sequence $\{x_n\}$ in $\mathbb{R}^n$ converges to a limit, if and only if for each $\varepsilon > 0$, there exists a positive integer p, such that*

$$\| x_m - x_n \| < \varepsilon, \forall m, n \geq p.$$

The following notions of limit infimum and limit supremum of sequences in $\mathbb{R}$ are important to understand the concept of continuity of real-valued functions. Let $\{x_n\}$ be any sequence in $\mathbb{R}$. Define

$$y_k := \inf\{x_n : n \geq k\} \text{ and } z_k := \sup\{x_n : n \geq k\}.$$

By the definitions, it is clear that $\{y_k\}$ and $\{z_k\}$ are nondecreasing and nonincreasing sequences, respectively. If the sequence $\{x_n\}$ is a bounded below or above, then by Proposition 1.6, the sequences $\{y_k\}$ and $\{z_k\}$ respectively, have a limit. The limit of $\{y_k\}$ is called lower limit or limit infimum of $\{x_n\}$ and is denoted by $\lim_{n \to \infty} \inf \{x_n\}$, while the limit of $\{z_k\}$ is called the upper limit or limit supremum of $\{x_n\}$ and is denoted by $\lim_{n \to \infty} \sup \{x_n\}$.

If a sequence $\{x_n\}$ is unbounded below, then we write $\lim_{n \to \infty} \inf \{x_n\} = -\infty$, while if the sequence $\{x_n\}$ is unbounded above, then we write $\lim_{n \to \infty} \sup \{x_n\} = \infty$.

Definition 1.26 (Compact set) *A set $X \subseteq \mathbb{R}^n$ is said to be a compact set, if it satisfies, any one of the following equivalent conditions:*

(i) X is closed and bounded.

(ii) **(Bolzano-Weierstrass property)** *Every sequence of points in X, has a limit point in X.*

(iii) **(Finite intersection property)** *For any family* $\{X_i, i \in I\}$ *of sets closed relative to X, if*

$$\bigcap_{i \in I} X_i = \phi \Rightarrow X_{i_1} \bigcap \cdots \bigcap X_{I_k} = \phi, \text{ for some } i_1, \ldots, i_k \in I.$$

(iv) **(Heine-Borel property)** *Every open cover of X admits a finite subcover. In other words for every family* $\{X_i, i \in I\}$ *of open sets such that* $\bigcup_{i \in I} X_i \subseteq X$, *there exists a finite subfamily* $\{X_{i_1}, \ldots, X_{i_k}\}$, *such that*

$$X_{i_1} \bigcup \cdots \bigcup X_{i_k} \subseteq X.$$

(v) *Every sequence of points in X, has a subsequence, that converges to a point of X.*

Now, we state the following proposition, that will be used frequently.

Proposition 1.8 *Let* $\{x_n\}$ *and* $\{y_n\}$ *be any two sequences in* $\mathbb{R}$. *Then,*

(i) *We have*

$$\sup\{x_n : n \geq 0\} \geq \limsup_{n \to \infty} x_n \geq \liminf_{n \to \infty} x_n \geq \{x_n : n \geq 0\}.$$

(ii) *The sequence* $\{x_n\}$ *converges if and only if*

$$\infty > \limsup_{n \to \infty} x_n = \liminf_{n \to \infty} x_n > \infty.$$

The common value of $\limsup_{n \to \infty} x_n$ *and* $\liminf_{n \to \infty} x_n$ *is the limit of the sequence* $\{x_n\}$.

(iii) *If* $x_n \leq y_n$, *for all n, then*

$$\limsup_{n \to \infty} x_n \leq \limsup_{n \to \infty} y_n$$

and

$$\liminf_{n \to \infty} x_n \leq \liminf_{n \to \infty} y_n.$$

(iv) *We have*

$$\limsup_{n \to \infty}(x_n + y_n) \leq \limsup_{n \to \infty} x_n + \limsup_{n \to \infty} y_n$$

and

$$\liminf_{n \to \infty}(x_n + y_n) \geq \liminf_{n \to \infty} x_n + \liminf_{n \to \infty} y_n.$$

Definition 1.27 (*Limit infimum and limit supremum of extended real-valued functions*) *Let* $f : \mathbb{R}^n \to \overline{\mathbb{R}}$ *be an extended real-valued function, then we define*

$$\liminf_{y \to x} = \sup_{\varepsilon > 0} \inf_{y \in \mathbb{B}_\varepsilon(x)} f(y)$$

and

$$\limsup_{y \to x} = \inf_{\varepsilon > 0} \sup_{y \in \mathbb{B}_\varepsilon(x)} f(y).$$

It is clear that

$$\liminf_{y \to x}(-f(y)) = -\limsup_{y \to x}(f(y))$$

and

$$\limsup_{y \to x}(-f(y)) = -\liminf_{y \to x}(f(y)).$$

Moreover, $\lim_{y \to x} f(y)$ *is said to exist, if*

$$\liminf_{y \to x}(f(y)) = \limsup_{y \to x}(f(y)).$$

The following results from Rockafellar and Wets [239], present characterizations for limit infimum and limit supremum, of an extended real-valued function.

Proposition 1.9 *Let* $f : \mathbb{R}^n \to \overline{\mathbb{R}}$ *be an extended real-valued function, then*

$$\liminf_{y \to x} = \min\{\alpha \in \overline{\mathbb{R}} : \exists x_n \to \bar{x} \ with \ f(x_n) \to \alpha\}$$

and

$$\limsup_{y \to x} = \max\{\beta \in \overline{\mathbb{R}} : \exists x_n \to \bar{x} \ with \ f(x_n) \to \beta\}.$$

Next, we state the following results for infimum and supremum operations from Rockafellar and Wets [239], that will be used in sequel.

Proposition 1.10 *Let* $f : \mathbb{R}^n \to \overline{\mathbb{R}}$ *and* $g : \mathbb{R}^n \to \overline{\mathbb{R}}$ *be extended real-valued functions and let* Ω, Ω_1 *and* Ω_2 *be subsets of* $\mathbb{R}^n$. *Then,*

(i) For $\alpha \geq 0$, *we have*

$$\inf_{x \in \Omega}(\alpha f(x)) = \alpha \inf_{x \in \Omega}(f(x))$$

and

$$\sup_{x \in \Omega}(\alpha f(x)) = \alpha \sup_{x \in \Omega}(f(x)).$$

(ii) *We have*

$$\sup_{x \in \Omega} f(x) + \sup_{x \in \Omega} g(x) \geq \sup_{x \in \Omega}(f(x) + g(x)) \geq \inf_{x \in \Omega}(f(x) + g(x))$$

$$\geq \inf_{x \in \Omega} f(x) + \inf_{x \in \Omega} g(x).$$

(iii) *If $\Omega_1 \subseteq \Omega_2$, then,*

$$\sup_{x_1 \in \Omega_1} f(x_1) \leq \sup_{x_2 \in \Omega_2} f(x_2)$$

and

$$\inf_{x_1 \in \Omega_1} f(x_1) \geq \inf_{x_2 \in \Omega_2} f(x_2).$$

Definition 1.28 *Let $X \subseteq \mathbb{R}^n$ be a nonempty set. Then, a function $f : X \to \overline{\mathbb{R}}$ is said to be*

(i) **Positively homogeneous**, *if for all $x \in X$ and all $r \geq 0$, we have*

$$f(rx) = rf(x).$$

It is clear that positive homogeneity is equivalent to the epigraph of f being a cone on $\mathbb{R}^{n+1}$.

(ii) **Subadditive**, *if for all $x, y \in X$, we have*

$$f(x + y) \leq f(x) + f(y).$$

(iii) **Sublinear**, *if it is a positively homogeneous and subadditive function.*

(iv) **Subodd**, *if for all $x \in \mathbb{R}^n / \{0\}$, we have*

$$f(x) \geq -f(-x)$$

or equivalently, if $f(x) + f(-x) \geq 0$. The function f is said to be odd, if f and $-f$ are both subodd.

Definition 1.29 *(Continuous function) Let $X \subseteq \mathbb{R}^n$ be a nonempty set and let $f : X \to \mathbb{R}$ be a real-valued function. Then, f is said to be continuous at $\bar{x} \in X$, if either of the following two conditions are satisfied*

(i) *For given $\varepsilon > 0$ there exists $\delta > 0$, such that for each $x \in X$, one has*

$$|f(x) - f(\bar{x})| < \varepsilon, \text{ whenever } \|x - \bar{x}\| < \delta.$$

(ii) *For each sequence $\{x_n\} \in X$ converging to $\bar{x}$, one has*

$$\lim_{n \to \infty} f(x_n) = f(\lim_{n \to \infty} x_n) = f(\bar{x}).$$

The function f is said to be continuous on X, if it is continuous at each point $\bar{x} \in X$.

Proposition 1.11 *The function $f : X \to \mathbb{R}$ is said to be continuous on X, if the following equivalent conditions are satisfied:*

(i) *The sets $\{x : x \in X, f(x) \geq \alpha\}$ and $\{x : x \in X, f(x) \geq \alpha\}$ are closed relative to X, for each real number α.*

(ii) *The sets $\{x : x \in X, f(x) < \beta\}$ and $\{x : x \in X, f(x) > \beta\}$ are open relative to X, for each real number β.*

(iii) *The epigraph of*

$$\text{epi}(f) = \{(x, \alpha) : x \in X, \alpha \in \mathbb{R}, f(x) \leq \alpha\}$$

and the hypograph of

$$\text{hyp}(f) = \{(x, \alpha) : x \in X, \alpha \in \mathbb{R}, f(x) \geq \alpha\}$$

are closed relative to $X \times \mathbb{R}$.

Now, we present the following theorem from Bazaraa *et al.* [17].

Theorem 1.1 (***Weierstrass theorem***) *Let $X \subseteq \mathbb{R}^n$ be a nonempty compact set and let $f : X \to \mathbb{R}$ be continuous on X, then, the problem*

$$\min\{f(x) : x \in X\}$$

attains its minimum.

Proof Since, X is both nonempty closed and bounded and f is continuous on X, therefore, f is bounded below on X. Moreover, there exists a greatest lower bound $\alpha = \inf\{f(x) : x \in X\}$. Let $0 < \varepsilon < 1$ and for each, $k = 1, 2, \ldots$, consider the set

$$X_k = \{x \in X : \alpha \leq f(x) \leq \alpha + \varepsilon^k\}.$$

By the definition of an infimum, the set $X_k \neq \phi$, for each k. Therefore, by selecting a point $x_k \in X_k$, for each $k = 1, 2, \ldots$, we may construct a sequence $\{x_k\} \subseteq X$. Since, X is bounded, there exists a convergent subsequence $\{x_k\}_K \to \bar{x}$ indexed by the set K. Since, X is closed, $\bar{x} \in X$. Since,

$$\alpha \leq f(x) \leq \alpha + \varepsilon^k, \forall k.$$

By continuity of f, we get

$$\alpha = \lim_{k \to \infty, k \in K} f(x_k) = f(\bar{x}).$$

Thus, we have established that there exists a solution $\bar{x}$, such that

$$f(\bar{x}) = \alpha = \inf\{f(x); x \in X\}.$$

Hence, $\bar{x}$ is a minimizing solution. This completes the proof.

Definition 1.30 (*Lower semicontinuous function*) *Let $X \subseteq \mathbb{R}^n$ be a nonempty set. Then a function $f : X \to \mathbb{R}$ is said to be lower semicontinuous at $\bar{x} \in X$, if either of the following two conditions is satisfied:*

(i) For each $\varepsilon > 0$, there exists $\delta > 0$, such that,

$$-\varepsilon < f(x) - f(\bar{x}), \text{ whenever } |x - \bar{x}| < \delta.$$

(ii) For each sequence $\{x_n\}$ in X, converging to $\bar{x}$,

$$\liminf_{n \to \infty} f(x_n) \geq f(\lim_{n \to \infty} x_n) = f(\bar{x}),$$

where $\liminf_{n \to \infty} f(x_n)$ denotes the infimum of the limit point of the sequence of real numbers $f(x^1), f(x^2), \ldots$

Example 1.1 *For example, the function $f : X = \mathbb{R} \to \mathbb{R}$, given by*

$$f(x) = \begin{cases} x, & \text{if } x \neq 1, \\ \frac{1}{3}, & \text{if } x = 1, \end{cases}$$

is lower semicontinuous on $\mathbb{R}$.

The following theorem relates the closedness of the epigraph and lower level sets with the lower semicontinuity of a function f. A simple proof can be found in Bertsekas *et al.* [23].

Theorem 1.2 *The function $f : X \to \mathbb{R}$ is lower semicontinuous on X, if the following equivalent conditions are satisfied:*

(i) The set $\{x : x \in X, f(x) \leq \alpha\}$ is closed relative to X, for each real number α.

(ii) The set $\{x : x \in X, f(x) > \beta\}$ is open relative to X, for each real number β.

(iii) The epigraph of f, i.e.,

$$\text{epi}(f) = \{(x, \alpha) : x \in X, \alpha \in \mathbb{R}, f(x) \leq \alpha\}$$

is closed relative to $X \times \mathbb{R}$.

Definition 1.31 (*Upper semicontinuous function*) *Let $X \subseteq \mathbb{R}^n$ be a nonempty set. Then a function $f : X \to \mathbb{R}$ is said to be upper semicontinuous at $\bar{x} \in X$, if either of the following two conditions are satisfied:*

(i) For each $\varepsilon > 0$, there exists $\delta > 0$, such that for each $x \in X$,

$$f(x) - f(\bar{x}) < \varepsilon, \text{ whenever } |x - \bar{x}| < \delta.$$

(ii) For each sequence $\{x_n\}$ in X, converging to $\bar{x}$,

$$\limsup_{n\to\infty} f(x_n) \le f(\lim_{n\to\infty} x_n) = f(\bar{x}),$$

where $\limsup\limits_{n\to\infty} f(x_n)$ *denotes the supremum of the limit point of the sequence of real numbers $f(x^1), f(x^2), \ldots$.*

Example 1.2 *For example, the function $f : X = \mathbb{R} \to \mathbb{R}$, given by*

$$f(x) = \begin{cases} x^2, & \text{if } x \ne 0, \\ \frac{1}{2}, & \text{if } x = 0 \end{cases}$$

is upper semicontinuous on $\mathbb{R}$.

The following theorem relates the closedness of the hypograph and upper level sets with the upper semicontinuity of a function f.

Theorem 1.3 *The function $f : X \to \mathbb{R}$ is lower semicontinuous on X, if the following equivalent conditions are satisfied:*

(i) The set $\{x : x \in X, f(x) \ge \alpha\}$ is closed relative to X, for each real number α.

(ii) The set $\{x : x \in X, f(x) < \beta\}$ is open relative to X, for each real number β.

(iii) The hypograph of f, i.e.,

$$\text{hyp}(f) = \{(x, \alpha) : x \in X, \alpha \in \mathbb{R}, f(x) \ge \alpha\}$$

is a closed relative to $X \times \mathbb{R}$.

1.4 Derivatives and Hessians

In optimization theory, derivatives and Hessians play a very important role. We recall the following definitions from Mangasarian [176].

Definition 1.32 (Partial derivative and gradient) *Let $X \subseteq \mathbb{R}^n$ be an open set and let $\bar{x} \in X$. Then the partial derivative of a function $f : X \to \mathbb{R}$ at a point $\bar{x}$, with respect to $x_i, i = 1, \ldots, n$, denoted by $\frac{\partial f(\bar{x})}{\partial x_i}$, is given by*

$$\frac{\partial f(\bar{x})}{\partial x_i} :=$$

$$\lim_{\delta \to 0} \frac{f(\bar{x}_1, \ldots, \bar{x}_{i-1}, \bar{x}_i + \delta, \bar{x}_{i+1}, \ldots, \bar{x}_n) - f(\bar{x}_1, \ldots, \bar{x}_{i-1}, \bar{x}_i, \bar{x}_{i+1}, \ldots, \bar{x}_n)}{\delta},$$

whenever the limit exists. The n-dimensional vector of the partial derivatives of f at $\bar{x}$ with respect to $x_1, x_2, \ldots, x_n$ is called the gradient of f at $\bar{x}$, denoted by $\nabla f(\bar{x})$ and is defined as

$$\nabla f(\bar{x}) := \left(\frac{\partial f(\bar{x})}{\partial x_1}, \ldots, \frac{\partial f(\bar{x})}{\partial x_n} \right).$$

Definition 1.33 *(**Differentiable function**) Let $X \subseteq \mathbb{R}^n$ be an open set and let $\bar{x} \in X$. Then a function $f : X \to \mathbb{R}$ is said to be differentiable at $\bar{x}$, if there exists a vector $\nabla f(x)$ called the gradient and a function $\alpha : \mathbb{R}^n \to \mathbb{R}$, such that*

$$f(x) = f(\bar{x}) + \langle \nabla f(\bar{x}), x - \bar{x} \rangle + \|x - \bar{x}\| \alpha(\bar{x}, x - \bar{x}), \forall x \in X,$$

where $\lim_{x \to \bar{x}} \alpha(\bar{x}, x - \bar{x}) = 0$. The function f is said to be differentiable on X, if it is differentiable at each $\bar{x} \in X$.

Definition 1.34 *(**Jacobian of a vector function**) Let $X \subseteq \mathbb{R}^n$ be an open set and let $\bar{x} \in X$. Suppose $f : X \to \mathbb{R}^m$ be a vector valued function such that the partial derivative $\frac{\partial f_i(\bar{x})}{\partial x_j}$ exists at $\bar{x}$, for each $i = 1, \ldots, m$ and $j = 1, \ldots, n$. Then the Jacobian (matrix) of f at $\bar{x}$ denoted by $J(f)(\bar{x})$ is a $m \times n$ matrix defined by*

$$J(f)(\bar{x}) = \begin{bmatrix} \frac{\partial f_1(\bar{x})}{\partial x_1} & \cdots & \frac{\partial f_1(\bar{x})}{\partial x_n} \\ \vdots & \vdots & \vdots \\ \frac{\partial f_m(\bar{x})}{\partial x_1} & \cdots & \frac{\partial f_m(\bar{x})}{\partial x_n} \end{bmatrix}.$$

Proposition 1.12 *(**Chain rule theorem**) Let $X \subseteq \mathbb{R}^n$ be an open set and let $\bar{x} \in X$. Suppose $f : X \to \mathbb{R}^m$ be a vector valued function and let $g : \mathbb{R}^m \to \mathbb{R}$ be a real-valued function. Suppose that f is differentiable at $\bar{x}$ and g is differentiable at $\bar{y} = f(\bar{x})$. Then the composite function*

$$\varphi(x) := g(f(x))$$

is also differentiable at $\bar{x}$, and

$$\nabla \varphi(\bar{x}) = \nabla g(\bar{y}) \nabla f(\bar{x}).$$

Definition 1.35 *(**Twice differentiable function and Hessian**) Let $X \subseteq \mathbb{R}^n$ be an open set and let $\bar{x} \in X$. Suppose $f : X \to \mathbb{R}^n$ be a vector valued function. Then f is said to be twice differentiable at $\bar{x}$, if there exists a vector $\nabla f(\bar{x}), n \times n$ symmetric matrix $\nabla^2 f(\bar{x})$ and a function $\alpha : X \to \mathbb{R}$, such that*

$$f(x) = f(\bar{x}) + \langle \nabla f(\bar{x}), x - \bar{x} \rangle + \langle x - \bar{x}, \nabla^2 f(\bar{x})(x - \bar{x}) \rangle + \|x - \bar{x}\|^2 \alpha(\bar{x}, x - \bar{x}), \forall x \in \mathbb{R}^n,$$

where $\lim_{x \to \bar{x}} \alpha(\bar{x}, x - \bar{x}) = 0$. The $n \times n$ matrix $\nabla^2 f(\bar{x})$, called the Hessian matrix of f at $\bar{x}$, is given by

$$\nabla^2 f(\bar{x}) := \begin{bmatrix} \frac{\partial^2 f(\bar{x})}{\partial x_1^2} & \cdots & \frac{\partial^2 f(\bar{x})}{\partial x_1 \partial x_n} \\ \vdots & \vdots & \vdots \\ \frac{\partial^2 f(\bar{x})}{\partial x_n \partial x_1} & \cdots & \frac{\partial^2 f(\bar{x})}{\partial x_n^2} \end{bmatrix}.$$

Remark 1.3 Let $X \subseteq \mathbb{R}^n \times \mathbb{R}^m$ be an open set and let $f : X \to \mathbb{R}$ be a real-valued function, differentiable at $(\bar{x}, \bar{y}) \in X$. Then, we define

$$\nabla_x f(\bar{x}, \bar{y}) := \left[\frac{\partial f(\bar{x}, \bar{y})}{\partial x_1}, \ldots, \frac{\partial f(\bar{x}, \bar{y})}{\partial x_n} \right]$$

and

$$\nabla_{\bar{x}} f(\bar{x}, \bar{y}) := \left[\frac{\partial f(\bar{x}, \bar{y})}{\partial \bar{x}_1}, \ldots, \frac{\partial f(\bar{x}, \bar{y})}{\partial x_m} \right].$$

If $f : X \to \mathbb{R}^m$ be a vector valued function, differentiable at $(\bar{x}, \bar{y}) \in X$, then, we define

$$\nabla_x f(\bar{x}, \bar{y}) = \begin{bmatrix} \frac{\partial f_1(\bar{x}, \bar{y})}{\partial x_1} & \cdots & \frac{\partial f_1(\bar{x}, \bar{y})}{\partial x_n} \\ \vdots & \vdots & \vdots \\ \frac{\partial f_m(\bar{x}, \bar{y})}{\partial x_1} & \cdots & \frac{\partial f_m(\bar{x}, \bar{y})}{\partial x_n} \end{bmatrix}$$

and

$$\nabla_y f(\bar{x}, \bar{y}) = \begin{bmatrix} \frac{\partial f_1(\bar{x}, \bar{y})}{\partial y_1} & \cdots & \frac{\partial f_1(\bar{x}, \bar{y})}{\partial y_n} \\ \vdots & \vdots & \vdots \\ \frac{\partial f_m(\bar{x}, \bar{y})}{\partial y_1} & \cdots & \frac{\partial f_m(\bar{x}, \bar{y})}{\partial y_n} \end{bmatrix}.$$

Now, we state one of the most fundamental theorems from classical theory of mathematical analysis.

Proposition 1.13 *(Mean value theorem)* Let $X \subseteq \mathbb{R}^n$ be an open convex set and let $f : X \to \mathbb{R}$ be a differentiable function on X. Then, for all $x, \bar{x} \in X$.

$$f(x) = f(\bar{x}) + \langle \nabla f(x + \delta(x - \bar{x})), x - \bar{x} \rangle,$$

for some real number $\delta, 0 < \delta < 1$.

Proposition 1.14 *(Taylor's theorem (second order))* Let $X \subseteq \mathbb{R}^n$ be an open convex set and let $f : X \to \mathbb{R}$ be a twice continuously differentiable function on X. Then, for all $x, \bar{x} \in X$,

$$f(x) = f(\bar{x}) + \langle \nabla f(\bar{x}), x - \bar{x} \rangle + \frac{\langle (x - \bar{x}), \nabla^2 f(x + \delta(x - \bar{x}))(x - \bar{x}) \rangle}{2},$$

for some real number $\delta, 0 < \delta < 1$.

Proposition 1.15 *(Implicit function theorem) Let $X \subseteq \mathbb{R}^n \times \mathbb{R}^m$ be an open set and let $f : X \to \mathbb{R}^m$ be a vector valued function. Suppose that f has a continuous partial derivative at $(\bar{x}, \bar{y}) \in X$, such that $f(\bar{x}, \bar{y}) = 0$ and $\nabla_y f(\bar{x}, \bar{y})$ are nonsingular. Then, there exists a ball $\mathbb{B}_\varepsilon(\bar{x}, \bar{y})$ with the radius $\varepsilon > 0$ in $\mathbb{R}^{n+m}$, an open set $\Omega \subseteq \mathbb{R}^n$ containing $\bar{x}$, and an m-dimensional vector function φ with continuous first partial derivative on Ω, such that $\nabla_y f(x, y)$ is nonsingular for all $(x, y) \in \mathbb{B}_\varepsilon(\bar{x}, \bar{y}), \bar{y} = \varphi(\bar{x})$ and*

$$f(x, \varphi(x)) = 0, \forall x \in \Omega.$$

1.5   Convex Sets and Properties

The notion of convexity of the sets is core of optimization theory. We start this section with the definition of a subspace of $\mathbb{R}^n$.

Definition 1.36 *(Subspace) A set $X \subseteq \mathbb{R}^n$ is said to be a subspace if for each $x, y \in X$ and $p, q \in \mathbb{R}$, one has*

$$px + qy \in X.$$

Each subspace of $\mathbb{R}^n$ contains the origin.

Definition 1.37 *(Affine set) Let x and y be any two points in $\mathbb{R}^n$. Then, the set of all points of the form*

$$z = (1 - \lambda)x + \lambda y = x + \lambda(y - x), \lambda \in \mathbb{R}$$

is called the line through x and y. A subset X of $\mathbb{R}^n$ is referred to as an affine set (or affine manifold) if it contains every line through any two points of it. Affine sets, which contain the origin are subspaces of $\mathbb{R}^n$. It is easy to prove that if the set X is affine, then there exist $x \in X$ and a subspace X_0 of X, such that $X = x + X_0$. The subspace X_0 is said to be parallel to the affine set X and is uniquely determined for a given nonempty affine set X.

The dimension of a nonempty affine set is defined as the dimension of the subspace parallel to it. The affine sets of dimensions $0, 1$ and 2 are points, lines and planes, respectively. Every affine subset of $\mathbb{R}^n$ can be characterized as the solution sets F (say) of the system of simultaneous linear equations in n variables, that is,

$$F := \{x \in \mathbb{R}^n : Ax = b\},$$

where A is an $m \times n$ real matrix and a vector $b \in \mathbb{R}^m$.

Definition 1.38 *(Hyperplanes) Any $(n-1)$-dimensional affine set in $\mathbb{R}$ is called a hyperplane or a plane, for short. Any hyperplane is a set of the form*

$$H = \{x \in \mathbb{R}^n : \langle \alpha, x \rangle = \beta\},$$

where $\alpha \in \mathbb{R}^n \backslash \{0\}$ and $\beta \in \mathbb{R}$.

The vector α is called a normal to the hyperplane H. Every other normal is either a positive or negative scalar multiple of α. Angle between two hyperplanes is the angle between their normal vectors. Every affine subset is the intersection of finite collection of hyperplanes. If $\bar{x} \in H$, then the hyperplane H can be expressed as

$$H = \{x \in \mathbb{R}^n : \langle \alpha, x \rangle = \langle \alpha, \bar{x} \rangle\} = \bar{x} + \{x \in \mathbb{R}^n : \langle \alpha, x \rangle = 0\}.$$

Therefore, H is an affine set parallel to $\{x \in \mathbb{R}^n : \langle \alpha, x \rangle = 0\}$. It is easy to see that the intersection of arbitrary collection of affine sets is again affine. This led to the concept of affine hull.

Definition 1.39 *(Half-spaces) Let α be a nonzero vector in $\mathbb{R}^n$ and β be a scalar. Then, the sets*

$$K^+ := \{x \in \mathbb{R}^n : \langle \alpha, x \rangle \geq \beta\} \text{ and } K^- := \{x \in \mathbb{R}^n : \langle \alpha, x \rangle \leq \beta\}$$

are called upper and lower closed half-spaces. The sets

$$H^+ := \{x \in \mathbb{R}^n : \langle \alpha, x \rangle > \beta\} \text{ and } H^- := \{x \in \mathbb{R}^n : \langle \alpha, x \rangle < \beta\}$$

are called upper and lower open half-spaces.

It is clear that all four sets are nonempty and convex sets. We note that, if we replace α and β, respectively by $\lambda\alpha$ and $\lambda\beta$, for some $\lambda \neq 0$, we will get the same quartet of half-spaces. Thus half-spaces depend only on the hyperplane

$$H = \{x \in \mathbb{R}^n : \langle \alpha, x \rangle = \beta\}.$$

Definition 1.40 *(Affine hull) Given any $X \subseteq \mathbb{R}^n$, the intersection of collections of affine sets containing X, is called affine hull of X, denoted by $\text{aff}(X)$. It can be proved that $\text{aff}(X)$ is the unique smallest affine set containing X.*

Definition 1.41 *(Supporting hyperplane) Let $X \subseteq \mathbb{R}^n$ be a nonempty set and $\bar{x} \in bd(X)$. A hyperplane*

$$H = \{x \in \mathbb{R}^n : \langle \alpha, x - \bar{x} \rangle = 0, \alpha \neq 0, \alpha \in \mathbb{R}^n\}$$

is called a supporting hyperplane to X at $\bar{x}$, if either

$$X \subseteq H^+, \text{that is}, \langle \alpha, x - \bar{x} \rangle \geq 0, \forall x \in X$$

or

$$X \subseteq H^-, \ i.e., \ \langle \alpha, x - \bar{x} \rangle \leq 0, \forall x \in X.$$

The hyperplane H is called a proper supporting hyperplane, if in addition to the above properties $X \nsubseteq H$ is also satisfied.

The following proposition provides a characterization for the affine hull of a set.

Proposition 1.16 *The affine hull of a set X, that is, $\mathrm{aff}(X)$ is a set consisting of all points of the form*

$$z : z = \lambda_1 x_1 + \ldots + \lambda_n x_n, \lambda_i \in \mathbb{R}, \sum_{i=1}^{k} \lambda_i = 1,$$

where k is a natural number.

Proof Let C be the set of collection of all the points of the above form. Let $x, y \in C$, then, we have

$$x = \sum_{i=1}^{k} \lambda_i x_i$$

and

$$y = \sum_{j=1}^{k} \mu_j y_j,$$

where $x_i, y_j \in X, \lambda_i, \mu_j \geq 0, \forall i, j = 1, \ldots, k$ and

$$\sum_{i=1}^{k} \lambda_i = 1 = \sum_{j=1}^{k} \mu_j.$$

Now, for any $\alpha \in \ ]0, 1[$, we get

$$\alpha x + (1 - \alpha)y = \sum_{i=1}^{k} \alpha \lambda_i x_i + \sum_{j=1}^{k} (1 - \alpha)\mu_j y_j$$

with $\sum_{i=1}^{k} \alpha \lambda_i + \sum_{j=1}^{k} (1 - \alpha)\mu_j = \alpha + (1 - \alpha) = 1$. Thus, $\alpha x + (1 - \alpha)y \in C$. Hence, C is an affine set. Therefore, $\mathrm{aff}(X) \subseteq C$.

On the other hand, it can be seen that if $x = \sum_{i=1}^{k} \lambda_i x_i, \lambda_i \geq 0, \forall i = 1, \ldots, k$ with $\sum_{i=1}^{k} \lambda_i = 1$ then $x \in \mathrm{aff}\,(X)$. For $k = 2$, the result follows by the definition of affine set and for $k = 3$, the result follows by induction. Hence, $C \subseteq \mathrm{aff}\,(X)$. Hence, we get $C = \mathrm{aff}\,(X)$.

Definition 1.42 (*Affine independence*) *Any k points $x_1, \ldots, x_k$ in $\mathbb{R}^n$ are said to be affinely independent, if $\{x_1, \ldots, x_k\}$ has dimension $k - 1$, that is the vectors $x_2 - x_1, \ldots, x_k - x_{k-1}$ are linearly independent.*

Definition 1.43 *(Line segment)* *Let x and y be any two points in $\mathbb{R}^n$. We define the following line segments joining any two points of it:*

(i) closed line segment $[x, y] := \{z : z = (1 - \lambda)x + \lambda y, \lambda \in [0,1]\}$;

(ii) open closed line segment $]x, y] := \{z : z = (1 - \lambda)x + \lambda y, \lambda \in]0,1]\}$;

(iii) closed open line segment $[x, y[:= \{z : z = (1 - \lambda)x + \lambda y, \lambda \in [0,1[\}$;

(iv) open line segment $]x, y[:= \{z : z = (1 - \lambda)x + \lambda y, \lambda \in]0,1[\}$.

Definition 1.44 *(Convex set)* *A set $X \subseteq \mathbb{R}^n$ is said to be convex, if it contains closed line segment joining any two points of it. In other words, X is a convex set, if*

$$\forall x, y \in X \Rightarrow \lambda x + (1 - \lambda)y \in X, \forall \lambda \in [0,1].$$

It is obvious that affine sets are convex sets. However, the converse is not true. For example, solid ellipsoid and cube in $\mathbb{R}^3$ are convex sets, but not affine. See Figure 1.2.

Next, we prove the following important results for convex sets. The proof follows along the lines of Mangasarian [176].

Theorem 1.4 *Let $\{X_i, i \in I\}$ be a family (finite or infinite) of convex sets in $\mathbb{R}^n$, then, their intersection $\bigcap_{i \in I} X_i$ is a convex set. Moreover, if C, D are convex sets in $\mathbb{R}^n$, then,*

$$C + D := \{x + y : x \in C, y \in D\}$$

and

$$\alpha C := \{\alpha x : x \in C\}, \alpha \in \mathbb{R},$$

are convex sets.

Proof Let $a, b \in \bigcap_{i \in I} X_i$. Then $a, b \in X_i$, for each $i \in I$. Now, for any $\lambda \in [0,1]$, $\lambda a + (1 - \lambda)b \in X_i$, for each $i \in I$. Therefore, $\lambda a + (1 - \lambda)b \in \bigcap_{i \in I} X_i$, for any $\lambda \in [0,1]$. Hence, $\bigcap_{i \in I} X_i$ is a convex set.

If C, D are convex sets and $a, b \in C + D$, then $a = x_1 + y_1$ and $b = x_2 + y_2$, where $x_1, x_2 \in C$ and $y_1, y_2 \in D$. Now, for each $\lambda \in [0,1]$, we have

$$\lambda a + (1 - \lambda)b = \lambda(x_1 + y_1) + (1 - \lambda)(x_2 + y_2)$$
$$= (\lambda x_1 + (1 - \lambda)x_2) + (\lambda y_1 + (1 - \lambda)y_2) \in C + D.$$

Hence, $C + D$ is a convex set. Similarly, we can prove that $\alpha C, \alpha \in \mathbb{R}$ is a convex set.

The following proposition from Mangasarian [176] will be used to prove the strict separation theorem.

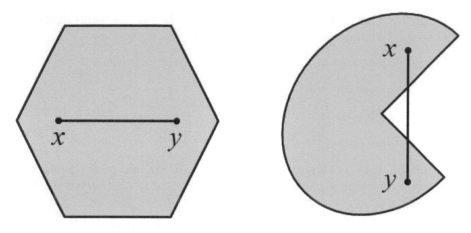

FIGURE 1.2: Convex and Nonconvex Sets

Proposition 1.17 *Let X and Y be two sets in $\mathbb{R}^n$ such that X is compact and Y is closed, then the sum $Z = X + Y$ is compact.*

Proof Let $\bar{z} \in \mathrm{cl}(Z)$. Then there exists a sequence $\{z_n\}$ in Z, which converges to $\bar{z}$. Then there exists a sequence $\{x_n\} \in X$ and $\{y_n\} \in Y$, such that $z_n = x_n + y_n$, for $n = 1, 2, \ldots$. Since, X is compact, there exists a subsequence $\{x_{n_i}\}$, which converges to $\bar{x} \in X$. Then, we have

$$\lim_{i \to \infty} z_{n_i} = \bar{z}$$

$$\lim_{i \to \infty} x_{n_i} = \bar{x} \in X.$$

Since, Y is closed, we have

$$\lim_{i \to \infty} y_{n_i} = \bar{z} - \bar{x} \in Y.$$

Therefore, $\bar{z} = \bar{x} + (\bar{z} - \bar{x}) \in Z$. Hence, Z is closed.

The following definition and characterization are from Mangasarian [176].

Definition 1.45 (*Convex hull*) *Let $X \subseteq \mathbb{R}^n$ be a given set. Then, the intersection of family of all convex sets containing X is the smallest convex set containing X. This set is called the convex hull of X and denoted by $\mathrm{co}(X)$. It can be shown that the convex hull of the set X can be expressed as*

$$\mathrm{co}\, X = \left\{ z : z = \lambda_1 x_1 + \ldots + \lambda_n x_n, x_i \in X, \lambda_i \in \mathbb{R}_+, \sum_{i=1}^{k} \lambda_i = 1 \right\}.$$

For example, see Figure 1.3.

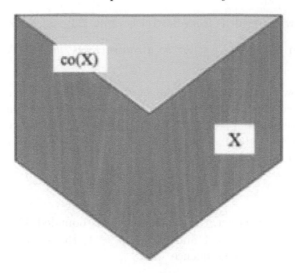

FIGURE 1.3: Convex Hull of a Set X

Definition 1.46 *(Convex combination) A point $x \in \mathbb{R}^n$, such that*

$$x = \sum_{i=1}^{k} \lambda_i a_i \text{ with } a_i \in X \subseteq \mathbb{R}^n, \lambda_i \geq 0, \sum_{i=1}^{k} \lambda_i = 1$$

is called a convex combination of $a_1, \ldots, a_n \in \mathbb{R}^n$.

The following theorem provides a characterization for convex sets. The proof can be found in Mangasarian [176], Rockafellar [238], and Tuy [273].

Theorem 1.5 *A subset $X \subseteq \mathbb{R}^n$ is convex if and only if convex combinations of points of X are contained in X. Equivalently, X is convex if and only if*

$$x_1, x_2 \ldots, x_k \in X, \lambda_1, \lambda_2, \ldots, \lambda_k \geq 0, \lambda_1 + \lambda_2 + \ldots + \lambda_k = 1$$

$$\Rightarrow \lambda_1 x_1 + \lambda_2 x_2 + \ldots + \lambda_k x_k \in X. \tag{1.1}$$

From the above discussion, it is clear that if X is a convex set, then any point of X can be expressed as a convex combination of m points of X, where $m \geq 1$ is arbitrary. However, for an arbitrary set X, if any point is a convex combination of m points of X, then $m \leq n + 1$. This is one of the most applicable results in optimization theory known as Caratheodory's theorem. There are several approaches to prove Caratheodory's theorem. See Mangasarian [176], Rockafellar [238], and Tuy [273].

Theorem 1.6 *(Caratheodory's theorem) Let $X \subseteq \mathbb{R}^n$ be a nonempty set. Let x be a convex combination of points of X. Then x can be expressed as a convex combination of $n + 1$ or fewer points of X.*

The following result is from Bertsekas *et al.* [23] and Rockafellar [238].

Proposition 1.18 *Let $X \subseteq \mathbb{R}^n$ be a compact set, then $\mathrm{co}(X)$ is a compact set.*

Proof To show that $\mathrm{co}(X)$ is a compact set, it is sufficient to show that every sequence in $\mathrm{co}(X)$ has a convergent subsequence, whose limit is in $\mathrm{co}(X)$. Consider a sequence $\{z_r\} \in \mathrm{co}(X)$. By Theorem 1.6, for all r there exist sequences $\{\lambda_i^r\}, \lambda_i^r \geq 0, i = 1, 2, \ldots, n+1, \sum_{r=1}^{n+1} \lambda_i^r = 1$ and $\{x_i^r\}, x_i^r \in X, r = 1, 2, \ldots, n+1$, such that

$$z_r = \sum_{r=1}^{n+1} \lambda_i^r x_i^r.$$

Since, X is a compact set, the sequence $\{x_i^r\}$ is bounded. Furthermore, since $\lambda_i^r \geq 0, i = 1, 2, \ldots, n+1$ and $\sum_{r=1}^{n+1} \lambda_i^r = 1$, the sequence, $\{\lambda_i^r\}$ is also bounded. Therefore, the sequence

$$\left\{ \lambda_1^r, \ldots, \lambda_{n+1}^r, x_1^r, \ldots, x_{n+1}^r \right\}$$

is bounded. From the Bolzano-Weierstrass theorem, the bounded sequence

$$\left\{ \lambda_i^1, \ldots, \lambda_i^{n+1}, x_i^1, \ldots, x_i^{n+1} \right\}$$

has a limit point say $\left\{ \lambda_1, \ldots, \lambda_{n+1}, x_1, \ldots, x_{n+1} \right\}$, which must satisfy

$$\lambda_i \geq 0, \sum_{r=1}^{n+1} \lambda_i = 1 \, and \, x_i \in X, \, \forall \, i = 1, \ldots, n+1.$$

Let us define $z = \sum_{i=1}^{n+1} \lambda_i x_i$. Thus, z is a limit point of the sequence $\{z_r\}$ and the sequence $z_r \to z \in \mathrm{co}(X)$. Hence, $\mathrm{co}(X)$ is a compact set.

Remark 1.4 *However, Proposition 1.18 does not hold true, if we assume that the set X is a closed set, rather than a compact set. The following example from Bertsekas et al. [23] illustrates the fact. Consider the closed set X, given by*

$$X = \{(0,1)\} \bigcup \left\{ (x_1, x_2) \in \mathbb{R}^2 : x_1 x_2 \geq 0, x_1 > 0, x_2 > 0 \right\}.$$

Then, the convex hull of X, is given by

$$X = \{(0,1)\} \bigcup \left\{ (x_1, x_2) \in \mathbb{R}^2 : x_1 > 0, x_2 > 0 \right\},$$

which is not a closed set.

Now, we give the definition of a radially continuous function.

Definition 1.47 (Radially continuous function) *Let X be a nonempty convex subset of $\mathbb{R}^n$. A function $f : X \to \mathbb{R}$ is said to be radially upper(lower) semicontinuous on X(also, known as upper (lower) hemicontinuous on X), if for every pair of distinct points $x, y \in X$, the function f is upper(lower) semicontinuous on the line segment $[x, y]$.*

Moreover, the function f is said to be radially continuous on X(also, known as hemicontinuous on X), if it is both radially upper semicontinuous and radially lower semicontinuous on X.

The following example from Ansari *et al.* [2] shows that a function may be radially continuous at a point, even not being continuous at that point.

Example 1.3 *Let $f : \mathbb{R}^2 \to \mathbb{R}$ be defined as*

$$f(x, y) := \begin{cases} \frac{2x^2 y}{x^4 + y^2}, & \text{if } (x, y) \neq (0, 0), \\ 0, & \text{if } (x, y) = (0, 0). \end{cases}$$

We note that the function f is continuous at every point of $\mathbb{R}^2$, except at $(0, 0)$. If we approach x along the path $y = mx^2$, then, we have

$$\lim_{(x,y) \to (0,0)} f(x, y) = \frac{2m}{1 + m^2},$$

which is different for different values of m. However, f is radially continuous, because if we approach $(0,0)$ along the line $y = mx$, then,

$$\lim_{(x,y) \to (0,0)} f(x, y) = \frac{2mx}{x^2 + m^2} = 0 - f(0, 0).$$

Definition 1.48 (Cone) *A subset $X \subseteq \mathbb{R}^n$ is called a cone, if it is closed under positive scalar multiplication, i.e., if*

$$x \in X, \lambda > 0 \Rightarrow \lambda x \in X.$$

The origin 0 itself may or may not be included in the set. However, the origin always lies in the closure of a nonempty cone. A set $\alpha + X$, which is a translation of a cone X by $\alpha \in \mathbb{R}^n$, is called a cone with apex α.

A cone X is said to be pointed, if it contains a line. In other words, X is said to be pointed, if

$$X \bigcap (-X) = \{0\}.$$

A subset $X \subseteq \mathbb{R}^n$ is a convex cone, if X is a convex set and cone, that is,

$$\forall x, y \in X \text{ and } \lambda, \mu \geq 0, \lambda x + \mu y \in X.$$

Subspaces of $\mathbb{R}^n$, closed and open half-spaces corresponding to a hyperplane through the origin are examples of convex cones. The nonnegative orthant of $\mathbb{R}^n$, that is,

$$\{x = (x_1, \ldots, x_n) : x_i \geq 0, i = 1, \ldots, n\}$$

is an example of a closed, convex and pointed cone. Like convex sets, intersection of an arbitrary collection of convex cones is a convex cone.

Theorem 1.7 *A subset $X \subseteq \mathbb{R}^n$ is a convex cone, if and only if it is closed under addition and positive scalar multiplication.*

Proof Let X be a cone and let $x, y \in X$. If X is convex, the vector $z = \frac{1}{2}(x + y) \in K$, hence, the vector $x + y = 2z \in X$. By definition of a cone X is closed under positive scalar multiplication. To prove the converse, we assume that cone X is closed under addition and positive scalar multiplication. Then for $x, y \in X$ and $0 < \lambda < 1, \lambda x \in X, (1 - \lambda)y \in X$ and therefore, $\lambda x + (1 - \lambda)y \in X$. Hence, X is convex.

The following corollaries are a direct consequence of the above theorem:

Corollary 1.1 *A subset $X \subseteq \mathbb{R}^n$ is a convex cone if and only if it contains all the positive linear combinations of its elements.*

Corollary 1.2 *Let $X \subseteq \mathbb{R}^n$ be an arbitrary set and let S be the set of all convex combinations of the elements of X. Then S is the smallest convex cone containing X. Furthermore, if X is a convex set, then the set*

$$S := \{\lambda x : x \in X, \lambda > 0\}$$

is the smallest convex cone, which includes X.

Definition 1.49 *(**Conic combination and polyhedral cone**) Let $S = \{x_1, x_2, \dots, x_n\}$ be a given set of vectors. Then, a vector*

$$x = \lambda_1 x_1 + \dots + \lambda_n x_n, \lambda_i \geq 0, i = 1, \dots, n$$

is called the conic combination of the vectors. The set of all conic combination of the elements of S is called the polyhedral cone generated by $x_1, \dots, x_n$.

Definition 1.50 *(**Conic hull**) The intersection of all the convex cones containing the given set X is called the conic hull of X and is denoted by $\mathrm{con}(X)$. In fact, $\mathrm{con}(X)$ is the smallest convex set containing the set X.*

Definition 1.51 *(**Polar or dual cone**) Let $X \subseteq \mathbb{R}^n$ be a nonempty set. Then the cone*

$$X^* := \{y : \langle y, x \rangle \leq 0, \forall x \in X\}$$

is called the polar cone to X. It is clear that polar cone X^* is the intersection of closed half-spaces and therefore, it is closed and convex, independent of the set closedness or convexity of the set X. It is easy to see that if X is a subspace, then X^* is equal to the orthogonal complement of X, that is, $X^* = X^\perp$.

The proof of the following proposition can be found in Bertsekas *et al.* [23].

Proposition 1.19 *Let $X \subseteq \mathbb{R}^n$ be a nonempty set. Then, we have*

(i) $X^* = (cl(X))^* = (co(X))^* = (con(X))^*$;

(ii) *(**Polar cone theorem**) In addition, if X is a cone, then*

$$(X^*)^* = cl(co(X)).$$

In particular, if X is closed and convex, we have $(X^*)^* = X$.

Definition 1.52 *(**Normal cone**) Let $X \subseteq \mathbb{R}^n$ be a nonempty closed convex set. A vector $\xi \in \mathbb{R}^n$ is said to be normal to X at $x \in X$, if*

$$\langle \xi, y - x \rangle \leq 0, \forall y \in X.$$

The set of all such vectors is called the normal cone to X at x and is denoted by $N_X(x)$.

Next, we state an important result about the interior of a convex set. The proof can be found in Bertsekas *et al.* [23].

Proposition 1.20 *A point $x \in X \subseteq \mathbb{R}^n$ is an interior point of X, if for every $y \in \mathbb{R}^n$, there exists $\alpha > 0$, such that*

$$x + \alpha(y - x) \in X.$$

We know that the interior of a nonempty convex set may be empty. For instance, the interior of a line or a triangle in $\mathbb{R}^3$ is an empty set. The interior of these convex sets is nonempty relative to the affine hull of the sets. In fact, a convex set has a nonempty interior relative to the smallest affine set containing it. This led to the introduction of the concept of relative interiors.

1.6 Relative Interiors

Definition 1.53 *(**Relative interior**) The relative interior of a convex set X, denoted by $ri(X)$ is defined as the interior of X, relative to the affine hull of X. In other words, $x \in ri(X)$ if there exists an open ball $\mathbb{B}_\varepsilon(x)$, centered at x, such that*

$$\mathbb{B}_\varepsilon(x) \bigcap aff(X) \subseteq X.$$

More precisely,

$$ri(X) := \{x \in aff(X) : \text{ there exists } \varepsilon > 0, \text{ such that } B_\varepsilon(x) \bigcap aff(X) \subseteq X\}.$$

A vector in the closure of X, which is not a relative interior point of X is called relative boundary of X. We know that if $X \subseteq \mathbb{R}^n$ is a convex set, then $\text{aff}(X) = \mathbb{R}^n$. Therefore, in this case interior and relative interior of the set X coincide, that is, $\text{int}(X) = \text{ri}(X)$.

For nonempty convex sets X and Y, we know that

$$X \subseteq Y \Rightarrow \text{int}(X) \subseteq \text{int}(Y) \text{ and } \text{cl}(X) \subseteq \text{cl}(X)(Y).$$

However, unlike the concept of closure and interior of nonempty convex sets, we have

$$X \subseteq Y \not\Rightarrow \text{ri}(X) \subseteq \text{ri}(Y),$$

which is a drawback of the concept of relative interiors. The following example from Dhara and Dutta [63] justifies the statement.

Example 1.4 *Let* $X = \{(0,0)\}$ *and* $Y = \{(0,y) \in \mathbb{R}^2 : y \geq 0\}$. *It is clear that* $X \subseteq Y$. *However,* $\text{ri}(X) = \{(0,0)\}$ *and* $\text{ri}(Y) = \{(0,y) \in \mathbb{R}^2 : y > 0\}$. *Thus,* $\text{ri}(X)$ *and* $\text{ri}(Y)$ *are nonempty and disjoint.*

The following proposition gives some basic facts about relative interiors and closures of a nonempty convex set. The proof can be found in Bertsekas *et al.* [23].

Proposition 1.21 *Let* $X \subseteq \mathbb{R}^n$ *be a nonempty convex set. Then the following hold:*

(i) *(**Nonemptiness**)* $\text{ri}(X)$ *and* $\text{cl}(X)$ *are nonempty convex set with same affine hull as* X.

(ii) *(**Line segment principle**) Let* $x \in \text{ri}(X)$ *and* $y \in \text{cl}(X)$. *Then,*

$$x + \lambda(y - x) \in \text{ri}(X), \forall \lambda \in [0,1[.$$

(iii) *(**Prolongation principle**)* $x \in \text{ri}(X)$ *if and only if every line segment in* X *with* x *has one end point that can be prolonged beyond* x *without leaving* X. *In other words, for all* $y \in X$, *there exists* $\lambda > 1$, *such that*

$$x + (1 - \lambda)(y - x) \in X.$$

The following result from Tuy [273] illustrates a very important property of relative interiors.

Proposition 1.22 *Let* $X \subseteq \mathbb{R}^n$ *be a nonempty convex set. Then* $\text{ri}(X)$ *is nonempty.*

Proof Let $S = \text{aff}(X)$ and $dim\ S = k$. Therefore, there exists $k + 1$ affinely independent elements $x_0, x_1, \ldots, x_k$ of X. We show that the point

$$a = \frac{1}{k+1} \sum_{i=1}^{k} x_i \in \text{ri}(X).$$

Let $x \in S$, then

$$x = \sum_{i=1}^{k+1} \lambda_i x_i \text{ with } \sum_{i=1}^{k+1} \lambda_i = 1.$$

Now, we consider

$$a + \alpha(x - a) = \frac{1}{k+1} \sum_{i=1}^{k+1} x_i \alpha + \alpha \left(\sum_{i=1}^{k+1} \lambda_i x_i - \frac{1}{k+1} \sum_{i=1}^{k+1} x_i \right)$$

$$= \sum_{i=1}^{k+1} \left((1-\alpha)\frac{1}{k+1} + \alpha\lambda_i \right) x_i = \sum_{i=1}^{k+1} \mu_i x_i$$

where $\mu_i = (1-\alpha)\frac{1}{k+1} + \alpha\lambda_i$ and $\sum_{i=1}^{k+1} \mu_i = 1$. For $\alpha > 0$ sufficiently small, we have

$$\mu_i > 0, i = 1, \ldots, k+1.$$

Hence, $a + \alpha(x - a) \in X$. Therefore, by Proposition 1.21, a is an interior point of X relative to $S = \text{aff}(X)$. Therefore, $a \in \text{ri}(X)$.

Corollary 1.3 *Let $X \subseteq \mathbb{R}^n$ be a nonempty convex set and let C be the nonempty set generated by $\{(1, x) : x \in C\}$. Then $\text{ri}(C)$ consists of the pairs (λ, x) such that $\lambda > 0$ and $x \in \lambda\text{ri}(C)$.*

The following corollaries from Rockafellar [238] will be needed in the sequel.

Corollary 1.4 *Let $C \subseteq \mathbb{R}^n$ be a convex set and let M be an affine set, which contains a point of $\text{ri}(C)$. Then,*

$$\text{ri}(M \cap C) = M \cap \text{ri}(C) \text{ and } \text{cl}(M \cap C) = M \cap \text{cl}(C).$$

Corollary 1.5 *Let C_1 be a convex set and let C_2 be the convex set contained in the $\text{cl}(C_1)$ but not contained in the relative boundary of C_1. Then*

$$\text{ri}(C_2) \subseteq \text{ri}(C_1).$$

Definition 1.54 (Recession cone) *Let $X \subseteq \mathbb{R}^n$ be a nonempty set. Then X is said to recede in the direction $y, y \neq 0$ if and only if*

$$x + \lambda y \in X, \forall \lambda \geq 0 \text{ and } x \in C.$$

The set of all vectors $y \in \mathbb{R}^n$, satisfying the condition, including $y = 0$, denoted by $0^+ X$, is called the recession cone of X. Directions in which X, recedes is referred to as direction of recessions of X.

Now, we present the following example of recession cone from Rockafellar [238].

Example 1.5 *We consider the following convex set in* $\mathbb{R}^n$, *given by*

$$X = \{(x_1, x_2)|x_1^2 + x_2^2 \leq 1\}.$$

Then the recession cone of X, *is given by*

$$0^+X := \{(x_1, x_2)|x_1 = x_2 = 0\} \bigcup \{(0, 0)\}.$$

An unbounded closed convex set contains at least one point at infinity, that is, it recedes in at least one direction.

The following theorem is from Rockafellar [238].

Theorem 1.8 *Let* $X \subseteq \mathbb{R}^n$ *be a nonempty closed convex set. Then* 0^+X *will be closed and it consists of all possible limits of sequences of the form* $\lambda_1 x_1, \lambda_2 x_2, \ldots$, *where* $x_i \in X$ *and* $\lambda_i \downarrow 0$. *In fact, for the convex cone* C *in* $X \subseteq \mathbb{R}^{n+1}$ *generated by* $\{(1, x) : x \in X\}$, *one has*

$$\text{cl}(C) = C \bigcup \{(0, x) : x \in 0^+X\}.$$

Proof By Corollary 1.3, the hyperplane $M = \{(1, x) : x \in \mathbb{R}^n\}$ must intersect with ri(C). Therefore, by the closure rule in Corollary 1.4, we have

$$M \cap \text{cl}(C) = \text{cl}(M \cap C) = M \cap C = \{(1, x) : x \in X\}.$$

Therefore, the cone

$$\bar{C} := \{(\lambda, x) : \lambda > 0, x \in \lambda x\} \bigcup \{(0, x) : x \in 0^+X\}$$

must contain cl(C), because of the maximality property. Again since, the cone $\bar{C}$ is contained in the half-space

$$H := \{(\lambda, x) : \lambda \geq 0\}$$

and meets int(H), therefore, by Corollary 1.5, ri($\bar{C}$) must be entirely contained in int(H). Hence, $ri(\bar{C})$ must be contained in C. Now, we have

$$\text{cl}(C) \subseteq \bar{C} \subseteq \text{cl}(ri(\bar{C})) \subseteq \text{cl}(C),$$

which implies that cl(C) = $\bar{C}$. Again, the set $\{(0, x) : x \in 0^+X\}$ is the intersection of cl(C) with $\{(0, x) : x \in \mathbb{R}^n\}$, therefore, it is closed set and consists of the limits of the sequences of the form $\lambda_1(1, x_1), \lambda_2(1, x_2), \ldots$, where $x_i \in X$ and $\lambda_i \downarrow 0$.

The following theorem provides a characterization of boundedness of a closed convex set in terms of its recession cone. The proof follows along the lines of Rockafellar [238].

Theorem 1.9 *Let* $X \subseteq \mathbb{R}^n$ *be a nonempty closed convex set. Then* X *is bounded if and only if its recession cone contains* 0^+X *consists of the zero vector alone.*

Proof If X is bounded, it certainly contains no half-line, so that $0^+X = \{0\}$. Suppose conversely that X is unbounded. Then it contains a sequence of nonzero vectors $x_1, x_2, \ldots$ whose Euclidean norms $|x_i|$ increase without bound. The vectors $\lambda_i x_i$, where $\lambda_i = \frac{1}{|x_i|}$, all belong to the unit sphere $S = \{x : |x| = 1\}$. Since, X is closed and bounded, there exists some subsequence of $\lambda_1 x_1, \lambda_2 x_2, \ldots$, will converge to a certain $y \in X$. By Theorem 1.8, this is a nonzero vector of 0^+X.

Definition 1.55 (Faces and vertex/extreme points) *Let $X \subseteq \mathbb{R}^n$ be a convex set. Then a face $\bar{X}$ of the set X is a convex subset of X, such that every closed line segment in X with relative interior points in $\bar{X}$ has both end points in $\bar{X}$. The empty set and the set X are faces of X. A face of zero dimension is called a vertex or extreme point. In other words, any point $x \in X$ is called a vertex (or extreme point) if there exist no two distinct points x_1, x_2, such that $x \in [x_1, x_2]$. In other words, if x is a vertex, then*

$$x \in [x_1, x_2] \Rightarrow x = x_1 \text{ or } x = x_2.$$

Definition 1.56 (Exposed faces and exposed points) *Let $X \subseteq \mathbb{R}^n$ be a convex set. Then the exposed faces of the set X are the set of the form $X \cap H$, where H is a nontrivial supporting hyperplane to X. An exposed point of X is an exposed face, which is a point. In other words, exposed points of a convex set X is a point, through which there is a supporting hyperplane, which contains no other points of X.*

The following proposition from Rockafellar [238] relates extreme points and the exposed points of a convex set.

Proposition 1.23 (Straszevicz's theorem) *Let $X \subseteq \mathbb{R}^n$ be a convex set. Then, the set of exposed points of X is a dense subset of the set of extreme points. Thus every exposed point is the limit of some sequence of exposed points.*

Remark 1.5 *A convex set $X \subseteq \mathbb{R}^n$ may have no vertices (for instance, the open ball $\mathbb{B}_\varepsilon(\bar{x})$ and the hyperplane $H = \{x \in \mathbb{R}^n : \langle \alpha, x \rangle = \beta\}$), finite number of vertices, (for instance, the set $\{x \in \mathbb{R}^n : x \geq, \langle e, x \rangle = 1\}$, where $e = (1, \ldots, 1)$ is n-vectors of 1s with $e_i = 1$ and $e_j = 0$, for $i \neq j$) or an infinite number of vertices, (for instance, the closed ball $\mathbb{B}_\varepsilon[\bar{x}]$ has an infinite number of vertices given by $\{x : x \in \mathbb{R}^n, \|x - \bar{x}\| = \varepsilon\}$.)*

Definition 1.57 (Polyhedron and polytope) *A set $X \subseteq \mathbb{R}^n$ which is the intersection of a finite number of closed half-spaces in $\mathbb{R}^n$ is called a polyhedron. A polyhedron is a convex hull of finite many points. A polyhedron, which is bounded is called a polytope. Hence, a polytope is a closed and bounded convex set. Polyhedral convex sets are more well behaved than ordinary convex sets due to lack of curvature. An extreme point (0-dimensional face) of a polyhedron is also called a vertex and 1-dimensional face is called an edge.*

Definition 1.58 (Simplex) *Let* $x_0, x_1, \ldots, x_k$ *be* $(k+1)$ *distinct point in* $\mathbb{R}^n$ *with* $k \leq n$. *If* $\{x_0, x_1, \ldots, x_k\}$ *are affinely independent, then its convex hull of the set, that is,*

$$\text{co } X := \left\{ z : z = \lambda_0 x_0 + \lambda_1 x_1, \ldots, \lambda_n x_n, x_i \in X, \lambda_i \in \mathbb{R}_+, \sum_{i=0}^{k} \lambda_i = 1 \right\}.$$

is called k-*simplex with vertices* $x_0, x_1, \ldots, x_k$. *In other words,* k-*simplex in* $\mathbb{R}^n$ *is a convex polyhedron having* $k+1$ *vertices. For instance, a* 0-*simplex is a point,* 1-*simplex is a line,* 2-*simplex is a triangle and* 3-*simplex is a tetrahedron.*

Now, we state the following result from Rockafellar [238].

Corollary 1.6 (Minkowski theorem) *Every closed and bounded convex set in* $\mathbb{R}^n$ *is the convex hull of its extreme points.*

1.7 Convex Functions and Properties

In this section, we will study convex functions and their properties. We start with the definition of a convex function.

Definition 1.59 (Convex function) *Let* $X \subseteq \mathbb{R}^n$ *be a nonempty convex set. Then a function* $f : X \to \mathbb{R}$ *is said to be convex function on* X, *if for all* $x, y \in X$, *we have*

$$f((1-\lambda)x + \lambda y) \leq (1-\lambda)f(x) + \lambda f(y), \forall \lambda \in [0,1]. \tag{1.2}$$

A convex function is proper if and only if its epigraph is nonempty and does not contain a vertical line. A function $f : X \to \mathbb{R}$ is said to be concave if $-f$ is convex on X. In other words, the function f is said to be concave, if

$$f((1-\lambda)x + \lambda y) \geq (1-\lambda)f(x) + \lambda f(y), \forall \lambda \in [0,1]. \tag{1.3}$$

An affine function i.e., the function of the form

$$f(x) = \langle c, x \rangle + \alpha, c, x \in \mathbb{R}^n$$

and $\alpha \in \mathbb{R}$. The function f is said to be strictly convex (strictly concave), if the above inequality (1.3) (respectively, (1.4)) is strict for $x \neq y$ and $\lambda \in]0,1[$. Every strictly convex (strictly concave) function is convex (concave), but the converse is not true. For example, the affine functions are convex (concave) but not strictly convex (strictly concave).

The following proposition presents the characterization of convex functions in terms of the epigraph of f.

Proposition 1.24 *Let $X \subseteq \mathbb{R}^n$ be a convex set. The function $f : X \to \mathbb{R}$ is convex if and only if* epi(f) *is a convex subset of $\mathbb{R}^{n+1}$.*

Proof Suppose that $f : X \to \mathbb{R}$ is a convex function. Let (x, α) and (y, β) be any two points of epi(f). By the convexity of f, we have

$$f((1 - \lambda)x + \lambda y) \le (1 - \lambda)f(x) + \lambda f(y).$$

$$\le (1 - \lambda)\alpha + \lambda \beta.$$

From which, it follows that

$$(1 - \lambda)(x, \alpha) + \lambda(y, \beta) \in \text{epi}(f), \forall \lambda \in [0, 1].$$

Hence, epi(f) is a convex subset of $\mathbb{R}^{n+1}$.

Suppose conversely that epi(f) is a convex subset of $\mathbb{R}^{n+1}$. To prove that f is a convex function on X, let $x, y \in X$. For $\lambda \in [0, 1]$, we have

$$(1 - \lambda)(x, f(x)) + \lambda(y, f(y)) \in \text{epi}(f),$$

which implies that,

$$f((1 - \lambda)x + \lambda y) \le (1 - \lambda)f(x) + \lambda f(y), \forall \lambda \in [0, 1].$$

Hence, f is a convex function on $\mathbb{R}^n$. This completes the proof.

An important property of convex and concave functions is that they are continuous on the interior of their domains. Though one may find various approaches to prove the result in Bertsekas *et al.* [23], Mangasarian [176], Rockafellar [238], and Tuy [273], we present a simple proof from Bazaraa *et al.* [17].

Proposition 1.25 *Let $X \subseteq \mathbb{R}^n$ be a nonempty set and let $f : X \to \mathbb{R}^n$ be convex. Then f is continuous on the interior of X.*

Proof Let $\bar{x} \in \text{int}(X)$. We have to prove that f is continuous at $\bar{x}$. Let $\varepsilon > 0$ be given. Now, we have to show that for given $\varepsilon > 0$, there exists $\delta > 0$ (depending on ε), such that

$$|f(x) - f(\bar{x})| \le \varepsilon, \text{ whenever } \|x - \bar{x}\| \le \delta.$$

First, we show that f is upper semicontinuous at $\bar{x}$, i.e.,

$$f(x) - f(\bar{x}) \le \varepsilon, \text{ whenever } \|x - \bar{x}\| \le \delta.$$

Since, $\bar{x} \in \text{int}(X)$, therefore, there exists a $\delta' > 0$ such that

$$|x - \bar{x}| \le \delta \Rightarrow x \in X.$$

Now, we define α as follows

$$\alpha = \max_{1 \le i \le n} \left\{ \max_{1 \le i \le n} [f(\bar{x} + \delta e_i) - f(\bar{x}), f(\bar{x} - \delta e_i) - f(\bar{x})] \right\}, \qquad (1.4)$$

where e_i is a vector of zeros except having 1 at the i th position. We note that $0 \leq \alpha \leq \infty$. Setting

$$\delta = \min \left(\frac{\delta'}{n}, \frac{\varepsilon \delta'}{n\alpha} \right). \tag{1.5}$$

Choose an x with $\|x - \bar{x}\| \leq \delta$. Let

$$z_i = \begin{cases} \delta' e_i, & \text{if } x_i - \bar{x}_i \geq 0, \\ -\delta' e_i, & \text{otherwise.} \end{cases}$$

Then $x - \bar{x} = \sum_{i=1}^n \lambda_i z_i$, where $\lambda_i \geq 0$, $i = 1, 2, ..., n$. Moreover,

$$\|x - \bar{x}\| = \delta' \left(\sum_{i=1}^n \lambda_i^2 \right)^{1/2} \tag{1.6}$$

Since, $\|x - \bar{x}\| \leq \delta$, from (1.7), it follows that $\lambda_i \leq 1/n$, for $i = 1, 2, ..., n$. Since, $0 \leq n\lambda_i \leq 1$, by the continuity of f, we have

$$f(x) = f\left(\bar{x} + \sum_{i=1}^n \lambda_i z_i \right) = f\left[\frac{1}{n} \sum_{i=1}^n \left(\bar{x} + \sum_{i=1}^n n\lambda_i z_i \right) \right]$$

$$\leq \frac{1}{n} \sum_{i=1}^n f\left[(1 - n\lambda_i) \bar{x} + n\lambda_i (\bar{x} + z_i) \right]$$

$$\leq \frac{1}{n} \sum_{i=1}^n \left[(1 - n\lambda_i) f(\bar{x}) + n\lambda_i f((\bar{x} + z_i)) \right].$$

Therefore,

$$f(x) - f(\bar{x}) \leq \sum_{i=1}^n \lambda_i \left[f(\bar{x} + z_i) - f(\bar{x}) \right].$$

By (1.5), we get

$$f(\bar{x} + z_i) - f(\bar{x}) \leq \alpha, \ \forall i.$$

Since, $\lambda_i \geq 0$, it follows that

$$f(x) - f(\bar{x}) \leq \alpha \sum_{i=1}^n \lambda_i. \tag{1.7}$$

From (1.7) and (1.8), it follows that $\lambda_i \leq \frac{\varepsilon}{n\alpha}$. Therefore, by (1.8), we get

$$f(x) - f(\bar{x}) \leq \varepsilon,$$

whenever

$$\|x - \bar{x}\| \leq \delta.$$

Thus, f is upper semicontinuous at $\bar{x}$.

Now, we have to prove that f is lower semicontinuous at $\bar{x}$, that is, we have to show that $f(\bar{x}) - f(x) \leq \varepsilon$. Let $y = 2\bar{x} - x$, then we note that $\|y - \bar{x}\| \leq \delta$. Therefore, as above

$$f(y) - f(\bar{x}) \leq \varepsilon. \tag{1.8}$$

Now, we have $\bar{x} = \frac{1}{2}y + \frac{1}{2}x$, therefore, by the continuity of f, we have

$$f(\bar{x}) \leq \frac{1}{2}f(y) + \frac{1}{2}f(x). \tag{1.9}$$

Combining, (1.9) and (1.10), it follows that $f(\bar{x}) - f(x) \leq \varepsilon$. Hence, the proof is complete.

Remark 1.6 *However, it should be noted that a convex function may not be continuous everywhere on their domain. Consider the following example, from Mangasarian [176]. Let $f : X \to \mathbb{R}$ be given by*

$$f(x) = \begin{cases} x^2, & \text{if } x > -1, \\ 2, & \text{if } x = -1, \end{cases}$$

where, $X := \{x : x \in \mathbb{R}, x > -1\}$. Then, it is clear that f is a convex function on X, but not continuous on X.

The following theorems illustrate that once or twice differentiable convex functions have their specific characterizations. We present a simple proof from Mangasarian [176].

Theorem 1.10 *Let $X \subseteq \mathbb{R}^n$ be an open convex set. Then $f : X \to \mathbb{R}$ is a convex function over X, if and only if*

$$f(y) - f(x) \geq \langle \nabla f(x), y - x \rangle, \forall x, y \in X.$$

Proof (Necessary) Assume that f is convex over X and let $x, y \in \mathbb{R}^n$ be such that $x \neq y$. For $\lambda \in]0, 1]$, we consider the function

$$g(\lambda) = \frac{f(x + \lambda(y - x)) - f(x)}{\lambda}.$$

Now, we will show that $g(\lambda)$ is monotonically increasing with λ. We consider any λ_1, λ_2 with $0 < \lambda_1 < \lambda_2$ and let

$$\bar{\lambda} = \frac{\lambda_1}{\lambda_2}, \quad \bar{y} = x + \lambda(y - x). \tag{1.10}$$

Now, by the convexity of f, we have

$$f(x + \bar{\lambda}(\bar{y} - x)) \leq \bar{\lambda}f(\bar{y}) + (1 - \bar{\lambda})f(x),$$

which can be rewritten as

$$\frac{f(x + \bar{\lambda}(\bar{y} - x)) - f(x)}{\bar{\lambda}} \leq f(\bar{y}) - f(x)). \tag{1.11}$$

Using (1.11) into (1.12), it may be shown that

$$\frac{f(x + \lambda_1(y - x)) - f(x)}{\lambda_1} \leq \frac{f(x + \lambda_2(y - x)) - f(x)}{\lambda_2}.$$

Hence, for $0 < \lambda_1 < \lambda_2 < 1$, we get

$$g(\lambda_1) \leq g(\lambda_2).$$

This will imply that

$$\langle \nabla f(x), y - x \rangle = \lim_{\lambda \to 0} g(\lambda) \leq g(1) = f(y) - f(x).$$

(Sufficiency) Let $x, y \in X$ and let $0 \leq \lambda \leq 1$. Since, X is convex, it follows that $(1 - \lambda)x + \lambda y \in X$. Now, we have

$$f(x) - f((1 - \lambda)x + \lambda y) \geq \lambda \langle \nabla f((1 - \lambda)x + \lambda y), y - x \rangle,$$

$$f(y) - f((1 - \lambda)x + \lambda y) \geq -(1 - \lambda) \langle \nabla f((1 - \lambda)x + \lambda y), y - x \rangle,$$

Multiplying the first inequality by $(1 - \lambda)$ and the second by λ and adding them, we get

$$(1 - \lambda)f(x) + \lambda f(y) \geq f((1 - \lambda)x + \lambda y), \forall \lambda \in [0, 1].$$

Hence, the proof is complete.

Remark 1.7 *(i) In a similar way, by using the definition of concave functions, we can prove that a function $f : X \to \mathbb{R}$ is concave over a convex set X, if and only if*

$$f(y) - f(x) \leq \langle \nabla f(x), y - x \rangle, \forall x, y \in X.$$

Again taking $x \neq y$ and $\lambda \in]0, 1[$, we can prove that a function $f : X \to \mathbb{R}$ is strictly convex (strictly concave) over a convex set X, if and only if

$$f(y) - f(x) > (<) \langle \nabla f(x), y - x \rangle.$$

(ii) Geometrically, the above theorem illustrates that, for a convex function, the linearization $f(x) + \langle \nabla f(x), y - x \rangle$ never overestimates $f(x)$ for any $x \in X$. Similarly, for a concave function, the linearization $f(x) + \langle \nabla f(x), y - x \rangle$ never underestimates $f(x)$ for any $x \in X$ (for example, see Figure 1.4).
(iii) Furthermore, it is clear from the above theorem that if $f : X = \mathbb{R}^n \to$ is a convex function and $\nabla f(x) = 0$, then x minimizes f over $\mathbb{R}^n$. This is classical sufficient optimality condition for unconstrained optimization known as Fermat's rule in one dimension.

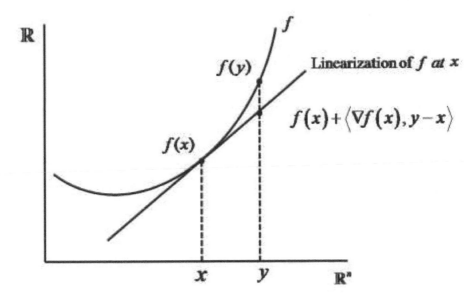

FIGURE 1.4: Linearization of a Convex Function f

Definition 1.60 (*Distance function*) *Let $C \subseteq \mathbb{R}^n$ be any nonempty set. Then the distance function of C, denoted by $d_C(.)$, is defined by*

$$d_C(x) = \inf\{\|x - c\| : c \subset C\}.$$

It is clear that if C is a closed set, then $x \in C$ if and only if $d_C(x) = 0$. Distance function $d_C(.)$ satisfies the following global Lipschitz property, i.e.,

$$|d_C(x) - d_C(y)| \leq \|x - y\|, \forall x, y \in \mathbb{R}^n.$$

If the set C is a convex set, then $d_C(.)$ is a convex function.

Definition 1.61 (*Indicator function*) *Let $C \subseteq \mathbb{R}^n$ be any set. Then indicator function of C denoted by $\delta(.|C)$, is defined as*

$$\delta(x|C) = \begin{cases} 0, & \text{if } x \in C, \\ \infty, & \text{if } x \notin C. \end{cases}$$

It is clear that the set C is a convex set if and only if $\delta(., C)$, is a convex function. It can be seen that the epigraph of an indicator function is a "half-cylinder" with the cross section as C.

Definition 1.62 (*Support function*) *The support function of a convex set $C \subseteq \mathbb{R}^n$ is a function $\sigma_C(.) : X \rightarrow \mathbb{R} \bigcup \{\infty\}$, defined by*

$$\sigma_C(x) = \sup\{\langle x, y \rangle : y \in X\}.$$

For instance, the support function of the convex set $C = [-1, 1]$ is the absolute value function $f(x) = |x|$.

Next, we state some important properties of the support function from Clarke [49].

Theorem 1.11 *Let C, D be nonempty closed convex subset of $\mathbb{R}^n$. Then,*

(i) $C \subseteq D \Leftrightarrow \sigma_C(x) \leq \sigma_D(x), \forall x \in \mathbb{R}^n$.

(ii) C is compact if and only if $\sigma_C(.) : (R)^n \to \mathbb{R}$ is finite valued on $\mathbb{R}^n$.

(iii) A given function $\sigma : \mathbb{R}^n \to \mathbb{R} \bigcup \{\infty\}$ is positively homogeneous, subadditive, lower semicontinuous function and not identically $+\infty$, if and only if there is a nonempty closed convex subset C of $\mathbb{R}^n$ such that $\sigma = \sigma_C$. Any such C is unique.

Definition 1.63 (Monotone function) *Let $f : X \subseteq \mathbb{R}^n \to \mathbb{R}^n$ be a vector valued function, then f is said to be monotone (strictly monotone) function on X, if*

$$\langle f(y) - f(x), y - x \rangle \geq (>)0, \forall x, y \in \mathbb{R}.$$

Theorem 1.12 *Let $X \subseteq \mathbb{R}^n$ be an open convex set. Then $f : X \to \mathbb{R}$ is a convex (concave) function over X, if and only if*

$$\langle \nabla f(y) - \nabla f(x), y - x \rangle \geq 0(\leq 0), \forall x, y \in X.$$

Proof We prove the theorem for the convex case. The proof for the concave case follows similarly.

(Necessary) Let f be a convex function and let $x, y \in X$. Then by Theorem 1.10, we have

$$f(y) - f(x) \geq \langle \nabla f(x), y - x \rangle$$

and

$$f(x) - f(y) \geq \langle \nabla f(y), x - y \rangle.$$

Adding these two inequalities, we get

$$\langle \nabla f(y) - \nabla f(x), y - x \rangle \geq 0.$$

(Sufficiency) Let $x, y \in X$. Since, X is convex, then,

$$(1 - \lambda)x + \lambda y \in X, \forall \lambda \in [0, 1].$$

Now, by mean value theorem, there exists $\bar{\lambda}, 0 < \bar{\lambda} < 1$, such that

$$f(y) - f(x) = \bar{\lambda} \langle \nabla f(x + \bar{\lambda}(y - x)), y - x \rangle. \tag{1.12}$$

By our assumption, we have

$$\langle \nabla f(x + \bar{\lambda}y - x) - \nabla f(x), y - x \rangle \geq 0,$$

which is equivalent to

$$\langle \nabla f(x + \bar{\lambda}(y - x)), y - x \rangle - \langle \nabla f(x), y - x \rangle \geq 0. \tag{1.13}$$

Hence, by (1.13) and (1.14), we get

$$f(y) - f(x) \geq \langle \nabla f(x), y - x \rangle, \ \forall x, y \in X.$$

Hence, f is a convex function over X.

Remark 1.8 *It is clear from the above theorem, that a differentiable function* $f : X \to \mathbb{R}$ *is convex on open convex set* X *if and only if* ∇f *is monotone on* X. *Similarly, we can show that differential function* $f : X \to \mathbb{R}$ *is strictly convex on open convex set* X *if and only if* ∇f *is strictly monotone on* X.

The following theorem presents a characterization for twice continuously differential convex fucntions.

Theorem 1.13 *Let* $X \subseteq \mathbb{R}^n$ *be an open convex set and let* f *be a twice differentiable function on* X. *Then* f *is a convex (concave) function over* X *if and only if for all* $x \in X$, *the Hessian matrix* $\nabla^2 f(x)$ *is positive (negative) semidefinite matrix on* X.

Proof (Necessary) By Proposition 1.14, for all $x, y \in X$, we have

$$f(y) = f(x) + \langle \nabla f(x), y - x \rangle + \frac{1}{2} \langle y - x, \nabla^2 f(x + \lambda(y - x))(y - x) \rangle,$$

for some $\lambda \in [0, 1]$. Since, $\nabla^2 f(x)$ is positive semidefinite, we must have

$$f(y) \geq f(x) + \langle \nabla f(x), y - x \rangle, \forall x, y \in X.$$

Hence, f is convex over X. (Sufficiency) Suppose f is convex and $\bar{x} \in X$. We have to prove that $\nabla^2 f(x)$ is a positive semidefinite matrix at $\bar{x} \in X$, that is, for each $y \subset \mathbb{R}^n$, we must have

$$\langle y, \nabla^2 f(\bar{x}) y \rangle \geq 0.$$

Since, X is open, there exists a $\bar{\lambda} > 0$, such that $\bar{x} + \lambda y \in X$, for $0 < \lambda < \bar{\lambda}$. Taking into consideration, the convexity and twice differentiability of f, for $0 < \lambda < \bar{\lambda}$, we have

$$f(\bar{x} + \lambda y) \geq f(\bar{x}) + \lambda \langle y, \nabla f(\bar{x}) y \rangle,$$

$$f(\bar{x} + \lambda y) \geq f(\bar{x}) + \lambda \langle \nabla f(\bar{x}), y \rangle + \frac{1}{2} \lambda^2 \langle y, \nabla^2 f(\bar{x}) y \rangle + \lambda^2 \alpha(\bar{x}, \lambda y)(\|y\|)^2.$$

Hence, we get

$$\frac{1}{2} \lambda^2 \langle y, \nabla^2 f(\bar{x}) y \rangle + \lambda^2 \alpha(\bar{x}, \lambda y)(\|y\|)^2 \geq 0, \ for \ 0 < \lambda < \bar{\lambda}.$$

Taking the limit as $\lambda \to 0$, we get

$$\langle y, \nabla^2 f(\bar{x})y \rangle \geq 0, \forall y \in \mathbb{R}^n.$$

This completes the proof for the convex case.

Taking into consideration the concavity of f and proceeding on the same line, we can prove the theorem for concave case.

Remark 1.9   *(i) The above theorem is useful in checking the convexity and concavity of the twice differentiable functions. Moreover a quadratic function*

$$f(x) = \frac{1}{2} \langle x, Qx \rangle + \langle x, a \rangle + b$$

where $x, a \in \mathbb{R}^n, b \in \mathbb{R}$ and Q is a $n \times n$ symmetric matrix, is convex on $\mathbb{R}^n$ if and only if Q is a positive definite matrix.

(ii) In a similar way, we can show that if $f : X \to \mathbb{R}$ be a twice differentiable function over an open convex set $X \subseteq \mathbb{R}^n$ and if the Hessian $\nabla^2 f(x)$ is positive definite for all $x \in X$, then f is a strictly convex function, but not conversely. However, if $X \subseteq \mathbb{R}^n$ is an open set and f is a strictly convex (strictly concave) function over X, then f is a convex (concave) function over X. Hence, for each $x \in X$, the Hessian matrix $\nabla^2 f(x)$ is positive (negative) semidefinite matrix but not necessarily positive definite on X. For example the function $f : \mathbb{R} \to \mathbb{R}$ given by $f(x) = x^4$ is strictly convex but $\nabla^2 f(x) = 12x^2$ is not positive definite, since $\nabla^2 f(0) = 0$.

Proposition 1.26 *Let $X \subseteq \mathbb{R}^n$ be a convex set. If the function $f : X \to \mathbb{R}$ is convex on X, then the lower level set $\Lambda(f, \alpha) := \{x \in X : f(x) \leq \alpha\}$ is a convex subset of X, for arbitrary $\alpha \in \mathbb{R}$.*

Proof Suppose f be convex on X and for arbitrary $\alpha \in \mathbb{R}$, let $x, y \in \Lambda(f, \alpha)$. Then by the convexity of f, we have

$$f((1 - \lambda)x + \lambda y) \leq (1 - \lambda)f(x) + \lambda f(y), \forall \lambda \in [0, 1].$$

$$\leq (1 - \lambda)\alpha + \lambda\alpha = \alpha.$$

Thus, for all $\lambda \in [0, 1]$, it follows that $(1 - \lambda)x + \lambda y \in \Lambda(f, \alpha)$. Hence, $\Lambda(f, \alpha)$ is a convex subset of X, for each $\alpha \in \mathbb{R}$.

Remark 1.10 *The converse of the above proposition is not true in general. For example, consider the function $f : \mathbb{R} \to \mathbb{R}$ given by $f(x) = x^3$. Clearly, f is not a convex function. However, the lower level set*

$$\Lambda(f, \alpha) = \left\{x : x \in \mathbb{R}, x^3 \leq \alpha\right\} = \left\{x : x \in \mathbb{R}, x \leq \alpha^{\frac{1}{3}}\right\}$$

is a convex set. This led to the generalization of the notion of convexity to quasiconvexity.

1.8   Generalized Convex Functions

The Italian mathematician Bruno de Finetti [62] introduced one of the fundamental generalized convex functions characterized by convex lower level sets, now known as quasiconvex functions. From an analytical point of view, we have the following definition:

Definition 1.64 (Mangasarian [176]) *Let* $f : X \to \mathbb{R}$ *be a real-valued function, defined on a convex set* X. *The function* f *is said to be quasiconvex on* X, *if*

$$f(\lambda x + (1 - \lambda) y) \leq \max (f(x), f(y)), \forall x, y \in X \text{ and } \lambda \in [0,1].$$

The function f *is said to be quasiconcave on* X, *if* $-f$ *is quasiconvex on* X. *A function* f *is said to be quasilinear on* X, *if it is both quasiconvex and quasiconcave on* X.

Example 1.6 *The function* $f(x) = x^3$ *is a quasilinear function on* $\mathbb{R}$.

If the above inequality is strict, we have the following notion of strictly quasiconvex functions.

Definition 1.65 (Strictly quasiconvex function) *Let* $f : X \to \mathbb{R}$ *be a real-valued function defined on a convex set* X. *Then, the function* f *is said to be strictly quasiconvex function on* X, *if for* $x \neq y$, $\lambda \in]0,1[$, *we have*

$$f(\lambda x + (1 - \lambda) y) < \max \{f(x), f(y)\}, \forall x, y \in X.$$

It is clear from the definitions that every strictly quasiconvex function is quasiconvex, but not conversely. However, the converse is not true as illustrated by the following example, from Cambini and Martein [36].

Example 1.7 *The function* $f(x) = \begin{cases} \frac{|x|}{x}, & \text{if } x \neq 0, \\ 0, & \text{if } x = 0 \end{cases}$ *is quasiconvex but not strictly quasiconvex.*

The following theorem shows that the quasiconvex functions, can be completely characterized in terms of lower level sets. The proof is along the lines of Mangasarian [176].

Theorem 1.14 *Let* $f : X \to \mathbb{R}$ *be a real-valued function defined on a convex set* X. *The function* f *is quasiconvex on* X, *if and only if the lower level set*

$$\Lambda(f, \alpha) = \{x \in X : f(x) \leq \alpha\}$$

is convex for each $\alpha \in \mathbb{R}$.

Proof Suppose that $\Lambda(f, \alpha)$ is convex for each $\alpha \in \mathbb{R}$. To show that f is *quasiconvex* on X, let $x, y \in X$ be such that $f(x) \geq f(y)$ and let $0 \leq \lambda \leq 1$. Setting, $f(x) = \alpha$, by convexity of $\Lambda(f, \alpha)$, we get

$$f(\lambda x + (1 - \lambda) y) \leq \alpha = f(x),$$

Hence, f is quasiconvex on X.

Suppose conversely that f is quasiconvex on X. Let α be any real number and let $x, y \in \Lambda_\alpha$. Without loss of generality, assume that $f(y) \leq f(x)$. Since, $x, y \in \Lambda_\alpha$, we get

$$f(y) \leq f(x) \leq \alpha.$$

Since, f is quasiconvex and X is convex, for $0 \leq \lambda \leq 1$, we have

$$f(\lambda x + (1 - \lambda) y) \leq f(x) \leq \alpha.$$

Hence, $\lambda x + (1 - \lambda) y \in \Lambda(f, \alpha)$ and $\Lambda(f, \alpha)$ is convex.

Taking into account the properties of quasiconvex and quasiconcave functions, the following theorem from Cambini and Martein [36] characterizes the properties of quasilinear functions.

Corollary 1.7 *Let $f : X \subseteq \mathbb{R}^n$ be a function defined on the convex set X. Then f is quasilinear on X if and only if one of the following conditions holds:*

(1) For $x, y \in X$, and $\lambda \in [0, 1]$, one has

$$\min \{f(x), f(y)\} \leq f((1 - \lambda)x + \lambda y) \leq \max \{f(x), f(y)\}$$

(ii) The lower level and upper level sets of f are convex.

(iii) Any restriction of f on a line segment is a nonincreasing or a nondecreasing function.

The following theorem states the relationship between convexity, strict convexity, quasiconvexity and strict quasiconvexity. The proof may be found in Cambini and Martein [36] and Mangasarian [176].

Theorem 1.15 *Let $X \subseteq \mathbb{R}^n$ be a convex set and $f : X \to \mathbb{R}$ any function. Then*

(i) If f is convex on S, then f is quasiconvex on S.

(ii) If f is strictly convex on S, then f is strictly quasiconvex on S.

(iii) If f is strictly quasiconvex on S, then f is quasiconvex on S.

It is obvious that the class of quasiconvex functions contains properly the classes of convex as well as strictly quasiconvex functions. However, there is no inclusion relation between the classes of convex and strictly quasiconvex functions.

The following theorem characterizes differentiable quasiconvex functions. The proof may be found in Cambini and Martein [36] and Mangasarian [176].

Theorem 1.16 *Let $X \subseteq \mathbb{R}^n$ be an open convex set and let $f : X \to \mathbb{R}$ be a differentiable function on X. Then, f is quasiconvex on X, if and only if the following implication holds:*

$$f(y) \leq f(x) \Rightarrow \langle \nabla f(x), y - x \rangle \leq 0, \forall x, y \in X. \tag{1.14}$$

However, if x is not a critical point, then on both sides of (1.15), we have a strict inequality.

Theorem 1.17 *Let $X \subseteq \mathbb{R}^n$ be an open convex set and let $f : X \to \mathbb{R}$ be a differentiable function on X. Then, f is quasiconvex on X, if and only if the following implication holds:*

$$f(y) < f(x), \nabla f(x) \neq 0 \Rightarrow \langle \nabla f(x), y - x \rangle < 0, \forall x, y \in X.$$

Proof For the proof, we refer to Cambini and Martein [36].

The following theorem from Cambini and Martein [36] gives a characterization for quasilinear functions. The characterization is based on the behavior of the differentiable quasilinear functions on the same level sets.

Theorem 1.18 *Let $f : X \to \mathbb{R}$ be a differentiable function on an open convex set X. Then f is quasilinear on X if and only if the following condition holds:*

$$f(x) = f(y) \Rightarrow \langle \nabla f(x), y - x \rangle = 0, \ \forall x, y \in X.$$

Mangasarian [176] introduced the concepts of pseudoconvex and pseudoconcave functions as generalizations of convex and concave functions, respectively. Analytically, pseudoconvex and pseudoconcave functions are defined as follows:

Definition 1.66 *Let $f : X \to \mathbb{R}$ be a differentiable function on open convex set X. The function f is said to be pseudoconvex at $x \in X$, if for all $y \in X$, one has*

$$\langle \nabla f(x), y - x \rangle \geq 0 \Rightarrow f(y) \geq f(x);$$

or equivalently,

$$f(y) < f(x) \Rightarrow \langle \nabla f(x), y - x \rangle < 0.$$

The function f is called pseudoconvex on X, if the above property is satisfied for all $x \in X$. The function f is called pseudoconcave on X, if $-f$ is pseudoconvex on X. The function f is called pseudolinear on X, if f is both pseudoconvex and pseudoconcave on X.

In the subsequent chapters, we study in detail about pseudolinear functions and their properties.

Definition 1.67 *Let $f : X \to \mathbb{R}$ be a differentiable function on an open convex set X. The function f is said to be strictly pseudoconvex at $x \in X$, if for all $y \in X$, $x \neq y$, one has*

$$f(y) \leq f(x) \Rightarrow \langle \nabla f(x), y - x \rangle < 0.$$

The function f is called strictly pseudoconvex on X, if the above property is satisfied for all $x \in X$.

It is obvious that every strictly pseudoconvex function is pseudoconvex, but the converse is not true, in general. For example, every constant function is pseudoconvex, but not strictly pseudoconvex.

The following theorem states the relationship between convex, pseudoconvex, and quasiconvex functions. For the proof, we refer to Mangasarian [176].

Theorem 1.19 *Let $f : X \to \mathbb{R}$ be a differentiable function on the open convex set X.*

(i) If f is convex on X, then f is pseudoconvex on X.

(ii) If f is strictly pseudoconvex on X, then f is strictly quasiconvex function on X, and therefore quasiconvex function on X.

However, the converse is not necessarily true. For example the function $f(x) = x + x^3$ is pseudoconvex, but not convex on $\mathbb{R}$. However, there is no relation between the classes of pseudoconvex and strictly quasiconvex functions. For example the function $f(x) = x^3$ is strictly quasiconvex on $\mathbb{R}$, but not pseudoconvex on $\mathbb{R}$. Moreover, the constant function is pseudoconvex but not strictly quasiconvex on $\mathbb{R}$.

Definition 1.68 (*Karamardian [131]*) *Let $X \subseteq \mathbb{R}^n$ be a nonempty set and $F : X \to \mathbb{R}^n$ be any map. Then F is said to be pseudomonotone, if for all $x, y \in X$, we have*

$$\langle F(x), y - x \rangle \geq 0 \Rightarrow \langle F(y), y - x \rangle \geq 0.$$

Now, we need the following lemma from Cambini and Martein [36].

Lemma 1.1 *Let $X \subseteq \mathbb{R}^n$ be a convex set and $f : X \to \mathbb{R}$ be a differentiable function on X. Suppose that ∇f is pseudomonotone on X and $x, y \in X$ such that*

$$\langle \nabla f(x), y - z \rangle \geq 0.$$

Then, the restriction of f on $[x, y]$ is nondecreasing.

The following theorem from Cambini and Martein [36] characterizes pseudoconvex functions in terms of the pseudomonotonicity of the gradient map.

Theorem 1.20 *Let $X \subseteq \mathbb{R}^n$ be a convex set and $f : X \to \mathbb{R}$ be a differentiable function on X. Then f is pseudoconvex on X, if and only if ∇f is pseudomonotone on X.*

Proof Assume that f is pseudoconvex on X and let $x, y \in X$. Suppose that

$$\langle \nabla f(x), y - x \rangle \geq 0.$$

Since, f is pseudoconvex on X, it is also quasiconvex on X, therefore, for all $x, y \in X$, we get

$$\langle \nabla f(y), x - y \rangle \leq 0,$$

which is equivalent to

$$\langle \nabla f(y), y - x \rangle \geq 0.$$

Hence, ∇f is pseudomonotone on X. Conversely, suppose that ∇f is pseudomonotone on X. Suppose contrary that there exists $x, y \in X$, such that

$$f(x) < f(y) \text{ and } \langle \nabla f(y), y - x \rangle \geq 0.$$

Then, by Lemma 1.1, f is nondecreasing on the restriction $[x, y]$. Therefore, we must have

$$f(x) \geq f(y),$$

which is a contradiction. This completes the proof.

The following theorem gives a very important characterization of pseudoconvex and strictly pseudoconvex functions.

Theorem 1.21 *Let $f : X \to \mathbb{R}$ be a differentiable function on the open convex set X. Then f is pseudoconvex (strictly pseudoconvex) if and only if for every $\bar{x} \in X$ and $u \in \mathbb{R}^n$, such that*

$$\langle u, \nabla f(\bar{x}) \rangle = 0,$$

the function $\varphi(t) = f(\bar{x} + tu)$ attains a local minimum (strict local minimum) at $t = 0$.

1.9 Separations and Alternative Theorems

It is obvious that the hyperplane H divides the space $\mathbb{R}^n$, into two half-spaces either closed or open. This led to the development of the idea of separation of spaces. We start the section with the concept of a separating plane.

Definition 1.69 (Separating plane) *The plane $\{x : x \in \mathbb{R}^n, \langle a, x \rangle = \alpha, \alpha \neq 0\}$ is said to separate two nonempty sets X and Y in $\mathbb{R}^n$, if*

$$\langle a, x \rangle \leq \alpha \leq \langle a, y \rangle, \forall x \in X \text{ and } \forall y \in Y.$$

The separation is said to be strict if

$$\langle a, x \rangle < \alpha < \langle a, y \rangle, \forall x \in X \text{ and } \forall y \in Y.$$

The sets X and Y are said to be separable, if such a plan exists.

It can be shown that two disjoint sets may not be separable and also two separable sets need not be disjoint. However, under the convexity assumptions on the set, we have certain theorems known as separation theorems, which imply that the disjoint convex sets are separable. The separation theorems play a fundamental role in establishing several optimality conditions. Following are some basic and useful separation theorems. There are several approaches to prove these results. See Mangasarian [176], Rockafellar [238], and Tuy [273].

First, we state the following lemma from Mangasarian [176].

Lemma 1.2 *(Separation of a point and a plane) Let X be a nonempty convex sets in $\mathbb{R}^n$ not containing the origin. Then, there exists a plane*

$$\{x : x \in \mathbb{R}^n, \langle a, x \rangle = 0, a \neq 0\}$$

separating X and 0. That is,

$$x \in X \Rightarrow \langle a, x \rangle \geq 0.$$

Using Lemma 1.2, we have the following separation theorem known as the fundamental separation theorem from Mangasarian [176].

Theorem 1.22 *(Separation of two planes) Let X and Y be two nonempty disjoint convex sets in $\mathbb{R}^n$. Then, there exists a plane*

$$\{x : x \in \mathbb{R}^n, \langle a, x \rangle = \gamma, a \neq 0\},$$

which separates them, that is,

$$x \in X \Rightarrow \langle a, x \rangle \leq \gamma$$

and

$$x \in Y \Rightarrow \langle a, x \rangle \geq \gamma.$$

Proof We consider the set

$$X - Y = \{z : z = x - y, x \in X, y \in Y\}.$$

By Theorem 1.1, the set $X - Y$ is convex. Since, $X \bigcap Y = \phi$, hence, $0 \notin X - Y$. By Lemma 1.2, there exists a plane $\{z : z \in \mathbb{R}^n, \langle a, z \rangle = 0, a \neq 0\}$, such that

$$z \in X - Y \Rightarrow \langle a, z \rangle \geq 0$$

or

$$x \in X, y \in Y \Rightarrow \langle a, x - y \rangle \geq 0.$$

Hence,

$$\alpha = \inf_{x \in X} \langle a, x \rangle \geq \sup_{y \in Y} \langle a, y \rangle = \beta.$$

Define

$$\gamma = \frac{\alpha + \beta}{2}.$$

Then,

$$x \in X \Rightarrow \langle a, x \rangle \leq \gamma$$

and

$$x \in Y \Rightarrow \langle a, x \rangle \geq \gamma.$$

This completes the proof.

The following corollary and lemma can be proved with the help of Theorem 1.22.

Corollary 1.8 *Let* $X \subseteq \mathbb{R}^n$ *be a nonempty convex set. If* $0 \notin \mathrm{cl}(X)$, *then there exists a hyperplane*

$$\{x : x \in \mathbb{R}^n, \langle a, x \rangle = \alpha, a \neq 0, a \in \mathbb{R}^n\},$$

which strictly separates X *and* 0 *conversely. In other words,*

$$0 \notin \mathrm{cl}(X) \Leftrightarrow \exists a \neq 0, a > 0 \text{ such that } x \in X \Rightarrow \langle a, x \rangle > \alpha.$$

The following lemma is very important to prove a strict separation theorem. The proof may be found in Bazaraa *et al.* [17] and Mangasarian [176].

Lemma 1.3 *Let* $X \subseteq \mathbb{R}^n$ *be a nonempty, closed and convex set. If* $0 \notin X$, *then, there exists a hyperplane*

$$\{x : x \in \mathbb{R}^n, \langle a, x \rangle = \alpha, a \neq 0, a \in \mathbb{R}^n\},$$

which strictly separates X *and* 0, *and vice-versa, that is,*

$$a \notin (X) \Leftrightarrow \exists a \neq 0, a > 0, \text{ such that } x \in X \Rightarrow \langle a, x \rangle > \alpha.$$

In other words, Lemma 1.3 may be restated as follows:

Let $X \subseteq \mathbb{R}^n$ *be a nonempty closed convex set. If* $0 \notin X$, *then, there exists a nonzero vector* a *and a scalar* α, *such that* $\langle a, y \rangle > \alpha$ *and* $\langle a, x \rangle \leq \alpha$, *for each* $x \in X$.

Using Corollary 1.6 and Lemma 1.3, we have the following strict separation theorem. We present a simple proof from Mangasarian [176].

Theorem 1.23 (Strict separation theorem) *Let* X *and* Y *be two nonempty disjoint convex sets in* $\mathbb{R}^n$ *with* X *compact and* Y *closed. Then, there exists a hyperplane*

$$\{x : x \in \mathbb{R}^n, \langle a, x \rangle = \alpha, a \neq 0, a \in \mathbb{R}^n\},$$

which strictly separates X *and* Y *and conversely. In other words,*

$$X \cap Y = \phi \Leftrightarrow \exists a \neq 0, \alpha > 0 \text{ such that } \langle a, x \rangle < \gamma, \forall x \in X \text{ and } \langle a, y \rangle > \gamma, \forall y \in Y.$$

Proof (Necessary) We consider the set

$$X - Y := \{z : z = x - y, x \in X, y \in Y\}.$$

By Theorem 1.1, the set $X - Y$ is convex. Since, $X \cap Y = \phi$, hence, $0 \notin X - Y$. By Lemma 1.1, there exists a plane $\{x : x \in \mathbb{R}^n, \langle a, x \rangle = \mu\}, a \neq 0, \mu > 0$, such that

$$z \in X - Y \Rightarrow \langle a, z \rangle > \mu > 0$$

or

$$x \in X, y \in Y \Rightarrow \langle a, x - y \rangle > \mu > 0.$$

Hence,

$$\alpha = \inf_{x \in X} \langle a, x \rangle \geq \sup_{y \in Y} \langle a, y \rangle + \mu > \sup_{y \in Y} \langle a, y \rangle = \beta.$$

Define

$$\gamma := \frac{\alpha + \beta}{2}.$$

Then, we have

$$x \in X \Rightarrow \langle a, x \rangle < \gamma$$

and

$$x \in Y \Rightarrow \langle a, x \rangle > \gamma.$$

(Sufficiency) Suppose to the contrary that $z \in X \cap Y$, then $\langle a, x \rangle < \gamma < \langle a, x \rangle$, which is a contradiction. This completes the proof.

The alternative theorems play a very important role in establishing the necessary optimality conditions for linear as well as nonlinear programming problems. Now, we state the two most applicable alternative theorems, that is, Farkas's and Gordan's alternative theorems. The proof can be found in Bazaraa *et al.* [17] and Mangasarian [176].

Theorem 1.24 *(Farkas's theorem) For each $p \times n$ matrix A and each fixed vector $b \in \mathbb{R}^n$, either*
(I) $Ax \leqq 0$, $b^T x > 0$ *has a solution* $x \in \mathbb{R}^n$;
or
(II) $A^T y = b$ *has a solution* $y \in \mathbb{R}^p$;
but never both.

Proof Assume that the system (II) has a solution. Then there exists $y \geq 0$, such that $A^T y = b$. Let there exist some $x \in \mathbb{R}^n$, such that $Ax \leq 0$. Then, we have

$$b^T x = y^T Ax \leq 0.$$

Hence, system (I) has no solution.

Now, we assume that the system (II) has no solution. Consider the set

$$X := \left\{ x \in \mathbb{R}^n : x = A^T y, y \geqq 0 \right\}.$$

It is clear that X is a closed convex set and that $b \in X$. Therefore, by Lemma 1.3, there exists vector $p \in \mathbb{R}^n$ and scalar α, such that $p^T b > \alpha$ and $p^T x \leq \alpha$, for all $x \in X$. Since, $0 \in X, \alpha \geq 0$, hence, $p^T b > 0$. Moreover,

$$\alpha \geq p^T x = p^T A^T y = y^T A p, \forall y \geq 0.$$

Since, $y \geq 0$, can be made arbitrary large, the last inequality implies that $Ap \leq 0$. Thus, we have constructed a vector $p \in \mathbb{R}^n$, such that $Ap \leq 0$ and $b^T p > 0$. Hence, the system (I) has a solution. This completes the proof.

Remark 1.11 *To understand geometrically, we may rewrite the system (I) and (II) as follows:*
(I*) $A_j x \leq 0, j = 1, 2, ..., p, \ bx > 0$;

(II*) $\sum_{j=1}^{p} A_{.j}^T y_j = \sum_{j=1}^{p} A_j^T y_j = b, y_j \geq 0$, where $A_{.j}^T$ denotes the jth column vector of A^T and A_J the jth row of A. Then, it is clear that system (I*) requires that there exists a vector $x \in \mathbb{R}^n$, that makes an obtuse angle $\left(\geq \frac{\pi}{2}\right)$ with the vectors A_1 to A_p and a strictly acute angle $\left(< \frac{\pi}{2}\right)$ with the vector b. The system (II*) requires that the vector b be a nonnegative linear combination of the vectors A_1 to A_p.

Theorem 1.25 *(Gordan's alternative theorem) Let A be an $p \times n$ matrix, then either*
(i) $Ax < 0$ has a solution $x \in \mathbb{R}^n$;
or
(ii) $A^T y = b, y \geq 0$ for some nonzero $y \in \mathbb{R}^p$, but never both.

Theorem 1.26 *(Motzkin's alternative theorem) Let A, B, and C be given $p^1 \times n, p^2 \times n$ and $p^3 \times n$ matrices, with A being nonvacuous. Then, either the system of inequalities (i) $Ax > 0$, $Bx \geq 0$, $Cx = 0$ has a solution $x \in \mathbb{R}^n$; or (ii) $A^T y_1 + B^T y_2 + CT y_3 = 0$, $y_1 \geq 0$, $y_3 \geq 0$ has a solution y_1, y_2, y_3 but never both.*

The following theorem by Slater [256] presents a fairly general theorem of the alternative. The proof can be found in Mangasarian [176].

Theorem 1.27 *(Slater's alternative theorem) Let A, B, C and D be given $p_1 \times n, p_2 \times n, p_3 \times n$, and $p_4 \times n$ matrices with A and B, being nonvacuous. Then, either*
(i) $Ax > 0, Bx \geq 0, Cx \geq 0$ has a solution $x \in \mathbb{R}^n$;
or
(ii) $A^T y_1 + B^T y_2 + CT y_3 + D^T y_4 = 0$, with $y_1 \geq 0$, $y_2 \geq 0, y_3 \geq 0$ or $y_1 \geq 0$, $y_2 > 0, y_3 \geq 0$ has a solution y_1, y_2, y_3, y_4 but never both.

The following theorem from Mangasarian [176] provides the alternative theorem for the most general systems:

Theorem 1.28 *Let A, B, C, and D be given $p_1 \times n, p_2 \times n, p_3 \times n,$ and $p_4 \times n$ matrices with A and B being nonvacuous. Then, either*
(i) $Ax \geq 0, Bx \geqq 0, Cx \geqq 0, Dx = 0$ or $Ax \geqq 0, Bx > 0, Cx \geqq 0, Dx = 0$ has a solution $x \in \mathbb{R}^n$
or
(ii) $A^T y_1 + B^T y_2 + C^T y_3 + D^T y_4 = 0$, with $y_1 \geq 0$, $y_2 \geqq 0, y_3 \geqq 0$ or $y_1 \geqq 0$, $y_2 > 0, y_3 \geqq 0$ has a solution y_1, y_2, y_3, y_4 but never both.

Now, we state the following alternative result for the system of convex and linear vector valued functions. For the proof, we refer to Mangasarian [176].

Lemma 1.4 *Let $X \subseteq \mathbb{R}^n$ be a convex set and $f_1 : X \to \mathbb{R}^{m_1}$, $f_2 : X \to \mathbb{R}^{m_2}$, and $f_3 : X \to \mathbb{R}^{m_3}$ be convex vector valued functions and $f_4 : X \to \mathbb{R}^{m_4}$ be a linear vector valued function. If*

$$\left\langle \begin{array}{l} f_1(x) < 0, f_2(x) \leq 0, f_3(x) \leqq 0 \\ f_4(x) = 0, \end{array} \right\rangle \text{ has no solution } x \in X,$$

then, there exist $p^1 \in \mathbb{R}^{m_1}, p^2 \in \mathbb{R}^{m_2}, p^3 \in \mathbb{R}^{m_3}, p^4 \in \mathbb{R}^{m_4}$, such that

$$\left\langle \begin{array}{l} p_1, p_2, p_3 \geqq 0, (p_1, p_2, p_3, p_4) \neq 0 \\ p_1 f_1(x) + p_2 f_2(x) + p_3 f_3(x) + p_4 f_4(x) \geq 0, \ \forall x \in X \end{array} \right\rangle.$$

1.10   Subdifferential Calculus

The nonsmooth phenomena occur naturally and frequently in optimization theory. For example, the well-known absolute value function $f(x) = |x|$ is an example of convex function, which is differentiable everywhere except at $x = 0$. Moreover, $x = 0$ is the global minimizer of f on $\mathbb{R}$. We know that for a function to be differentiable, both the left and right handed derivative must exist and be equal. But this is not the case with nondifferentiable functions. Therefore, the characterization of the convex function in terms of derivative (gradient) is not applicable for nondifferentiable convex functions. This led to the generalization of the notion of gradients to subgradients. However, for nondifferentiable convex functions, the one sided derivatives exist universally. In this section, we will study subdifferentials for convex functions, Dini, Dini-Hadamard derivatives, and the Clarke subdifferentials for locally Lipschitz functions.

1.10.1   Convex Subdifferentials

We start this section, with an important notion of the Lipschitzian property, that plays a very important role in the theory of nonsmooth analysis:

Definition 1.70 (Locally Lipschitz function) *A function $f : X \subseteq \mathbb{R}^n \to \mathbb{R}$ is said to be locally Lipschitz (of rank M) near $z \in X$, if there exist a positive constant M and a neighborhood N of z, such that for any $x, y \in N$, one has*

$$|f(y) - f(x)| \le M \|y - x\|.$$

The function f is said to be a locally Lipschitz on X if the above condition is satisfied for all $z \in X$.

The following theorem presents that if a convex function is bounded above in a neighborhood of some point, then it is bounded as well as a locally Lipschitz in that neighborhood. The proof is along the lines of Roberts and Verberg [237].

Theorem 1.29 *Let $X \subseteq \mathbb{R}^n$ be an open convex set. Let $f : X \to \mathbb{R}$ be a convex function on X and bounded above on a neighborhood of some point of X. Then f is bounded as well as Lipschitz near every $x \in X$.*

Proof Without loss of generality, we assume that f is bounded above by M on the set $B_\varepsilon(0) \subseteq X$. Let $x \in X$ be arbitrary. First, we prove that f is bounded on a neighborhood of x. Choose $\mu > 0$ such that $y = \mu x \in X$. If $\lambda = \frac{1}{\mu}$, then the set

$$\Omega := \{\nu : \nu = (1 - \lambda)\bar{x} + \lambda y, \bar{x} \in B_\varepsilon(0)\}$$

is a neighborhood of $x = \lambda y$ with radius $(1 - \lambda)\varepsilon$. Now, for all $\nu \in \Omega$, by the convexity of f, we have

$$f(\nu) \le (1 - \lambda) f(\bar{x}) + \lambda f(y) \le K + \lambda f(y).$$

Therefore, f is bounded above on a neighborhood of x. If z is any point in the $B_{(1-\lambda)\varepsilon}(x,)$, there exist another such point $\bar{z}$ such that $x = (z + \bar{z})/2$. Therefore,

$$f(x) \le \frac{1}{2} f(z) + \frac{1}{2} f(\bar{z}),$$

which implies that

$$f(z) \ge 2f(x) - f(\bar{z}) \ge 2f(x) - M - \lambda f(y).$$

Thus f is bounded below near x, hence, we have established that f is bounded near x.

Now, to prove that f is Lipschitz near x. Suppose that $|f| \le N$ on the set $B_{2\delta}(x)$, where $\delta > 0$. Let x_1 and x_2 be any two distinct points in $B_{2\delta}$. Setting

$$x_3 = x_2 + \frac{\delta}{\alpha}(x_2 - x_1),$$

where $\alpha = \|x_2 - x_1\|$. It is clear that $x_3 \in B_{2\delta}(x)$. Solving for x_2, we get

$$x_2 = \frac{\delta}{\alpha + \delta} x_1 + \frac{\alpha}{\alpha + \delta} x_3.$$

Therefore, by convexity of f, we get

$$f(x_2) \leq \frac{\delta}{\alpha + \delta} f(x_1) + \frac{\alpha}{\alpha + \delta} f(x_3).$$

Then,

$$f(x_2) - f(x_1) \leq \frac{\alpha}{\alpha + \delta} [f(x_3) - f(x_1)] \leq \frac{\alpha}{\delta} |f(x_3) - f(x_1)|.$$

Since, $|f| \leq N$ and $\alpha = \|x_2 - x_1\|$, it results that

$$f(x_2) - f(x_1) \leq \frac{2N}{\delta} |x_2 - x_1|.$$

Since, the role of x_1 and x_3 can be changed, we conclude that f is Lipschitz near x.

Definition 1.71 *(Directional derivatives)* *Let $f : \mathbb{R}^n \to \bar{\mathbb{R}}$ be a function and $x \in \mathbb{R}^n$ be a point, where f be finite. Then,*

(i) The right sided directional derivative of f at x in direction of the vector $d \in \mathbb{R}^n$, is defined as the limit

$$f'_+(x; d) = \lim_{\lambda \to 0^+} \frac{f(x + \lambda d) - f(x)}{\lambda},$$

if it exists, provided that $+\infty$ and $-\infty$ are allowed as limits.

(ii) The left sided directional derivative of f at x in direction of the vector $d \in \mathbb{R}^n$, is defined as the limit

$$f'_-(x; d) = \lim_{\lambda \to 0^-} \frac{f(x + \lambda d) - f(x)}{\lambda}.$$

if it exists, provided that $+\infty$ and $-\infty$ are allowed as limits.

It is easy to see that

$$f'_+(x; -d) = -f'_-(x; d), \ \forall d \in \mathbb{R}^n.$$

The two sided directional derivative $f'(x; d)$ is said to exist, if $f'_+(x; d)$ and $f'(x; -d)$ both exist and are equal, i.e.,

$$f'_+(x; d) = f'(x; -d) = f'(x; d).$$

We know that if f is differentiable at x, then the two sided directional derivative exists and is finite. Moreover,

$$f'(x; d) = \langle \nabla f(x), d \rangle, \forall d \in \mathbb{R}^n,$$

where $\nabla f(x)$ is the gradient of f at x.

The following theorem from Rockafellar [238] establishes that for a convex function, the one sided directional derivatives always exist.

Theorem 1.30 *Let $f : \mathbb{R}^n \to \bar{\mathbb{R}}$ be a convex function and $x \in \mathbb{R}^n$ be a point, where f be finite. Then, for each $d \in \mathbb{R}^n$, the difference quotient*

$$\frac{f(x + \lambda d) - f(x)}{\lambda}$$

is a nondecreasing function of $\lambda > 0$, so that $f'_+(x;d)$ and $f'_-(x;d)$ both exist and

$$f'_+(x;d) = \inf_{\lambda > 0} \frac{f(x + \lambda d) - f(x)}{\lambda},$$

$$f'_-(x;d) = \sup_{\lambda < 0} \frac{f(x + \lambda d) - f(x)}{\lambda}.$$

Moreover, $f'(x;d)$ is a positively homogeneous convex function of d, with $f'(x;0) = 0$ and

$$f'_+(x;d) \geq -f'(x;-d), \, \forall d \in \mathbb{R}^n.$$

To deal with nondifferentiable convex functions, the usual notion of gradient has been replaced by the notion of subgradients.

Definition 1.72 (Subgradient) *Let $f : X \subseteq \mathbb{R}^n \to \bar{\mathbb{R}}$ be a convex function and let $x \in \mathbb{R}^n$ be a point, where f be finite. Then, a vector $\xi \in \mathbb{R}^n$ is said to be subgradient of f at x, if*

$$f(y) \geq f(x) + \langle \xi, y - x \rangle, \, \forall y \in \mathbb{R}^n.$$

We note that the affine function $f(x) + \langle \zeta, y - x \rangle$ is a supporting hyperplane to the epigraph of f at $(x, f(x))$ with slope ξ. Moreover, at a point, where the function is not differentiable, there can be an infinite number of such supporting hyperplanes. The collection of slopes of all these supporting hyperplanes of the function f at a point x is called subdifferential at x and is denoted by $\partial f(x)$, that is,

$$\partial f(x) = \{\xi \in \mathbb{R}^n : f(y) - f(x) \geq \langle \xi, y - x \rangle, \forall y \in \mathbb{R}^n\}.$$

The subdifferential $\partial f(x)$ of a convex function f at a point x is the image of a set valued map $\partial f : \mathbb{R}^n \to 2^{\mathbb{R}^n}$, known as subdifferential mapping. For more details about ∂f as a set valued map or multifunction, we refer to John [122] and Rockafellar [238].

We know that if C is a convex set, then the indicator function $\delta_C : \mathbb{R}^n \to \bar{\mathbb{R}}$ is a proper convex function. Then the subdifferential of δ_C at $x \in C$, is given by

$$\partial \delta_C(x) = \{\xi \in \mathbb{R}^n : \delta_C(x) - \delta_C(y) \geq \langle \xi, y - x \rangle, \forall y \in \mathbb{R}^n\}$$
$$= \{\xi \in \mathbb{R}^n : 0 \geq \langle \xi, y - x \rangle, \forall x \in \mathbb{R}^n\},$$

which is nothing but the normal cone to the set C at x, i.e., $N_C(x)$. Hence,

$$\partial \delta_C(x) = N_C(x).$$

The following proposition summarizes some basic properties of the convex subdifferential. The proof may be found in Bazaraa *et al.* [17], Bertsekas *et al.* [23], and Rockafellar [238].

Proposition 1.27 *Let* $f : \mathbb{R}^n \to \mathbb{R}$ *be a convex function. Then*

1. $\partial f(x)$ *is nonempty, convex and compact set for all* $x \in \mathbb{R}^n$.

2. $\partial f(x) = \left\{ \xi \in \mathbb{R}^n : f'_+(x;d) \geq \langle \xi, d \rangle, \forall d \in \mathbb{R}^n \right\}$.

3. $f'_+(x;.)$ *is the support function of the set* $\partial f(x)$ *at* $x \in \mathbb{R}^n$.

4. *If* f *is differentiable at* x *with gradient* $\nabla f(x)$, *then unique subgradient is the gradient, i.e.,* $\partial f(x) = \{ \nabla f(x) \}$.

Now, we present the mean value theorem for convex subdifferentiable functions. The proof may be found in Hiriart-Urruty and Lemarechal [105].

Theorem 1.31 *(Mean value theorem) Let* $f : \mathbb{R}^n \to \mathbb{R}$ *be a convex function. Then, for* $x, y \in \mathbb{R}^n$, *there exists* $z \in]x, y[$ *and* $\xi \in \partial f(z)$, *such that*

$$f(y) - f(x) = \langle \xi, y - x \rangle.$$

In other words,

$$f(y) - f(x) \in \langle \partial f(z), y - x \rangle.$$

The following theorem from Rockafellar [238] states that for a proper convex function f, the multifunction ∂f is a monotone map.

Theorem 1.32 *Let* $f : \mathbb{R}^n \to \bar{\mathbb{R}}$ *be a proper convex function. Then the subdifferential is a monotone map, i.e.,*

$$\langle \xi_1 - \xi_2, x - y \rangle \geq 0, \ \forall \xi_1 \in \partial f(x), \xi_2 \in \partial f(y).$$

Moreover, ∂f *is a maximal monotone map, that is, its graph is not properly contained in the graph of any other monotone map.*

Now, we state without proof, the following theorem on subdifferential calculus. The proof may be found in Bertsekas *et al.* [23] and Rockafellar [238].

Theorem 1.33 *Let* $X \subseteq \mathbb{R}^n$ *be a nonempty convex set and* $f, f_i : X \to \mathbb{R}, i = 1, 2, ..., m$ *be convex functions. Then,*

(i) *For* $\lambda > 0$ *and* $x \in X$, $\partial(\lambda f)(x) = \lambda \partial f(x)$.

(ii) $\partial(f_1 + f_2 + ... + f_m)(x) = \partial f_1(x) + \partial f_2 + ... + \partial f_m(x)$.

1.10.2 Dini and Dini-Hadamard Derivatives

For convex functions, the one sided directional derivatives exist universally. However, this is not the case with the nonconvex functions. For nonconvex functions, the limit in the definitions of directional derivatives may not exist. This led to the generalizations of the notion of directional derivatives to the Dini and Dini-Hadamard directional derivtives.

Definition 1.73 *Let* $f : X \to \mathbb{R}, X \subseteq \mathbb{R}^n$ *be an open convex set. Let* $x \in \mathbb{R}^n, \nu \in \mathbb{R}^n$ *with* $\nu^T \nu = 1$. *Dini derivatives of* f *in the direction* ν *are defined as follows:*

$$D^+ f(x, \nu) := \lim_{n \to \infty} \sup_{\{t_n\}} \left\{ \frac{f(x + t_n \nu) - f(x)}{t_n} : 0 < t_n < \frac{1}{n} \right\};$$

$$D_+ f(x, \nu) := \lim_{n \to \infty} \inf_{\{t_n\}} \left\{ \frac{f(x + t_n \nu) - f(x)}{t_n} : 0 < t_n \leq \frac{1}{n} \right\};$$

$$D^- f(x, \nu) := \lim_{n \to \infty} \sup_{\{t_n\}} \left\{ \frac{f(x - t_n \nu) - f(x)}{-t_n} : 0 < t_n \leq \frac{1}{n} \right\};$$

$$D_- f(x, \nu) := \lim_{n \to \infty} \inf_{\{t_n\}} \left\{ \frac{f(x - t_n \nu) - f(x)}{-t_n} : 0 < t_n \leq \frac{1}{n} \right\},$$

where $D^+ f(x; \nu)$ *is the upper right derivative,* $D_+ f(x; \nu)$ *is the lower right derivative,* $D^- f(x; \nu)$ *is the upper left derivative and* $D_- f(x; \nu)$ *is the lower left derivative evaluated at* x *in the direction* ν. *Limits can be infinite in the above definition. By using the definitions, we may prove that*

(1) Dini derivatives always exist (finite or infinite) for any function f *and satisfy*

$$D^+ f(x; \nu) \geq D_+ f(x; \nu), D^- f(x; \nu) \geq D_- f(x; \nu).$$

Since, $D^- f(x; \nu) = -D_+ f(x; -\nu), D_- f(x; \nu) = -D^+ f(x; -\nu)$, *therefore, it is quite obvious to deal with the directional Dini derivatives* $D^+ f(x; \nu)$ *and* $D_+ f(x; \nu)$.

(2) If $D^+ f(x; \nu) = D_+ f(x; \nu)$, *then, the common value, written as* $f'_+(x, \nu)$ *is clearly the right derivative of* f *at* x *in the direction* ν, *given by*

$$f'_+ x; \nu = \lim_{t \to 0^+} \frac{f(x + t\nu) - f(x)}{t}.$$

Moreover, if $f'_+(x; \nu)$ exists and is finite, then the function f is called differentiable (or Dini differentiable) at x in the direction ν. The function f is called Dini directionally differentiable at the point x, if f is Dini differentiable at x, for every $\nu \in \mathbb{R}^n$. If $f : \mathbb{R} \to \bar{\mathbb{R}}$ be any function, then $D^+ f(x; 1), D^- f(x; 1), D_+ f(x; 1)$, and $D_- f(x; 1)$ are denoted by $D^+ f(x), D^- f(x), D_+ f(x)$, and $D_- f(x)$, respectively. Indeed, we have

$$D^+ f(x) := \lim\sup_{t \to x^+} \frac{f(t) - f(x)}{t - x} \text{ and } D^+ f(x) := \lim\inf_{t \to x^+} \frac{f(t) - f(x)}{t - x}.$$

A continuous function may not have even a one sided directional derivative at a point but it may have the Dini directional derivative at that point. The following example from Ansari *et al.* [2] illustrates the fact:

Example 1.8 *Let us consider the function* $f : \mathbb{R} \to \mathbb{R}$, *defined by*

$$f(x) := \begin{cases} |x| \ |cos(\frac{1}{x})|, & \text{if } x \neq 0, \\ 0, & \text{if } x = 0. \end{cases}$$

Since, $|cos\left(\frac{1}{x}\right)| \leq 1$ *for all* $x \neq 0$, *we have*

$$\lim_{x \to 0} f(x) = 0 = f(0).$$

Therefore, f *is continuous at* $x = 0$. *It is clear that* f *is also continuous at all the other points of* $\mathbb{R}$, *so* f *is continuous function. However, inspection of the difference quotient reveals that*

$$\limsup_{x \to 0+} \frac{f(x) - f(0)}{x - 0} = 1, \quad \text{while } \liminf_{x \to 0+} \frac{f(x) - f(0)}{x - 0} = 0,$$

hence, $f'_+(0)$ *does not exist. Similarly,* $f'_-(0)$ *does not exist.*
      However, we have

$$D^+ f(0) = 1, D^- f(0) = 0, D_+ f(0) = 1, D^- f(0) = 0,$$

Moreover, we can see that $f'(x)$ *exists everywhere, except at* $x = 0$, *and all the four Dini derivatives have the same value.*

It is known from Giorgi and Komlosi [88] that for a function $f : I \to \mathbb{R}$, defined on real interval I, if the Dini derivatives $D^+ f(0), D^- f(0), D_+ f(0), D_- f(0)$, are finite at each point of I, then f is differentiable, almost everywhere on I. However, if all of the Dini derivatives are not finite, then continuity is in general not ensured. The following example illustrates the fact.

Example 1.9 *(Giorgi and Komlosi [88]) Consider the function* $f : \mathbb{R} \to \mathbb{R}$, *defined by*

$$f(x) := \begin{cases} 0, & \text{if } x > 0 \text{ and } x \text{ is irrational,} \\ 1, & \text{if } x < 0 \text{ or } x \text{ is rational.} \end{cases}$$

Then, for $x = 0$, *we have*

$$D^+ f(x) = 0, D^- f(x) = 0, D_+ f(x) = -\infty, D_- f(x) = 0.$$

Moreover, it is easy to see that the function f *is not continuous at* $x = 0$.

The important feature of the Dini derivatives is that they always exist and admit useful calculus rules. The following theorem summarizes some elementary properties and calculus rules for the Dini upper (Dini lower) directional derivatives. The proof follows directly from the definitions of the Dini upper (Dini lower) directional derivatives. For further study, we refer to Mc Shane [181].

Theorem 1.34 *Let $f, g : \mathbb{R}^n \to \mathbb{R}$ be real-valued functions. The following assertion holds:*

*(1) **(Homogeneity)** $D^+ f(x; v)$ is positively homogeneous in v, that is,*

$$D^+ f(x; rv) = r D^+ f(x; v), \forall r > 0.$$

*(2) **(Scalar multiplication)** For $r > 0, D^+(rf)(x; v) = r D^+ f(x; v)$, and for $r < 0, D^+(rf)(x; v) = r D_- f(x; v)$.*

*(3) **(Sum rule)***

$$D^+(f + g)(x; v) \le D^+(x; v) + D^+ g(x; v),$$

provided that the sum on the right hand side exists.

*(4) **(Product rule)***

$$D^+(fg)(x; v) \le D^+[g(x)f](x; v) + D^+[f(x)g](x; v),$$

provided that the sum on the right hand side exists, the functions f and g are continuous at x and that one of the following conditions is satisfied: $f(x) \ne 0; g(x) \ne 0; D^+ f(x; v)$ is finite; and $D^+ g(x; v)$ is finite.

*(5) **(Quotient rule)***

$$\left(D^+ \left(\frac{f}{g}\right)\right)(x; v) \le \frac{D^+[g(x)f](x; v) + D^+[-f(x)g](x; v)}{[g(x)]^2},$$

provided that the right hand side exists and the function g is continuous at x.

In addition, if the functions f and g are directionally differentiable at x, then, the inequalities in the last three assertions become equalities.

Similarly, the properties and calculus rules for the Dini lower directional derivatives can be obtained.

Next, we present the following theorem, which shows that the Dini upper and Dini lower directional derivatives can be used conveniently for characterizing an extremum of a function.

Theorem 1.35 *For a function $f : \mathbb{R}^n \to \mathbb{R}$, the following assertions hold:*

(a) *If $f(x) \leq f(x + t\nu)$ (respectively, $f(x) \geq f(x + t\nu)$,) for all $t > 0$ are sufficiently small, then $D_+f(x;\nu) \geq 0$ (respectively, $D^+f(x;\nu) \leq 0$). In particular, if f is directionally differentiable at x and $f(x) \leq f(x + t\nu)$ (respectively, $f(x) \geq f(x+t\nu)$), for all $t > 0$ are sufficiently small, then, $f'(x;\nu) \geq 0$ (respectively, $f'(x;\nu) \leq 0$).*

(b) *If $D^+f(x + t\nu;\nu) \geq 0$ for all $x, \nu \in \mathbb{R}^n$ and $t \in \,]0,1[$ and if the function $t \mapsto f(x + t\nu)$ is continuous on $[0,1]$, then, $f(x) \leq f(x + \nu)$.*

Proof (a) From the definitions of the Dini lower and Dini upper directional derivatives, the proof follows.

(b) On the contrary, suppose that $f(x) > f(x + \nu)$ for some $x, \nu \in \mathbb{R}^n$. We consider the function $h : [0,1] \to \mathbb{R}$, defined by

$$h(t) = f(x + t\nu) - f(x) + t(f(x) - f(x + \nu)).$$

Clearly, h is continuous on $[0,1]$ and $h(0) = h(1) = 0$. Hence, there exists $\bar{t} \in [0,1[$, such that h has a maximum value at $\bar{t}$. Set $y := x + \bar{t}\nu$. Then,

$$h(\bar{t}) \geq h(\bar{t} + t), \forall t \in [0, 1 - \bar{t}].$$

Hence, we have

$$f(y + t\nu) - f(y) \leq t(f(x + \nu) - f(x)),$$

for all $t > 0$ are sufficiently small.

Dividing the above inequality by t and taking the limit superior as $t \to 0^+$, it follows that

$$D^+f(y;\nu) = \limsup_{t \to 0^+} \frac{f(y + t\nu) - f(y)}{t} \leq f(x + \nu) - f(x) < 0,$$

which is a contradiction to our assumption. This completes the proof.

We know that the mean value theorem plays a key role in the classical theory of the differential calculus. The following mean value theorem for upper semicontinuous functions of one variable from Diewert [65] provided an extension for nondifferentiable functions.

Theorem 1.36 (Diewert's mean value theorem) *If $f : [a,b] \to \mathbb{R}$ is an upper semicontinuous function, then, there exists $c \in [a,b[$, such that*

$$D_+f(c) \leq D^+f(c) \leq \frac{f(b) - f(a)}{b - a}.$$

Proof Let $\alpha = \frac{f(b)-f(a)}{b-a}$. Let us define a function $h : [a,b] \to \mathbb{R}$ by

$$h(t) = f(t) - \alpha t.$$

Since f is upper semicontinuous on $[a, b]$, it follows that h is also upper semicontinuous on the compact set $[a, b]$. By Berge's maximum theorem [22], there exists a point $c \in [a, b]$ such that h attains its maximum at c. Then, we get

$$h(c) \geq h(t), \forall t \in [a, b].$$

That is,

$$f(t) - f(c) \leq \alpha(t - c), \forall t \in [a, b].$$

If $c \in [a, b[$, then from the above inequality, we get

$$\frac{f(t) - f(c)}{t - c} \leq \alpha, \forall\, t \in\,]c, b].$$

From the above inequality, it follows that

$$D_+ f(c) \leq D_+ f(c) \leq \alpha.$$

This completes the proof.

The following mean value theorem for lower semicontinuous functions can be proved by using the same argument as in Theorem 1.36.

Corollary 1.9 *Let $f : [a, b] \to \mathbb{R}$ be a lower semicontinuous function. Then, there exists $c \in [a, b[$, such that*

$$D^+ f(c) \geq D_+ f(c) \geq \frac{f(b) - f(a)}{b - a}.$$

The following mean value theorem for continuous functions may be obtained by using Theorem 1.36 and Corollary 1.8.

Corollary 1.10 *If $f : [a, b] \to \mathbb{R}$ is a continuous function, then there exist $c, d \in [a, b[$ such that*

$$D^+ f(c)(b - a) \leq f(b) - f(a) \leq D_+ f(d)(b - a).$$

Now, we state the following important result. The proof is based on Theorem 1.36.

Corollary 1.11 *Let $f : [a, b] \to \mathbb{R}$ be an upper semicontinuous function. If for all $c \in [a, b[, D^+ f(c) \geq 0$ (respectively, $D^+ f(c) \leq 0$), then f is a nondecreasing (respectively, nonincreasing) function on $[a, b]$.*

Proof Assume that $D^+ f(c) \geq 0$, for all $c \in [a, b[$. On the contrary, suppose that f is strictly decreasing on $[a, b]$. Then there exist $t_1, t_2 \in [a, b[$ such that $t_1 \leq t_2$ and $f(t_1) > f(t_2)$. By Theorem 1.36 there exists $t \in [t_1, t_2[$ such that

$$D^+ f(t) \leq \frac{f(t_2) - f(t_1)}{t_2 - t_1},$$

which implies that $D^+ f(t) < 0$, which is a contradiction to our assumption. By using Corollary 1.8, we can show that f is nonincreasing on $[a, b]$ if $D^+ f(c) \leq 0$ for all $c \in [a, b[$.

We now give Diewert's mean value theorem for functions defined on a convex subset of $\mathbb{R}^n$. The proof is along the lines of Ansari *et al.* [2].

Theorem 1.37 *(Diewert's mean value theorem)* *Let X be a nonempty convex subset of $\mathbb{R}^n$ and $f : X \to \mathbb{R}$ be a real-valued function. Then, for every pair of distinct points x and y in X, the following assertions hold.*

(i) *If f is radially upper semicontinuous on X, then there exists $z \in [x, y[$, such that*

$$D^+ f(z; y - x) \leq f(y) - f(x).$$

(ii) *If f is radially lower semicontinuous on X, then there exists $w \in [x, y[$, such that*

$$D_+ f(w; y - x) \geq f(y) - f(x).$$

(iii) *If f is radially continuous on X, then there exists $\nu \in [x, y[$ such that*

$$D^+ f(\nu; y - x) \leq f(y) - f(x) \leq D_+ f(\nu; y - x).$$

Moreover, if the Dini upper directional derivative $f^D(\nu; y - x)$ is continuous in ν on the line segment $[x, y[$, then there exists a point $w \in [x, y[$, such that

$$D_+ f(w; y - x) = f(y) - f(x).$$

Proof Let x and y be two distinct points in X. Define a function $h : [0, 1] \to \mathbb{R}$ by

$$h(t) = f(x + t(y - x)).$$

(a) If f is radially upper semicontinuous on X, then h is an upper semicontinuous function on $[0, 1]$. By Theorem 1.36, there exists $\bar{t} \in [0, 1[$ such that

$$D^+ h(\bar{t}) \leq h(1) - h(0).$$

If we set $w = x + \bar{t}(y - x)$, then we have

$$D^+ h(\bar{t}) = D^+ f(w; y - x),$$

which lead to the results.

(b) By using Corollary 1.8, the proof follows.

(c) By using Corollary 1.9, the proof follows. Moreover, if $D^+ f(\nu; y - x)$ is continuous in ν on the line segment $[, y[$, then by the intermediate value theorem, there exists a point $w \in [x, y[$ such that

$$D^+ f(w; y - x) = f(y) - f(x).$$

A generalization of the Dini (upper and lower) directional derivative is the Dini-Hadamard (upper and lower) directional derivative.

Definition 1.74 *Let $f : \mathbb{R}^n \to \bar{\mathbb{R}}$ be a function and $x \in \mathbb{R}^n$ be a point where f is finite.*

(a) *The Dini-Hadamard upper directional derivative of f at x in the direction $nu \in \mathbb{R}^n$ is defined by*

$$f^{DH}(x; \nu) := \limsup_{u \to \nu, t \to 0+} \frac{f(x + tu) - f(x)}{t}.$$

(b) *The Dini-Hadamard lower directional derivative of f at x in the direction $\nu \in \mathbb{R}^n$ is defined by*

$$f_{DH}(x; \nu) := \liminf_{u \to \nu, t \to 0+} \frac{f(x + tu) - f(x)}{t}.$$

If $f^{DH}(x; \nu) = f_{DH}(x; \nu)$, then we denote it by $f^{DH*}(x; \nu)$, that is,

$$f^{DH*}(x; \nu) = \lim_{u \to \nu, t \to 0+} \frac{f(x + tu) - f(x)}{t}.$$

From the definitions of the Dini upper(lower) directional derivative and the Dini-Hadamard upper (lower) directional derivative, we can easily obtain the following relations:

$$(-f)^{DH}(x; \nu) = -f_{DH}(x; \nu), \quad (-f)_{DH}(x; \nu) = -f^{DH}(x; \nu),$$

$$f_{DH}(x; \nu) \leq f_D(x; \nu) \leq f^D(x; \nu) \leq f^{DH}(x; \nu).$$

The following example illustrates that the Dini (upper and lower) directional derivative and the Dini-Hadamard(upper and lower) directional derivative are different.

Example 1.10 *Let $f : \mathbb{R}^2 \to \mathbb{R}$ be a function defined by*

$$f(x_1, x_2) = \begin{cases} 0, & \text{if } x_2 = 0, \\ x_1 + x_2, & \text{if } x_2 \neq 0. \end{cases}$$

Let $x = (0, 0)$ and $\nu = e_1 = (1, 0)$. Then, by an easy calculation, we get

$$f^{DH}(x; d) = 1 \text{ and } f^D(x; d) = 0.$$

The following result shows that the Dini upper (lower) directional derivative and the Dini-Hadamard upper (lower) directional derivative at a point coincide, if f is locally Lipschitz around that point.

Theorem 1.38 *Let f be locally Lipschitz around a point $x \in \mathbb{R}^n$. Then, for every $d \in \mathbb{R}^n$,*

$$f^{DH}(x; d) = f^D(x; d) \text{ and } f_{DH}(x; d) = f_D(x; d).$$

Proof Let $N(x)$ be a neighborhood of x and f be Lipschitz continuous on $N(x)$ with Lipschitz constant k. Let $d \in \mathbb{R}^n$ be arbitrary. Then there exist $\delta > 0$ and $\tau > 0$ such that τ for all $v \in \mathbb{R}^n$ and $t \in \mathbb{R}$ satisfying the conditions $\|v - d\| < \delta$ and $0 < t < \tau$, we have $x + td, x + tv \in N(x)$. Consequently,

$$|f(x + tv) - f(x + td)| \leq kt\|v - d\|.$$

Therefore,

$$\limsup_{v \to d, t \to 0+} \frac{f(x + td) - f(x + tv)}{t} = 0.$$

By applying the properties of lim sup, we obtain

$$f^{DH}(x; d) = \limsup_{v \to d, t \to 0+} \frac{f(x + td) - f(x)}{t}$$

$$\leq \limsup_{v \to d, t \to 0+} \frac{f(x + td) - f(x)}{t} + \limsup_{v \to d, t \to 0+} \frac{f(x + tv) - f(x + td)}{t}$$

$$= f^D(x; d).$$

Since $f^D(x; d) \leq f^{DH}(x; d)$, we have the equality.

Similarly, we can prove that $f_D(x; d) = f_{DH}(x; d)$.

Next, we state the following result about upper (lower) semicontinuity of the Dini-Hadamard derivatives.

Theorem 1.39 *Let $f : \mathbb{R}^n \to \mathbb{R}$ be a function. Then $f_{DH}(x; d)$ (respectively, $f^{DH}(x; d)$) is a lower (respectively, upper) semicontinuous function of d.*

For further details about the Dini and Dini-Hadamard derivatives, we refer to Giorgi and Komlosi [88, 89, 90] and Schirotzek [245].

1.10.3   Clarke Subdifferentials

The concept of subdifferentials for general nonconvex, locally Lipschitz functions was introduced by Clarke [49]. To deal with the problems in which the smoothness of the data is not necessarily postulated, Clarke [49] has developed the following basic concepts:

Definition 1.75 *Let $f : \mathbb{R}^n \to \mathbb{R}$ be a locally Lipschitz function near x. The Clarke generalized directional derivative of f at x in the direction of a vector $\nu \in \mathbb{R}^n$, denoted by $f^0(x; \nu)$, is defined as*

$$f^0(x; \nu) := \limsup_{\substack{y \to x \\ t \downarrow 0}} \frac{f(y + t\nu) - f(y)}{t}.$$

Definition 1.76 *Let $f : \mathbb{R}^n \to \mathbb{R}$ be a locally Lipschitz function on X. The Clarke generalized subdifferential of f at x, denoted by $\partial^c f(x)$, is defined as*

$$\partial^c f(x) := \left\{ \xi \in \mathbb{R}^n : f^0(x; \nu) \geq \langle \xi, \nu \rangle, \forall \nu \in \mathbb{R}^n \right\}.$$

In fact, a locally Lipschitz function f is not differentiable everywhere. Rademacher's theorem (see Evans and Gariepy [76]) states that a locally Lipschitz function is differentiable almost everywhere (a.e.) in the sense of Lebesgue measure, i.e., a set of points where f is not differentiable, forms a set of measure zero.

Theorem 1.40 *Let* $f : \mathbb{R}^n \to \mathbb{R}$ *be a locally Lipschitz function near* x. *Let* $S \subseteq \mathbb{R}^n$ *be a set with Lebesgue measure zero. Then, at any* $x \in \mathbb{R}^n$, *the Clarke generalized subdifferential is given by*

$$\partial^c f(x) = \mathrm{co}\left\{\xi \in \mathbb{R}^n : \xi = \lim_{i \to \infty} \nabla f(x_i), x_i \to x, x_i \notin \Omega_f \bigcup S\right\},$$

where Ω_f *denotes the set of points, where* f *is not differentiable.*

The following proposition from Clarke [49], summarizes some basic properties for the Clarke generalized gradient.

Proposition 1.28 *Let* $f : \mathbb{R}^n \to \mathbb{R}$ *be a locally Lipschitz function near* x, *with Lipschitz constant* M. *Then*

1. *The function* $f^0(x;.)$ *is a finite, positively homogeneous function of* ν *and satisfies*

$$\left|f^0(x;\nu)\right| \leq M\|\nu\|.$$

2. $f^0(x;\nu)$ *is Lipschitz of rank* M *as a function of* ν *and upper semicontinuous as a function of* $(x;\nu)$.

3. $f^0(x;-\nu) = (-f)^0(x;\nu)$.

The following proposition states some important properties of the Clarke generalized subdifferential.

Proposition 1.29 *Let* $f : \mathbb{R}^n \to \mathbb{R}$ *be a locally Lipschitz function near* x *with Lipschitz constant* M. *Then,*

1. *For each* x, $\partial^c f(x)$ *is nonempty, convex and compact subset of* $\mathbb{R}^n$. *Moreover,*

$$\|\xi\| \leq M, \quad \forall \xi \in \partial^c f(x).$$

2. *For every* $v \in \mathbb{R}^n$,

$$f^0(x;\nu) = \max\left\{\langle \xi, v \rangle : \xi \in \partial^c f(x)\right\}.$$

That is, $f^0(x;.)$ *is the support function of the set* $\partial^c f(x)$.

The following lemma states that the Clarke subdifferential for locally Lipschitz functions are closed sets.

Lemma 1.5 *Let $f : X \to \mathbb{R}$ be a locally Lipschitz function at $x \in K$. If $\{x_n\}$ and $\{\zeta_n\}$ are two sequences in $\mathbb{R}^n$ such that, $\zeta_n \in \partial^c f(x_n)$, for all n and if $x_n \to x$ and ζ is a cluster point of $\{\zeta_n\}$, then $\zeta \in \partial^c f(x)$.*

The following theorem states that for convex functions, the Clarke and convex subdifferentials coincide. The proof follows along the lines of Clarke [49].

Theorem 1.41 *Let $X \subseteq \mathbb{R}^n$ be an open convex set and $f : X \to \mathbb{R}$ is convex and locally Lipschitz with rank M on X. Then $\partial^c f(x)$ coincides with $\partial f(x)$, for each $x \in X$. Moreover, $f^0(x; \nu)$ coincides with $f'(x; \nu)$, for each $x \in X$ and $\nu \in \mathbb{R}^n$.*

Proof From Theorem 1.30, it is clear that if f is a convex function, then $f'_+(x; \nu)$ exists for each $\nu \in \mathbb{R}^n$ and $f'_+(x; .)$ is the support function of ∂f at x. Therefore, we need to prove only that

$$f'_+(x; \nu) = f^0(x; \nu), \ \forall \nu \in \mathbb{R}^n.$$

Now, the Clarke generalized gradient $f^0(x; \nu)$, may be written as

$$f^0(x; \nu) = \lim_{\varepsilon \downarrow 0} \sup_{\|\bar{x} - x\| \leq \varepsilon\delta} \sup_{0 < \lambda < \varepsilon} \frac{f(\bar{x} + \lambda\nu) - f(\bar{x})}{\lambda},$$

where δ is any positive real number. By Theorem 1.30, for a convex function, the difference quotient

$$\frac{f(\bar{x} + \lambda\nu) - f(\bar{x})}{\lambda}$$

is a nondecreasing function of λ. Hence,

$$f^0(x; \nu) = \lim_{\varepsilon \downarrow 0} \sup_{\|\bar{x} - x\| \leq \varepsilon\delta} \frac{f(\bar{x} + \varepsilon\nu) - f(\bar{x})}{\varepsilon}.$$

Now, by the locally Lipschitz property of f, for any $\bar{x} \in B_x(\varepsilon)$, we have

$$\left| \frac{f(\bar{x} + \varepsilon\nu) - f(\bar{x})}{\varepsilon} - \frac{f(x + \varepsilon\nu) - f(x)}{\varepsilon} \right| \leq 2\delta M.$$

Therefore,

$$f^0(x; \nu) \leq \lim_{\varepsilon \downarrow 0} \frac{f(x + \varepsilon v) - f(x)}{\varepsilon} + 2M\delta = f'_+(x; \nu) + 2M\delta.$$

Since, δ is arbitrary, we get $f^0(x; \nu) \leq f'_+(x; v)$. Moreover, $f'_+(x; \nu) \leq f^0(x; \mu)$ is always true. Hence, the equality follows.

The following theorem provides a characterization for convexity in terms of the Clarke subdifferential.

Theorem 1.42 *Let $f : X \subseteq \mathbb{R}^n \to \mathbb{R}$ be a locally Lipschitz on open convex set X. Then f is a convex function on X if and only if $\partial^c f$ is a monotone map on X.*

Next, we present some basic calculus rules, which will be used frequently. Before, that we give the definition of a strictly differentiable function.

Definition 1.77 (Strictly differentiable function) *Let $f : X \subseteq \mathbb{R}^n \to \mathbb{R}$ be any function. Then, strict directional derivative of f at the point x in the direction $\nu \in \mathbb{R}^n$, is given by*

$$f^s (x; \nu) = \lim_{\substack{\bar{x} \to x \\ \lambda \downarrow 0}} \frac{f (\bar{x} + \lambda \nu) - f (\bar{x})}{\lambda},$$

provided the limit exists. The function f is said to admit a strict derivative at $x \in X$, denoted by $D_s f (x)$, if for each $\nu \in \mathbb{R}^n$ the following holds

$$f^s (x; \nu) = \langle D_s f (x), \nu \rangle.$$

It is known that if f is a continuously differentiable function at x, then it is strictly differentiable at x.

The following proposition characterizes the relationship between a locally Lipschitz property and strict differentiability.

Proposition 1.30 *Let $X \subseteq \mathbb{R}^n$ be an open set and $f : X \to \mathbb{R}$ be any map. Let ξ be any vector in $\mathbb{R}^n$. Then the following statements are equivalent:*

1. f is strictly differentiable at x and $D_s f (x) = \xi$.

2. f is locally Lipschitz around x and $f^s (x; v) = \langle \xi, v \rangle, \forall v \in \mathbb{R}^n$.

Proposition 1.31 *Let $f : \mathbb{R}^n \to \mathbb{R}$ be a locally Lipschitz function near x. Then, for any scalar s, one has*

$$\partial^c (\lambda f) (x) = \lambda \partial^c f (x).$$

Proposition 1.32 *Let $f_i : \mathbb{R}^n \to \mathbb{R}, i = 1, ..., m$ be a finite family of locally Lipschitz functions near x. Then,*

$$\partial^c (f_1 + f_2 + ... + f_m) (x) \subseteq \partial^c f_1 (x) + \partial^c f_2 (x) + ... + \partial^c f_m (x).$$

Moreover, the equality holds if all but at most one of the function f_i are strictly differentiable at x.

Proposition 1.33 *Let $f_i : \mathbb{R}^n \to \mathbb{R}, i = 1, ..., m$ be a finite family of locally Lipschitz function near x. Then, for any scalar $\lambda_i, i = 1, ..., m$, one has*

$$\partial^c \left(\sum_{i=1}^m \lambda_i f_i \right) (x) \subseteq \lambda_i \sum_{i=1}^m \partial^c f_i (x).$$

Moreover, the equality holds, if all but at most one of the function f_i are strictly differentiable at x.

The following notion of regular functions helps to sharpen the subdifferential calculus rules by turning inclusions to equality.

Definition 1.78 *Let $f : \mathbb{R}^n \to \mathbb{R}$ be a locally Lipschitz function near x. Then f is said to be regular at x, if*

(i) *For all $\nu \in \mathbb{R}^n$, the one sided directional derivative $f'_+(x; \nu)$ exists.*

(ii) *For all $\nu \in \mathbb{R}^n, f'_+(x; \nu) = f^0(x; \nu)$.*

Under the regularity assumption, Propositions 1.32 and 1.33 take the following form.

Proposition 1.34 *Let $f_i : \mathbb{R}^n \to \mathbb{R}, i = 1, ..., m$ be a finite family of locally Lipschitz function near x. Suppose, each $f_i, i = 1, ..., m$ be regular at $x \in X$. Then,*

$$\partial^c (f_1 + f_2 + ... + f_m)(x) = \partial^c f_1(x) + \partial^c f_2(x) + ... + \partial^c f_m(x).$$

In addition, if each λ_i is nonnegative, then,

$$\partial^c \left(\sum_{i=1}^m \lambda_i f_i \right)(x) = \lambda_i \sum_{i=1}^m \partial^c f_i(x).$$

Proposition 1.35 *Let $f, g : \mathbb{R}^n \to \mathbb{R}$ be a locally Lipschitz function near x and suppose that $g(x) \neq 0$. Then $\frac{f}{g}$ is locally Lipschitz near x and one has*

$$\partial^c \left(\frac{f}{g} \right)(x) \subseteq \frac{g(x)\partial^c f(x) - f(x)\partial^c g(x)}{(g(x))^2}.$$

If in addition $f(x) \geq 0, g(x) > 0$ and if f and $-g$ are regular at x, then the equality holds and $\frac{f}{g}$ is regular at x.

Next, we state the following important chain rule for the Clarke subdifferential.

Proposition 1.36 *(Chain rule) Let $h : \mathbb{R}^n \to \mathbb{R}^m$ and $g : \mathbb{R}^m \to \mathbb{R}$ be the given functions, such that each h be strictly differentiable at x and g is Lipschitz near $h(x)$. Then $f = g \circ h$ is Lipschitz near x and one has*

$$\partial^c f(x) \subseteq \partial^c g(h(x)) \circ D_s h(x).$$

The equality holds, if g(or $-g$) is regular at $h(x)$.

Next, we state the following well-known mean value theorem. The proof may be found in Clarke [49] and Schirotzek [245].

Theorem 1.43 *(Lebourg mean value theorem) Let x and y be points in X and suppose that $f : X \subseteq \mathbb{R}^n \to \mathbb{R}$ be locally Lipschitz on an open set containing the line segment $[x, y]$. Then, there exists a point $u \in]x, y[$, such that*

$$f(y) - f(x) \in \langle \partial^c f(u), y - x \rangle,$$

where $]x, y[$ denotes the line segment joining x and y excluding the end points.

The following proposition describes the role of the distance function in exact penalization. The proof may be found in Clarke [49].

Definition 1.79 *Let $C \subseteq \mathbb{R}^n$ be a nonempty set. The Clarke tangent cone of C denoted by $T_C(x)$ is defined as*

$$T_C(x) := \left\{ y \in \mathbb{R}^n : d_C^0(x, y) = 0 \right\}$$

where $d_C(.)$ denotes the distance function related to C. It is easy to show that $T_C(x)$ is a closed convex cone containing 0.

The normal cone to the set C at x is defined as a cone polar to the tangent cone $T_C(x)$, as follows.

Definition 1.80 *(Clarke normal cone) Let $C \subseteq \mathbb{R}^n$ be a nonempty set. The Clarke normal cone of C at $x \in C$, denoted by $N_C(x)$, is defined as*

$$N_C(x) := \left\{ \xi \in \mathbb{R}^n : \langle \xi, \nu \rangle \leq 0, \forall \nu \in T_C(x) \right\}.$$

It follows that, $N_C(x)$ is the closed convex cone generated by $\partial^c d_C(x)$. Moreover, if C is a convex set, then $N_C(x)$ coincides with the normal cone in the sense of convex analysis.

The following proposition provides the characterization of a normal cone in terms of a generalized gradient.

Proposition 1.37 *Let $C \subseteq \mathbb{R}^n$ be a nonempty set and let $x \in C$, then,*

$$N_C(x) = \mathrm{cl} \left\{ \bigcup_{\lambda \geq 0} \lambda \partial^c d_C(x) \right\}.$$

Proposition 1.38 *Let $f : X \subseteq \mathbb{R}^n \to \mathbb{R}$ be locally Lipschitz of rank M on X and let $\bar{x}$ belong to a set $C \subseteq X$. Suppose that f attains a minimum at $\bar{x}$. Then for any $\bar{M} \geq M$, the function $g(y) = f(y) + \bar{M} d_C(y)$ attains a minimum over X at $\bar{x}$. If $\bar{M} > M$ and C is closed, then any other point minimizing g over X lies within C.*

1.11   Optimality Criteria

In the simplest case, an optimization problem consists of maximizing or minimizing a real-valued single objective function, by systematically choosing input values from within an allowed set, known as the feasible set and computing the value of the function. Depending upon the feasible set, the optimization problems have been classified as unconstrained and constrained optimization problems.

1.11.1   Unconstrained Minimization Problem

Consider an optimization problem of the form

$$(UMP) \quad \min f(x)$$

$$\text{subject to } x \in \mathbb{R}^n,$$

where $f : \mathbb{R}^n \to \mathbb{R}$ is a given function. The problem (UMP) is referred to as an unconstrained minimization problem. Unconstrained minimization problems arise seldom in practical applications. However, optimality conditions for constrained minimization problems become logical extension of the optimality conditions for unconstrained minimization problems.

Definition 1.81 *(Local and global minimum) A vector $\bar{x} \in \mathbb{R}^n$ is said to be a local minimum for (UMP), if there exists an $\varepsilon > 0$, such that*

$$f(\bar{x}) \leq f(x), \ \forall x \in B_\varepsilon(\bar{x}) \cap \mathbb{R}^n.$$

It is said to be global minimum, if

$$f(\bar{x}) \leq f(x), \ \forall x \in \mathbb{R}^n.$$

It is clear that a global minimum is also a local minimum.

Theorem 1.44 *Suppose that $f : \mathbb{R}^n \to \mathbb{R}$ is differentiable at $\bar{x}$ and there exists a vector $d \in \mathbb{R}^n$ such that $\langle \nabla f(\bar{x}), d \rangle < 0$. Then there exists some $\bar{\lambda} > 0$, such that*

$$f(\bar{x} + \lambda d) < f(\bar{x}), \ \forall \lambda \in \,]0, \bar{\lambda}[\,.$$

Proof Since, f is differentiable at $\bar{x}$, we have

$$f(\bar{x} + \lambda d) = f(\bar{x}) + \lambda \langle \nabla f(\bar{x}), d \rangle + \lambda \|d\| \varphi(\bar{x}, \lambda d), \qquad (1.15)$$

where $\varphi(\bar{x}, \lambda d) \to 0$ as $\lambda \to 0$. Now, the equation (1.16) can be rewritten as

$$\frac{f(\bar{x} + \lambda d) - f(\bar{x})}{\lambda} = \langle \nabla f(\bar{x}), d \rangle + \|d\| \varphi(\bar{x}, \lambda d). \qquad (1.16)$$

By our assumption, $\langle \nabla f (\bar{x}), d \rangle < 0$ and $\varphi (\bar{x}, \lambda d) \to 0$ as $\lambda \to 0$. Therefore, there exists some $\bar{\lambda} > 0$, such that

$$\langle \nabla f (\bar{x}), d \rangle + \|d\| \, \varphi (\bar{x}, \lambda d) < 0, \ \forall \lambda \in \,]0, \bar{\lambda}[\, .$$

Hence, by (1.17), the required result follows.

Now, we state the following well-known optimality conditions. The proof may be found in Luenberger [173].

Proposition 1.39 *Suppose that $\bar{x}$ is a local minimum for (UMP) and f is differentiable at $\bar{x}$. Then,*

$$\nabla f (\bar{x}) = 0.$$

Proof Suppose to the contrary that $\nabla f (\bar{x}) \neq 0$. Then setting $d = -\nabla f (\bar{x})$, we get

$$\langle \nabla f (\bar{x}), d \rangle = - \|\nabla f (\bar{x})\|^2 < 0.$$

Therefore, by Theorem 1.38, there exists $\bar{\lambda} > 0$, such that

$$f (\bar{x} + \lambda d) < f (\bar{x}), \ \forall \lambda \in \,]0, \bar{\lambda}[\, ,$$

a contradiction to our assumption, that $\bar{x}$ is a local minimum for (UMP). Hence, $\nabla f (\bar{x}) = 0$.

Proposition 1.40 *Suppose that $\bar{x}$ is a local minimum for (UMP) and f is a twice differentiable function at $\bar{x}$. Then, $\nabla^2 f (\bar{x})$ is positive semidefinite.*

Proof Since, f is a twice differentiable function at $\bar{x}$, therefore, for arbitrary $d \in \mathbb{R}^n$, we have

$$f (\bar{x} + \lambda d) = f (\bar{x}) + \lambda \langle \nabla f (\bar{x}), d \rangle + \frac{1}{2} \lambda^2 \langle d, \nabla^2 f (\bar{x}) d \rangle + \lambda^2 \|d\|^2 \varphi (\bar{x}, \lambda d),$$
$$(1.17)$$

where $\varphi (\bar{x}, \lambda d) \to 0$ as $\lambda \to 0$. Since, $\bar{x}$ is a local minimum, by Proposition 1.39, we get

$$\nabla f (\bar{x}) = 0. \tag{1.18}$$

Using (1.19), the equation (1.18) can be rewritten as

$$\frac{f (\bar{x} + \lambda d) - f (\bar{x})}{\lambda^2} = \frac{1}{2} \langle d, \nabla^2 f (\bar{x}) d \rangle + \|d\|^2 \varphi (\bar{x}, \lambda d). \tag{1.19}$$

Since, $\bar{x}$ is a local minimum for (UMP), $f (\bar{x} + \lambda d) \geq f (\bar{x})$, for λ sufficiently small. Therefore, from (1.20), it follows that

$$\frac{1}{2} \langle d, \nabla^2 f (\bar{x}) d \rangle + \|d\|^2 \varphi (\bar{x}, \lambda d) \geq 0,$$

for λ sufficiently small.

By taking the limit as $\lambda \to 0$, it follows that

$$\langle d, \nabla^2 f(\bar{x}) d \rangle \geq 0.$$

Since, d is arbitrary, hence, $\nabla^2 f(\bar{x})$ is positive semidefinite.

The following results provide an optimality condition for the unconstrained optimization problem (UMP) in terms of the Dini-Hadamard derivative.

Theorem 1.45 *If $x \in \mathbb{R}$ is a local minimum of the function (UMP), then, we have*

$$f_{DH}(x; d) \geq 0, \forall d \in \mathbb{R}^n.$$

Proof Since $x \in \mathbb{R}^n$ is a local minimum point of the function f, then for each $d \in \mathbb{R}^n$, each v in the neighborhood of d and each $t > 0$ sufficiently small, we have

$$\frac{f(x + tv) - f(x)}{t} \geq 0.$$

Thus,

$$f_{DH}(x; d) = \liminf_{v \to d, t \to 0+} \frac{f(x + tv) - f(x)}{t} \geq 0.$$

This completes the proof.

In terms of the Dini derivatives, we have the following optimality conditions for the unconstrained optimization problem (UMP).

Corollary 1.12 *If $x \in \mathbb{R}^n$ is a local minimum of the function $f : \mathbb{R}^n \to \mathbb{R}$, then,*

$$D^+ f(x; d) \geq 0, \forall d \in \mathbb{R}^n.$$

Next, we state the following necessary and sufficient optimality conditions, which describe one of the most important properties possessed by convex as well as pseudoconvex functions, that is, every stationary point is a global minimum.

Theorem 1.46 *Suppose that f is a differentiable convex function at $\bar{x}$. Then $\bar{x}$ is a global minimum for (UMP) if and only if*

$$\nabla f(\bar{x}) = 0.$$

Proof If $\bar{x}$ is a global minimum for (UMP), then it is local minimum for (UMP). Therefore, by Proposition 1.39, we get

$$\nabla f(\bar{x}) = 0.$$

Now, suppose conversely that $\nabla f(\bar{x}) = 0$. Therefore, for each $x \in \mathbb{R}^n$, we have

$$\langle \nabla f(\bar{x}), x - \bar{x} \rangle = 0. \tag{1.20}$$

By the convexity of f at $\bar{x}$, we get

$$f(x) - f(\bar{x}) \geq \langle \nabla f(\bar{x}), x - \bar{x} \rangle, \forall x \in \mathbb{R}^n. \tag{1.21}$$

From (1.21) and (1.22), we get

$$f(x) \geq f(\bar{x}), \forall x \in \mathbb{R}^n.$$

Hence, $\bar{x}$ is a global minimum for (UMP).

Remark 1.12 *If f is a differentiable pseudoconvex function at $\bar{x}$, then, from (1.21) and the pseudoconvexity of f at $\bar{x}$, we get*

$$f(x) \geq f(\bar{x}), \forall x \in \mathbb{R}^n.$$

Next, we present the following theorem, which states that for a strictly quasiconvex function, every local minimizer is a global one.

Theorem 1.47 *Let $f : \mathbb{R}^n \to \mathbb{R}$ be a differentiable function. Let $\bar{x} \in \mathbb{R}^n$ be a local minimizer of (UMP). If f is a strictly quasiconvex function at $\bar{x}$, then $\bar{x}$ is a global minimum of (UMP).*

Proof If $\bar{x}$ is a local minimum, then there exists an open ball $\mathbb{B}_\varepsilon(\bar{x})$, such that

$$f(\bar{x}) \leq f(x), \ \forall x \in B_\varepsilon(\bar{x}) \cap \mathbb{R}^n.$$

On the contrary, suppose that there exist $\hat{x} \in \mathbb{R}, \hat{x} \notin B_\varepsilon(\bar{x})$, such that

$$f(\hat{x}) < f(\bar{x}).$$

By the strict quasiconvexity of f, we get

$$f(\lambda \hat{x} + (1 - \lambda) \bar{x}) < f(\bar{x}), \text{ for } \lambda \in]0, 1[. \tag{1.22}$$

For $\lambda < \frac{\delta}{\|\hat{x} - \bar{x}\|}$, we have that

$$(1 - \lambda) \bar{x} + \lambda \hat{x} \in B_\varepsilon(\bar{x}) \cap \mathbb{R}^n.$$

Then, we have

$$f(\bar{x}) \leq f((1 - \lambda) \bar{x} + \lambda \hat{x}), \text{ for } 0 < \lambda < \frac{\delta}{\|\hat{x} - \bar{x}\|},$$

which is a contradiction to (1.23). This completes the proof.

The following theorem states that the set of global minimizers of a quasiconvex function is a convex set.

Theorem 1.48 *Let $f : \mathbb{R}^n \to \mathbb{R}$ be a quasiconvex function on the convex set X. Let $\bar{X}$ be the set of all global minimizers of f. Then $\bar{X}$ is a convex set.*

Proof If $\bar{X} = \phi$, then the result is trivially true. Let α be the minimum value of f on $\mathbb{R}^n$. By the definition of global minimizers, we have

$$\bar{X} = \{x \in X : f(x) = \alpha\} = \{x \in X : f(x) \leq \alpha\}.$$

The convexity of $\bar{X}$ follows by Theorem 1.14.

The following theorem gives the optimality condition for the unconstrained minimization problem (UMP), if the objective function f is convex.

Theorem 1.49 *Let the function $f : \mathbb{R}^n \to \mathbb{R}$ be a convex function in the unconstrained minimization problem (UMP). Then $\bar{x} \in \mathbb{R}^n$ is a global minimizer of (UMP) if and only if*

$$0 \in \partial f(\bar{x}).$$

Proof Suppose that $\bar{x} \in \mathbb{R}^n$ is a global minimizer of (UMP). Therefore, by the definition of minimizer, we have

$$f(x) - f(\bar{x}) \geq 0, \ \forall x \in \mathbb{R}^n.$$

By the definition of subdifferential, we get

$$0 \in \partial f(\bar{x}).$$

Again, using the definition of subdifferentials, the converse can be proved.

1.11.2   Constrained Minimization Problems

Consider the optimization problem

$$(CMP) \quad \min f(x)$$

$$\text{subject to } x \in X \subseteq \mathbb{R}^n,$$

where $f : \mathbb{R}^n \to \mathbb{R}$ is a given function. The problem (CMP) is referred to as a constrained minimization problem.

Definition 1.82 *(Local and global minimum) A vector $\bar{x} \in X$ is said to be a local minimum for (CMP), if there exists an $\varepsilon > 0$, such that*

$$f(\bar{x}) \leq f(x), \ \forall x \in B_\varepsilon(\bar{x}) \cap X.$$

It is said to be a global minimum, if

$$f(\bar{x}) \leq f(x), \ \forall x \in X.$$

From Figure 1.5, it is clear that a global minimum is also a local minimum for (CMP), but not conversely.

When X is a convex set, the following optimality conditions for the problem (CMP) holds. The proof can be found in Mangasarian [176], Luenberger [173], and Bazaraa *et al.* [17].

Proposition 1.41 *Suppose that X is a convex set and for some $\varepsilon > 0$ and $\bar{x} \in X$, the function f is a continuously differentiable function over $B_\varepsilon(\bar{x})$. Then, the following statement holds:*

1. *If $\bar{x}$ is a local minimum for (CMP), then*

$$\langle \nabla f(\bar{x}), x - \bar{x} \rangle \geq 0, \ \forall x \in X. \tag{1.23}$$

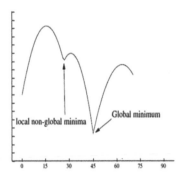

FIGURE 1.5: Local and Global Minima for CMP

2. *In addition if f is a convex function over X and (1.24) holds, then $\bar{x}$ is a global minimum for (CMP).*

The constrained minimization problem (CMP) can be converted to an unconstrained minimization (UMP) by using indicator functions. For (CMP), associated unconstrained minimization problem may be formulated as:

$$\min \bar{f}(x)$$

$$\text{subject to } x \in \mathbb{R}^n,$$

where $\bar{f} : \mathbb{R}^n \to \bar{\mathbb{R}}$ is a function given by $\bar{f}(x) = f(x) + \delta_X(x)$, that is,

$$\bar{f}(x) = \left\{ \begin{array}{ll} f(x), & \text{if} \quad x \in X, \\ \infty, & \text{if} \quad \text{otherwise.} \end{array} \right.$$

Using indicator functions, the optimality condition for constrained optimization problem (CMP), may be given as follows.

Theorem 1.50 *Let X be a convex set and the function $f : X \subseteq \mathbb{R}^n \to \mathbb{R}$ be a convex function in the constrained minimization problem (CMP). Then $\bar{x} \in \mathbb{R}^n$ is a minimizer of (CMP) if and only if*

$$0 \in \partial(f + \delta_X)(\bar{x}).$$

Next, we state the following theorem, which gives the necessary optimality condition for (CMP), when the objective function is locally Lipschitz.

Theorem 1.51 *Suppose that $f : \mathbb{R}^n \to \mathbb{R}$ be locally Lipschitz of rank M near $\bar{x}$ and let $\bar{x}$ belong to a set $C \subseteq \mathbb{R}^n$. Suppose that f attains a minimum over C at $\bar{x}$. Then*

$$0 \in \partial^c f(\bar{x}) + N_C(\bar{x}).$$

Proof Let N be a neighborhood of $\bar{x}$ upon which f is Lipschitz of rank M. Since C and $C \cap N$ have the same normal cone at $\bar{x}$, we may assume that $C \subseteq X$. By Proposition 1.38, $\bar{x}$ is a local minimizer of the function $f(x) + M d_C(x)$. Therefore,

$$0 \in \partial^c (f + M d_C)(\bar{x}) \subseteq \partial^c f(\bar{x}) + M \partial d_C(\bar{x}).$$

By Proposition 1.37, we get

$$0 \in \partial f(\bar{x}) + N_C(\bar{x}).$$

.

1.11.3   Scalar Optimization Problems and Optimality Criteria

The optimization problems, where the constrained set X is described explicitly by inequality constraints, arise frequently in optimization theory. Analytically, a single objective (scalar) optimization problem with explicit inequality constraints may be formulated as follows:

$$(SOP) \quad \min f(x)$$

$$\text{subject to } g_j(x) \leq 0, \ j = 1, 2 ..., m,$$

where $f : X \to \mathbb{R}, g_j : X \to \mathbb{R}, j = 1, 2, ..., m$ are differentiable functions on an open set $X \subseteq \mathbb{R}^n$. Let $S = \{x : x \in X, g(x) \leq 0\}$ denote the set of all feasible solutions of the problem (SOP). Suppose that $J(\bar{x}) = \{j \in \{1, 2, ..., m\} : g_j(\bar{x}) = 0\}$ denotes the set of active constraint index set at $\bar{x}$.

Optimality criteria form the foundation of mathematical programming both theoretically and computationally. The conditions that must be satisfied at the optimum point are called necessary optimality conditions. If any point does not satisfy the necessary optimality condition, it cannot be optimum. However, not every point that satisfies the necessary condition is optimal. The well-known necessary optimality conditions for a mathematical programming problem are Fritz John [85] and Kuhn-Tucker [151] type necessary optimality conditions. The latter condition is referred to as Karush-Kuhn-Tucker optimality conditions to give credit to Karush [135], who had derived these conditions in his Master's thesis in 1939 (see Boyd and Vandenberghe [29]).

Now, we state the following Fritz John necessary optimality conditions. For the proof we refer to Mangasarian [176].

Theorem 1.52 *(Fritz John necessary optimality conditions) Let $\bar{x}$ be a local minimum of the (SOP). Then, there exist multipliers $\lambda_0 \in \mathbb{R}$ and $\mu \in \mathbb{R}^m$, such that*

$$\lambda_0 \nabla f(\bar{x}) + \sum_{j=1}^{m} \mu_j \nabla g_j(\bar{x}) = 0, \tag{1.24}$$

$$\mu_j g_j(\bar{x}) = 0, \quad j = 1, 2, ..., m, \tag{1.25}$$

$$\lambda_0 \geq 0, \mu \geq 0 \; and \; (\lambda, \mu) \neq 0. \tag{1.26}$$

The conditions (1.25)–(1.26) are known as Fritz John condition and the condition (1.27) is known as complementary slackness condition.

We note that in case $\lambda_0 = 0$, the objective function f disappears and we have a degenerate case. To avoid degeneracy, we apply some regularity conditions on the constraints. These regularity conditions, which ensure that $\lambda_0 \neq 0$, are referred to as constraint qualifications.

Next, we state Slater's constraint qualification, which is frequently used in nonlinear optimization.

Definition 1.83 *(Slater's constraint qualification) The problem (SOP) is said to satisfy the Slater constraint qualification if each g_j is convex (or pseudoconvex) and there exists a feasible point $\bar{x} \in S$, such that*

$$g_j(\bar{x}) < 0, \; j = 1, 2, ..., m.$$

Besides Slater's constraint qualification, there are several other constraint qualifications for (SOP), such as the Mangasarian Fromovitz constraint qualification, Karlin's constraint qualification, linear independent constraint qualification, strict constraint qualification, and others. For further details about different constraint qualifications for (SOP) and relations between them, we refer to Mangasarian [176].

If $\lambda_0 \neq 0$, we can take $\lambda_0 = 1$. The Fritz John condition then reduces to the famous Karush-Kuhn-Tucker optimality conditions stated as follows:

Theorem 1.53 *(Karush-Kuhn-Tucker necessary optimality conditions) Let $\bar{x}$ be a local minimum of the (SOP) and a suitable constraint qualification is satisfied at $\bar{x}$. Then, there exist multiplier $\mu \in \mathbb{R}^m$, such that*

$$\nabla f(\bar{x}) + \sum_{j=1}^{m} \mu_j \nabla g_j(\bar{x}) = 0,$$

$$\mu_j g_j(\bar{x}) = 0, \; j = 1, 2, ..., m.$$

$$\mu_j \geq 0, j = 1, 2, ..., m.$$

Of course, one would like to have the same criterion be both necessary as well as sufficient. The validity of the Karush-Kuhn-Tucker necessary optimality condition does not guarantee the optimality of $\bar{x}$. However, these conditions become sufficient under somewhat ideal conditions which are rarely satisfied in practice. The above necessary optimality conditions become sufficient if f and g have some kind of convexity. In the presence of convexity it is very convenient to find an optimal solution for the (SOP).

Now, we state the Karush-Kuhn-Tucker sufficient optimality conditions for (SOP).

Theorem 1.54 (Karush-Kuhn-Tucker sufficient optimality condition) *Let $\bar{x}$ be a feasible solution for the (SOP) and let the functions $f : X \to \mathbb{R}$ and g_j, $j = 1, 2, ..., m$ be convex and continuously differentiable at $\bar{x}$. If there exist multipliers $0 \le \mu \in \mathbb{R}^m$, such that*

$$\nabla f(\bar{x}) + \sum_{j=1}^{m} \mu_j \nabla g_j(\bar{x}) = 0,$$

$$\mu_j g_j(\bar{x}) = 0, \quad j = 1, 2, ..., m.$$

then, $\bar{x}$ is a global minimum for the (SOP).

1.11.4 Multiobjective Optimization and Pareto Optimality

In practice, we usually encounter with the optimization problems, having several conflicting objectives rather than a single objective, known as multiobjective optimization problems. Multiobjective optimization provides a flexible modeling framework that allows for simultaneous optimization of more than one objective over a feasible set. These problems occur in several areas of modern research such as in analyzing design trade-offs, selecting optimal product or process designs, or any other application, where we need an optimal solution with trade-offs between two or more conflicting objectives.

Because these objectives conflict naturally, a trade-off exists. The set of trade-off designs that cannot be improved upon according to one criterion without hurting another criterion is known as the Pareto set.

Analytically, a nonlinear multiobjective optimization problem (MOP) may be formulated as follows:

$$(MOP) \quad \min f(x) = (f_1(x), f_2(x), ..., f_k(x))$$

$$\text{subject to} \quad g_j(x) \le 0, \quad j = 1, 2, ..., m,$$

where $f : X \to \mathbb{R}^k, g_j : X \to \mathbb{R}$, $j = 1, 2, ..., m$ are differentiable functions on an open set X. Let $S = \{x : x \in X, g(x) \le 0\}$ denote the set of feasible solutions of the problem (MOP). Suppose that $J(\bar{x}) = \{j \in \{1, 2, ..., m\} : g_j(\bar{x}) = 0\}$ denotes the set of active constraint index set at $\bar{x}$.

Remark 1.13 *For $k = 1$, the problem (MOP) reduces to the scalar optimization problem (SOP).*

Edgeworth [71] and Pareto [219] have given the definition of the standard optimality concept via the usage of utility functions. Then it has been extended to the classical notion of Pareto efficiency/optimality defined via an ordering cone. The allocation of resources is *Pareto optimal,* often called the *Pareto efficient,* if it is not possible to change the allocation of resources in such a way as to make some people better off without making others worse off.

Definition 1.84 (Pareto optimal) *A feasible point $\bar{x}$, is said to be Pareto optimal (Pareto efficient) solution for (MOP), if there does not exist another feasible point x, such that*

$$f(x) \leq f(\bar{x})$$

or equivalently,

$$f_i(x) \leq f_i(\bar{x}), \forall i = 1, 2, ..., k, i \neq r,$$

$$f_r(x) < f_r(\bar{x}), \text{ for some } r.$$

All Pareto optimal points lie on the boundary of the feasible criterion space (see Chen *et al.* [44]). Often, algorithms provide solutions that may not be Pareto optimal, but may satisfy other criteria, making them significant for practical applications. For instance, weakly Pareto optimality is defined as follows:

Definition 1.85 *A feasible point $\bar{x}$, is said to be weakly Pareto optimal (weakly Pareto efficient) solution for (MOP), if there does not exist another feasible point x, such that*

$$f(x) < f(\bar{x}),$$

or equivalently,

$$f_i(x) < f_i(\bar{x}), \forall \ i = 1, 2, ..., k.$$

A point is weakly Pareto optimal if there is no other point that improves all of the objective functions, simultaneously. In contrast, a point is Pareto optimal if there is no other point that improves at least one objective function without detriment to another function. Pareto optimal points are weakly Pareto optimal, but weakly Pareto optimal points are not Pareto optimal. All Pareto optimal points may be categorized as being either proper or improper.

Kuhn and Tucker [151] noticed that some of the Pareto optimal solutions had some undesirable properties. To avoid such properties they divided the class of Pareto optimal solutions into properly and improperly Pareto optimal solutions. Proper Pareto optimal solutions are those Pareto optimal solutions, that do not allow the unbounded trade-offs between the objectives. Although there are several definitions of proper Pareto optimal solutions, we present the easiest one by Geoffrion [86].

Definition 1.86 (Proper Pareto optimality) *A feasible solution $\bar{x}$ is said to be proper Pareto optimal (properly efficient) solution of (MOP), if it is efficient and if there exists a scalar $M > 0$, such that for each i,*

$$\frac{f_i(x) - f_i(\bar{x})}{f_r(\bar{x}) - f_r(x)} \leq M,$$

for some r, such that

$$f_r(x) > f_r(\bar{x}), \text{ whenever } x \in S \text{ with } f_i(x) < f_i(\bar{x}).$$

If a Pareto optimal point is not proper, it is improper.

The above quotient is referred to as a trade-off, and it represents the increment in objective function r, resulting from a decrement in objective function i. The definition requires that the trade-off between each function and at least one other function be bounded in order for a point to be proper Pareto optimal. In other words, a solution is properly Pareto optimal if there is at least one pair of objectives, for which a finite decrement in one objective is possible only at the expense of some reasonable increment in the other objective.

1.11.5 Necessary and Sufficient Conditions for Pareto Optimality

Now, we state the following Fritz John type necessary optimality conditions. The proof may be found in Da Cunha and Polak [59].

Theorem 1.55 *(Fritz John type necessary optimality conditions)* Let $f : X \to \mathbb{R}^k$ and $g : X \to \mathbb{R}^m$ be continuously differentiable functions at $\bar{x} \in X$. Then a necessary condition for $\bar{x}$ to be an efficient solution for (MOP) is that there exist multipliers $\lambda \in \mathbb{R}^k$ and $\mu \in \mathbb{R}^m$, such that

$$\sum_{i=1}^{k} \lambda_i \nabla f_i (\bar{x}) + \sum_{j=1}^{m} \mu_j \nabla g_j (\bar{x}) = 0,$$

$$\mu_j g_j(\bar{x}) = 0, \quad j = 1, 2, ..., m,$$

$$\lambda \geqq 0, \quad \mu \geqq 0, (\lambda, \mu) \neq 0.$$

We know that for scalar optimization problem (SOP), the multiplier (λ) of the objective function in the Karush-Kuhn-Tucker type optimality conditions is assumed to be positive rather than being nonnegative. Likewise, in the case of multiobjective optimization problems (MOP), we further need some constraint qualification, so that all the multipliers of the objective functions are not equal to zero.

Next, we present the Kuhn-Tucker constraint qualification.

Definition 1.87 *(Kuhn-Tucker constraint qualification)* Suppose that the constraint functions $g_j, j = 1, 2, ..., m$ of the problem (MOP) be continuously differentiable at $\bar{x} \in S$. The problem (MOP) is said to satisfy the Kuhn-Tucker constraint qualification at $\bar{x}$, if for any $d \in \mathbb{R}^n$ such that

$$\langle \nabla g_j (\bar{x}) , d \rangle \leq 0, \ \forall j \in J (\bar{x}) ,$$

there exists a vector function $\alpha : [0,1] \to \mathbb{R}^n$, which is continuously differentiable at 0 and some real number $\gamma > 0$, such that

$$\alpha (0) = \bar{x}, \ g (\alpha (t)) \leq 0, \forall t \in [0, 1] \ and \ \alpha' (t) = \gamma d.$$

Next, we present the Karush-Kuhn-Tucker necessary optimality conditions:

Theorem 1.56 *(Karush-Kuhn-Tucker necessary condition for Pareto optimality)* *Let $f : X \to \mathbb{R}^k$ and $g : X \to \mathbb{R}^m$ be continuously differentiable functions at $\bar{x} \in X$. Suppose that the Kuhn-Tucker constraint qualification is satisfied at $\bar{x}$. Then, a necessary condition for $\bar{x}$ to be an efficient solution for (MOP) is that there exist multipliers $\lambda \in \mathbb{R}^k$ and $\mu \in \mathbb{R}^m$, such that*

$$\sum_{i=1}^{k} \lambda_i \nabla f_i(\bar{x}) + \sum_{j=1}^{m} \mu_j \nabla g_j(\bar{x}) = 0, \tag{1.27}$$

$$\mu_j g_j(\bar{x}) = 0, \quad j = 1, 2, ..., m, \tag{1.28}$$

$$\lambda \geq 0, \ \lambda \neq 0, \ \mu \geq 0,$$

Proof Suppose that $\bar{x}$ is an efficient solution for (MOP). Consider the following system

$$\left. \begin{array}{l} \langle \nabla f_i(\bar{x}), d \rangle < 0, \ \forall i = 1, 2, ..., k \\ \langle \nabla g_j(\bar{x}), d \rangle \leq 0, \ \forall j \in J(\bar{x}). \end{array} \right\} \tag{1.29}$$

First, we prove that the system (1.30) has no solution $d \in \mathbb{R}^n$. On the contrary suppose that $d \in \mathbb{R}^n$ solves the above system. Then from the Kuhn-Tucker constraint qualification, we know that there exists a function $\alpha : [0, 1] \to \mathbb{R}^n$ which is continuously differentiable at 0 and some real number $\gamma > 0$ such that

$$\alpha(0) = \bar{x}, \ g(\alpha(t)) \leq 0, \forall t \in [0, 1] \ and \ \alpha'(t) = \gamma d.$$

Since, the functions $f_i, i = 1, 2, ..., k$ are continuously differentiable, we can approximate $f_i(\alpha(t))$ linearly as

$$f_i(\alpha(t)) = f_i(\bar{x}) + \langle \nabla f_i(\bar{x}), \alpha(t) - \bar{x} \rangle + \|\alpha(t) - \bar{x}\| \varphi(\alpha(t), \bar{x})$$
$$= f_i(\bar{x}) + \langle \nabla f_i(\bar{x}), \alpha(t) - \alpha(0) \rangle + \|\alpha(t) - \alpha(0)\| \varphi(\alpha(t), \alpha(0))$$
$$= f_i(\bar{x}) + t \left\langle \nabla f_i(\bar{x}), \frac{\alpha(0 + t) - \alpha(0)}{t} \right\rangle + \|\alpha(t) - \alpha(0)\| \varphi(\alpha(t), \alpha(0)),$$

where, $\varphi(\alpha(t), \alpha(0)) \to 0$ as $\|\alpha(t) - \alpha(0)\| \to 0$. As $t \to 0$, $\|\alpha(t) - \alpha(0)\| \to 0$ and

$$\frac{\alpha(0 + t) - \alpha(0)}{t} \to \alpha'(0) = \gamma d.$$

Using the assumption $\langle \nabla f_i(\bar{x}), d \rangle < 0$, for all $i = 1, 2, ..., k$ and $t \geq 0$, we have $f_i(\alpha(t)) < f_i(\bar{x})$, for all $i = 1, 2, ..., k$ and for sufficiently small t. This contradicts the Pareto optimality of $\bar{x}$. Hence, the system (1.30) has no solution $d \in \mathbb{R}^n$. Therefore, invoking the Motzkin's theorem (Theorem 1.26), there exist multipliers $\lambda_i \geq 0, \ i = 1, 2, ..., k, \ \lambda \neq 0$ and $\mu_j \geq 0, j \in J(\bar{x})$ such that

$$\sum_{i=1}^{k} \lambda_i \nabla f_i(\bar{x}) + \sum_{j \in J(\bar{x})} \mu_j \nabla g_j(\bar{x}) = 0,$$

Setting $\mu_j = 0, \forall j \in \{1, 2, ..., m\} \setminus J(\bar{x})$, the equality (1.28) of the theorem follows.

If $g_j(\bar{x}) < 0$, for some $j = 1, 2, ..., m$, then by the above setting $\mu_j = 0$, the equality (1.29) holds.

When objective and constraint functions are convex, we have the following Karush-Kuhn-Tucker type sufficient optimality conditions (see, Miettinen [184]).

Theorem 1.57 (*Karush-Kuhn-Tucker sufficient conditions for Pareto optimality*) *Let $\bar{x}$ be a feasible solution for the (MOP) and let $f_i : X \to \mathbb{R}, i = 1, 2, ..., k$ and $g_j : X \to \mathbb{R}, j = 1, 2, ..., m$ be continuously differentiable and convex function at $\bar{x}$. If there exist multipliers $0 < \lambda \in \mathbb{R}^k$ and $0 \leq \mu \in \mathbb{R}^m$, such that*

$$\sum_{i=1}^{k} \lambda_i \nabla f_i(\bar{x}) + \sum_{j=1}^{m} \mu_j \nabla g_j(\bar{x}) = 0, \qquad (1.30)$$

$$\mu_j g_j(\bar{x}) = 0, \quad j = 1, 2, ..., m, \qquad (1.31)$$

then, $\bar{x}$ is an efficient solution for the (MOP).

Proof Suppose there exists vectors λ and μ such that the conditions of the theorem are satisfied. Define a function $F : \mathbb{R}^n \to \mathbb{R}$ as $F(x) = \sum_{i=1}^{k} \lambda_i f_i(x)$, where $x \in S$. Since, each $f_i, i = 1, 2, ..., k$ is convex at $\bar{x}$, and $\lambda > 0$, therefore, the function F is also convex at $\bar{x}$.

Now, from (1.31) and (1.32), we get

$$F(\bar{x}) + \sum_{j=1}^{m} \mu_j \nabla g_j(\bar{x}) = 0,$$

$$\mu_j g_j(\bar{x}) = 0, \quad j = 1, 2, ..., m.$$

Therefore, by Theorem 1.47, it follows that F attains its minimum at $\bar{x}$. Therefore, we get

$$F(\bar{x}) \leq F(x), \forall x \in S.$$

In other words,

$$\sum_{i=1}^{k} \lambda_i f_i(\bar{x}) \leq \sum_{i=1}^{k} \lambda_i f_i(x), \ \forall x \in S. \qquad (1.32)$$

On the contrary, suppose that $\bar{x}$ is not Pareto optimal. Then, there exists some point $x \in S$, such that

$$f_i(x) \leq f_i(\bar{x}), \forall i = 1, 2, ..., k, i \neq r,$$

$$f_r(x) < f_r(\bar{x}), \text{ for some } r.$$

Since, each $\lambda_i > 0$, $i = 1, 2, ..., k$, it results that

$$\sum_{i=1}^{k} \lambda_i f_i(x) \leq \sum_{i=1}^{k} \lambda_i f_i(\bar{x}),$$

which is a contradiction to (1.33). Hence $\bar{x}$ is an efficient solution for the (MOP).

In optimization theory, the notion of convexity is just a convenient sufficient condition. In fact, most of the time it is not necessary and it is a rather rigid assumption, often not satisfied in real-world applications. In many cases, nonconvex functions provide a more accurate representation of reality. For instance, their presence may ensure that usual first order necessary optimality conditions are also sufficient or that a local minimum is also a global one. This led to the introduction of several generalizations of the classical notion of convexity. See Mishra and Giorgi [189], Mishra *et al.* [202, 203], and the references cited therein.

The following theorem shows that the necessary Karush-Kuhn-Tucker optimality conditions become sufficient under suitable generalized convexity assumptions.

Theorem 1.58 (Miettinen [184]) *Let $\bar{x}$ be a feasible point for the (MOP). Suppose that $f_i, i = 1, 2, ..., p$ be pseudoconvex at $\bar{x}$ and g_j, $j = 1, 2, ..., m$ be quasiconvex at $\bar{x}$. If there exist $\lambda_i \in \mathbb{R}$, $i = 1, 2, ..., p$ and $\mu_j \in \mathbb{R}$, $j = 1, 2, ..., m$, such that*

$$\sum_{i=1}^{p} \lambda_i \nabla f_i(\bar{x}) + \sum_{j=1}^{m} \mu_j \nabla g_j(\bar{x}) = 0,$$

$$\mu_j g_j(\bar{x}) = 0, \quad j = 1, 2, ..., m,$$

$$\lambda \geq 0, \quad \sum_{i=1}^{p} \lambda_i = 1, \quad \mu \geqq 0.$$

Then, $\bar{x}$ is an optimal solution of (MOP).

1.11.6 Nondifferentiable Optimality Conditions

Here, we present necessary and sufficient conditions for Pareto optimality, if the objective and constraint functions of the problem (MOP) are not necessarily differentiable. We assume that the functions are locally Lipschitz and use the Clarke subdifferential and their properties to provide optimality conditions.

First, we state the following Fritz John necessary conditions for Pareto optimality. The proof may be found in Miettinen [184].

Theorem 1.59 *Let $\bar{x}$ be a feasible solution for the (MOP). Assume that the functions $f_i : X \to \mathbb{R}, i = 1, 2, ..., k$ and $g_j : X \to \mathbb{R}, j = 1, 2, ..., m$ be locally Lipschitz at $\bar{x}$. Then $\bar{x}$ is a Pareto optimal solution for (MOP) if there exists multipliers $\lambda \in \mathbb{R}^p$ and $\mu \in \mathbb{R}^m$, such that*

$$0 \in \sum_{i=1}^{k} \lambda_i \partial^c f_i(\bar{x}) + \sum_{j=1}^{m} \mu_j \partial^c g_j(\bar{x}),$$

$$\mu_j g_j(\bar{x}) = 0, \quad j = 1, 2, ..., m,$$

$$\lambda \geqq 0, \ \mu \geqq 0, (\lambda, \mu) \neq 0.$$

Remark 1.14 *For $k = 1$, the above Fritz John optimality condition for (MOP) reduces to that for (SOP). See for example, Clarke [49] or Kiwiel [143].*

To move from the Fritz John to the Karush-Kuhn-Tucker optimality condition, we need constraint qualifications. The constraint qualifications for nondifferentiable case are different from those for differentiable case. Frequently used constraint qualifications for nondifferentiable (MOP)s are Calmness, Mangasarian Fromovitz, and Cottle constraint qualifications. For details about these and other constraint qualifications, we refer to Dolezal [69], Ishizuka and Shimijhu [109].

Here, we present the Cottle constraint qualification.

Definition 1.88 *(Cottle constraint qualification) Suppose that the functions $f_i : X \to \mathbb{R}, i = 1, 2, ..., k$ and $g_j : X \to \mathbb{R}, j = 1, 2, ..., m$ are locally Lipschitz at $\bar{x}$. Then, (MOP) is said to satisfy the Cottle constraint qualification at $\bar{x}$, if either*

$$g_j(\bar{x}) < 0, j = 1, 2, ..., m$$

or

$$0 \in \text{co} \left\{ \partial^c g_j(\bar{x}) : g_j(\bar{x}) = 0 \right\}.$$

Under Cottle constraint qualification and convexity assumptions on the functions, the following Karush-Kuhn-Tucker necessary and sufficient optimality conditions hold for nondifferentiable (MOP):

Theorem 1.60 *(Miettinen [184]) Let $\bar{x}$ be a feasible solution for the (MOP) and the Cottle constraint qualification be satisfied at $\bar{x}$. Let $f_i : X \to \mathbb{R}, i = 1, 2, ..., k$ and $g_j : X \to \mathbb{R}, j = 1, 2, ..., m$ be locally Lipschitz and convex function at $\bar{x}$. A necessary and sufficient optimality condition for $\bar{x}$ to be an efficient solution is that there exist multipliers $\lambda \in \mathbb{R}^p$ and $\mu \in \mathbb{R}^m$, such that*

$$0 \in \sum_{i=1}^{k} \lambda_i \partial^c f_i(\bar{x}) + \sum_{j=1}^{m} \mu_j \partial^c g_j(\bar{x}),$$

$$\mu_j g_j(\bar{x}) = 0, \quad j = 1, 2, ..., m,$$

$$\lambda \geq 0, \ \sum_{i=1}^{k} \lambda_i = 1, \ \mu \geqq 0.$$

1.12 Duality

One of the most interesting, useful and fundamental aspects of linear as well as nonlinear programming is duality theory. It provides a theoretical foundation for many optimization algorithms. Duality can be used to directly solve nonlinear programming problems as well as to derive lower bounds in other high-level search techniques of the solution quality. In constrained optimization, it is often used in a number of constraint decomposition schemes such as separable programming and in space decomposition algorithms such as branch and bound.

Wolfe [285] used the Karush-Kuhn-Tucker conditions to formulate a dual program for a nonlinear optimization problem in the spirit of duality in linear programming, that is, with the aim of defining a problem whose objective value gives lower bound on the optimal value of the original or primal problem and whose optimal solution yields an optimal solution for the primal problem under certain regularity conditions. See Mishra *et al.* [203]. For (MOP), the Wolfe dual model (WD) may be formulated as follows:

$$(WD) \quad \max f(u) + \sum_{j=1}^{m} \mu_j g_j(u)$$

$$\text{subject to } \sum_{i=1}^{p} \lambda_i \nabla f_i(u) + \sum_{j=1}^{m} \mu_j \nabla g_j(u) = 0,$$

$$(\lambda, \mu) \geq 0, u \in X \subseteq \mathbb{R}^n.$$

Mangasarian [176] has pointed out that whereas some results such as sufficiency and converse duality hold for (P) and Wolfe dual (WD), if f is only pseudoconvex and g quasiconvex. However, by means of a counterexample, Mangasarian [176] also showed that weak and strong duality theorems do not hold for these functions for Wolfe dual (WD). We present an example from Mond [206]:

$$(P1) \quad \min x + x^3$$

$$\text{subject to } x \geq 1.$$

It is easy to show that $x = 1$ is the optimal solution and the optimal value is 2.

The Wolfe dual for the problem (P1) may be formulated as

$$(WD1) \quad \max u + u^3 + \mu(1 - u)$$

$$\text{subject to } 3u^2 + 1 - \mu \geq 0, \ \mu \geq 0.$$

It is obvious that the objective function of (WD1) tends to $+\infty$, when $u \to -\infty$.

One of the reasons that in Wolfe duality the convexity cannot be weakened to pseudoconvexity is that unlike for convex functions, the sum of two pseudoconvex functions is not pseudoconvex. However, Wolfe duality holds if the Lagrangian $f(u) + \mu^T g(u)$, $\mu \geq 0$ is pseudoconvex.

In order to weaken the convexity assumption, Mond and Weir [207] proposed a new type of dual based on the Wolfe dual (WD). Mond-Weir type dual (MWD) to the above nonlinear multiobjective optimization problem (MOP) is given by:

$$(MWD) \quad \max f(u)$$

$$\text{subject to} \sum_{i=1}^{p} \lambda_i \nabla f_i(u) + \sum_{j=1}^{m} \mu_j \nabla g_j(u) = 0,$$

$$\mu_j g_j(u) \geq 0, \quad j = 1, ..., m,$$

$$(\lambda, \mu) \geq 0, u \in X \subseteq \mathbb{R}^n.$$

In a nonsmooth setting, using the Clarke subdifferential, the Mond-Weir dual may be formulated as follows:

$$(MWD) \quad \max f(u)$$

$$\text{subject to} \sum_{i=1}^{p} \lambda_i \partial^c f_i(u) + \sum_{j=1}^{m} \mu_j \partial^c g_j(u) = 0,$$

$$\mu_j g_j(u) \geq 0, \quad j = 1, ..., m,$$

$$(\lambda, \mu) \geq 0, u \in X \subseteq \mathbb{R}^n.$$

For further details about different types of duals and duality theory, we refer to Craven [52], Craven and Glover [55], Egudo and Mond [73], Preda [223], Mishra *et al.* [202, 203], and the references therein.

Chapter 2

Pseudolinear Functions: Characterizations and Properties

2.1 Introduction

Convexity plays a central role in optimization theory. To provide a more accurate representation, modelling and solutions of several real world problems, various generalizations of convexity have been introduced. Mangasarian [176] has introduced pseudoconvex and pseudoconcave functions as generalizations of convex and concave functions, respectively. Kortanek and Evans [149] investigated the properties of the functions, which are both pseudoconvex as well as pseudoconcave. Many authors extended the study of this class of functions, which were later termed as pseudolinear functions. Chew and Choo [47] derived first and second order characterizations for differentiable pseudolinear functions. The class of pseudolinear functions includes many important classes of functions, for example, the linear and quadratic fractional functions, which arise in many practical applications (see [34, 246]). Rapcsak [233] characterized the general form of gradient of twice continuously differentiable pseudolinear functions on Riemannian manifolds. Komlosi [147] presented the characterizations of the differentiable pseudolinear functions using a special property of the normalized gradient.

Kaul *et al.* [139] introduced a new class of functions, known as semilocally pseudolinear functions, as an extension of the class of differentiable pseudolinear functions. Aggarwal and Bhatia [1] introduced the notion of pseudolinear functions in terms of Dini derivatives, which generalized the class of semilocally pseudolinear functions. Lu and Zhu [171] have established some characterizations for locally Lipschitz pseudolinear functions using the Clarke subdifferential on Banach spaces. Recently, Lalitha and Mehta [155, 156] introduced the notion of h-pseudolinear functions and studied the properties of this class of pseudolinear functions. Very recently, Smietanski [257] derived the characterizations of the class of directionally differentiable pseudolinear functions. Szilagyi [267] introduced a new class of pseudolinear functions, known as weakly pseudolinear functions, where the differentiability of the functions are assumed only at a point of the domain not over the whole domain.

In this chapter, we will study the properties of different types of pseudolinear functions such as differentiable pseudolinear, semilocally pseudolinear, Dini pseudolinear, directionally differentiable pseudolinear, locally Lipschitz pseudolinear, h-pseudolinear and weakly pseudolinear functions and present the characterizations for these classes of pseudolinear functions.

2.2   Differentiable Pseudolinear Functions

Let $X \subseteq \mathbb{R}^n$ be a nonempty set equipped with the Euclidean norm $\|\cdot\|$. Throughout the chapter, let $x \lambda y := (1 - \lambda)x + \lambda y, \lambda \in [0, 1]$, unless otherwise specified.

We recall the following definiiton from Chapter 1.

Definition 2.1 *Let $f : X \to \mathbb{R}$ be a differentiable function on an open convex set X. The function f is said to be*

(i) *pseudoconvex at $x \in X$, if for all $y \in X$,*

$$\langle \nabla f(x), y - x \rangle \geq 0 \Rightarrow f(y) \geq f(x);$$

(ii) *pseudoconcave at $x \in X$, if for all $y \in X$,*

$$\langle \nabla f(x), y - x \rangle \leq 0 \Rightarrow f(y) \leq f(x).$$

The function f is said to be pseudoconvex (pseudoconcave) on X, if it is pseudoconvex (pseudoconcave) at every $x \in X$. Moreover, the function f is said to be pseudolinear on X, if f is both pseudoconvex and pseudoconcave on X.

To be precise; let $f : X \to \mathbb{R}$ be a differentiable function on an open convex subset $X \subseteq \mathbb{R}^n$, the function f is said to be pseudolinear on X, if for all $x, y \in X$, one has

$$\langle \nabla f(x), y - x \rangle \geq 0 \Rightarrow f(y) \geq f(x)$$

and

$$\langle \nabla f(x), y - x \rangle \leq 0 \Rightarrow f(y) \leq f(x).$$

Example 2.1

(i) *Every linear fractional function f of the form*

$$f(x) = \frac{ax + b}{cx + d},$$

where $a, c, x \in \mathbb{R}^n, b, d \in \mathbb{R}$ and $cx + d$ keeps the same sign for every $x \in X := \{x \in \mathbb{R}^n : cx + d \neq 0\}$ is pseudolinear function over the set X, with respect to the proportional function $p(x, y) = \frac{cx+d}{cy+d}$.

(ii) The function $x + x^3$ is a pseudolinear function on $\mathbb{R}$.

The following results from Chew and Choo [47] characterize the class of differentiable pseudolinear functions.

Theorem 2.1 *Let $f : X \to \mathbb{R}$, where $X \subseteq \mathbb{R}^n$, be an open convex set. Suppose that f is a differentiable pseudolinear function on X. Then for all $x, y \in X$,*

$$\langle \nabla f(x), y - x \rangle = 0 \Leftrightarrow f(y) = f(x).$$

Proof Suppose f is pseudolinear on X. Then, for any $x, y \in X$, we have

$$\langle \nabla f(x), y - x \rangle \geq 0 \Rightarrow f(y) \geq f(x)$$

and

$$\langle \nabla f(x), y - x \rangle \leq 0 \Rightarrow f(y) \leq f(x).$$

Combining these two inequalities, we get

$$\langle \nabla f(x), y - x \rangle = 0 \Rightarrow f(y) = f(x), \forall x, y \in X.$$

Now, we prove that

$$f(y) = f(x) \Rightarrow \langle \nabla f(x), y - x \rangle = 0, \forall x, y \in X.$$

For that, we first show that

$$f(x\lambda y) = f(x), \forall \lambda \in [0, 1].$$

When $\lambda = 0$ and $\lambda = 1$, it is obvious. Now, we will prove it for $\lambda \in]0, 1[$.
If $f(x\lambda y) > f(x)$ then by the pseudoconvexity of f, we have

$$\langle \nabla f(x\lambda y), x - x\lambda y \rangle < 0.$$

We know that $y - x\lambda y = y - (x + \lambda(y - x)) = (1 - \lambda)(y - x)$. Therefore, we have

$$x - x\lambda y = -\lambda(y - x) = \frac{-\lambda}{1 - \lambda}(y - x\lambda y).$$

This yields

$$-\frac{\lambda}{1 - \lambda} \langle \nabla f(x\lambda y), y - x\lambda y \rangle < 0,$$

which implies that

$$\langle \nabla f(x\lambda y), y - x\lambda y \rangle > 0.$$

By pseudoconvexity of f, we have

$$f(y) \geq f(x\lambda y),$$

which is a contradiction to the fact that

$$f(x\lambda y) > f(x) = f(y).$$

Similarly, if $f(x\lambda y) < f(x)$, then by the pseudoconcavity of f, we get

$$f(y) \leq f(x\lambda y),$$

again a contradiction. Hence, for any $\lambda \in]0,1[$,

$$f(x\lambda y) = f(x).$$

Thus,

$$\langle \nabla f(x), y - x \rangle = \lim_{\lambda \to 0^+} \frac{f(x\lambda y) - f(x)}{\lambda} = 0.$$

Theorem 2.2 *Let* $f : X \to \mathbb{R}$, *where* $X \subseteq \mathbb{R}^n$, *be an open convex set. Then* f *is a differentiable pseudolinear function on* X *if and only if for all* $x, y \in X$, *there exists a function* $p : X \times X \to \mathbb{R}_{++}$, *such that*

$$f(y) = f(x) + p(x,y) \langle \nabla f(x), y - x \rangle. \tag{2.1}$$

Proof Let f be pseudolinear on X. We need to construct a function $p : X \times X \to \mathbb{R}_{++}$, such that for all $x, y \in X$,

$$f(y) = f(x) + p(x,y) \langle \nabla f(x), y - x \rangle.$$

If $\langle \nabla f(x), y - x \rangle = 0$, for any $x, y \in X$, we define $p(x,y) = 1$. From Theorem 2.1, we get $f(y) = f(x)$ and thus, (2.1) holds.
If $\langle \nabla f(x), y - x \rangle \neq 0$, for any $x, y \in X$, we define

$$p(x,y) = \frac{f(y) - f(x)}{\langle \nabla f(x), y - x \rangle}. \tag{2.2}$$

Evidently, $p(x,y)$ satisfies (2.1). Now, we have to show that $p(x,y) > 0$, for all $x, y \in X$. If $f(y) > f(x)$, then by pseudoconcavity of f, we have $\langle \nabla f(x), y - x \rangle > 0$. From (2.1), we get $p(x,y) > 0$. Similarly, if $f(y) < f(x)$, then by pseudoconvexity of f, we have

$$\langle \nabla f(x), y - x \rangle < 0.$$

From (2.2), we get $p(x,y) > 0$, for all $x, y \in X$.
Conversely, suppose that for any $x, y \in X$, there exists a function $p : X \times X \to \mathbb{R}_{++}$, such that (2.1) holds. If for any $x, y \in X, \langle \nabla f(x), y - x \rangle \geq 0$, then, from (2.1), it follows that

$$f(y) - f(x) = p(x,y) \langle \nabla f(x), y - x \rangle \geq 0.$$

Hence, f is pseudoconvex on X. Similarly, if $\langle \nabla f(x), y - x \rangle \leq 0$, for any $x, y \in X$, from (2.1), it follows that f is pseudoconcave on X.

Theorem 2.3 *Let* $f : X \to \mathbb{R}$, *where* $X \subseteq \mathbb{R}^n$, *be an open convex set. Then* f *is once continuously differentiable pseudolinear function on* X *if and only if for all* $x, y \in X$,

$$\langle \nabla f(x), y - x \rangle = 0 \Rightarrow f(x\lambda y) = f(x), \forall \lambda \in [0,1]. \tag{2.3}$$

Proof Suppose f is pseudolinear on X. Let $x, y \in X$ be such that $\langle \nabla f(x), y - x \rangle = 0$. Then,

$$\langle \nabla f(x), x\lambda y - x \rangle = \lambda \langle \nabla f(x), y - x \rangle = 0, \forall \lambda \in [0, 1].$$

Now, the result follows by Theorem 2.1.

Conversely, suppose that (2.3) holds. On the contrary, suppose that f is not pseudolinear on X, then, there exist two points $x, y \in X$ such that $f(x) = f(y)$ but $\langle \nabla f(x), y - x \rangle \neq 0$. It is evident from (2.3), that the alternative case $\langle \nabla f(x), y - x \rangle = 0$ and $f(x) \neq f(y)$ cannot occur. Define the function $f_{x,y} : [0, 1] \to \mathbb{R}$ by

$$f_{x,y}(\lambda) := f(x\lambda y), \lambda \in [0, 1].$$

We get that $f_{x,y}$ is once continuously differentiable function, such that $f_{x,y}(0) = f_{x,y}(1)$ and $f'_{x,y}(0) > 0$. Therefore, $f_{x,y}$ assumes a local maximum at some point $\bar{\lambda} \in [0, 1]$. Now, we have

$$0 = f'_{x,y}(\bar{\lambda}) = \langle \nabla f(x\bar{\lambda}y), x\bar{\lambda}y - x \rangle.$$

Hence, $\langle \nabla f(x\bar{\lambda}y), x\bar{\lambda}y - x \rangle = \bar{\lambda} \langle \nabla f(x\bar{\lambda}y), y - x \rangle - 0$. From condition (2.3), we have

$$f(x) = f(x\lambda y), \forall 0 \leq \lambda \leq \bar{\lambda}.$$

From which, it follows that

$$\langle \nabla f(x), y - x \rangle = 0,$$

which is a contradiction. This completes the proof.

Remark 2.1 *From Theorem 2.1 and Theorem 2.3, it is evident that if f : $X \to \mathbb{R}$ is a differentiable pseudolinear function on an open convex set $X \subseteq \mathbb{R}^n$ and $\nabla f(x) = 0$ at any arbitrary point $x \in X$, then f is a constant function on X. Moreover, Chew and Choo [47] have pointed out that a real differentiable pseudolinear function defined on an interval of the real line must be either a constant function or strictly monotonic (increasing or decreasing) function, where the derivative does not vanish at any point of the interval.*

In fact, the following corollary from Cambini and Martein [36] formalizes the fact explicitly.

Corollary 2.1 *Let $f : X \subseteq \mathbb{R}^n$ be a differentiable function on an open convex set X. Then f is pseudolinear on X, if the following properties hold:*

(i) *If f is pseudolinear on X and $\nabla f(x) = 0$, then, f is a constant function on X.*

(ii) *If f is pseudolinear on X, then f is also quasilinear on X.*

(iii) *f is pseudolinear on X if and only if derivative of any of its nonconstant restriction on a line is constant in sign.*

Motivated by the work of Avriel and Schaible [10], Chew and Choo [47] derived the following characterizations for twice continuously differentiable pseudolinear functions.

Theorem 2.4 *A twice continuously differentiable function $f : X \to \mathbb{R}$ is pseudolinear on a convex set X if and only if there exists a function $h : [0,1] \times X \times X \to \mathbb{R}$, such that $h(\cdot, x, y)$ is continuous over $[0,1]$ and for all (λ, x, y) in the domain of h, one has*

$$(y-x)^T \left[\nabla^2 f(x\lambda y) + h(\lambda, x, y)\nabla f(x\lambda y)^T \nabla f(x\lambda y)\right](y-x) = 0. \quad (2.4)$$

Proof Suppose f is pseudolinear on X. For any x and y in X and $\lambda \in]0,1[$, we define

$$h(\lambda, x, y) = \begin{cases} 0, & \text{if } \langle \nabla f(x\lambda y), (y-x)\rangle = 0, \\ -\frac{(y-x)^T \nabla^2 f(x\lambda y)(y-x)}{[\langle \nabla f(x\lambda y), (y-x)\rangle]^2}, & \text{if } \langle \nabla f(x\lambda y), (y-x)\rangle \neq 0. \end{cases} \quad (2.5)$$

By Theorem 2.1, we note that $\langle \nabla f(x), (y-x)\rangle = 0$ if and only if $\langle \nabla f(x\lambda y), (y-x)\rangle = 0$, which implies that f is a constant on the line segment joining x and y. Therefore, we have

$$(y-x)^T \nabla^2 f(x\lambda y)(y-x) = 0.$$

Hence, (2.4) holds.

If $\langle \nabla f(x), (y-x)\rangle = 0$, then $h(\lambda, x, y) = 0$, for any $\lambda \in [0,1]$ and so $h(\cdot, x, y)$ is continuous. If $\langle \nabla f(x), (y-x)\rangle \neq 0$, then $\langle \nabla f(x\lambda y), (y-x)\rangle \neq 0$. It follows from (2.5) and the assumption that f is twice continuously differentiable that $h(\cdot, x, y)$ is continuous.

Conversely, assume that there exists such a functional h with the given properties. For $x, y \in X$, let

$$H(x, y) = \max\left(\max_{0\leq\lambda\leq1} h(\lambda, x, y), 1\right), \quad (2.6)$$

and

$$L(x, y) = \min\left(\min_{0\leq\lambda\leq1} h(\lambda, x, y), -1\right). \quad (2.7)$$

Now, for any $\lambda \in T_{x,y} := \{\lambda \in \mathbb{R} : x\lambda y \in X\}$, define

$$q(\lambda) = e^{H(x,y)f(x\lambda y)} \quad (2.8)$$

and

$$r(\lambda) = e^{L(x,y)f(x\lambda y)}. \quad (2.9)$$

Then,

$$q'(\lambda) = H(x, y)e^{H(x,y)f(x\lambda y)}\nabla f(x\lambda y)(y-x)$$

and

$$q''(\lambda) = H(x,y)^2 e^{H(x,y)f(x\lambda y)} \nabla f(x\lambda y)(y-x)^2$$
$$+H(x,y)e^{H(x,y)f(x\lambda y)}(y-x)^T \nabla^2 f(x\lambda y)(y-x).$$

It is evident from (2.5) and (2.6) that $q''(\lambda) \geq 0$, for all $\lambda \in [0,1]$. Therefore, q is convex on $[0,1]$. In particular,

$$q(1) - q(0) \geq q'(0),$$

that is,

$$e^{H(x,y)f(y)} - e^{H(x,y)f(x)} \geq H(x,y)e^{H(x,y)f(x)}\nabla f(x)(y-x). \tag{2.10}$$

Similarly, by considering the function r defined in (2.9), we can show that

$$e^{L(x,y)f(y)} - e^{L(x,y)f(x)} \geq L(x,y)e^{L(x,y)f(x)}\nabla f(x)(y-x). \tag{2.11}$$

Now, if $\langle \nabla f(x), (y-x)\rangle \geq 0$, then by (2.10), it follows that f is pseudoconvex on X. If $\langle \nabla f(x), (y-x)\rangle \leq 0$, then by (2.11), it follows that f is pseudoconcave on X. Hence, f is pseudolinear on X.

Theorem 2.5 *Let $f : X \to \mathbb{R}$ be a differentiable pseudolinear function, defined on open set $X \subseteq \mathbb{R}^n$. Let $F : \mathbb{R} \to \mathbb{R}$ be differentiable with $F'(t) > 0$ or $F'(t) < 0, \forall t \in \mathbb{R}$. Then, the composite function $F \circ f$ is also pseudolinear on X.*

Proof Let $g(x) = F(f(x))$, for all $x \in X$. We prove the result for $F'(t) > 0$, for $t \in \mathbb{R}$. We have

$$\langle \nabla g(x), y - x\rangle = \left\langle F'(f(x))\nabla f(x), y - x \right\rangle.$$

Since, $F'(t) > 0$, for all $t \in \mathbb{R}$, therefore,

$$\langle \nabla g(x), y - x\rangle \geq 0 (\leq 0) \Rightarrow \langle \nabla f(x), y - x\rangle \geq 0 (\leq 0).$$

By the pseudolinearity of f, it follows that

$$f(y) \geq f(x)(f(y) \leq f(x)).$$

Since, F is strictly increasing, we get

$$g(y) \geq g(x)(g(y) \leq g(x)).$$

For $F'(t) < 0$, for $t \in \mathbb{R}$, the result follows, similarly.
Hence, g is pseudolinear on X.

Remark 2.2 *The above theorem may be used to construct a pseudolinear function from known functions. For example, if $g : X \to \mathbb{R}$ be a pseudolinear function on open convex set X, then, the function $f(x) = e^{g(x)}$ is also a pseudolinear function on X.*

Komlosi [147] has shown that the pseudolinearity can be characterized in terms of the normalized gradients. We present a simple proof along the lines of Cambini and Martein [36].

Corollary 2.2 *If $f : X \to \mathbb{R}$ is continuously differentiable and pseudolinear on an open convex set X, where it has nonvanishing gradient, then,*

$$\text{for } x, y \in X, f(x) = f(y) \Rightarrow \frac{\nabla f(x)}{\|\nabla f(x)\|} = \frac{\nabla f(y)}{\|\nabla f(y)\|}. \tag{2.12}$$

Proof Suppose that f is pseudolinear on X. Let us define

$$\Omega_1 := \{d \in \mathbb{R}^n : \langle \nabla f(x), d \rangle = 0\} \, and \, \Omega_2 := \{d \in \mathbb{R}^n : \langle \nabla f(y), d \rangle = 0\}.$$

Then, we have $\Omega_1 = \Omega_2$. To see it, let $d \in \Omega_1$, then, from (2.1) and for every t, such that $x + td \in X$, we have

$$f(x + td) = f(x) = f(y).$$

Again from (2.1), it follows that

$$\langle \nabla f(y), x + td - y \rangle = 0 \quad and \quad \langle \nabla f(x), y - x \rangle = 0,$$

which imply that $\langle \nabla f(y), d \rangle = 0$. Hence, $d \in \Omega_1$ and thus $\Omega_1 \subseteq \Omega_2$. Similarly, we can show that $\Omega_2 \subseteq \Omega_1$.

Since, $\Omega_1 = \Omega_2$, it follows that

$$\frac{\nabla f(x)}{\|\nabla f(x)\|} = \pm \frac{\nabla f(y)}{\|\nabla f(y)\|}.$$

Let us put $u = \frac{\nabla f(y)}{\|\nabla f(y)\|}$ and assume that

$$\frac{\nabla f(x)}{\|\nabla f(x)\|} = -u.$$

For a suitable $t \in \,]0, \varepsilon[$, choose point $z_1 = x + tu, z_2 = y + tu$, such that

$$f(z_1) < f(x) \quad and \quad f(z_2) > f(y).$$

Therefore, by the continuity of f, there exists $\lambda \in \,]0, 1[$, such that

$$f(z) = f(x) = f(y), \text{ where } z = z_1 \lambda z_2.$$

From (2.1), it follows that

$$\langle u, z - y \rangle = 0.$$

On the other hand,

$$\langle u, z - y \rangle = \langle u, (\lambda(x - y) + tu) \rangle = t \|u\|^2 > 0,$$

so that $f(y) \neq f(z)$ and this is a contradiction. Consequently, we have

$$\frac{\nabla f(x)}{\|\nabla f(x)\|} = \frac{\nabla f(y)}{\|\nabla f(y)\|}.$$

Conversely, suppose that (2.12) holds. Let $x, y \in X$ and for $t \in [0,1]$, we define $\varphi(t) = f(x + t(y - x))$. If $\varphi\prime(t)$ is constant in the sign, then $\varphi(t)$ is quasilinear on the line segment $[0,1]$.

Suppose $\varphi'(t)$ is not constant in sign, then there exist $t_1, t_2 \in]0,1[$, such that $\varphi(t_1) = \varphi(t_2)$ with $\varphi'(t_1)\varphi'(t_2) < 0$. Setting $z_1 = x + t_1(y - x)$, $z_2 = x + t_2(y - x)$. Since,

$$f(z_1) = \varphi(t_1) = \varphi(t_2) = f(z_2).$$

Therefore, we have

$$\varphi\prime(t_2) = (1 - t_2) \langle \nabla f(z_2), y - x \rangle$$

$$= (1 - t_2) \left\langle \nabla f(z_1) \frac{\|\nabla f(z_2)\|}{\|\nabla f(z_1)\|}, y - x \right\rangle$$

$$= \frac{1 - t_2}{1 - t_1} \varphi'(t_1) \frac{\|\nabla f(z_2)\|}{\|\nabla f(z_1)\|}.$$

Since, $\varphi'(t_1)\varphi'(t_2) < 0$, we get a contradiction. It follows that the restriction of the function over every line segment contained in X is quasiliner, so that f is quasilinear and also pseudolinear as $\nabla f(x) = 0, \forall x \in X$. This completes the proof.

This shows that the normalized gradient is constant on the solution set of a pseudolinear program, where the objective function has a nonvanishing gradient on the feasible set.

If the function f is defined on the whole space, the above corollary can be strengthened as follows:

Corollary 2.3 *A nonconstant function f is pseudolinear on the whole space $\mathbb{R}^n$, if and only if its normalized gradient map is constant on $\mathbb{R}^n$.*

Geometrically, the above theorem states that the level set of a nonconstant pseudolinear function, defined on the whole space $\mathbb{R}^n$ are parallel hyperplanes. Conversely, if the level sets of a differentiable function, with noncritical points are hyperplanes, then the function is pseudolinear.

Using the local analysis, Komlosi [147] has shown that the above condition characterizes pseudolinearity for once differentiable functions.

Now, in Sections 2.3 and 2.4 we assume that the function $f : X \rightarrow \mathbb{R}, X \subseteq \mathbb{R}^n$ has a nonvanishing gradient over X.

2.3   Local Characterizations of Pseudolinear Functions

Using implicit function technique, Komlosi [147] has studied the local analysis of pseudolinear functions. Komlosi [147] has defined locally pseudolinear functions as follows:

Definition 2.2 *(Locally pseudolinear function)(Komlosi [147]) Suppose that the function f is defined and differentiable around $a \in \mathbb{R}^n$. The function f is said to be locally pseudoconvex at a, if there exists a neighborhood N of a, such that*

$$\text{for } x \in N, f(x) < f(a) \Rightarrow \langle \nabla f(a), x - a \rangle < 0$$

and

$$\text{for } x \in N, f(x) = f(a) \Rightarrow \langle \nabla f(a), x - a \rangle \leq 0.$$

The function f is said to be locally pseudoconcave at a, if $-f$ is locally pseudoconvex at a. The function f is said to be locally pseudolinear at a, if f is both locally pseudoconvex and locally pseudoconcave at a.

To be precise; any function $f : X \to \mathbb{R}$ is locally pseudolinear at a, if there exists a neighborhood N of a, such that

$$x \in N, f(x) < f(a) \Rightarrow \langle \nabla f(a), x - a \rangle < 0,$$

$$x \in N, f(x) = f(a) \Rightarrow \langle \nabla f(a), x - a \rangle = 0,$$

and

$$x \in N, f(x) > f(a) \Rightarrow \langle \nabla f(a), x - a \rangle > 0.$$

The following theorem from Komlosi [147] provides the basis for the global analysis of pseudolinear functions.

Theorem 2.6 *Let the differentiable function $f : X \to \mathbb{R}$ be defined on the open convex set X. The function f is pseudolinear on X if and only if it is locally pseudolinear at each point of X.*

Proof It is clear that global pseudolinearity of f implies locally pseudolinearity of f. To prove the converse, we assume that f is locally pseudolinear at each point of X. Let $x, y \in X$, such that $f(x) = f(y)$. Then, by Theorem 2.1, to prove the pseudolinearity of f on X, it is sufficient to show that $\langle \nabla f(y), x - y \rangle = 0$. On the contrary, suppose that $\langle \nabla f(y), x - y \rangle > 0$. It results that the direction $x - y$ is a strict ascent direction of f at y. Therefore, by the continuity of f, there exists a real number $0 < t < 1$ such that $z = tx + (1 - t)y$ is the first point in $[x, y]$, where the function f attains a maximum over the line segment $[x, y]$.

For every $w \in [x, z]$, one has

$$\langle \nabla f(z), x - z \rangle = \langle \nabla f(z), y - z \rangle = \langle \nabla f(z), w - z \rangle = 0,$$

and

$$f(w) < f(z),$$

which is a contradiction to our assumption that f is locally pseudolinear at z. Similarly, assuming $\langle \nabla f(y), x - y \rangle < 0$, we arrive again at a contradiction.

The following theorem can be easily proved by using Definition 2.2.

Theorem 2.7 *Let $\nabla f(a) \neq 0$ at $a \in X$. Then, $f : X \to \mathbb{R}$ is locally pseudolinear at a if and only if there exists a neighborhood N of a, such that the following condition holds:*

$$\text{for } x \in N, f(x) = f(a) \Rightarrow \langle \nabla f(a), x - a \rangle - 0. \tag{2.13}$$

Komlosi [147] employed this result to characterize the local pseudolinearity using the implicit functions. Let $f : X \to \mathbb{R}$ be a continuously differentiable function. Following the notations of Komlosi [147], let the vectors $d, c_1, \ldots, c_{n-1}$ form an orthonormal basis in $\mathbb{R}^n$. The basis is said to be admissible for $f(x)$ at a, if

$$\langle \nabla f(a), d \rangle \neq 0.$$

Let us form the $n \times (n - 1)$ type matrix C from the basis vector $\{c_1, c_2, \ldots, c_{n-1}\}$ in the following way:

$$C = [c_1, c_2 \ldots c_{n-1}].$$

Let $x \in \mathbb{R}^n$ be arbitrary and let

$$u = C^T x \in \mathbb{R}^{n-1}, v = d^T x \in \mathbb{R}.$$

It is clear that v and u are the coordinates of x, in the basis $\{d, C\}$, which means that

$$x = vd + Cu.$$

From now on, the vector x will be identified with the pair (v, u). The following form of the implicit function theorem (see Komlosi [147]) will be used:

Proposition 2.1 *(Implicit function theorem) Let $f : X \to \mathbb{R}$ be a continuously differentiable function. Consider the equation $f(v, u) = f(a)$, where $a = (v_o, u_o)$. Let the basis $\{d, C\}$ be admissible for $f(x)$ at $x = a$. The implicit function theorem states that there exist a neighborhood N of a, neighborhood M of u_o and a unique function $h(u)$, defined and differentiable on M, such that*

$$f(h(u), u) = f(a) \text{ and } (h(u), u) \in M, \forall u \in M,$$

and for all $x \in N$, we have $\langle \nabla f(x), d \rangle \neq 0$. Moreover,

$$x \in N, f(x) = f(a) \Rightarrow x = (h(u), u), \text{ for some } u \in M.$$

An immediate consequence of the implicit function theorem is the following lemma:

Lemma 2.1 *In view of the implicit function theorem, for each $x \in N$ and $u \in M$, one has*

$$\langle \nabla f(a), x - a \rangle = \nabla f(a)^T d \left[h(u) - h(u_o) - \langle \nabla h(u_o), u - u_o \rangle \right], \qquad (2.14)$$

where $x = (h(u), u)$.

Proof Let $x = (h(u), u)$. It means that $x = h(u)d + Cu$. It is clear that

$$\langle \nabla f(a), x - a \rangle = \nabla f(a)^T d(h(u) - h(u_o)) + \nabla f(a)^T C(u - u_o). \qquad (2.15)$$

Computing the gradient of $\nabla h(u)$, from the implicit relation

$$f(h(u), u) = f(a),$$

we get

$$\nabla h(u)^T = -\frac{\nabla f(x)^T C}{\nabla f(x)^T d}, \qquad (2.16)$$

where $u \in M$ and $x = \langle h(u), u \rangle$.

From (2.16), it follows that

$$\nabla f(a)C = \left(-\nabla f(a)^T d \right) \nabla h(u_o)^T. \qquad (2.17)$$

From (2.15) and (2.17), the result follows.

Lemma 2.2 *In view of the implicit function theorem, condition (2.13) holds if and only if $h(u)$ is linear on M.*

Proof To prove the necessary part, we assume that condition (2.13) holds. Let $u \in M$ be arbitrary and $x = (h(u), u)$. Then $x \in N$ and $f(x) = f(a)$. In view of implication (2.13) and identity (2.14), we get

$$0 = \langle \nabla f(a), x - a \rangle = \nabla f(a)^T d(h(u) - h(u_o)) + \langle \nabla h(u_o), u - u_o \rangle, u \in M.$$

Since, $\nabla f(a)^T d \neq 0$, we get

$$h(u) = h(u_o) + \langle \nabla h(u_o), u - u_o \rangle, u \in M. \qquad (2.18)$$

Taking into account the identity (2.14), the sufficient part follows from (2.18).

The next theorem from Komlosi [147] is an immediate consequence of Theorem 2.7 and Lemma 2.2.

Theorem 2.8 *Let $f : X \to \mathbb{R}$ be a continuously differentiable function satisfying condition $\nabla f(a) \neq 0$. Then, function f is locally pseudolinear at a if and only if the implicit function $h(u)$ is linear on M.*

The following theorem form Komlosi [147] is a local variant of Theorem 6 of Rapcsak [233].

Theorem 2.9 *Let $f : X \to \mathbb{R}$ be a continuously differentiable function satisfying $\nabla f(a) \neq 0$. Then f is locally pseudolinear at a, if and only if there exist a function $c : X \to \mathbb{R}$ defined on a neighborhood N of a and a nonzero vector $g \in \mathbb{R}^n$ such that $c(x) \neq 0$, for each $x \in N$ and*

$$f(x) = f(a) \Rightarrow \nabla f(x) = c(x)g. \tag{2.19}$$

Proof From Theorem 2.8, it is sufficient to prove that the implicit function $h(u)$ is linear if and only if the implication (2.19) holds true.

To prove the necessary part, we assume that $h(u)$ is linear and that

$$h(u) = h(u_o) + a^T(u - u_o), u \in M.$$

Using (2.16), we get

$$a^T = -\frac{\nabla f(x)^T C}{\nabla f(x)^T d}$$

and it follows that

$$\nabla f(x)^T C = \left(-\nabla f(x)^T d \right) a^T, \tag{2.20}$$

where $x = (h(u), u)$. We consider the nonsingular matrix $D = [d, C]$. From (2.20), we get

$$\nabla f(x)^T D = \nabla f(x)^T d(1 - a^T).$$

From which, it follows that

$$\nabla f(x) = c(x)g,$$

where $c(x) = \nabla f(x)^T d$ and $g^T = \left[1 - a^T \right] D^{-1}$.

Since, for any $x \in N$, such that $f(x) = f(a)$, we have $x = (h(u), u)$, for some $u \in M$.

To prove the sufficient part, we assume that the implication (2.19) holds. From (2.16), we have

$$\nabla h(u)^T = -\frac{\nabla f(x)^T C}{\nabla f(x)^T d} = -\frac{g^T C}{g^T d} = a^T = constant.$$

Hence, the linearity of $h(u)$ on M follows.

Taking into account Theorems 2.6 and 2.9, the following characterizations of once differentiable pseudolinear functions can be deduced.

Theorem 2.10 *Let $f : X \to \mathbb{R}$, be a continuously differentiable function on an open convex set X and $\nabla f(x) \neq 0$, for each $x \in X$. Then, f is pseudolinear on X, if and only if there exists two functions $c : X \to \mathbb{R}$ and $g : f(X) \to \mathbb{R}^n$ such that for each $x \in X$, we have $c(x) \neq 0$, and*

$$\nabla f(x) = c(x)g(f(x)). \tag{2.21}$$

The following examples from Komlosi [147] justify the significance of Theorem 2.10.

Example 2.2 *Consider the function $f : \mathbb{R}^n \to \mathbb{R}$, given by*

$$f(x) = \frac{a^T x + \alpha}{b^T x + \beta},$$

on the open convex set $X = \left\{ x \in \mathbb{R}^n : b^T x + \beta > 0 \right\}, a, b \in \mathbb{R}^n; \alpha, \beta \in \mathbb{R}$. *Then, we can show that*

$$\nabla f(x) = c(x)g(f(x)),$$

where $c(x) = (b^T x + \beta)^{-1}$ *and* $g(t) = a - tb$. *It follows by Theorem 2.10, that* f *is pseudolinear on* X.

Example 2.3 *Consider the composite function*

$$F(x) = \phi(f(x)),$$

where $f : X \to \mathbb{R}$ *is continuously differentiable and pseudolinear on the open convex set* X *and* $\phi(t), t \in \mathbb{R}$ *is continuously differentiable on* $f(X)$. *From (2.21), we have*

$$\nabla F(x) = \left[\phi'(f(x)) \right] \nabla (f(x)).$$

Suppose $f(x)$ *is not a singleton, it follows that* $F(x)$ *is pseudolinear on* X *if and only if* $\phi'(t) \neq 0$ *on* $f(X)$.

2.4   Second Order Characterizations of Pseudolinear Functions

Komlosi [147] has investigated the Hessian $\nabla^2 h(u)$ of the implicit function $h(u)$, to obtain the second order information on twice differentiable pseudolinear functions. Differentiating twice the implicit relation $f(h(u), u) = f(a)$, we have

$$\left(-\nabla f(a)^T d \right) \nabla^2 h(u_\circ) = C^T \left[H_1 - H_2 + H_3 \right] C, \qquad (2.22)$$

where

$$H_1 = \nabla^2 f(a),$$

$$H_2 = \frac{1}{\nabla f(a)^T d} \left(\nabla f(a) d^T \nabla^2 f(a) + \nabla^2 f(a) d \nabla f(a)^T \right),$$

$$H_3 = \frac{d^T \nabla^2 f(a) d}{\left(\nabla f(a)^T d \right)^2} \nabla f(a) \nabla f(a)^T.$$

Let us introduce the following notation:

$$Q_C f(a; d) = \left(-\nabla f(a)^T d\right) \nabla^2 h(u_o),\qquad(2.23)$$

which emphasizes the dependence of $h(u)$ on the basis $\{d, C\}$. The matrix $Q_C f(a; d)$ is called quasi-Hessian matrix of $f(x)$ at a, by Komlosi [144, 145] as it characterizes pseudoconvexity/pseudoconcavity in the same manner as the Hessian $\nabla^2 f(a)$ does for convexity. The quasi-Hessian matrix is not unique. Any admissible basis $\{d, C\}$ determines a quasi-Hessian matrix. It is shown in Theorem 3 by Komlosi [144] that, all the quasi-Hessian matrices are similar.

Theorem 2.11 *Let $f : X \to \mathbb{R}$ be a twice continuously differentiable function, such that $\nabla f(a) \neq 0$ and the basis $\{d, C\}$ is admissible for $f(x)$ at $x = a$. Then, f is pseudolinear at a if and only if there exists a neighborhood N of a such that*

$$Q_C f(x; d) = 0, \forall x \in N.$$

Proof Since

$$Q_C f(x; d) = \left(-\nabla f(x)^T d\right) \nabla^2 h(u),$$

for any $x \in N$, whenever N is the neighborhood of a in the statement of the implicit function theorem, the condition $Q_C f(x; d) = 0$ is equivalent to the linearity of the implicit function $h(u)$, which is equivalent to the local pseudolinearity of f at a.

The next result by Komlosi [147], presents a new proof for the sufficient part of Theorem 6 of Rapcsak [233] using the quasi-Hessian matrix.

Theorem 2.12 *Assume that $f : X \to \mathbb{R}$ is twice continuously differentiable on open convex set X. Let*

$$\nabla f(x) = c(x)g(f(x)),\qquad(2.24)$$

where $c(x)$ and $g(t)$ are continuously differentiable functions and $\nabla f(x) \neq 0$, for each $x \in X$. Then, f is pseudolinear on X.

Proof Let $a \in X$ be arbitrary. Let us choose $d = \frac{\nabla f(a)}{\|\nabla f(a)\|}$ and let $\{d, C\}$ be a basis in $\mathbb{R}^n$. We can check that $C^T H_2 C = 0$ and $C^T H_3 C = 0$. Then, we have

$$Q_C f(a, d) = C^T \nabla^2 f(a) C.$$

Computing $\nabla^2 f(a)$ from (2.24), we get

$$\nabla^2 f(a) = \left[\frac{\nabla c(a)}{c(a)} + c(a)g'f(a)\right] \nabla f(a)^T.$$

Since, $\nabla f(a)$ is orthogonal to C, it follows that

$$C^T \nabla f(a) = 0.$$

Hence, we have

$$Q_C f(a; d) = C^T \nabla^2 f(a) C = 0.$$

Let N be a neighborhood of a, where $\nabla f(x)^T d \neq 0$ holds. Therefore, the base $\{d, C\}$ is admissible for f at each $x \in N$ and thus the quasi-Hessian $Q_C f(x; d)$ exists for all $x \in N$. Without loss of generality, we may assume that N satisfies all the conditions of the implicit function theorem. Let $x' \in N$ be arbitrary and proceeding as above for x' and $d' = \dfrac{\nabla f(x')}{\|\nabla f(x')\|}$ in place of a and d, we get $Q_{C'} f(x', d') = 0$. Since, quasi-Hessian matrices are similar, we get $Q_C f(x', d') = 0$, hence, from Theorem 2.11, that f is locally pseudolinear at a. Since, a is arbitrary, by Theorem 2.6, f is pseudolinear on X.

Komlosi [147] studied the eigenvalues and eigenvectors of the Hessian $\nabla^2 f$ of the twice continuously differentiable pseudolinear function f. Let $u_1, u_2, \ldots, u_n$ be the orthonormal eigenvectors of the symmetric matrix $\nabla^2 f$ and let $h_1, h_2, \ldots, h_n \in \mathbb{R}$ be the corresponding eigenvalues of the twice continuously differentiable function f.

The following result from Komlosi [147] is needed for the characterization of the Hessian of a pseudolinear function.

Lemma 2.3 *The quasi-Hessian $Q_C f(a; d)$ is the zero matrix if and only if the Hessian $\nabla^2 f(a)$ possesses the following property:*
Property-(Z) *one of the following conditions holds:*

(Z1) $\nabla^2 f(a) = 0$;

(Z2) $\nabla^2 f(a)$ *has exactly one nonzero single eigenvalue, to be denoted by h_k, and*

$$\nabla f(a) = \alpha u_k, \alpha \neq 0.$$

(Z3) $\nabla^2 f(a)$ *has exactly one positive and one negative single eigenvalue, to be denoted by h_k and h_p and*

$$\nabla f(a) = \alpha u_k + \beta u_p, \alpha \neq 0, \beta \leq 0, \tag{2.25}$$

$$h_p (\nabla f(a)^T u_k)^2 = -h_k \left(\nabla f(a)^T u_p \right)^2. \tag{2.26}$$

Proof To prove the necessary part, we assume that $Q_C f(a; d)$ is the zero matrix. Since, $\nabla f(a) \neq 0$, there exists an index k such that $\nabla f(a)^T u_k \neq 0$. Let us put

$$C_k = [u_1, \ldots, u_{k-1}, u_{k+1}, \ldots, u_n]$$

and consider the quasi-Hessian associated with the admissible basis $\{u_k, C_k\}$, denoted by the symbol $Q_k f(a)$. Since $Q_k f(a)$ is similar to $Q_C f(a; d)$, therefore $Q_k f(a)$ is also the zero matrix.

Now, we compute the quadratic form $z^T Q_k f(a) z, z \in \mathbb{R}^{n-1}$. The computation is simplified by the fact that

$$u_j^T \nabla^2 f(a) u_k = 0, \quad \text{for any } j \neq k.$$

We consider the following matrices:

$$U = [u_1, u_2, \ldots, u_n],$$
$$E_k = [e_1, \ldots, e_{k-1}, e_{k+1}, \ldots, e_n],$$

where e_j stands for the j-th unit vector of $\mathbb{R}^n$. From (2.22), we get

$$Q_k f(a) = E_k^T U^T \left[\nabla^2 f(a) + \frac{h_k}{(\nabla f(a)^T u_k)^2} \nabla f(a) \nabla f(a)^T \right] U E_k. \qquad (2.27)$$

Let $z \in \mathbb{R}^{n-1}$ be arbitrary and following the notation:

$$E_k z = (z_1, \ldots, z_n)^T \in \mathbb{R}^n.$$

It is clear that $z_k = 0$. By using (2.27), we get

$$z^T Q_k f(a) z = \sum_{j \neq k} h_j z_j^2 + \frac{h_k}{(\nabla f(a)^T u_k)^2} \left[\sum_{j \neq k} (\nabla f(a)^T u_j) z_j \right]^2. \qquad (2.28)$$

We consider (2.28) for some special choices of z. Put $z = E_k^T e_j \in \mathbb{R}^{n-1}, j \neq k$. Since $E_k z = e_j$, from (2.28), we have

$$z^T Q_k f(a) z = h_j + h_k \frac{(\nabla f(a)^T u_j)^2}{(\nabla f(a)^T u_k)^2}, j \neq k. \qquad (2.29)$$

Since $Q_k f(a) = 0$, from (2.29), we get

$$h_j = -h_k \frac{(\nabla f(a)^T u_j)^2}{(\nabla f(a)^T u_k)^2}, j \neq k. \qquad (2.30)$$

If $h_k = 0$, then by (2.30) all the eigenvalues of $\nabla^2 f(a)$ are zero, which imply that (Z1) holds.

If $h_k \neq 0$, but the other eigenvalues are all zero, then, by (2.30), we get $\nabla f(a)^T u_j = 0$, for all $j \neq k$, which implies that (Z2) holds.

If $h_k \neq 0$ and $h_p \neq 0$, then by (2.30), they must have opposite sign. Since, $h_p \neq 0$ implies $\nabla f(a)^T u_p \neq 0$, therefore (2.30) remains valid replacing k by p in it. It follows that h_k and h_p are the only nonzero eigenvalues of $\nabla^2 f(a)$ and they have the opposite sign. Since $\nabla f(a)^T u_j = 0$, for each $j \neq k, p$ (2.25) holds with $\alpha = \nabla f(a)^T u_k$ and $\beta = \nabla f(a)^T u_p$. Replacing j by p in (2.30), we get (2.26) and thus (Z3) holds.

To prove the sufficient part, assume that $\nabla^2 f(a)$ possesses the Property-(Z). It is clear by the definition of quasi-Hessian $Q_C f(a; d)$ that condition (Z1) implies $Q_C f(a; d) = 0$.

Now suppose that condition (Z2) or (Z3) holds. Then the quasi-Hessian matrix $Q_k f(a)$ exists. Let $z \in \mathbb{R}^{n-1}$ be arbitrary. Next, we show that $z^T Q_k f(a) z = 0$, which proves that $Q_k f(a) = 0$ and hence, $Q_C f(a; d) = 0$.

If condition (Z2) is satisfied then both terms on the right hand side of (2.28) are zero and so $z^T Q_k f(a) z = 0$, for every $z \in \mathbb{R}^{n-1}$.

If condition (Z3) is fulfilled, then for $z \in \mathbb{R}^{n-1}$, (2.28) gives

$$z^T Q_k f(a) z = \left[h_p + h_k \frac{(\nabla f(a)^T u_p)^2}{(\nabla f(a)^T u_k)^2} \right] z_p^2 = 0.$$

Using the Property-(Z), Komlosi [147] has given the following characterization for the Hessian of a pseudolinear function:

Theorem 2.13 *Let $f : X \to \mathbb{R}$ be twice continuously differentiable function on the open convex set X. Suppose that $\nabla f(x) \neq 0$ on X. Then f is pseudolinear on X, if and only if the Hessian $\nabla^2 f(x)$ of $f(x)$ possesses Property-(Z) at each point of X.*

**Example 2.4 *(Komlosi [147])* *Consider the following separable function*

$$f(x_1, x_2, \ldots, x_n) = f_1(x_1) + f_2(x_2) + \cdots + f_n(x_n),$$

where the functions $f_j(x_j), j = 1, 2, \ldots, n$ are twice continuously differentiable functions. The function $f(x), x \in \mathbb{R}^n$ is pseudolinear on the open convex set $X \subseteq \mathbb{R}^n$ if and only if $f(x)$ has any one of the following forms:

(i) $f(x) = a^T x + b, a \in \mathbb{R}^n, b \in \mathbb{R}$;

(ii) $f(x) = f_k(x_k), f_k'(x_k) \neq 0, f_k''(x_k) \neq 0$;

(iii) $f(x) = f_k(x_k) + f_p(x_p)$, *where $f_k(x_k) = m \ln |a x_k + b|$ and $f_p(x_p) = -m \ln |c x_p + d|$, $a, b, c, d, m \in \mathbb{R}$; $a, b, m \neq 0$.*

To see it, let f be pseudolinear on the open convex set X. Since, pseudolinearity of f is of radial character, the pseudolinearity of f implies the pseudolinearity of $f_j(x_j), x_j \in X_j$ for all $j = 1, 2, \ldots, n$, where X_j is the projection of X onto the j–th axis. Therefore, $f_j'(x_j) = 0$ everywhere on X_j or $f_j'(x_j) \neq 0$, everywhere on X_j. Therefore, the Property-(Z) is manifested on X either by condition (Z3) or by (Z2) or by (Z1) but never in mixed form.

If $f(x)$ satisfies Condition (Z1) on X, then $f(x)$ is of type (i).

If $f(x)$ satisfies Condition (Z2) on X, then all the function $f_j(x_j)$ but one are constant and so $f(x)$ is of type (ii).

If $f(x)$ satisfies Condition (Z3) on X, then all the function $f_j(x_j)$ but two are constant and thus, we may take that

$$f(x) = f_k(x_k) + f_p(x_p).$$

Also, $f_k'(x_k), f_k''(x_k), f_p'(x_p), f_p''(x_p) \neq 0$. Moreover, from (Z2), we have

$$\frac{f_k''(x_k)}{\left[f_k'(x_k) \right]^2} = -\frac{f_p''(x_p)}{\left[f_p'(x_p) \right]^2},$$

which can be rewritten as

$$\frac{d}{dx_k}\frac{1}{f'_k(x_k)} = -\frac{d}{dx_p}\frac{1}{f'_p(x_p)} = \frac{1}{c} = \text{constant.}$$

From, which we get

$$f_k(x_k) = m\ln|ax_k + b|$$

and

$$f_p(x_p) = -m\ln|cx_p + d|.$$

The sufficiency of the above statement follows by testing the Property-(Z) in case of (i), (ii), and (iii), respectively.

2.5 Semilocally Pseudolinear Functions

An important generalization of convex functions, termed as semilocally convex functions has been introduced by Ewing [77]. Kaul and Kaur [136] introduced the notion of semilocally pseudoconvex and semilocally quasiconvex functions as generalization of pseudoconvex and quasiconvex functions, respectively.

To understand the notion of semilocal pseudolinearity, we recall the following definitions and preliminary results from Kaul *et al.* [138].

Definition 2.3 *(Semidifferentiable function) Let $f : X \to \mathbb{R}$, where $X \subseteq \mathbb{R}^n$ and $x, y \in X$. The right differential of f at x in the direction of $y - x$ is defined by*

$$f'_+(x; y - x) = \lim_{\lambda \to 0^+} \frac{f(x + \lambda(y - x)) - f(x)}{\lambda}$$

provided the limit exists. We take $f'_+(x; y - x) = 0$, whenever $x = y$.

If the right differential of f at each point $x \in X$, for any direction $y - x, y \in X$ exists, then we say that f is semidifferentiable on X.

Definition 2.4 *(Locally star-shaped set) A subset $X \subseteq \mathbb{R}^n$ is said to be locally star-shaped at $\bar{x} \in X$, if corresponding to $\bar{x}$ and each $x \in X$, there exists a maximal positive number $\alpha(\bar{x}, x) \leq 1$, such that*

$$(1 - \lambda)\bar{x} + \lambda x \in X, 0 < \lambda < \alpha(\bar{x}, x).$$

If $\alpha(\bar{x}, x) = 1$, for each $x \in X$, then X is star-shaped at $\bar{x}$. If X is star-shaped at every $\bar{x} \in X$, then X is convex.

The set X is said to be locally star-shaped, if it is locally star-shaped at *each of its points, i.e., for every $x, y \in X$, there exists a maximal positive number $\alpha(x, y) \leq 1$, such that*

$$(1 - \lambda)x + \lambda y \in X, 0 < \lambda < \alpha(x, y).$$

Kaul and Kaur [136] have pointed out that there do exist sets, which are locally star-shaped at each of their points, but which are not convex. For example the set $\{x : x \in \mathbb{R} \setminus \{0\}, x^3 \leq 1\}$ is locally star-shaped at each of its points but is not convex.

Definition 2.5 (Ewing [77]) *Let $f : X \to \mathbb{R}$, where $X \subseteq \mathbb{R}^n$ is locally star-shaped. The function f is said to be semilocally convex (slc) on X, if for every $x, y \in X$, there exists a positive number $d(x, y) \leq \alpha(x, y)$, such that*

$$f((1 - \lambda)x + \lambda y) \leq (1 - \lambda)f(x) + \lambda f(y), 0 < \lambda < d(x, y).$$

If $d(x, y) = \alpha(x, y) = 1$, for every $x, y \in X$, then f is convex on X.

Definition 2.6 (Semilocally pseudoconvex) (Kaul et al. [136]) *A function $f : X \to \mathbb{R}$ is said to be semilocally pseudoconvex (slpc) on X, if for every $x, y \in X$, the right differential exists and*

$$f'_+(x; y - x) \geq 0 \Rightarrow f(y) \geq f(x).$$

The function $f : X \to \mathbb{R}$ is said to be semilocally pseudoconcave (slpv) on X, if $-f$ is pseudoconvex (slpc) on X.

Remark 2.3 (Kaul and Kaur [136]) *If the function $f : X \to \mathbb{R}, X \subseteq \mathbb{R}^n$ is pseudoconvex on X, then f is semilocally pseudoconvex (slpc) on X, but not conversely. For example, the function $f : \mathbb{R} \to \mathbb{R}$ defined by $f(x) = |x|$ is semilocally pseudoconvex (slpc) on $\mathbb{R}$ but not pseudoconvex on $\mathbb{R}$.*

Definition 2.7 (Semilocally pseudolinear) (Kaul et al. [136]) *A function $f : X \to \mathbb{R}, X \subseteq \mathbb{R}^n$ is said to be semilocally pseudolinear (slpl) on X, if f is both semilocally pseudoconvex and semilocally pseudoconcave on X.*

To be precise; $f : X \to \mathbb{R}, X \subseteq \mathbb{R}^n$ is semilocally pseudolinear (slpl) on X, if for every $x, y \in X$, the right differential exists and

$$f'_+(x; y - x) \geq 0 \Rightarrow f(y) \geq f(x),$$
$$f'_+(x; y - x) \leq 0 \Rightarrow f(y) \leq f(x).$$

Remark 2.4 *From the definition of semilocally pseudoconvex (semilocally pseudoconcave) functions and Remark 2.3, it is clear that every pseudolinear function f on $X \subseteq \mathbb{R}^n$ is semilocally pseudolinear (slpl) on X, but the converse of this is not true. The following example from Kaul et al. [138] illustrates the fact.*

Example 2.5 *Consider the function* $f : X = [0, \infty[\rightarrow \mathbb{R}$, *defined by*

$$f(x) = \begin{cases} x, & \text{if } 0 \leq x < 1, \\ 3x - 2, & \text{if } x \geq 1. \end{cases}$$

Clearly, f is not differentiable at $x = 1$ and therefore f is not pseudolinear. We can check that the right differential exists and the function f is semilocally pseudolinear (slpl) on X. Hence, the class of semilocally pseudolinear functions (slpl) properly contains the class of pseudolinear functions.

The following characterizations and properties of semilocally pseudolinear functions are from Kaul *et al.* [138].

Theorem 2.14 *Let $f : X \rightarrow \mathbb{R}$, where $X \subseteq \mathbb{R}^n$ is locally star-shaped. If f is semilocally pseudolinear (slpl) on X, then for every pair of points $x, y \in X$,*

$$f'_+(x; y - x) = 0 \Leftrightarrow f(y) = f(x).$$

Proof Since f is semilocally pseudolinear (slpl) on X, then, for any $x, y \in X$, we have

$$f'_+(x; y - x) \geq 0 \Rightarrow f(y) \geq f(x)$$

and

$$f'_+(x; y - x) \leq 0 \Rightarrow f(y) \leq f(x).$$

Combining these two inequalities, we get

$$f'_+(x; y - x) = 0 \Rightarrow f(y) = f(x), \forall x, y \in X.$$

To prove the converse implication, we must prove that

$$f(y) = f(x) \Rightarrow f'_+(x; y - x) = 0, \forall x, y \in X.$$

Suppose to the contrary, that

$$f(y) = f(x), \tag{2.31}$$

but $f'_+(x; y - x) \neq 0$. Let

$$f'_+(x; y - x) > 0. \tag{2.32}$$

Since $x, y \in X$ and X are locally star-shaped sets, therefore, there exists a maximal positive number $\alpha(x, y) \leq 1$, such that

$$x\lambda y \in X, 0 < \lambda < \alpha(x, y),$$

where $x\lambda y := (1 - \lambda)x + \lambda y$.

Consider the function

$$\varphi(x, y, \lambda) = f(x\lambda y) - f(x).$$

Clearly, $\varphi(x, y, 0) = 0$. Therefore, the right differential of $\varphi(x, y, \lambda)$ with respect to λ at $\lambda = 0$ is

$$\lim_{\lambda \to 0^+} \frac{\phi(x, y, \lambda) - \phi(x, y, 0)}{\lambda} = \lim_{\lambda \to 0^+} \frac{f(x\lambda y) - f(x)}{\lambda} = f'_+(x; y - x) > 0,$$

by (2.32). Therefore, $\varphi(x, y, \lambda) > 0$, if λ is in some open interval $]0, e_1(x, y)[$. Taking $e(x, y) = \min \{e_1(x, y), \alpha(x, y)\}$, we have

$$f(x\lambda y) > f(x), 0 < \lambda < e(x, y). \tag{2.33}$$

From (2.31) and (2.33), we get $f(x\lambda y) > f(y), 0 < \lambda < e(x, y)$. Since, f is semilocally pseudoconvex,

$$f(y) < f(x\lambda y) \Rightarrow f'_+(x\lambda y; y - x\lambda y) < 0, 0 < \lambda < e(x, y). \tag{2.34}$$

Proceeding similarly as in the derivation of (2.33) and using (2.34), we have

$$f'_+(x\lambda y; y - x\lambda y) < 0 \Rightarrow f((x\lambda y)ky) < f(x\lambda y), \tag{2.35}$$

$0 < \lambda < e(x, y)$ and $0 < k < e_1(x\lambda y, y)$.

Taking the limit as $\lambda \to 0$, we get

$$f(xky) \leq f(x), 0 < k < e_1(x, y). \tag{2.36}$$

Choosing $\bar{e}(x, y) = \min \{e_1(x, y), e(x, y)\}$, from (2.36), it follows that

$$f(xky) \leq f(x), 0 < k < \bar{e}(x, y), \tag{2.37}$$

but (2.37) contradicts (2.33). Hence, $f'_+(x; y - x) > 0$ is not possible. Similarly, using semilocal pseudoconcavity of f, we can prove that $df^+(x, y - x) < 0$ is also not possible.

Hence, $f'_+(x; y - x) = 0$.

The following example illustrates that the converse of Theorem 2.14 is not necessarily true.

Example 2.6 (Kaul et al. [138]) *Consider the function* $f : X = \mathbb{R}_{++} \setminus \{2\} \to \mathbb{R}$ *defined by*

$$f(x) = \begin{cases} 3x, & \text{if } 0 < x < 2, \\ -x, & \text{if } x > 2. \end{cases}$$

It is clear that

$$f'_+(x; y - x) = 0 \Leftrightarrow f(y) = f(x).$$

But, the function f *is not semilocally pseudolinear, as it is not semilocally pseudoconvex on* X.

Theorem 2.15 *Let $f : X \to \mathbb{R}$ be a function, where $X \subseteq \mathbb{R}^n$ is locally star-shaped. Then f is semilocally pseudolinear on X, if and only if there exists a positive real-valued function $p : X \times X \to \mathbb{R}_{++}$, such that*

$$f(y) = f(x) + p(x,y)f'_+(x; y - x), \tag{2.38}$$

for any $x, y \in X$.

Proof Let f be semilocally pseudolinear on X. We need to construct a function $p : X \times X \to \mathbb{R}_{++}$, such that for all $x, y \in X$,

$$f(y) = f(x) + p(x,y)f'_+(x; y - x).$$

If $f'_+(x; y - x) = 0$, for any $x, y \in X$, we define $p(x, y) = 1$. From Theorem 2.14, we get $f(y) = f(x)$ and thus, (2.38) holds.

If $f'_+(x; y - x) \neq 0$, for any $x, y \in X$, we define

$$p(x,y) = \frac{f(y) - f(x)}{f'_+(x; y - x)}. \tag{2.39}$$

Evidently, $p(x, y)$ satisfies (2.38). Now, we have to show that $p(x, y) > 0$, for all $x, y \in X$. If $f(y) > f(x)$, then, by semilocal pseudoconcavity of f, we have $f'_+(x; y-x) > 0$. From (2.38), we get $p(x, y) > 0$. Similarly, if $f(y) < f(x)$, then by semilocally pseudoconvexity of f, we have

$$f'_+(x; y - x) < 0.$$

From (2.39), we get $p(x, y) > 0$, for all $x, y \in X$.

Conversely, suppose that for any $x, y \in X$, there exists a function $p : X \times X \to \mathbb{R}_{++}$, such that (2.38) holds. If for any $x, y \in X, f'_+(x; y - x) \geq 0$, from (2.38), it follows that

$$f(y) - f(x) = p(x,y)f'_+(x; y - x) \geq 0,$$

and hence, f is semilocally pseudoconvex on X.

Similarly, if $f'_+(x; y - x) \leq 0$, for any $x, y \in X$, from (2.38), it follows that $f(y) - f(x) \leq 0$. Hence, f is semilocally pseudoconcave on X.

Theorem 2.16 *Let $f : X \to \mathbb{R}$ be a function, where $X \subseteq \mathbb{R}^n$ is locally star-shaped. If f is semilocally pseudolinear then for any $x, y \in X$, there exists a maximal positive number $d(x, y)$, such that*

$$f'_+(x; y - x) = 0 \Rightarrow f(x) = f(x\lambda y), 0 < \lambda < d(x, y),$$

where $x\lambda y := x + \lambda(y - x)$.

Proof Let f be semilocally pseudolinear on X and $x, y \in X$, such that

$$f'_+(x; y - x) = 0. \tag{2.40}$$

Since X is locally star-shaped, therefore, there exists a maximal positive number $d(x, y) \leq 1$, such that

$$(1 - \lambda)x + \lambda y \in X, 0 < \lambda < d(x, y).$$

That is,

$$x\lambda y \in X, 0 < \lambda < d(x, y).$$

We know that,

$$f'_+(x; x\lambda y - x) = \lambda f'_+(x; y - x).$$

From (2.40), we get

$$f'_+(x; x\lambda y - x) = 0, 0 < \lambda < d(x, y).$$

Using Theorem 2.14, we have

$$f(x) = f(x\lambda y), 0 < \lambda < d(x, y).$$

2.6   Dini Derivatives and Pseudolinear Functions

In this section, we will study the pseudolinear functions and their properties, in terms of the Dini derivatives instead of right derivatives. First, we present the definition of pseudoconcave functions in terms of the Dini derivative, as follows:

Definition 2.8 (Diewert [65]) *Let* $f : X \to \mathbb{R}, X \subseteq \mathbb{R}^n$, *be an open convex set. Then* f *is said to be pseudoconcave over* X, *if for every* $\bar{x} \in X, v \in \mathbb{R}^n$ *satisfying*

$$v^T v = 1, t > 0, \bar{x} + tv \in X, D^+ f(\bar{x}; v) \leq 0 \Rightarrow f(\bar{x} + tv) \leq f(\bar{x}).$$

Following Diewert [65], Aggarwal and Bhatia [1] defined pseudoconvex functions, in terms of the Dini derivatives, as follows:

Definition 2.9 *Let* $f : X \to \mathbb{R}, X \subseteq \mathbb{R}^n$ *an open convex set. Then* f *is said to be pseudoconvex over* X, *if for every* $\bar{x} \in X, v \in \mathbb{R}^n$, *satisfying*

$$v^T v = 1, t > 0, \bar{x} + tv \in X, D_+ f(\bar{x}; v) \geq 0 \Rightarrow f(\bar{x} + tv) \geq f(\bar{x}).$$

Aggarwal and Bhatia [1] defined pseudolinear functions in terms of the Dini derivatives, as follows:

Definition 2.10 *Let* $f : X \to \mathbb{R}, X \subseteq \mathbb{R}^n$ *be an open convex set. Then,* f *is said to be pseudolinear in terms of the Dini derivatives over* X, *if* f *is both pseudoconvex and pseudoconcave in terms of the Dini derivatives over* X.

To be precise; any function $f : X \to \mathbb{R}, X \subseteq \mathbb{R}^n$ an open convex set is said to be pseudolinear in terms of the Dini derivatives, if for every $\bar{x} \in X, v \in \mathbb{R}^n$ satisfying $v^T v = 1, t > 0, \bar{x} + tv \in X$, one has

$$D_+ f(\bar{x}; v) \geq 0 \Rightarrow f(\bar{x} + tv) \geq f(\bar{x}),$$
$$D^+ f(\bar{x}; v) \leq 0 \Rightarrow f(\bar{x} + tv) \leq f(\bar{x}).$$

Remark 2.5 *It is clear from Definition 2.10, that every semilocally pseudolinear function is pseudolinear in terms of the Dini derivatives, but the converse may not be true. Aggarwal and Bhatia [1] illustrated this fact by the following example. Consider the function $f : X = [-2, 1[$ defined as follows:*

$$f(x) = \begin{cases} 1, & \text{if } -2 \leq x \leq 1, \\ 0, & \text{if } -1 \leq x \leq 0, \\ 1/2^{n+1}, & \text{if } 1/2^{n+1} \leq x \leq 1/2^n, n = 0, 1, 2, \ldots. \end{cases}$$

The Dini derivatives at different points of the domain are as follows:

$$D^+ f(-1; 1) = -\infty, D_+ f(-1; 1) = -\infty,$$
$$D^+ f(0; 1) = 1, D_+ f(0; 1) = 1/2.$$

Then, it is evident that

$$D^+ f(-1; 1) < 0 \Rightarrow f(-1 + t) < f(-1), t > 0, (-1 + t) \in]-1, 0],$$
$$D_+ f(0; 1) > 0 \Rightarrow f(0 + t) > f(0), t > 0, 0 + t \in]0, 1[.$$

Therefore, from Definition 2.10, the pseudolinearity of the function f in terms of the Dini derivatives follows. Moreover, it is clear that f does not have the right derivative at $x = 0$ and hence f is not semilocally pseudolinear over X.

Now, in this section, we assume all the functions and the Dini derivatives to be finite.

The following theorem by Aggarwal and Bhatia [1] characterizes pseudolinear functions in terms of the Dini derivatives.

Theorem 2.17 *Let $f : X \to \mathbb{R}, X \subseteq \mathbb{R}^n$ be an open convex set. Then, the following statements are equivalent:*

(a) *f is pseudolinear over X;*

(b) *There exists functions $p, q : X \times X \to \mathbb{R}$ such that $p(\bar{x}, \bar{x}+tv) > 0, q(\bar{x}, \bar{x}+ tv) > 0$ and*

$$f(\bar{x}+tv) = f(\bar{x}) + p(\bar{x}, \bar{x}+tv)D^+ f(\bar{x}; v) + q(\bar{x}, \bar{x}+tv)D_+ f(\bar{x}; v), \quad (2.41)$$

for any $\bar{x} \in X, v \in \mathbb{R}^n$ satisfying $v^T v = 1, t > 0, \bar{x} + tv \in X$.

Proof $(a) \Rightarrow (b)$ Let $\bar{x} \in X, v \in \mathbb{R}^n$ satisfying $v^T v = 1, t > 0, \bar{x} + tv \in X$ and f be pseudolinear over X. Therefore,

$$D_+ f(\bar{x}, v) \geq 0 \Rightarrow f(\bar{x} + tv) \geq f(\bar{x}),$$
$$D^+ f(\bar{x}, v) \leq 0 \Rightarrow f(\bar{x} + tv) \leq f(\bar{x}). \tag{2.42}$$

Since, the right derivative of f may not exist at every point of X, the following two cases may arise:

Case (1): When the right derivative at $\bar{x}$ does not exist, i.e.,

$$D^+ f(\bar{x}; v) \neq D_+ f(\bar{x}; v).$$

Case (2): When the right derivative at $\bar{x}$ does exist, i.e.,

$$df^+(x, v) = D^+ f(x, v) = D_+ f(x, v).$$

We consider Case (1), then from (2.42), we have

(i) $D^+ f(\bar{x}; v) < 0 \Rightarrow D_+ f(\bar{x}; v) < 0$ and $f(\bar{x} + tv) < f(\bar{x})$;

(ii) $D^+ f(\bar{x}; v) = 0 \Rightarrow D_+ f(\bar{x}; v) < 0$ and $f(\bar{x} + tv) < f(\bar{x})$;

(iii) $D_+ f(\bar{x}; v) > 0 \Rightarrow D^+ f(\bar{x}; v) > 0$ and $f(\bar{x} + tv) > f(\bar{x})$;

(iv) $D_+ f(\bar{x}; v) = 0 \Rightarrow D^+ f(\bar{x}; v) > 0$ and $f(\bar{x} + tv) > f(\bar{x})$.

Since $D_+ f(x; v) \leq D^+ f(x; v)$, for all $x \in X$ and moreover, for any $x \in X, v \in \mathbb{R}^n$ satisfying $v^T v = 1, t > 0, x + tv \in X$, if $f(x + tv) = f(x)$ then $D^+ f(x; v) = D_+ f(x; v) = 0$.

For each of the above four possibilities, we now establish statement (b). For possibilities (i) and (iii), we have

$$p(\bar{x}, \bar{x} + tv) = \frac{f(\bar{x} + tv) - f(\bar{x})}{D^+ f(x; v)}$$

and

$$q(\bar{x}, \bar{x} + tv) = \frac{f(\bar{x} + tv) - f(\bar{x})}{D_+ f(x; v)}.$$

From which, the results follow. For possibility (ii), we can define $p(\bar{x}, \bar{x} + tv)$ to be any positive real number and

$$q(\bar{x}, \bar{x} + tv) = \frac{f(\bar{x} + tv) - f(\bar{x})}{D_+ f(x; v)}.$$

Similarly, in case of possibility (iv), we define $q(\bar{x}, \bar{x} + tv)$ to be any positive real number and

$$p(\bar{x}, \bar{x} + tv) = \frac{f(\bar{x} + tv) - f(\bar{x})}{D^+ f(x; v)}.$$

Thus, in all the four possibilities, the statement (b) holds.

In Case (2), relation (2.41) can be written as

$$f(\bar{x} + tv) = f(\bar{x}) + \bar{p}(\bar{x}, \bar{x} + tv) D^+ f(\bar{x}; v),$$

where $\bar{p}(\bar{x}, \bar{x} + tv) = p(\bar{x}, \bar{x} + tv) + q(\bar{x}, \bar{x} + tv)$ and can be defined as

$$\bar{p}(\bar{x}, \bar{x} + tv) = \begin{cases} \frac{f(\bar{x}+tv)-f(\bar{x})}{D^+ f(x;v)}, & D^+ f(x; v) \neq 0, \\ M, & D^+ f(x; v) = 0, \end{cases}$$

where, M is a positive real number. The positiveness of the function $\bar{p}$ can be proved exactly in the same manner as proved for the functions p and q in Case (1).

$(b) \Rightarrow (a)$: Assume that there are functions p and q defined over $X \times X$ such that $p(\bar{x}, \bar{x} + tv) > 0, q(\bar{x}, \bar{x} + tv) > 0$ and (2.41) holds.

Then $D^+ f(\bar{x}; v) \leq 0$ and the fact that $D_+ f(\bar{x}; v) \leq D^+ f(\bar{x}; v)$ implies that $f(\bar{x}+tv) \leq f(\bar{x})$, that is f is pseudoconcave over X. Moreover, $D_+ f(\bar{x}; v) \geq 0$ and again the fact that $D^+ f(\bar{x}; v) \geq D_+ f(\bar{x}; v)$ implies that $f(\bar{x}+tv) \geq f(\bar{x})$, that is f is pseudoconvex over X. Hence, f is pseudolinear over X, in terms of the Dini derivatives.

2.7 Pseudolinearity in Terms of Bifunctions

Several authors have studied the notions of nondifferentiable pseudoconvexity/pseudoconcavity using the concept of generalized directional derivatives and subdifferentials. See Luc [172], Daniilidis and Hadjisavvas [60], and Hadjisavvas [98]. To extend generalized convexity to nondifferentiable functions, Komlosi [146] and Sach and Penot [242] used the concept of bifunctions, which are substitutes to directional derivatives. Recently, Lalitha and Mehta [155] used Komlosi's [146] concept to define the notion of pseudolinearity in terms of a bifunction h. Lalitha and Mehta [155] introduced the class of h-pseudolinear functions, as functions which are both h-pseudoconvex and h-pseudoconcave and obtained various characterizations for these types of functions.

To understand the notion of h-pseudolinearity, we recall the following definitions and preliminary results from Lalitha and Mehta [155]. In this section, we assume that X is a convex subset of $\mathbb{R}^n$. An alternative definition of the Dini upper and lower directional derivatives are as folllows.

Definition 2.11 *Let $f : \mathbb{R}^n \to \overline{\mathbb{R}}$ be a function and $x \in \mathbb{R}^n$, be a point where f is finite.*

(a) The Dini upper directional derivative of f at the point $x \in \mathbb{R}^n$, in the

direction $d \in \mathbb{R}^n$, *is defined as:*

$$D^+ f(x, d) = \limsup_{t \to 0^+} \frac{f(x + td) - f(x)}{t} = \inf_{s > 0} \sup_{0 < t < s} \frac{f(x + td) - f(x)}{t}.$$

(b) The Dini lower directional derivative of f at the point $x \in \mathbb{R}^n$, in the direction $d \in \mathbb{R}^n$, is defined as:

$$D_+ f(x, d) = \limsup_{t \to 0^+} \frac{f(x + td) - f(x)}{t} = \inf_{s > 0} \sup_{0 < t < s} \frac{f(x + td) - f(x)}{t}.$$

It is obvious from the definitions that for each fixed $x \in \mathbb{R}^n$, $D^+ f(x; d)$ and $D_+ f(x; d)$ are positively homogeneous. For basic elementary properties and calculus rules about the Dini upper (lower) derivative, we refer to Chapter 1 of this book and Giorgi and Komlosi [88, 89, 90].

Usually, generalized derivatives of a function might be considered in a unified way as a bifunction $h(x; .)$ of the given point $x \in X$, and the given direction $d \in \mathbb{R}^n$. All the generalized derivatives have the common feature, that they are positively homogeneous functions of the direction $d \in \mathbb{R}^n$.

Let $h : X \times \mathbb{R}^n \to \overline{\mathbb{R}}$ be a bifunction associated with the function $f : X \to \mathbb{R}$, such that for each $x \in X$,

$$h(x; .) \ is \ \text{positively homogeneous} \qquad (2.43)$$

and

$$h(x; 0) = 0. \qquad (2.44)$$

Definition 2.12 (Sach and Penot [242]) *Let $h : X \times \mathbb{R}^n \to \overline{\mathbb{R}}$ be a bifunction associated with the function $f : X \to \mathbb{R}$. Then, f is said to be h-convex if for all $x, y \in X$ with $x \neq y$, we have*

$$f(y) - f(x) \geq h(x; y - x).$$

Sach and Penot [242] have proved that if f is a convex function and h is a bifunction associated with the function f such that it satisfies (2.43) and the condition

$$h(x; d) \leq D^+ f(x; d), x \in X, d \in \mathbb{R}^n,$$

then, f is h-convex.

Lalitha and Mehta [155] defined pseudolinearity for a nondifferentiable case in terms of a bifunction, in the following way:

Definition 2.13 *Let $h : X \times \mathbb{R}^n \to \overline{\mathbb{R}}$ be a bifunction associated with the function $f : X \to \overline{\mathbb{R}}$. Then, f is said to be*

(i) h-pseudoconvex, if for all $x, y \in K$ with $x \neq y$

$$f(y) < f(x) \Rightarrow h(x; y - x) < 0;$$

(ii) h-pseudoconcave, if for all $x, y \in K$ with $x \neq y$

$$f(y) > f(x) \Rightarrow h(x; y - x) > 0;$$

(iii) h-pseudolinear, if it is both h-pseudoconvex as well as h-pseudoconcave.

To be precise; the function $f : X \to \overline{\mathbb{R}}$ is said to be h-pseudolinear, if for all $x, y \in X$ with $x \neq y$, we have

$$f(y) < f(x) \Rightarrow h(x; y - x) < 0$$

and

$$f(y) > f(x) \Rightarrow h(x; y - x) > 0.$$

Definition 2.14 (Sach and Penot [242]) Let $h : X \times \mathbb{R}^n \to \overline{\mathbb{R}}$ be a bifunction associated with the function $f : X \to \mathbb{R}$. Then, f is said to be h-quasiconvex, if for all $x, y \in X$ with $x \neq y$, we have

$$f(y) \geq f(x) \Rightarrow h(x; y - x) \geq 0.$$

We recall the following definition from Chapter 1.

Definition 2.15 A function $f : \mathbb{R}^n \to \mathbb{R}$ is said to be radially upper semicontinuous (radially lower semicontinuous) on a convex subset X of $\mathbb{R}^n$, if f is upper semicontinuous on every line segment $[x, y]$ in X, where $[x, y] = \{z : z = ty + (1 - t)x, 0 \leq t \leq 1\}$. The function f is said to be radially continuous on X, if f is both radially upper semicontinuous and radially lower semicontinuous on X.

Lalitha and Mehta [155] have given the following two examples of h-pseudolinear functions, which are not pseudolinear.

Example 2.7 Let $f : \mathbb{R} \to \mathbb{R}$ be defined as

$$f(x) = \begin{cases} x^2 + 3x + 1, & \text{if } x \geq 0, \\ x, & \text{if } x < 0. \end{cases}$$

Then, we have

$$D^+ f(x; d) = D_+ f(x; d) = \begin{cases} (2x + 3)d, & \text{if } x > 0, \forall d \text{ or } x = 0, d > 0, \\ -\infty, & \text{if } x = 0, d < 0, \\ d, & \text{if } x < 0, \forall d. \end{cases}$$

Taking $h(x; d) = D^+ f(x; d)$, for each $x, d \in \mathbb{R}$, we can see that f is h-pseudolinear on $\mathbb{R}$.

Example 2.8 *Let $f : \mathbb{R} \to \mathbb{R}$ be defined as*

$$f(x) = \begin{cases} 5x/2, & \text{if } x \leq 0, \\ x + \frac{1}{2^{2n}}, & \text{if } \frac{1}{2^{2n+1}} < x \leq \frac{1}{2^{2n}}, n = 0,1,2,.... \\ 4x - \frac{1}{2^{2n+1}}, & \text{if } \frac{1}{2^{2n+2}} < x \leq \frac{1}{2^{2n+1}}, n = 0,1,2,.... \\ 4x - 2, & \text{if } x > 1, n = 0,1,2,.... \end{cases}$$

Then, f is h-pseudolinear function on $\mathbb{R}$, where

$$h(x;d) = \begin{cases} 5d/2, & \text{if } x \leq 0, \forall d, \\ d, & \text{if } \frac{1}{2^{2n+1}} < x < \frac{1}{2^{2n}}, \forall d, n = 0,1,2,.... \\ 4d, & \text{if } \frac{1}{2^{2n+2}} < x < \frac{1}{2^{2n+1}}, \forall d, n = 0,1,2,.... \\ \max\{d, 4d\}, & \text{if } x = \frac{1}{2^{2n}}, \forall d, n = 0,1,2,..... \\ \min\{d, 4d\}, & \text{if } x = \frac{1}{2^{2n+1}}, \forall d, \\ 4d, & \text{if } x > 1, \forall d. \end{cases}$$

It is clear that $D_+ f(0; d) < h(0; d) < D^+ f(0; d), \forall d > 0$.

The following theorem of Lalitha and Mehta [155] gives a necessary condition for the function to be h-pseudolinear.

Theorem 2.18 *Let $f : X \to \mathbb{R}$ be a radially continuous function on open convex set X and $h : X \times \mathbb{R}^n \to \bar{R}$ be a bifunction associated to f such that it satisfies (2.43)–(2.44) and the condition*

$$D_+ f(x; .) \leq h(x; .) \leq D^+ f(x; .), \forall x \in X. \tag{2.45}$$

If the function f is h-pseudolinear on X, then, for any $x, y \in X$,

$$h(x; y - x) = 0 \Leftrightarrow f(x) = f(y). \tag{2.46}$$

Proof Let $x, y \in X$ be such that $x \neq y$. Assume that $h(x; y - x) = 0$. Then, by h-pseudoconvexity and h-pseudoconcavity of f, it follows that

$$f(x) = f(y).$$

Conversely, suppose that $f(x) = f(y)$, for some $x, y \in X$. If $x = y$ then by (2.44), we get $h(x; y - x) = 0$. Let $x \neq y$. Now, we first prove that

$$f(x) = f(z), \forall z \in \,]x, y[\, .$$

On the contrary, suppose that $f(x) > f(z)$, for some $z \in \,]x, y[$. By h-pseudoconcavity of f, we have $h(z; x - z) > 0$. Therefore, from (2.45), we get $D^+ f(z; x - z) > 0$, that is, there exists a sequence $\lambda_n \downarrow 0$, $\lambda_n \in \,]0,1[$ such that $f(z + \lambda_n(x-z)) > f(z)$. Since, f is radially continuous on X, there exists some $\lambda_n \in \,]0,1[$, such that $f(z) < f(z_n) < f(x)$, where $z_n := z + \lambda_n(x - z)$. Since, $f(x) = f(y)$, therefore, we get $f(z_n) < f(y)$. Using the h-pseudoconcavity of

f, we get $h(z_n; y - z_n) > 0$. Using (2.43), we have $h(z_n; z - z_n) > 0$, which by h-pseudoconvexity of f implies that $f(z_n) \le f(z)$, which is a contradiction. Now, assume that $f(x) < f(z)$. Using h-pseudoconcavity of f, we get $h(x, z - z) > 0$. From (2.45), it follows that $D^+ f(x; z - x) > 0$, that is there exists a sequence $t_n \downarrow 0, t_n \in]0, 1[$, such that

$$f(x + t_n(z - x)) > f(z).$$

Then by radial continuity of f, there exists some $t_n \in]0, 1[$, such that $f(x) < f(x_n) < f(z)$, where $x_n = x + t_n(z - x)$. As $f(y) = f(x) < f(x_n)$, by h-pseudoconvexity of f, it follows that $h(x_n; y - x_n) < 0$. By positive homogeneity of h we get

$$h(x_n; z - x_n) < 0.$$

By h-pseudoconcavity of f it follows that $f(x_n) \ge f(z)$, which is a contradiction.

Therefore, $f(x) = f(z), \forall z \in]x, y[$ and hence,

$$D^+ f(x; y - x) = D_+ f(x; y - x) = 0.$$

Then, relation (2.45) yields that $h(x; y - x) = 0$.

Lalitha and Mehta [155] proved the converse of the above theorem assuming that the function f is continuous.

Theorem 2.19 *Let $f : X \to \mathbb{R}$ be a continuous function on open convex set X and $h : X \times \mathbb{R}^n \to \overline{\mathbb{R}}$ be a positively homogeneous function and satisfy (2.43)–(2.45). If condition (2.46) holds for all $x, y \in X$, then f is h-pseudolinear on X.*

Proof Let $f(x) < f(y)$, for some $x, y \in X$. Then by (2.46), it follows that $h(x; y - x) \neq 0$. Let us assume $h(x; y - x) > 0$, then by using (2.45), it follows that $D^+ f(x; y - x) > 0$. Hence, there exist some $\lambda \in]0, 1[$, such that $f(x + \lambda(y - x)) > f(x)$. Now as f is continuous on X and $f(y) < f(x) < f((1 - \lambda)x + \lambda y)$, there exists some $\bar{\lambda} \in]0, 1[$, such that $f(z) = f(x)$, where $z = (1 - \bar{\lambda})x + \bar{\lambda}y$. Then, by assumption (2.46), we get

$$h(x; z - x) = 0.$$

Using (2.43), we get $h(x; y - x) = 0$, which is a contradiction to the assumption that $h(x; y - x) > 0$. Hence, $h(x; y - x) < 0$. Therefore, f is h-pseudoconvex on X. Similarly, we can prove that f is h-pseudoconcave on X.

The following example from Lalitha and Mehta [155] illustrates the fact that the continuity assumption in Theorem 2.16 cannot be relaxed.

Example 2.9 *Let $f : \mathbb{R} \to \mathbb{R}$ be defined as*

$$f(x) = \begin{cases} e^x, & \text{if } x < 0, \\ -x - 1, & \text{if } x \ge 0. \end{cases}$$

Then, we have

$$D^+ f(x; d) = D_+ f(x; d) = \begin{cases} e^x d, & \text{if } x < 0, \forall d, \\ \infty, & \text{if } x = 0, d < 0, \\ -d, & \text{if } x = 0, d \geq 0 \text{ or } x > 0, \forall d. \end{cases}$$

Choosing $h(x; d) = D^+ f(x; d)$, *for each* $x, d \in \mathbb{R}$, *we can see that condition (2.46) is trivially true. However, the function* f *is not* h-*pseudolinear on* $\mathbb{R}$, *because for* $y = 0$ *and* $x = -1$, $f(y) < f(x)$, *whereas* $h(x; y - x) > 0$.

Theorem 2.20 *Let* $f : X \to \mathbb{R}$ *be a radially continuous function on open convex set* X *and* $h : X \times \mathbb{R}^n \to \bar{\mathbb{R}}$ *is a bifunction associated to* f, *such that it satisfies (2.43)–(2.45). Then,* f *is* h-*pseudolinear on* X, *if and only if there exists a function* $p : X \times X \to \mathbb{R}_{++}$, *such that for each* $x, y \in X$, *we have*

$$f(y) = f(x) + p(x, y)h(x; y - x). \tag{2.47}$$

Proof We suppose that f is h-pseudolinear on X. If $h(x; y - x) = 0$, for some $x, y \in X$, by Theorem 2.18, we get $f(x) = f(y)$. Defining $p(x, y) = 1$ it follows that (2.47) holds. If $h(x; y - x) \neq 0$, then, by Theorem 2.18, we get $f(x) \neq f(y)$. Now, we define

$$p(x, y) = \frac{f(y) - f(x)}{h(x, y - x)}. \tag{2.48}$$

Now, we prove that $p(x, y) > 0$. If $f(y) > f(x)$, then h-pseudoconcavity of f implies that $h(x, y - x) > 0$, hence, by (2.48), we get the result. Similarly, if $f(y) < f(x)$, the result follows by h-pseudoconvexity of f. The converse implication is trivially true. $\qquad \square$

The following theorem by Lalitha and Mehta [155] provides another characterization for h-pseudolinear functions.

Theorem 2.21 *Let* $f : X \to \mathbb{R}$ *be a radially continuous function on open convex set* X *and* $h : X \times \mathbb{R}^n \to \bar{\mathbb{R}}$ *is a bifunction associated to* f, *such that it satisfies (2.43)–(2.45). If* f *is* h-*pseudolinear on* X, *then, for each* $x, y \in X$, *we have*

$$h(x; y - x) = 0 \Leftrightarrow f(x) = f((1 - \lambda)x + \lambda y), \forall \lambda \in [0, 1]. \tag{2.49}$$

Proof For $x, y \in X$, let $h(x; y - x) = 0$. Then, for all $\lambda \in \,]0, 1]$, by the assumption (2.43), it follows that

$$h(x; (1 - \lambda)x + \lambda y - x) = \lambda h(x; y - x) = 0.$$

The result follows by using Theorem 2.18. $\qquad \square$

The following example by Lalitha and Mehta [155, 156] provides an illustration for Theorem 2.21.

Example 2.10 *Consider the function $f : \mathbb{R}^2 \to \mathbb{R}$, defined by*

$$f(x) = \begin{cases} x_1 + x_2, & \text{if } x_1 + x_2 > 0, \\ \frac{(x_1+x_2)}{2}, & \text{if } x_1 + x_2 \le 0. \end{cases}$$

Then, we have

$$D^+ f(x; d) = D_+ f(x; d) = \begin{cases} d_1 + d_2, & \text{if } x_1 + x_2 > 0, \\ \max\{d_1 + d_2, \frac{(d_1+d_2)}{2}\}, & \text{if } x_1 + x_2 = 0, \\ \frac{(d_1+d_2)}{2}, & \text{if } x_1 + x_2 < 0, \end{cases}$$

where $x = (x_1, x_2)$ and $d = (d_1, d_2)$.

   Evidently, f is a continuous h-pseudolinear function with

$$h(x; d) = D^+ f(x; d) = D_+ f(x; d), \forall x, d \in \mathbb{R}^2.$$

For $x - (1, 2), y = (2, 1)$, we have $h(x, y - x) = 0$ and it is clear that

$$f(x) = f((1 - \lambda)x + \lambda y), \forall \lambda \in [0, 1].$$

Remark 2.6 (Lalitha and Mehta [155]) *The converse of Theorem 2.21 is not necessarily true. The function considered in Example 2.9 satisfies condition (2.49), but f is not a h-pseudolinear function on $\mathbb{R}$. It is clear by Theorem 2.19, that the converse of Theorem 2.21 holds, if f is a continuous function on $\mathbb{R}$.*

The following theorem has been established by Sach and Penot [242].

Theorem 2.22 *Let $h : X \times \mathbb{R}^n \to \overline{\mathbb{R}}$ be a bifunction associated to f such that h is positively homogeneous and subodd in the second argument. Further suppose that*

$$h(x; d) \le D^+ f(x; d), \forall x \in X \text{ and } d \in \mathbb{R}^n.$$

If f is h-pseudoconvex on X, then f is h-quasiconvex on X.

2.8   Directionally Differentiable Pseudolinear Functions

Smietanski [257] has extended the notion of pseudolinearity of Chew and Choo [47] for nondifferentiable functions as follows:

Definition 2.16 (Lower and upper Dini pseudolinearity) *A function $f : X \to \mathbb{R}$ is said to be lower Dini pseudolinear on C if for all $x, y \in C$, we have*

$$f(x) < f(y) \Rightarrow D_+ f(y; x - y) < 0 \text{ and } D_+ f(x; y - x) > 0.$$

The function f is called upper Dini pseudolinear if for all $x, y \in C$, we have

$$f(x) < f(y) \Rightarrow D^+ f(y; x - y) < 0 \text{ and } D^+ f(x; y - x) > 0.$$

Smietanski [257] has remarked that pseudolinear functions can alterna-tively be defined with Clarke derivatives, upper f° and lower $f_\circ$, the Dini-Hadamard derivatives (hypo-derivatives): upper f^{DH} and lower f_{DH} and oth-ers. With any of the directional derivatives (Dini, Clarke, and Dini-Hadamard) considered above (upper and lower), denoted by f^*, Smietanski [257] intro-duced the notion of ∗-pseudolinearity, as follows:

Definition 2.17 (∗-*pseudolinearity*) *A function* $f : X \to \mathbb{R}$ *is said to be* ∗-*pseudolinear on* C, *if for all* $x, y \in C$, *we have*

$$f(x) < f(y) \Rightarrow f_*(y; x - y) < 0 \text{ and } f_*(x; y - x) > 0.$$

Example 2.11 *Let* $X = \mathbb{R}$ *and* $f : X \to \mathbb{R}$, *be defined as*

$$f(x) := \begin{cases} \left|x - \frac{1}{2}\sin x\right|, & \text{if } x \in \left]-\infty, -\frac{\pi}{2}\right[\cup \left]\frac{\pi}{2}, \infty\right[, \\ \left(\frac{\pi}{2} - \frac{1}{2}\sin\frac{\pi}{2}\right), & \text{if } x \in \left]\frac{-\pi}{2}, \frac{\pi}{2}\right[. \end{cases}$$

Then, the function f *is a nondifferentiable pseudolinear function.*

Remark 2.7 *From Chapter 1, we know that the function* $f : X \subseteq \mathbb{R}^n \to \mathbb{R}$ *is called directionally differentiable at* $x \in X$, *if for all direction* $\nu \in \mathbb{R}^n$, *the Dini directional derivative* $f'_+(x; \nu)$ *exists and is finite. Moreover, then,*

$$D^+ f(x; \nu) = D_+ f(x; \nu) = f'_+(x; \nu)$$

holds. It is known from Preda [227], if a function is directionally differentiable, then it is semidifferentiable but the converse is not true. Hence, a direction-ally differentiable pseudolinear function is a semilocal pseudolinear function. However, the converse, may not be true.

The following characterizations of ∗-pseudolinear functions are variants of the corresponding results from Smietanski [257] on normed vector spaces.

Proposition 2.2 *Let* f *be a directionally differentiable and radially lower semicontinuous function. Moreover, let* $f_*(x; .)$ *be a homogeneous function. Then the following statements (a)–(c) are equivalent:*

(a) f *is* ∗-*pseudolinear on* C;

(b) *for any* $x, y \in C$, $f'_+(x, y - x) = 0 \Leftrightarrow f(x) = f(y)$;

(c) *there exists a real-valued function* $p : C \times C \to \mathbb{R}$, *such that*

$$p(x, y) > 0 \text{ and } f(y) = f(x) + p(x, y)f'_+(x; y - x), \forall x, y \in C. \quad (2.50)$$

Proof $(a) \Rightarrow (b)$ Suppose that f is $*$-pseudolinear on C. Now, we prove that

$$f(x) = f(y) \Rightarrow f'_+(x, y - x) = 0.$$

The inverse implication follows from the definition of directionally differentiable $*$-pseudolinear functions. Let $x, y \in C$, be such that $f(x) = f(y)$. We first prove that

$$f(x\lambda y) = f(x), \forall \lambda \in]0, 1[.$$

If

$$f(x\lambda y) > f(x), \forall \lambda \in]0, 1[,$$

then, by $*$-pseudoconvexity of f, we get

$$f_*(x\lambda y; x - x\lambda y) < 0.$$

We note that

$$y - x\lambda y = (1 - \lambda)(y - x) = \frac{-(1 - \lambda)}{\lambda}(x - x\lambda y),$$

therefore, we have

$$f_*(x\lambda y, y - x\lambda y) = f_*(x\lambda y; \frac{-(1 - \lambda)}{\lambda}(x - x\lambda y)).$$

By the homogeneity of $f_*(x; .)$ we get

$$f_*(x\lambda y; y - x\lambda y) > 0.$$

Since, f is $*$-pseudoconcave, we get $f(y) \geq f(x\lambda y)$, which is a contradiction. Similarly, we can show that $f(x\lambda y) < f(x)$ leads us to a contradiction. This proves that

$$f(x\lambda y) = f(x), \forall \lambda \in]0, 1[.$$

Then,

$$f'_+(x; y - x) = \lim_{\lambda \downarrow 0} \frac{f(x + \lambda(y - x)) - f(x)}{\lambda} = \lim_{\lambda \downarrow 0} \frac{f(x\lambda y) - f(x)}{\lambda}.$$

$(b) \Rightarrow (c)$ Let (b) holds. We need to construct a function p, for which condition (2.50) holds. When $f'_+(x, y - x) = 0$ for $x, y \in C$, we define $p(x, y) = 1$. If $f'_+(x, y - x) \neq 0$, we define

$$p(x, y) = \frac{f(y) - f(x)}{f'_+(x; y - x)}.$$

Assume that $f(y) > f(x)$. If $f(x\lambda y) \leq f(y)$, for some $\lambda \in]0, 1[$, then by the radial lower semicontinuity of f there is $\mu \in]0, 1[$, such that $\mu \leq \lambda$ and $f(x\mu y) = f(x)$ Now, by statement (b), we have

$$f'_+(x; y - x) = f'_+\left(x; \frac{x\mu y - x}{\mu}\right) = 0,$$

which is a contradiction. Hence, $f(x\lambda y) > f(x), \lambda \in \,]0,1[$ and

$$f'_+(x; y - x) = \lim_{\lambda \downarrow 0} \left(\frac{f(x\lambda y) - f(x)}{\lambda} \right) \geq 0.$$

The strict inequality holds since $f'_+(x; y-x) \neq 0$. This implies that $p(x,y) > 0$. Similarly, we can show that $p(x,y) > 0$, if $f(y) < f(x)$.

The inclusion $(c) \Rightarrow (a)$ is obviously true.

Proposition 2.3 *Let f be a directionally differentiable function on X. Then f is $*$-pseudolinear on C, if and only if for all $x, y \in C$, the following condition holds:*

$$f'_+(x; y - x) = 0 \Rightarrow f(x) = f(x\lambda y), \forall \lambda \in [0,1]. \qquad (2.51)$$

Proof Suppose that f is $*$-pseudolinear on C. Let $x, y \in C$ be the points, such that

$$f'_+(x; y - x) = 0.$$

Then,

$$f'_+(x; x\lambda y - x) = \lambda f'_+(x; y - x) = 0.$$

By Proposition 2.2, it follows that

$$f(x) = f(x\lambda y), \forall \lambda \in [0,1].$$

Conversely, suppose that (2.51) holds. If f is not $*$-pseudolinear on C, then there exist points $x, y \in C$ such that $f(x) = f(y)$ but $f'_+(x, y - x) \neq 0$. In view of condition (2.51), the alternative case $f'_+(x, y-x) = 0$ but $f(x) \neq f(y)$ cannot occur. Define the function $f_{x,y}(\lambda)$ on $C_{x,y} = \{\lambda \in \mathbb{R} : x\lambda y \in C\}$ by setting

$$f_{x,y}(\lambda) := f(x\lambda y).$$

Now, we suppose that $f'_+(x; y - x) > 0$. It means that $f_{x,y}$ is directionally differentiable and $f'_{x,y}(0; 1) > 0$. Therefore,

$$f_{x,y}(\lambda) > f_{x,y}(0),$$

for sufficiently small $\lambda > 0$. Since, $f_{x,y}(0) = f_{x,y}(1)$, then $f_{x,y}$ attains a local maximum at some point $\bar{\lambda} \in \,]0,1[$. Therefore,

$$0 = f'_{x,y}(\bar{\lambda}; 1) = \bar{\lambda} f'_+(x\bar{\lambda}y; y - x).$$

Hence,

$$f'_+(x\bar{\lambda}y; x\bar{\lambda}y - x) = \bar{\lambda} f'_+(x\bar{\lambda}y; y - x) = 0.$$

By condition (2.51), we have

$$f(x) = f(x\lambda y), \forall \lambda \in [0, \bar{\lambda}].$$

It follows that

$$f'_+(x; y - x) = 0,$$

which is a contradiction.

2.9 Locally Lipschitz Pseudolinear Functions and Clarke Subdifferentials

In the previous chapter, we studied the properties and calculus rules of the Clarke subdifferential. Now, we use the tools of Clarke subdifferentials to study the properties of locally Lipschitz pseudolinear functions.

The following definition is an extension of the concept of pseudolinearity to the nondifferentiable case (see Aussel [8], Lu and Zhu [171]).

Definition 2.18 *Let* $f : X \to \mathbb{R}$ *be a locally Lipschitz function on an open convex set* X. *Then,* f *is said to be pseudoconvex at* $x \in X$, *if for all* $y \in X$, *one has*

$$f(y) < f(x) \Rightarrow \langle \zeta, y - x \rangle < 0, \forall \zeta \in \partial^c f(x)$$

or equivalently,

$$\exists \zeta \in \partial^c f(x) : \langle \zeta, y - x \rangle \geq 0 \Rightarrow f(y) \geq f(x).$$

The function f *is said to be pseudoconvex on* X, *if it is pseudoconvex for each* $x \in X$.

The function f *is said to be pseudoconcave on* X, *if* $-f$ *is pseudoconvex on* X. *Moreover, the function* $f : X \to \mathbb{R}$ *is said to be pseudolinear on* X, *if* f *is both pseudoconvex and pseudoconcave on* X.

To be precise; let $f : X \to \mathbb{R}$ be a locally Lipschitz function on an open convex set X. The function f is said to be *pseudolinear* at $x \in X$, if for all $y \in X$, one has

$$\exists \zeta \in \partial^c f(x) : \langle \zeta, y - x \rangle \geq 0 \Rightarrow f(y) \geq f(x)$$

and

$$\exists \zeta \in \partial^c f(x) : \langle \zeta, y - x \rangle \leq 0 \Rightarrow f(y) \leq f(x).$$

The function in the following example is taken from Zhao and Tang [299].

Example 2.12 *Let* $X = \mathbb{R}$. *Let* $f : X \to \mathbb{R}$ *be defined as* $f(x) = 2x + |x|$. *Then, clearly one can see that:*

$$\partial^c f(x) = \begin{cases} 3, & \text{if } x > 0, \\ [1,3], & \text{if } x = 0, \\ 1, & \text{if } x < 0. \end{cases}$$

Evidently, the f *is pseudolinear on* $\mathbb{R}$.

Theorems 2.23 and 2.24, which follow, are variants of Theorems 3.1 and 3.2 in Lu and Zhu [171].

Theorem 2.23 *Let X be an open convex set. If $f : X \to \mathbb{R}$ is a locally Lipschitz pseudolinear function on X, then $f(x) = f(y)$ if and only if there exists $\zeta \in \partial^c f(x)$, such that*

$$\langle \zeta, y - x \rangle = 0, \forall x, y \in X.$$

Proof Suppose that, for any $x, y \in X$, there exists $\zeta \in \partial^c f(x)$, such that $\langle \zeta, y - x \rangle = 0$. By the pseudoconvexity of f, there exists $\zeta \in \partial^c f(x)$, such that

$$\langle \zeta, y - x \rangle \geq 0 \Rightarrow f(y) \geq f(x). \tag{2.52}$$

By the pseudoconcavity of f, there exists $\zeta \in \partial^c f(x)$, such that

$$\langle \zeta, y - x \rangle \leq 0 \Rightarrow f(y) \leq f(x). \tag{2.53}$$

From (2.52) and (2.53), we get

$$\langle \zeta, y - x \rangle = 0 \Rightarrow f(y) = f(x).$$

Conversely, for any $x, y \in X$, we suppose that $f(x) = f(y)$. We have to prove that there exists $\zeta \in \partial^c f(x)$, such that $\langle \zeta, y - x \rangle = 0$. We will show that

$$f(x\lambda y) = f(x), \forall \lambda \in [0, 1]. \tag{2.54}$$

When $\lambda = 0$ and $\lambda = 1$, it is obvious. Now, we will prove it for $\lambda \in]0, 1[$. If $f(x\lambda y) > f(x)$, then by the pseudoconvexity of f, we have

$$\left\langle \zeta', x - x\lambda y \right\rangle < 0, \forall \zeta' \in \partial^c f(x\lambda y). \tag{2.55}$$

We know that $y - x\lambda y = y - (x + \lambda(y - x)) = (1 - \lambda)(y - x)$. Therefore, we have

$$x - x\lambda y = -\lambda(y - x) = -\frac{\lambda}{1 - \lambda}(y - x\lambda y). \tag{2.56}$$

From (2.55) and (2.56), we get

$$\left\langle \zeta', y - x\lambda y \right\rangle > 0, \forall \zeta' \in \partial^c f(x\lambda y).$$

From the pseudoconvexity of f, we have

$$f(y) \geq f(x\lambda y),$$

a contradiction, to the fact that

$$f(x\lambda y) > f(x) = f(y).$$

Similarly, if $f(x\lambda y) < f(x)$, then by the pseudoconcavity of f, we get

$$f(y) \leq f(x\lambda y),$$

again a contradiction. Hence, for any $\lambda \in [0,1]$,

$$f(x\lambda y) = f(x).$$

Since, X is an open convex set, $[x\lambda y, x] \subseteq K$, for all $\lambda \in]0,1[$. By Theorem 1.43, there exist $\bar{\lambda} \in]0, \lambda[$ and $\zeta_{\bar{\lambda}} \in \partial^c f(x\bar{\lambda}y)$, such that

$$0 = f(x\bar{\lambda}y) - f(x) = \bar{\lambda}\langle \zeta_{\bar{\lambda}}, y - x \rangle,$$

that is,

$$\langle \zeta_{\bar{\lambda}}, y - x \rangle = 0 \tag{2.57}$$

Since f is locally Lipschitz and $\zeta_{\bar{\lambda}} \in \partial^c f(x\bar{\lambda}y)$, by Proposition 1.29, $\zeta_{\bar{\lambda}}$ is a bounded sequence. Without loss of generality, we may assume that $\lim_{\lambda \to 0} \zeta_{\bar{\lambda}} = \zeta$. Since, $x\bar{\lambda}y \to x$ as $\bar{\lambda} \to 0$, by Lemma 1.5 and (2.57), it follows that there exists $\zeta \in \partial^c f(x)$, such that $\langle \zeta, y - x \rangle = 0$.

Theorem 2.24 *Let X be an open convex set and let $f : X \to \mathbb{R}$ be a locally Lipschitz function on X. Then, the function f is pseudolinear on X, if and only if there exists a function $p : X \times X \to \mathbb{R}_{++}$, such that for every $x, y \in X$, there exists $\zeta \in \partial^c f(x)$, such that*

$$f(y) = f(x) + p(x,y)\langle \zeta, y - x \rangle. \tag{2.58}$$

Proof Let f be pseudolinear on X. We need to construct a function $p : X \times X \to \mathbb{R}_{++}$, such that for all $x, y \in X$, there exists $\zeta \in \partial^c f(x)$, such that

$$f(y) = f(x) + p(x,y)\langle \zeta, y - x \rangle.$$

If $\langle \zeta, y - x \rangle = 0$, for some $\zeta \in \partial^c f(x)$ and for any $x, y \in X$, we define $p(x,y) = 1$. From Theorem 2.23, we get $f(y) = f(x)$ and thus, (2.49) holds.

If $\langle \zeta, y - x \rangle \neq 0$, for some $\zeta \in \partial^c f(x)$ and for any $x, y \in X$, we define

$$p(x,y) = \frac{f(y) - f(x)}{\langle \zeta, y - x \rangle}. \tag{2.59}$$

Evidently, $p(x,y)$ satisfies (2.58). Now, we have to show that $p(x,y) > 0$, for all $x, y \in X$. If $f(y) > f(x)$, then by pseudoconcavity of f, we have $\langle \zeta, y - x \rangle > 0$, for all $\zeta \in \partial^c f(x)$. From (2.59), we get $p(x,y) > 0$. Similarly, if $f(y) < f(x)$, then by pseudoconvexity of f, we have $\langle \zeta, y - x \rangle < 0$, for all $\zeta \in \partial^c f(x)$. From (2.59), we get $p(x,y) > 0$, for all $x, y \in K$.

Conversely, suppose that for any $x, y \in X$, there exists a function $p : X \times X \to \mathbb{R}_{++}$, such that for every $x, y \in X$, there exists $\zeta \in \partial^c f(x)$ such that (2.58) holds. If for some $\zeta \in \partial^c f(x)$ and for any $x, y \in X, \langle \zeta, y - x \rangle \geq 0$, from (2.54), it follows that

$$f(y) - f(x) = p(x,y)\langle \zeta, y - x \rangle \geq 0,$$

and hence, f is pseudoconvex on X. Similarly, if $\langle \zeta, y - x \rangle \leq 0$, for some $\zeta \in \partial^c f(x)$ and for any $x, y \in X$ from (2.58), it follows that f is pseudoconcave on X.

The following results are from Mishra *et al.* [200].

Remark 2.8 *For any two nonzero vectors $a, b \in \mathbb{R}^n$, we know that*

$$\langle a, b \rangle \leq \|a\| \, \|b\|$$

and the equality holds if and only if $a = \lambda b$, for some $\lambda > 0$. Evidently, the equality does not hold for $\lambda \leq 0$. We note that a, b are not zero by assumption, so $\lambda \neq 0$. Again, $\lambda \not< 0$, because then $\langle a, b \rangle = \langle \lambda b, b \rangle = \lambda \|b\|^2 < 0$, which is impossible, since $\langle a, b \rangle = \|a\| \, \|b\| > 0$.

Proposition 2.4 *Let $X \subseteq \mathbb{R}^n$ be an open convex set and $f : X \to \mathbb{R}$ be a locally Lipschitz pseudolinear function on X. Let $z \in X$. Then the set*

$$A := \{x \in X : f(x) = f(z)\}$$

is convex.

Proof Suppose $y, z \in A$, then, we have $f(x) = f(y) = f(z)$. By Theorem 2.23, there exists $\zeta \in \partial^c f(x)$, such that

$$\langle \zeta, y - x \rangle = 0.$$

Then, for every $\lambda \in [0, 1]$, there exists $\zeta \in \partial^c f(x)$, such that

$$\langle \zeta, ((1 - \lambda) x + \lambda y) - x \rangle = \lambda \langle \zeta, y - x \rangle = 0.$$

Thus, using Theorem 2.23, again, we infer that

$$f((1 - \lambda)x + \lambda y) = f(x) = f(z), \forall \lambda \in [0, 1].$$

Consequently, $(1 - \lambda)x + \lambda y \in A, \forall \lambda \in [0, 1]$, so, A is convex.

Remark 2.9 *From Proposition 2.4, it is evident that if $f : X \to \mathbb{R}$ is a locally Lipschitz pseudolinear function on an open convex set $X \subseteq \mathbb{R}^n$ and $0 \in \partial^c f(x)$ at any arbitrary point $x \in X$, then f is a constant function on X.*

The following proposition from Mishra *et al.* [200] provides a very important characterization for the locally Lipschitz pseudolinear functions.

Proposition 2.5 *Let $f : X \to \mathbb{R}$ be a locally Lipschitz pseudolinear function on an open convex set X. Let $x \in X$ be arbitrary. Then for any $\xi, \xi' \in \partial^c f(x)$, there exists $\lambda > 0$, such that*

$$\xi' = \lambda \xi.$$

Proof If $\xi = 0$, then for any $y \in X$, we have $\langle \xi, y - x \rangle = 0$, hence $f(y) = f(x)$. Consequently, f is a constant function on X. This implies that $\partial^c f(x) = \{0\}$, so that $\xi = \xi' = 0$ and we can take $\lambda = 1$. The same argument is valid if $\xi' = 0$.

So, we suppose that $\xi \neq 0, \xi' \neq 0$. Assume that there does not exist any $\lambda > 0$ such that $\xi' = \lambda \xi$. Then there exists $v \in \mathbb{R}^n$ such that $\langle \xi, v \rangle > 0$ and

$\left\langle \xi^{'}, v \right\rangle < 0$. Since X is open, we can choose $t > 0$, such that $y_{\circ} = x + tv \in K$. Note that $\langle \xi, y_{\circ} - x \rangle > 0$ and $\left\langle \xi^{'}, y_{\circ} - x \right\rangle < 0$. So $y_{\circ}$ belongs to the open set

$$B := X \bigcap \{ y \in \mathbb{R}^n : \langle \xi, y - x \rangle > 0 \} \bigcap \left\{ y \in \mathbb{R}^n : \left\langle \xi^{'}, y - x \right\rangle < 0 \right\}.$$

Hence, B is nonempty. For every $y \in B$, one has

$$\langle \xi, y - x \rangle > 0 \Rightarrow f(y) \geq f(x),$$
$$\langle \xi, y - x \rangle < 0 \Rightarrow f(y) \leq f(x).$$

Hence, f is constant on B. But this means that $\partial^c f(y) = \{0\}$ for every $y \in B$. Since, $y \in X$ and f is pseudolinear on X, it follows that f is constant everywhere on X. Hence, $\partial^c f(x) = \{0\}$, which is a contradiction to our assumption.

2.10 Weakly Pseudolinear Functions

Szilagyi [267] introduced the notion of weakly pseudolinear functions and established the interrelations between this class of functions and other classes of generalized convex functions. Here, the differentiability of the functions are assumed only at a point of the domain not over the whole domain.

Let $\phi \neq X \subseteq \mathbb{R}^n$ be an open convex set, and let $f : X \to \mathbb{R}$ be a continuous function on X and differentiable only at the point $\bar{x} \in X$. The following definitions and preliminaries are from Szilagyi [267]:

Definition 2.19 *A function* $f : X \to \mathbb{R}$ *is said to be weakly pseudoconvex (WPX) (weakly pseudoconcave (WPV)) for X at $\bar{x}$, if for every $x \in X$, the following implication holds:*

$$\langle \nabla f(\bar{x}), x - \bar{x} \rangle - 0 \Rightarrow f(x) \geq f(\bar{x}),$$
$$(\langle \nabla f(\bar{x}), x - \bar{x} \rangle = 0 \Rightarrow f(x) \leq f(\bar{x})).$$

The function f is called weakly pseudolinear (WPL) for X at $\bar{x}$, if f is both weakly pseudoconvex (WPX) and weakly pseudoconcave (WPV).

To be precise; a function $f : X \to \mathbb{R}$ *is said to be weakly pseudolinear (WPL) for X at $\bar{x}$, if for every $x \in X$, the following implication holds:*

$$\langle \nabla f(\bar{x}), x - \bar{x} \rangle = 0 \Rightarrow f(x) = f(\bar{x}).$$

Definition 2.20 *A function* $f : X \to \mathbb{R}$ *is said to be weakly strictly pseudoconvex (WSPX) (weakly strictly pseudoconcave (WSPV)) for X at $\bar{x}$, if for every $x \in X$, $x \neq \bar{x}$, the following implication holds:*

$$\langle \nabla f(\bar{x}), x - \bar{x} \rangle = 0 \Rightarrow f(x) > f(\bar{x}),$$
$$(\langle \nabla f(\bar{x}), x - \bar{x} \rangle = 0 \Rightarrow f(x) < f(\bar{x})).$$

Definition 2.21 *A function $f : X \to \mathbb{R}$ is said to be weakly quasiconvex (WQX) (weakly quasiconcave (WQV)) for X at $\bar{x}$, if for every $x \in X$, the following implication holds:*

$$\langle \nabla f(\bar{x}), x - \bar{x} \rangle > 0 \Rightarrow f(x) \geq f(\bar{x}),$$
$$(\langle \nabla f(\bar{x}), x - \bar{x} \rangle < 0 \Rightarrow f(x) \leq f(\bar{x})).$$

Definition 2.22 *A function $f : X \to \mathbb{R}$ is said to be orthoisoval (OI) (antiorthoisoval (AOI)) for X at $\bar{x}$, if for every $x \in X, x \neq \bar{x}$, the following implication holds:*

$$\langle \nabla f(\bar{x}), x - \bar{x} \rangle = 0 \Rightarrow f(x) = f(\bar{x}),$$
$$(\langle \nabla f(\bar{x}), x - \bar{x} \rangle < 0 \Rightarrow f(x) \neq f(\bar{x})).$$

Definition 2.23 *A function $f : X \to \mathbb{R}$ is said to be isoorthogonal (IO) for X at $\bar{x}$, if for every $x \in X, x \neq \bar{x}$, the following implication holds:*

$$f(x) = f(\bar{x}) \Rightarrow \langle \nabla f(\bar{x}), x - \bar{x} \rangle = 0.$$

Definition 2.24 *A function $f : X \to \mathbb{R}$ is said to be isonegative (IN) (isopositive (IP)) for X at $\bar{x}$, if for every $x \in X, x \neq \bar{x}$, the following implication holds:*

$$f(x) = f(\bar{x}) \Rightarrow \langle \nabla f(\bar{x}), x - \bar{x} \rangle < 0$$
$$(f(x) = f(\bar{x}) \Rightarrow \langle \nabla f(\bar{x}), x - \bar{x} \rangle > 0).$$

Definition 2.25 *A function $f : X \to \mathbb{R}, X \subseteq \mathbb{R}^n$ is an open convex set, differentiable at $\bar{x}$ is said to be quasiconvex (QX) (quasiconcave (QV)) for X at $\bar{x}$, if for every $x \in X$, the following implication holds:*

$$\langle f(\bar{x}), x - \bar{x} \rangle > 0 \Rightarrow f(x) > f(\bar{x})$$
$$(\langle f(\bar{x}), x - \bar{x} \rangle > 0 \Rightarrow f(x) > f(\bar{x})).$$

The function f is called quasilinear for X at $\bar{x}$, if it is both quasiconvex (QX) and quasiconcave (QV) for X at $\bar{x}$.

To be precise: A function $f : X \to \mathbb{R}, X \subseteq \mathbb{R}^n$ is an open convex set, differentiable at $\bar{x}$ is said to be quasilinear (QL) for X at $\bar{x}$, if for every $x \in X$, the following implication holds:

$$\langle f(\bar{x}), x - \bar{x} \rangle > 0 \Rightarrow f(x) > f(\bar{x})$$
$$(\langle f(\bar{x}), x - \bar{x} \rangle > 0 \Rightarrow f(x) > f(\bar{x})).$$

Definition 2.26 (*Functional class of differentiable at a point functions*) *Let $\mathfrak{R}$ denote the following set of relations:*

$$\mathfrak{R} = \{>, \geq, <, \leq, =, \neq\}.$$

For a given pair of relations (R_1, R_2), a function differentiable at a point $\bar{x}$ is called a function of (R_1, R_2)-type for X at $\bar{x}$, if the following implication holds for all $x \in X$:

$$\langle \nabla f(\bar{x}), x - \bar{x} \rangle \, R_1 0 \Rightarrow f(x) R_2 f(\bar{x}). \tag{2.60}$$

The functional class (2.60) of differentiable at a point functions includes several classes of functions for different combinations of R_1 and R_2. For example, pseudoconvex functions are $(\geq, \geq)$-type functions of this class.

2.11 Properties of the Functional Class of Differentiable at a Point Functions

In this section, the properties of the elements and the interrelations between the elements of the functional class of differentiable at a point functions will be studied. Szilagyi [267] stated and proved these theorems for the case, where generalized convexity properties are investigated at a point, not on the set. Let $\phi \neq X \subseteq \mathbb{R}^n$ be an open convex set, and let $f : X \to \mathbb{R}$ be a continuous function on X and differentiable only at the point $\bar{x} \in X$, for the set X. Now onward, for the sake of brevity, we omit the phrase "for the set X."

Theorem 2.25 *The function f is quasilinear at $\bar{x}$, if and only if f is isoorthogonal at $\bar{x}$, that is,*

$$f(x) = f(\bar{x}) \Rightarrow \langle \nabla f(\bar{x}), x - \bar{x} \rangle = 0, \ \forall x \in X. \tag{2.61}$$

Proof To prove the necessary part, let f be quasilinear at $\bar{x}$ and let $f(x) = f(\bar{x})$. Then, $f(x) \leq f(\bar{x})$ and since f is quasiconvex at $\bar{x}$, we get $\langle \nabla f(\bar{x}), x - \bar{x} \rangle \leq 0$. Similarly, $f(x) > f(\bar{x})$ and since f is quasiconcave at $\bar{x}$, we get that $\langle \nabla f(\bar{x}), x - \bar{x} \rangle \geq 0$. Hence, we get

$$\langle \nabla f(\bar{x}), x - \bar{x} \rangle = 0.$$

To prove the sufficient part, we first prove the following lemma:

Lemma 2.4 *If f has property (2.61), then f is quasiconvex at $\bar{x}$.*

Proof Suppose to the contrary that f is not quasiconvex at $\bar{x}$. Then there exists an $x_1 \in X, x_1 \neq \bar{x}$ with $f(x_1) \leq f(\bar{x})$ such that $\langle \nabla f(\bar{x}), x_1 - \bar{x} \rangle > 0$. We claim that there is no x_2 with $f(x_2) > f(\bar{x})$ and $\langle \nabla f(\bar{x}), x_2 - \bar{x} \rangle > 0$. If possible there exist such an x_2, then $x_2 \neq x_1$. Take the interval between x_1 and x_2, that is $\lambda x_1 + (1 - \lambda)x_2, \lambda \in]0, 1]$. By the continuity of f there is a $\bar{\lambda} \in]0, 1]$ such that $f(\bar{\lambda} x_1 + (1 - \bar{\lambda})x_2) = f(\bar{x})$. Since, f satisfies (2.61), we get

$$\langle \nabla f(\bar{x}), (\bar{\lambda} x_1 + (1 - \bar{\lambda})x_2) - \bar{x} \rangle = 0.$$

However,

$$\langle \nabla f(\bar{x}), (\bar{\lambda}x_1 + (1 - \bar{\lambda})x_2) - \bar{x} \rangle = \langle \nabla f(\bar{x}), (\bar{\lambda}(x_1 - \bar{x}) + (1 - \bar{\lambda})(x_2 - \bar{x})) \rangle$$
$$= \bar{\lambda} \langle \nabla f(\bar{x}), x_1 - \bar{x} \rangle + (1 - \bar{\lambda}) \langle \nabla f(\bar{x}), x_2 - \bar{x} \rangle > 0,$$

which is a contradiction. Therefore, for every x_2 with $\langle \nabla f(\bar{x}), x_2 - \bar{x} \rangle > 0$, we have $f(x_2) \leq f(\bar{x})$, which is also a contradiction, if x_2 is close enough to $\bar{x}$. Thus f is quasiconvex at $\bar{x}$. Hence, the lemma is proved.

Taking $-f$ instead of f, we get that if f satisfies (2.61), then it is quasiconcave. Hence, f is quasilinear at $\bar{x}$ and sufficiency is established.

Theorem 2.26 *Let f be differentiable and quasiconvex at $\bar{x}$ and assume that $\nabla f(\bar{x}) \neq 0$. Then f is pseudoconvex at $\bar{x}$.*

Proof To prove that f is pseudoconvex at $\bar{x}$, we have to prove that for all $x \in X$,

$$\langle \nabla f(\bar{x}), x - \bar{x} \rangle \geq 0 \Rightarrow f(x) \geq f(\bar{x}).$$

Suppose contrary that, f is not pseudoconvex at $\bar{x}$. Therefore, there exists an $x_1 \in X$ such that $\langle \nabla f(\bar{x}), x_1 - \bar{x} \rangle \geq 0$ but $f(x_1) < f(\bar{x})$. Since, f is quasiconvex at $\bar{x}$, hence, we must have

$$\langle \nabla f(\bar{x}), x_1 - \bar{x} \rangle = 0.$$

Since, $\nabla f(\bar{x}) \neq 0$, it follows that in every neighborhood N of $\bar{x}$, there exists a $y \in N$, such that

$$\langle \nabla f(\bar{x}), y - \bar{x} \rangle \neq 0.$$

Since, f is quasiconvex at $\bar{x}$, from $\langle \nabla f(\bar{x}), y - \bar{x} \rangle > 0$, it follows that $f(y) > f(\bar{x}) > f(x_1)$, which is a contradiction to the continuity of f on X, if $y \in N$ and N is small enough. Hence, there exists a neighborhood N_1 of $\bar{x}$, such that $\langle \nabla f(\bar{x}), y - \bar{x} \rangle \leq 0$, for all $y \in N_1$.

Therefore, we can choose a neighborhood N_1 of x_1 and a vector $x \neq 0$, such that the vector $y_1 = x_1 + z \in N_1$ and $\langle \nabla f(\bar{x}), y_1 - \bar{x} \rangle < 0$. Let $y_2 = x_1 - z$, then $y_2 \in N$. Hence, we have

$$\langle \nabla f(\bar{x}), y_2 - \bar{x} \rangle \leq 0.$$

However, we have

$$0 > \langle \nabla f(\bar{x}), y_1 - \bar{x} \rangle = \langle \nabla f(\bar{x}), x_1 - \bar{x} \rangle + \langle \nabla f(\bar{x}), z \rangle = \langle \nabla f(\bar{x}), z \rangle.$$

Hence, we have

$$0 \geq \langle \nabla f(\bar{x}), y_2 - \bar{x} \rangle = \langle \nabla f(\bar{x}), x_1 - \bar{x} \rangle - \langle \nabla f(\bar{x}), z \rangle > 0,$$

which is a contradiction.

Theorem 2.27 *Let $\nabla f(\bar{x}) \neq 0$. Then f is pseudolinear at $\bar{x}$, if and only if f is isoorthogonal at $\bar{x}$.*

Proof To prove the necessary part, we assume that f be pseudolinear at $\bar{x}$. Then, f is pseudoconvex and pseudoconcave at $\bar{x}$, hence, f is quasiconvex and quasiconcave at $\bar{x}$ that is quasilinear at $\bar{x}$. Thus the necessary part follows from that of Theorem 2.25.

To prove the sufficient part, we assume that the implication (2.60) holds. By Lemma 2.4, we get that f is quasiconvex at $\bar{x}$. Since, f is quasiconvex at $\bar{x}$ and $\nabla f(\bar{x}) \neq 0$, by Theorem 2.26, it follows that f is pseudoconvex at $\bar{x}$. Similarly, taking $-f$ instead of f, we have that f is pseudoconcave at $\bar{x}$. Hence, f is pseudolinear at $\bar{x}$.

Theorem 2.28 *The function f be weakly pseudolinear at $\bar{x}$, if and only if it is orthoisoval at $\bar{x}$, that is,*

$$\langle \nabla f(\bar{x}), x - \bar{x} \rangle = 0 \Rightarrow f(x) = f(\bar{x}), \forall x \in X. \tag{2.62}$$

Proof To prove the necessary part, let f be weakly pseudolinear at $\bar{x}$ and let

$$\langle \nabla f(\bar{x}), x - \bar{x} \rangle = 0.$$

Since, f is weakly pseudoconvex at $\bar{x}$, it follows that

$$f(x) \geq f(\bar{x}).$$

By the weakly pseudoconcavity of f at $\bar{x}$, we get

$$f(x) \leq f(\bar{x}).$$

Hence, we get $f(x) = f(\bar{x})$ and hence (2.62) holds.

To prove the sufficiency, let f has the property (2.62) and let $\langle \nabla f(\bar{x}), x - \bar{x} \rangle = 0$. By the implication (2.62), we get that $f(x) = f(\bar{x})$, so $f(x) \geq f(\bar{x})$ and $f(x) \geq f(\bar{x})$. Therefore, f is both weakly pseudoconvex and weakly pseudoconcave at $\bar{x}$. Hence, f is weakly pseudolinear at $\bar{x}$.

From Theorem 2.25 and Theorem 2.28, we get the following theorem:

Theorem 2.29 *The function f is pseudolinear at $\bar{x}$, if and only if it is quasilinear at $\bar{x}$ and weakly pseudolinear at $\bar{x}$, that is f is pseudolinear at $\bar{x}$, if and only if*

$$f(x) = f(\bar{x}) \Leftrightarrow \langle \nabla f(\bar{x}), x - \bar{x} \rangle = 0, \forall x \in X.$$

Remark 2.10 *In Theorem 2.1, the above result has been proved by Chew and Choo [47] with the assumption that the function f is differentiable on an open convex set containing $\bar{x}$.*

The following corollaries and propositions are the immediate consequences of the above definitions.

Corollary 2.4 *The function f is quasiconvex and pseudoconvex at $\bar{x}$, if and only if*

$$f(x) \leq f(\bar{x}) \Leftrightarrow \langle \nabla f(\bar{x}), x - \bar{x} \rangle \leq 0, \forall x \in X.$$

TABLE 2.1:   Geometrical Meaning of the
Functional Class

Property at $\bar{x}$	Geometrical Meaning at $\bar{x}$
SPX	SMIP on CHS
PS	MIP on CHS and SMIP on OHS
IN	SMIP or SMAP on CHS
WPL	$f = $ constant on HP
IP	SMIP or SMAP on HP
WSPX	SMIP on HP
WPX	MIP on HP

Corollary 2.5 *The function f is quasiconcave and pseudoconcave at $\bar{x}$, if and only if*

$$f(x) \geq f(\bar{x}) \Leftrightarrow \langle \nabla f(\bar{x}), x - \bar{x} \rangle \geq 0, \forall x \in X.$$

Proposition 2.6 *If the function f is strictly pseudoconvex at $\bar{x}$, then f is isonegative at $\bar{x}$. If f is strictly pseudoconcave at $\bar{x}$, then f is isopositive at $\bar{x}$.*

Proposition 2.7 *The fact that the function f is strictly pseudoconvex at $\bar{x}$, does not imply that f is isonegative at $\bar{x}$.*

Proof Let $f(x) = $ constant, $X = \mathbb{R}$ and $\bar{x}$ be an arbitrary point in X. In this case f is pseudoconvex at $\bar{x}$, since $\nabla f(\bar{x}) = 0$ and $\langle \nabla f(\bar{x}), x - \bar{x} \rangle = 0$ and $f(x) = f(\bar{x})$ for every $x \in X$, but, of course, it is isonegative at $\bar{x}$.

Proposition 2.8 *The fact that the function f is isonegative at $\bar{x}$, does not imply that f is pseudoconvex at $\bar{x}$.*

Proof Let $f(x) = -x^2, X = \mathbb{R}$ and $\bar{x} = 0$. Then, $\nabla f(\bar{x}) = 0$ and $f(x) \neq f(\bar{x})$, if $x \neq \bar{x}$. Therefore, f is isonegative at $\bar{x}$. On the other hand, f is not pseudoconvex at $\bar{x}$, since, for all $x \neq \bar{x}, \langle \nabla f(\bar{x}), x - \bar{x} \rangle = 0$ and $f(x) < f(\bar{x})$.
    Similar results are valid between pseudoconcavity and isopositivity.

Remark 2.11 *Table 2.1, summarizes the geometrical interpretation of the elements of the functional class:*

Abbreviations:
MIP-$\bar{x}$ is a minimum point; SMIP-$\bar{x}$ is strict MIP; SMAP-$\bar{x}$ is a strict maximum point; CHS is the closed half-spaces $\{x : \langle \nabla f(\bar{x}), x - \bar{x} \rangle \geq 0\}$; OHS is the open half-spaces $\{x : \langle \nabla f(\bar{x}), x - \bar{x} \rangle > 0\}$; HP is the hyperplane $\{x : \langle \nabla f(\bar{x}), x - \bar{x} \rangle = 0\}$.

Remark 2.12 *Szilagyi [267] has pointed out that the functional classes introduced in this section can be defined for functions nondifferentiable at $\bar{x}$, but having some generalized differentiability property at $\bar{x}$, such as directional differentiability in every direction, the Dini derivatives, Clarke derivatives, and so on, at $\bar{x}$.*

Chapter 3

Constrained Pseudolinear Optimization: Characterizations of Solution Sets

3.1 Introduction

The characterizations and the properties of the solution set related to an optimization problem having multiple optimal solutions are of fundamental importance in understanding the behavior of solution methods. Mangasarian [177] has presented simple and elegant characterizations for the solution set of convex extremum problems with one solution known. These results have been further extended to various classes of optimization problems such as infinite dimensional convex optimization problems (Jeyakumar and Wolkowicz [117], Jeyakumar [114]), generalized convex optimization problems (Burke and Ferris [32], Penot [220], Jeyakumar *et al.* [115]), and convex vector optimization problems (Jeyakumar *et al.* [116]). Jeyakumar and Yang [119] have obtained the characterizations for the solution set of a differentiable pseudolinear programming problem by using the basic properties of pseudolinear functions. Lu and Zhu [171] have established some characterizations for locally Lipschitz pseudolinear functions and the solution set of a pseudolinear programming problem using the Clarke subdifferential on Banach spaces. Recently, Lalitha and Mehta [155, 156] introduced the notion of h-pseudolinear functions and derived the characterizations for the solution set of h-pseudolinear programming problems. Very recently, Smietanski [257] has derived some characterizations of directionally differentiable pseudolinear functions, which are not necessarily differentiable. Using these characterizations, Smietanski [257] obtained several characterizations for the solution sets of constrained non-smooth optimization problems involving directionally differentiable pseudolinear functions.

In Chapter 2, we studied the properties of different types of pseudolinear functions, both differentiable and nondifferentiable. In this chapter, we will study the characterizations of the solution sets of constrained optimization problems by using the properties of different types of pseudolinear functions studied in the previous chapter, such as differentiable pseudolinear, h-pseudolinear, Dini pseudolinear, and locally Lipschitz pseudolinear functions.

In this chapter, we consider the following constrained optimization problem:

$$(P) \quad \min \ f(x)$$

$$\text{subject to } x \in C,$$

where $f : X \to \mathbb{R}, X \subseteq \mathbb{R}^n$ is an open set and C is a convex subset of X. Throughout the chapter, we assume that the solution set

$$\bar{S} := \arg\min_{x \in C} f(x) := \{x \in C : f(x) \le f(y), \forall y \in C\}$$

is nonempty. Let $x\lambda y := (1 - \lambda)x + \lambda y, \lambda \in [0,1]$, unless otherwise specified.

Now, we present the characterizations for the solution set of the problem (P) under various pseudolinearity assumptions on the function f.

3.2 Characterizations of the Solution Set of (P) Using Differentiable Pseudolinear Functions

In this section, we present the characterizations for the solution set of (P) under the assumption that the function $f : X \rightharpoonup \mathbb{R}$ is a differentiable pseudolinear function on C, derived by Jeyakumar and Yang [119].

Remark 3.1 *Let $\bar{x} \in \bar{S}$. Then, it is clear that $x \in \bar{S}$ if and only if $f(x) = f(\bar{x})$. Hence, by Theorem 2.1, f is pseudolinear on $\bar{S}$ if and only if for each $\bar{x} \in \bar{S}$, we have*

$$\bar{S} = \{x \in C : \langle \nabla f(\bar{x}), x - \bar{x} \rangle = 0\}.$$

Remark 3.2 *It is evident from Theorem 2.1, that if f is a differentiable pseudolinear function, then, the solution set $\bar{S}$ of the problem (P) is a convex set. Indeed, suppose $x, \bar{x} \in \bar{S}$, then, we have $f(x) = f(\bar{x})$. By Theorem 2.1, $f(x) = f(\bar{x})$ if and only if $\langle \nabla f(\bar{x}), x - \bar{x} \rangle = 0$. Now, we note that*

$$\langle \nabla f(\bar{x}), x\lambda\bar{x} - \bar{x} \rangle = (1 - \lambda)\langle \nabla f(\bar{x}), x - \bar{x} \rangle = 0, \forall \lambda \in [0,1].$$

Therefore, by Theorem 2.1, we get $f(x\lambda\bar{x}) = f(\bar{x}) = f(x), \forall \lambda \in [0,1]$. Hence, for each $x, \bar{x} \in \bar{S}$, we have $x\lambda\bar{x} \in \bar{S}, \forall \lambda \in [0,1]$.

The following first order characterizations of the solution set of (P) using differentiable pseudolinear functions in terms of any of its solution points are due to Jeyakumar and Yang [119].

Theorem 3.1 *Let $\bar{x} \in \bar{S}$. Then, $\bar{S} = \tilde{S} = \hat{S} = S^*$, where*

$$\tilde{S} := \{x \in C : \langle \nabla f(x), \bar{x} - x \rangle = 0\}, \tag{3.1}$$

$$\hat{S} := \{x \in C : \langle \nabla f(\bar{x}), \bar{x} - x \rangle = 0\}, \tag{3.2}$$

$$S^* := \{x \in C : \langle \nabla f(x\lambda\bar{x}), \bar{x} - x \rangle = 0, \forall \lambda \in [0,1]\}. \tag{3.3}$$

Proof Clearly, the point $x \in \bar{S}$ if and only if $f(x) = f(\bar{x})$. By Theorem 2.1, $f(x) = f(\bar{x})$ if and only if

$$\langle \nabla f(x), \bar{x} - x \rangle = 0 = \langle \nabla f(\bar{x}), x - \bar{x} \rangle.$$

Therefore, $\bar{S} = \tilde{S} = \hat{S}$. To prove that $\bar{S} = S^*$. Now, by the definitions, we note that $S^* \subseteq \tilde{S} = \bar{S}$. To prove the converse implication, let $x \in \bar{S} = \tilde{S}$. Then, we have

$$\langle \nabla f(x), \bar{x} - x \rangle = 0.$$

Therefore, for each $\lambda \in [0,1]$, we have

$$\langle \nabla f(x), x\lambda\bar{x} - x \rangle = \lambda \langle \nabla f(x), \bar{x} - x \rangle = 0.$$

By Theorem 2.1, we get

$$f(x) = f(x\lambda\bar{x}).$$

Again, applying Theorem 2.1, we have

$$\langle \nabla f(x\lambda\bar{x}), x - x\lambda x \rangle = 0, \forall \lambda \in [0,1].$$

Now, we note that

$$\langle \nabla f(x\lambda\bar{x}), \bar{x} - x \rangle = (-1/\lambda) \langle \nabla f(x\lambda\bar{x}), x - x\lambda\bar{x} \rangle = 0, \forall \lambda \in \,]0,1].$$

When, $\lambda = 0$, we have

$$\langle \nabla f(x\lambda\bar{x}), \bar{x} - x \rangle = 0.$$

Therefore,

$$\langle \nabla f(x\lambda\bar{x}), \bar{x} - x \rangle = 0, \forall \lambda \in [0,1].$$

Thus, $x \in S^*$ and hence, $\bar{S} \subseteq S^*$.

Corollary 3.1 *Let $\bar{x} \in \bar{S}$. Then, $\bar{S} = \tilde{S}_1 = \hat{S}_1 = S_1^*$, where*

$$\tilde{S}_1 := \{x \in C : \langle \nabla f(x), \bar{x} - x \rangle \ge 0\}, \tag{3.4}$$

$$\hat{S}_1 := \{x \in C : \langle \nabla f(\bar{x}), \bar{x} - x \rangle \ge 0\}, \tag{3.5}$$

$$S_1^* := \{x \in C : \langle \nabla f(x\lambda\bar{x}), \bar{x} - x \rangle \ge 0, \forall \lambda \in [0,1]\}. \tag{3.6}$$

Proof By Theorem 3.1, it is clear that $\bar{S} \subseteq \tilde{S}_1$. To prove the converse implication, we assume that $x \in \tilde{S}_1$, that is

$$x \in C \ and \ \langle \nabla f(x), \bar{x} - x \rangle \ge 0.$$

Then, by Theorem 2.2, there exists a function $p : C \times C \to \mathbb{R}^n_{++}$, such that

$$f(\bar{x}) = f(x) + p(x, \bar{x}) \langle \nabla f(x), \bar{x} - x \rangle \ge f(x),$$

which implies that $x \in \bar{S}$. Hence, $\bar{S} = \tilde{S}_1$. Similarly, we can show that $\bar{S} = \hat{S}_1$. Now to prove that $\bar{S} = S_1^*$, we note that

$$\bar{S} = S^* \subseteq S_1^* \subseteq \tilde{S}_1 = \bar{S}.$$

This completes the proof.

The following corollary due to Jeyakumar and Yang [119] provides a second order characterization for the solution set of a pseudolinear program.

Corollary 3.2 *Let f be twice continuously differentiable and pseudolinear function on C. Let $\bar{x} \in \bar{S}$. Then,*

$$\bar{S} := \{x \in C : \langle \nabla f(x\lambda\bar{x}), x - \bar{x} \rangle = 0, \forall \lambda \in [0,1]\}$$

$$\bigcap \{x \in C : (\bar{x} - x) \langle \nabla^2 f(x\lambda\bar{x}), \bar{x} - x \rangle = 0, \forall \lambda \in [0,1]\}.$$

Proof To prove the result, it is sufficient to show that

$$\bar{S} = S^* \subseteq \{x \in C : (\bar{x} - x) \langle \nabla^2 f(x\lambda\bar{x}), \bar{x} - x \rangle = 0\}.$$

Let $x \in \bar{S} = S^*$. Then,

$$\langle \nabla f(x\lambda\bar{x}), x - \bar{x} \rangle = 0, \forall \lambda \in [0,1].$$

Since, f is pseudolinear on C, it follows from Theorem 2.4, that for each $\lambda \in [0,1]$, we have

$$(\bar{x} - x)^T \langle \nabla^2 f(x\lambda\bar{x}), \bar{x} - x \rangle = 0.$$

Theorem 3.2 *Let f be continuously differentiable and pseudolinear function on C. Let $\bar{x} \in \bar{S}$. Then, $\bar{S} = S_2^* = S_2^\sharp$, where*

$$S_2^* := \{x \in C : \langle \nabla f(\bar{x}), \bar{x} - x \rangle = \langle \nabla f(x), x - \bar{x} \rangle\}, \tag{3.7}$$

$$S_2^\sharp := \{x \in C : \langle \nabla f(\bar{x}), \bar{x} - x \rangle \geq \langle \nabla f(x), x - \bar{x} \rangle\}. \tag{3.8}$$

Proof We first prove that $\bar{S} \subseteq S_2^*$. Let $x \in \bar{S}$, from Theorem 3.1, we have

$$\langle \nabla f(x), \bar{x} - x \rangle = 0, \langle \nabla f(\bar{x}), \bar{x} - x \rangle) = 0.$$

Then, we get
$$\langle \nabla f(x), x - \bar{x} \rangle = \langle \nabla f(\bar{x}), \bar{x} - x \rangle.$$

Thus, $x \in S_2^*$. Again, it is clear that, $S_2^* \subseteq S_2^\sharp$. Now, we prove that $S_2^\sharp \subseteq \bar{S}$. Let $x \in S_2^\sharp$. Then, we have

$$\langle \nabla f(\bar{x}), \bar{x} - x \rangle \geq \langle \nabla f(x), x - \bar{x} \rangle. \tag{3.9}$$

Suppose that $x \notin \bar{S}$. Then, $f(x) > f(\bar{x})$. By the pseudoconcavity of f, we have
$$\langle \nabla f(\bar{x}), x - \bar{x} \rangle > 0.$$

From (3.9), we have

$$\langle \nabla f(x), \bar{x} - x \rangle > 0.$$

By Theorem 2.2, there exists a function $p : X \times X \to \mathbb{R}_{++}$, such that

$$f(\bar{x}) = f(x) + p(x, \bar{x})) \langle \nabla f(x), \bar{x} - x \rangle > f(x),$$

which is a contradiction. Hence, $x \in \bar{S}$.

The following theorem due to Jeyakumar and Yang [119] relates the solution set $\bar{S}$ to the set of normalized gradients.

Theorem 3.3 *Let f be continuously differentiable and pseudolinear function on an open convex set containing C. Let $\bar{x} \in \bar{S}$. Suppose that $\nabla f(x) \neq 0$ on C. Then,*

$$\bar{S} \subseteq \left\{ x \in C : \frac{\nabla f(x)}{\|\nabla f(x)\|} = \frac{\nabla f(\bar{x})}{\|\nabla f(\bar{x})\|} \right\} \subseteq \{ x \in C : \langle \nabla f(x), \bar{x} - x \rangle \leq 0 \}.$$

Proof Following Corollary 2.2, the inclusion

$$\bar{S} \subseteq \left\{ x \in C : \frac{\nabla f(x)}{\|\nabla f(x)\|} = \frac{\nabla f(\bar{x})}{\|\nabla f(\bar{x})\|} \right\}$$

follows easily. To establish the second inclusion, suppose that $x \in C$ and

$$\frac{\nabla f(x)}{\|\nabla f(x)\|} = \frac{\nabla f(\bar{x})}{\|\nabla f(\bar{x})\|}. \tag{3.10}$$

Assume that

$$\langle \nabla f(x), \bar{x} - x \rangle > 0.$$

Then, we have

$$\left\langle \frac{\nabla f(x)}{\|\nabla f(x)\|}, \bar{x} - x \right\rangle > 0.$$

Therefore, from (3.10), we have

$$\left\langle \frac{\nabla f(\bar{x})}{\|\nabla f(\bar{x})\|}, \bar{x} - x \right\rangle > 0.$$

From which, it follows that

$$\langle \nabla f(\bar{x}), \bar{x} - x \rangle > 0,$$

which is a contradiction to the necessary optimality condition of (P) at $\bar{x}$, that is,

$$\langle \nabla f(\bar{x}), x - \bar{x} \rangle \geq 0, \forall x \in C.$$

Therefore,

$$\langle \nabla f(\bar{x}), \bar{x} - x \rangle \leq 0.$$

This completes the proof.

The following example from Jeyakumar and Yang [119] illustrates that the inclusion in Theorem 3.3 can be strict.

Example 3.1 *Consider, the following pseudolinear programming problem:*

$$\min f(x) := x + (1/2)\sin x$$

$$\text{subject to } \ x \in C := \{x \in \mathbb{R} : x \geq 0\}.$$

Then, f is pseudolinear and $\bar{S} = \{0\}$. We note that

$$\bar{S} = \{0\} \subset \left\{ x \in C : \frac{\nabla f(x)}{\|\nabla f(x)\|} = \frac{\nabla f(\bar{x})}{\|\nabla f(\bar{x})\|} \right\}$$

$$= \{2k\pi, k = 0, 1, 2, ...\} \subset \{x \in C : \langle \nabla f(x), \bar{x} - x \rangle \leq 0\} = C.$$

The above inclusions are all strict.

3.3   Solution Sets of Linear Fractional Programs

In this section, we present the characterization and boundedness of the solution set of a linear fractional programming problem derived by Jeyakumar and Yang [119].

Consider the following linear fractional programming problem:

$$(FP) \quad \min \left(\frac{c^T x + \alpha}{d^T x + \beta} \right)$$

$$\text{subject to } \ x \in \mathbb{R}^n : Ax \geq b, x \geq 0,$$

where $c, d \in \mathbb{R}^n$ and $\alpha, \beta \in \mathbb{R}$ and the feasible set

$$C := \{x \in \mathbb{R}^n : Ax \geq b, x \geq 0\}$$

is a convex subset of $\mathbb{R}^n$. Let $\bar{S} \neq \phi$ denote the solution set of the problem (FP). Let $d^T x + \beta > 0$ on C. Let $Q = cd^T - dc^T$, and $q^T = \alpha d^T - \beta c^T$.

Applying Theorem 3.1, we have the following characterizations for the solution set of (FP).

Theorem 3.4 *Assume that $\bar{x} \in \bar{S}$. Then, $\bar{S} = \hat{S} = \tilde{S}$, where*

$$\hat{S} := \left\{ x \in C : \bar{x}^T Q x = q^T (\bar{x} - x) \right\};$$

$$\tilde{S} := \left\{ x \in C : (-\bar{x} - x)^T Q (\bar{x} - x) = 2q^T (\bar{x} - x) \right\}.$$

The following theorem due to Jeyakumar and Yang [119] gives the characterization for the boundedness of the solution set of (FP).

Theorem 3.5 *Assume that* $\bar{x} \in \bar{S}$. *Then,* $\bar{S}$ *is bounded if and only if*

$$\tilde{Z} := \left\{ z \in \mathbb{R}^{\kappa} : Az \geq 0, 0 \neq z \geq 0, \bar{x}^T Q z + q^T z = 0 \right\} \neq \phi.$$

Proof Let $\tilde{Z} \neq \phi$. Suppose that $\bar{S}$ is unbounded. Then, there exists a sequence $\{z_i\} \subseteq \bar{S}$, such that $z_i \neq 0$ and $\|z_i\| \to \infty$. Using Theorem 3.4, we get $\bar{S} = \hat{S}$ and so any limiting point z of $\frac{z_i}{\|z_i\|} \in \tilde{Z}$ which is a contradiction and hence, $\bar{S}$ is bounded. Conversely, assume that $\bar{S}$ is bounded. Let $\bar{x} \in \bar{S}$. On the contrary, suppose that $z \in \tilde{Z}$. Then, for any positive scalar λ, we have

$$x(\lambda) = \bar{x} + \lambda z \in \bar{S}.$$

Then, we have

$$\bar{x}^T Q(\bar{x} + \lambda z) = \lambda \bar{x}^T Q z = -\lambda q^T z = q^T(\bar{x} - (\bar{x} + \lambda z)).$$

Thus, $\|x(\lambda)\| \to \infty$ as $\lambda \to \infty$, which is a contradiction to the assumption that $\bar{S}$ is bounded.

Remark 3.3 *It is evident that the set $\tilde{Z}$ depends on a solution point and $\tilde{Z} \bigcup \{0\}$ is the recession cone of $\tilde{S}$. Therefore, by Theorem 1.9 in Chapter 1, the set $\bar{S}$ is bounded if and only if $\tilde{Z} \neq \phi$.*

3.4 Characterizations of the Solutions Set of (P) Using *-Pseudolinear Functions

In the previous chapter, we studied the properties and characterizations about directionally differentiable *-pseudolinear functions. Now, using the properties of directionally differentiable *-pseudolinear functions, we present the characterizations for the solution set of (P).

Throughout this section, we assume that $f : X \to \mathbb{R}$ be directionally differentiable on C.

Theorem 3.6 *Let f be *-pseudolinear on C and let $\bar{x} \in \bar{S}$. Then $\bar{S} = \tilde{S}_1 = \hat{S}_1 = S_1^{\sharp}$, where*

$$\tilde{S}_1 := \{x \in C : f'_+(x; \bar{x} - x) = 0\},$$

$$\hat{S}_1 := \{x \in C : f'_+(\bar{x}; x - \bar{x}) = 0\},$$

$$S_1^{\sharp} := \{x \in C : f'_+(x\lambda\bar{x}; \bar{x} - x) = 0, \forall \lambda \in [0,1]\}.$$

Proof Clearly, the point $x \in \bar{S}$ if and only if $f(\bar{x}) = f(x)$. Since f is *-pseudolinear on C, by Proposition 2.2, we get

$$f'_+(x; \bar{x} - x) = 0 = f'_+(\bar{x}; x - \bar{x}).$$

Therefore, $\bar{S} = \tilde{S}_1 = \hat{S}_1$.

Now, to prove that $\bar{S} = S_1^\sharp$. Let $x \in \bar{S}$. Then for each $\lambda \in \,]0, 1[$, we have

$$f'_+(x; x\lambda\bar{x} - x) = \lambda f'_+(x; \bar{x} - x) = 0.$$

By Proposition 2.2, we get $f(x) = f(x\lambda\bar{x})$. Again, from this equality, we get

$$0 = f'_+(x\lambda\bar{x}; x - x\lambda\bar{x}) = \lambda f'_+(x; x - \bar{x}).$$

Therefore, $f'_+(x\lambda\bar{x}; \bar{x} - x) = 0, \forall \lambda \in \,]0, 1[$. Hence, $x \in S_1^\sharp$ and therefore, $\bar{S} \subseteq S_1^\sharp$. Conversely, let $x \in S_1^\sharp$. In particular, taking $\lambda = 0$, we get $f'_+(x; \bar{x} - x) = 0$ and thus, $x \in \bar{S}$.

Hence $x \in S_1^\sharp \subseteq \bar{S}$.

Remark 3.4 *It is evident from Theorem 3.6 that f is $*$-pseudolinear on C if and only if for each $\bar{x} \in C$, we have*

$$\bar{S} := \{x \in C : f'_+(x; \bar{x} - x) = 0\}.$$

Theorem 3.7 *Let f be $*$-pseudolinear on C and let $\bar{x} \in \bar{S}$. Then $\bar{S} = \tilde{S}_2 = \hat{S}_2 = S_2^\sharp$, where*

$$\tilde{S}_2 := \{x \in C : f'_+(x; \bar{x} - x) \geq 0\},$$

$$\hat{S}_2 := \{x \in C : f'_+(\bar{x}; x - \bar{x}) \geq 0\},$$

$$S_2^\sharp := \{x \in C : f'_+(x\lambda\bar{x}; \bar{x} - x) \geq 0, \forall \lambda \in [0, 1]\}.$$

Proof By Theorem 3.6, the inclusion $\bar{S} \subseteq \tilde{S}_2$ follows. To prove the converse inclusion, suppose that $x \in C$ satisfies $f'_+(x; \bar{x} - x) \geq 0$. Since, f is $*$-pseudolinear, it follows that

$$f(\bar{x}) = f(x) + p(x, \bar{x}) f'_+(x; \bar{x} - x).$$

Hence, $f(\bar{x}) \geq f(x)$. As $\bar{x}$ is a solution of (P), therefore, $f(\bar{x}) \leq f(x)$, which implies that $f(\bar{x}) = f(x)$, that is $x \in \bar{S}$. Hence $\tilde{S}_2 \subseteq \bar{S}$. Likewise, we can prove that $\bar{S} = \hat{S}_2$.

To prove that $\bar{S} = S_2^\sharp$ we note that $S_2^\sharp \subseteq \tilde{S}_2 = \bar{S}$. Now, let $x \in \bar{S}$, then from Theorem 3.6, we have $f'_+(x\lambda\bar{x}; \bar{x} - x), \forall \lambda \in [0, 1]$ and hence, $\bar{S} \subseteq S_2^\sharp$.

Theorem 3.8 *Let f be $*$-pseudolinear on C and let $\bar{x} \in \bar{S}$. Then $\bar{S} = \tilde{S}_3 = \hat{S}_3$, where*

$$\tilde{S}_3 := \left\{x \in C : f'_+(\bar{x}; \bar{x} - x) = f'_+(x; x - \bar{x})\right\},$$

$$\hat{S}_3 := \left\{x \in C : f'_+(\bar{x}; \bar{x} - x) \geq f'_+(x; x - \bar{x})\right\}.$$

Proof First, we prove that $\bar{S} \subseteq \tilde{S}_3$. Let $x \in \bar{S}$. It follows from Theorem 3.6, that

$$f'_+(x; \bar{x} - x) = 0 \ and \ f'(\bar{x}; x - \bar{x}) = 0.$$

Then, $f'_+(x; \bar{x} - x) = 0 = f'_+(\bar{x}; x - \bar{x})$. Hence, $x \in \tilde{S}_3$, that is $\bar{S} \subseteq \tilde{S}_3$.

Now, we show that $\hat{S}_3 \subseteq \bar{S}$. Let $x \in \hat{S}_3$ then, we have $f'_+(\bar{x}; \bar{x} - x) \geq f'_+(x; x - \bar{x})$. Suppose that $x \notin \bar{S}$. Then $f(\bar{x}) > f(x)$. By *-pseudoconcavity of f, we have $f'_+(\bar{x}; \bar{x} - x) > 0$ and thus $f'_+(x; \bar{x} - x) > 0$. Using *-pseudoconvexity of f, we get $f(\bar{x}) \geq f(x)$, which is a contradiction. Hence, $x \in \bar{S}$, that is the inclusion $\hat{S}_3 \subseteq \bar{S}$ is true. Again, the inclusion $\tilde{S}_3 \subseteq \hat{S}_3$ is obvious. Hence, we have

$$\bar{S} \subseteq \tilde{S}_3 \subseteq \hat{S}_3 \subseteq \bar{S}.$$

The following example from Smietanski illustrates the significance of Theorem 3.6.

Example 3.2 *Consider the following optimization problem*

$$(P1) \quad \min f(x)$$

$$\text{subject to } x \in C := \{X \in \mathbb{R} : -\pi \leq x \leq \pi\},$$

where

$$f(x) = \begin{cases} |x - \dfrac{1}{2}\sin x|, & \text{if } x \in \left(-\infty, -\dfrac{\pi}{2}\right) \cup \left(\dfrac{\pi}{2}, \infty\right), \\ \dfrac{\pi}{2} - \dfrac{1}{2}\sin\dfrac{\pi}{2}, & \text{if } x \in \left(-\dfrac{\pi}{2}, \dfrac{\pi}{2}\right). \end{cases}$$

*Then, f is *-pseudolinear on C and*

$$\bar{S} = \left[-\dfrac{\pi}{2}, \dfrac{\pi}{2}\right].$$

Now, it is easy to see that

$$\bar{S} = \hat{S}_1 = \tilde{S}_1 = S_1^\sharp, \forall x \in \left[-\dfrac{\pi}{2}, \dfrac{\pi}{2}\right].$$

3.5 Characterization of the Solution Set of (P) Using h-Pseudolinear Functions

In the previous chapter, we studied the characterizations and properties of h-pseudolinear functions. In this section, we present the characterization for the solution set of the problem (P) assuming that the function $f : X \to \mathbb{R}$ is an h-pseudolinear function, where $h : X \times \mathbb{R} \to \bar{\mathbb{R}}$ is a bifunction associated with f.

Definition 3.1 (Sach and Penot [242]) *A function f is called h-quasiconvex on C, if for all $x, y \in C$, we have*

$$x \neq y, f(y) \leq f(x) \Rightarrow h(x; y - x) \leq 0.$$

Theorem 3.9 (Sach and Penot [242]) *Suppose that $h(x; .)$ is subodd and positively homogeneous in the second argument. Further, suppose that*

$$h(x; d) \leq D^+ f(x; d), \forall x \in C \text{ and } d \in \mathbb{R}^n. \tag{3.11}$$

If f is h-pseudoconvex on C, then, f is quasiconvex as well as h-quasiconvex on C, i.e.,

$$\forall x, y \in C, x \neq y, f(y) \leq f(x) \Rightarrow h(x; y - x) \leq 0.$$

Proof If f is not quasiconvex, then, there exist points $x, y, z \in C$ with $y \in]x, z[$, such that

$$f(y) > f(x) \text{ and } f(y) > f(z).$$

Now, the h-pseudoconvexity of f ensures that

$$h(y; x - y) < 0 \text{ or } h(y; x - z) < 0,$$

$$h(y; z - y) > 0 \text{ or } h(y; z - x) < 0.$$

Since, h is subodd, we arrive at a contradiction. To prove the h-quasiconvexity of f on X, we take $x, y \in X$, $x \neq y$, such that $f(y) \leq f(x)$. By quasiconvexity of f, we get $f(x + t(y - x)) \leq f(x), \forall t \in [0, 1]$, hence, we get $D^+ f(x; y - x) < 0$. Therefore, by (3.11), we get $h(x; y - x) < 0$.

The following theorem from Lalitha and Mehta [157] is a variant of Theorem 3.9.

Theorem 3.10 *Suppose that $-h$ is subodd and positively homogeneous in the second argument. Further, suppose that*

$$D_+ f(x; d) \leq h(x; d), \forall x \in C \text{ and } d \in \mathbb{R}^n.$$

If f is h-pseudoconcave on C, then, f is quasiconcave and h-quasiconcave on C.

The following theorem from Lalitha and Mehta [157] establishes a characterization for h-pseudolinear functions. However, the approach is different from that in Theorem 2.18.

Theorem 3.11 *Let h be odd in the second argument. If f is h-pseudolinear on C, then for each $x, y \in C$,*

$$h(x; y - x) = 0 \Leftrightarrow f(x) = f(y).$$

Proof Suppose that $x, y \in C$. If $x = y$, the result is trivially true. Assume that $x \neq y$. If $h(x; y - x) = 0$, then by h-pseudoconvexity and h-pseudoconcavity of f, it follows that

$$f(x) = f(y).$$

Conversely, suppose that f is h-pseudolinear on C. Then, by Theorems 3.9 and 3.10, it follows that f is both h-quasiconvex and h-quasiconcave on C. Hence, we get $h(x; y - x) = 0$.

The following example from Lalitha and Mehta [157] illustrates the necessity of the oddness assumption in Theorem 3.11.

Example 3.3 *Let $X = \mathbb{R}$ and $f : X \to \mathbb{R}$ be defined as*

$$f(x) := \begin{cases} 1, & \text{if } x \text{ is irrational,} \\ -1, & \text{if } x \text{ is rational.} \end{cases}$$

Define h as

$$h(x; d) := \begin{cases} -|d|, & \text{if } x \text{ is irrational,} \\ |d|, & \text{if } x \text{ is rational.} \end{cases}$$

It is easy to see that all the assumptions of Theorem 3.11 are satisfied except that h is odd in the second argument. We observe that

$$f(1) = (-1), \text{ but } h(1; -2) \neq 0.$$

Now, we present the characterization for the solution set of the problem (P) under the assumption that the function f is h-pseudolinear on C. The proofs are along the lines of Lalitha and Mehta [157].

Theorem 3.12 *The solution set $\bar{S}$ of the problem (P) is a convex set.*

Proof Let $x, y \in \bar{S}$. Then, $f(x) = f(y)$. From Theorem 3.11, we get

$$h(x; y - x) = 0 \text{ and } h(y; x - y) = 0.$$

Now, for $\lambda \in [0, 1]$, we have

$$h(x; x\lambda y - x) = h(x; \lambda(y - x)) = \lambda h(x; y - x) = 0.$$

By Theorem 3.11, we get $f(x\lambda y) = f(x), \forall \lambda \in [0, 1]$ and therefore, $x\lambda y \in \bar{S}, \forall \lambda \in [0, 1]$.

Hence, $\bar{S}$ is a convex set.

Theorem 3.13 *Let $x \in \bar{S}$ then $\bar{S} = \tilde{S}_1 = \hat{S}_1 = S_1^{\sharp}$, where*

$$\tilde{S}_1 := \{x \in C : h(x; \bar{x} - x) = 0\},$$

$$\hat{S}_1 := \{x \in C : h(\bar{x}; x - \bar{x}) = 0\},$$

$$S_1^{\sharp} := \{h(x\lambda \bar{x}; \bar{x} - x) = 0, \forall \lambda \in [0, 1]\}.$$

Proof The point $x \in \bar{S}$ if and only if $f(\bar{x}) = f(x)$. Since, f is h-pseudolinear on C, by Theorem 3.11, we get

$$h(x; \bar{x} - x) = 0 = h(\bar{x}; x - \bar{x}).$$

Hence, $\bar{S} = \tilde{S}_1 = \hat{S}_1$. Now, to prove that $\bar{S} = S_1^\sharp$. Let $x \in \bar{S}$. By Theorem 3.12, $x\lambda\bar{x} \in \bar{S}, \forall \lambda \in [0,1]$. Then, for each $\lambda \in [0,1]$, we have $f(x) = f(x\lambda\bar{x})$. Hence, by Theorem 3.11, we get

$$h(x\lambda\bar{x}; x - x\lambda\bar{x}) = 0.$$

Now, for $\lambda \in]0,1[$, we have

$$h(x\lambda\bar{x}; \bar{x} - x) = \frac{-1}{\lambda} h(x\lambda\bar{x}; x - x\lambda\bar{x}) = 0.$$

When $\lambda = 0$, we have

$$h(x\lambda\bar{x}; \bar{x} - x) = 0.$$

Hence, $h(x\lambda\bar{x}; \bar{x} - x) = 0, \forall \lambda \in [0,1]$. Therefore, $x \in S_1^\sharp$ and hence $\bar{S} \subseteq S_1^\sharp$. Conversely, let $x \in S_1^\sharp$. In particular, taking $\lambda = 0$, we get $h(x; \bar{x} - x) = 0$ and thus, $x \in \bar{S}$. Hence, $S_1^\sharp \subseteq \bar{S}$.

Theorem 3.14 *Let* $\bar{x} \in \bar{S}$, *then* $\bar{S} = \tilde{S}_2 = \hat{S}_2 = S_2^\sharp$, *where*

$$\tilde{S}_2 := \{ x \in C : h(x; \bar{x} - x) \geq 0 \};$$

$$\hat{S}_2 := \{ x \in C : h(\bar{x}; x - \bar{x}) \leq 0 \};$$

$$S_2^\sharp := \{ x \in C : h(x\lambda\bar{x}; \bar{x} - x) \geq 0, \forall \lambda \in [0,1] \}.$$

Proof By Theorem 3.13, the inclusion $\bar{S} \subseteq \tilde{S}_2$ follows. To prove the converse inclusion assume that $x \in \tilde{S}_2$, that is, $x \in C$ and $h(x; \bar{x} - x) \geq 0$. Since, f is h-pseudoconvex, it follows that $f(x) \leq f(\bar{x})$. As $\bar{x}$ is a solution of (P), hence $f(\bar{x}) \leq f(x)$. Therefore, $f(\bar{x}) = f(x)$, that is $x \in \bar{S}$. Similarly, using h-pseudoconcavity of f, we can prove that $\bar{S} = \hat{S}_2$. To prove that $\bar{S} = S_2^\sharp$, we note that $S_2^\sharp \subseteq \tilde{S}_2 = \bar{S}$. Now, let $x \in \bar{S}$, then, from Theorem 3.13, we have $h(x\lambda\bar{x}; \bar{x} - x) = 0, \forall \lambda \in [0,1]$ and hence, $\bar{S} \subseteq S_2^\sharp$.

Theorem 3.15 *Let* $\bar{x} \in \bar{S}$, *then,* $\bar{S} = \tilde{S}_3 = \hat{S}_3$, *where*

$$\tilde{S}_3 := \{ x \in C : h(x; \bar{x} - x) = h(\bar{x}; x - \bar{x}) \},$$

$$\hat{S}_3 := \{ x \in C : h(x; \bar{x} - x) \geq h(\bar{x}; x - \bar{x}) \}.$$

Proof First, we prove that $\bar{S} \subseteq \tilde{S}_3$. Let $x \in \bar{S}$. It follows from Theorem 3.13, that

$$h(\bar{x}; x - \bar{x}) = 0 \text{ and } h(x; \bar{x} - x) = 0.$$

Then,

$$h(x; \bar{x} - x) = 0 = h(x; \bar{x} - x).$$

Hence, $x \in \tilde{S}_3$, that is, $\bar{S} \subseteq \tilde{S}_3$.

Now, let $x \in \hat{S}_3$, then, we have $h(\bar{x}; x - \bar{x}) \leq h(x; \bar{x} - x)$. Suppose that $x \notin \bar{S}$. Then $f(x) > f(\bar{x})$. By h-pseudoconcavity of f, we have $h(\bar{x}; x - \bar{x}) > 0$ and thus $h(x; \bar{x} - x) > 0$. Using h-pseudoconvexity of f, we get $f(\bar{x}) \geq f(x)$, which is a contradiction. Thus, $x \in \bar{S}$ and we have $\hat{S}_3 \subseteq \bar{S}$.

The inclusion $\hat{S}_3 \subseteq \hat{S}_3$ is obvious. Hence, we have $\bar{S} \subseteq \tilde{S}_3 \subseteq \bar{S} \subseteq \bar{S}$.

3.6 Characterization of the Solution Set of (P) Using Locally Lipschitz Pseudolinear Functions and Clarke Subdifferential

In this section, we present the characterization of the solution set of (P) under the assumption that the function $f : X \to \mathbb{R}$ is a locally Lipschitz pseudolinear function on C.

We recall the following definition of pseudomonotone maps and their characterization from Floudas and Pardalos [83].

Definition 3.2 *A multimapping* $F : \mathbb{R}^n \to 2^{\mathbb{R}^n}$, *is said to be pseudomonotone, if for every pair of distinct points* $x, y \in C$, *one has*

$$\exists \xi \in \partial^c f(x) : \langle \xi, y - x \rangle \geq 0 \Rightarrow \langle \zeta, y - x \rangle \geq 0, \forall \zeta \in \partial^c f(y).$$

Proposition 3.1 *(Floudas and Pardalos [83])* *Let* $f : \mathbb{R}^n \to \mathbb{R}$ *be a locally Lipschitz function. Then,* f *is pseudoconvex if and only if* $\partial^c f$ *is pseudomonotone.*

Now, we present the characterizations for the solution set of (P) using the Clarke subdifferential. The following theorems and corollaries are variants of the corresponding results from Lu and Zhu [171] on Banach spaces.

Theorem 3.16 *Let* $\bar{x} \in \bar{S}$, *then* $\bar{S} = \tilde{S} = \hat{S} = S^*$, *where*

$$\tilde{S} := \{x \in C : \exists \xi \in \partial^c f(x) : \langle \xi, \bar{x} - x \rangle = 0\}; \tag{3.12}$$

$$\hat{S} := \{x \in C : \exists \zeta \in \partial^c f(\bar{x}) : \langle \zeta, \bar{x} - x \rangle = 0\}; \tag{3.13}$$

$$S^* := \{x \in C : \exists \varsigma \in \partial^c f(x\lambda\bar{x}) : \langle \varsigma, \bar{x} - x \rangle = 0, \forall \lambda \in [0,1]\}. \tag{3.14}$$

Proof The point $x \in \bar{S}$ if and only if $f(x) = f(\bar{x})$. By Theorem 2.23, $x \in \bar{S}$ if and only if there exists $\xi \in \partial^c f(x)$, such that $\langle \xi, \bar{x} - x \rangle = 0$, hence, $\bar{S} = \tilde{S}$. To prove that $\tilde{S} = \hat{S}$, let $x \in \tilde{S}$, then there exists $\xi \in \partial^c f(x)$, such that $\langle \xi, \bar{x} - x \rangle = 0$. By Theorem 2.23, we get $f(x) = f(\bar{x})$, which implies that there exists $\zeta \in \partial^c f(\bar{x})$, such that $\langle \zeta, x - \bar{x} \rangle = 0$, that is, $\langle \zeta, \bar{x} - x \rangle = 0$ and we get $\tilde{S} \subseteq \hat{S}$. Similarly, we can show that $\hat{S} \subseteq \tilde{S}$. Hence, $\tilde{S} = \hat{S}$.

Now, to show that $\bar{S} = S^*$. Let $x \in \tilde{S}$, then $f(x) = f(\bar{x})$ which implies that there exists $\xi \in \partial^c f(x)$, such that

$$\langle \xi, \bar{x} - x \rangle = 0.$$

From Equation (2.54) of Theorem 2.23, for every $\lambda \in \,]0, 1]$ and $z := x\lambda\bar{x}$, we have $f(z) = f(x)$. Moreover, from Theorem 2.23, there exists $\varsigma \in \partial^c f(z)$, such that

$$\langle \varsigma, z - x \rangle = 0.$$

Now,

$$\langle \varsigma, \bar{x} - x \rangle = \frac{1}{\lambda} \langle \varsigma, z - x \rangle = 0, \forall \lambda \in \,]0, 1].$$

When $\lambda = 0$, the following

$$\langle \varsigma, \bar{x} - x \rangle = 0$$

is obviously true. Therefore,

$$\langle \varsigma, \bar{x} - x \rangle = 0, \forall \lambda \in \,]0, 1].$$

Hence, $x \in S^*$ and $\bar{S} \subseteq S^*$.

Conversely, let $x \in S^*$, then $\forall \lambda \in (0, 1], z = x\lambda\bar{x}$, there exists $\varsigma \in \partial^c f(z)$, such that $\langle \varsigma, \bar{x} - x \rangle = 0$. Hence, $\frac{1}{\lambda} \langle \varsigma, z - x \rangle = 0$, that is,

$$\langle \varsigma, z - x \rangle = 0 \Rightarrow f(x) = f(z) \Rightarrow \exists \xi \in \partial^c f(x) : \langle \xi, z - x \rangle = 0,$$

$$\Rightarrow \lambda \langle \xi, \bar{x} - x \rangle = 0 \Rightarrow \langle \xi, \bar{x} - x \rangle = 0 \Rightarrow f(x) = f(\bar{x}).$$

Hence, $S^* \subseteq \bar{S}$. If $\lambda = 0$, then $S^* = \tilde{S} \subseteq \bar{S}$. Hence, $S^* \subseteq \bar{S}$ is always true.

Theorem 3.17 *Let* $\bar{x} \in \bar{S}$, *then* $\bar{S} = \tilde{S}_1 = \hat{S}_1 = S_1^*$, *where*

$$\tilde{S}_1 := \{ x \in C : \exists \xi \in \partial^c f(x) : \langle \xi, x - \bar{x} \rangle \geq 0 \}, \tag{3.15}$$

$$\hat{S}_1 := \{ x \in C : \exists \xi \in \partial^c f(\bar{x}) : \langle \zeta, \bar{x} - x \rangle \geq 0 \}, \tag{3.16}$$

$$S_1^* := \{ x \in C : \exists \varsigma \in \partial^c f(x\lambda\bar{x}) : \langle \varsigma, \bar{x} - x \rangle \geq 0, \forall \lambda \in [0, 1] \}. \tag{3.17}$$

Proof Clearly, from Theorem 3.16, $\bar{S} \subseteq \tilde{S}_1$ holds. Suppose that $x \in \tilde{S}_1$ and there exists $\xi \in \partial^c f(x)$ such that $\langle \xi, \bar{x} - x \rangle \geq 0$. By pseudoconvexity of f, we get $f(\bar{x}) \geq f(x)$, but $\bar{x}$ is a minimum of f on X, hence, $x \in \bar{S}$. Therefore, $\tilde{S}_1 \subseteq \bar{S}$. Thus $\bar{S} = \tilde{S}_1$. Similarly, using pseudoconcavity of f we can prove that $\bar{S} = \hat{S}_1$.

To prove that $\bar{S} = S_1^*$. Indeed, for $\lambda \in (0, 1], z := x\lambda\bar{x}$ and $\partial^c f$ is pseudomonotone, we have

$$\langle \varsigma, \bar{x} - x \rangle \geq 0 \Rightarrow \langle \varsigma, z - x \rangle \geq 0 \Rightarrow \langle \xi, z - x \rangle \geq 0 \Rightarrow \langle \xi, \bar{x} - x \rangle \geq 0.$$

When $\lambda = 0$, $S_1^* = \tilde{S}_1$. Hence, $S_1^* \subseteq \tilde{S}_1 = \bar{S}$. It is obvious that $S^* \subseteq S_1^*$. By Theorem 3.16, we have $\bar{S} = S^* \subseteq S_1^*$. Hence, we get $\bar{S} = S_1^*$.

Theorem 3.18 *Let $\bar{x} \in \bar{S}$, then $\bar{S} = \tilde{S}_2 = \hat{S}_2$, where*

$$\tilde{S}_2 := \{x \in C : \exists \xi \in \partial^c f(x), \zeta \in \partial^c f(\bar{x}) : \langle \zeta, \bar{x} - x \rangle = \langle \xi, x - \bar{x} \rangle \}; \quad (3.18)$$

$$\hat{S}_2 := \{x \in C : \exists \xi \in \partial^c f(x), \varsigma \in \partial^c f(\bar{x}) : \langle \varsigma, \bar{x} - x \rangle \geq \langle \xi, x - \bar{x} \rangle \}. \quad (3.19)$$

Proof First, we prove that $\bar{S} \subseteq \tilde{S}_2$. Let $x \in \bar{S}$. It follows from Theorem 3.16, that $\langle \xi, x - \bar{x} \rangle = 0$ and $\langle \zeta, \bar{x} - x \rangle = 0$. Then

$$\langle \zeta, \bar{x} - x \rangle = 0 = \langle \xi, x - \bar{x} \rangle.$$

Thus $x \in \tilde{S}_2$. The inclusion $\tilde{S}_2 \subseteq \hat{S}_2$ is clearly true. Finally, we prove that $\hat{S}_2 \subseteq \bar{S}$. Suppose that $x \subset \hat{S}_2$. Then x satisfies

$$\langle \zeta, \bar{x} - x \rangle \geq \langle \xi, x - \bar{x} \rangle,$$

which imply that

$$\langle \zeta, x - \bar{x} \rangle \leq \langle \xi, \bar{x} - x \rangle. \quad (3.20)$$

Suppose that $x \notin \bar{S}$ then $f(x) > f(\bar{x})$. By pseudoconcavity of f we have $\langle \zeta, x - \bar{x} \rangle > 0$. From (3.20), we have $\langle \xi, \bar{x} - x \rangle > 0$. By the pseudoconvexity of f we get $f(\bar{x}) \geq f(x)$, which is a contradiction. Hence, $x \in \bar{S}$.

[180]

Chapter 4

Characterizations of Solution Sets in Terms of Lagrange Multipliers

4.1 Introduction

The dual characterizations of the solution set of an optimization problem having multiple optimal solutions are of fundamental importance. It is useful in characterizing the boundedness of the solution sets as well as in understanding the behavior of solution methods. It is well known that the Lagrange multiplier is a key to identify the optimal solutions of a constrained optimization problem. In 1988, Mangasarian [177] presented simple and elegant characterizations of the solution set for a convex minimization problem over a convex set, when one solution is known. Jeyakumar *et al.* [115] have established that the Lagrangian function of an inequality constrained convex optimization problem is constant on the solution set. They employed this property of Lagrangian to establish several Lagrange multiplier characterizations of the solution set of a convex optimization problem. Dinh *et al.* [67] presented several Lagrange multiplier characterizations of a pseudolinear optimization problem over a closed convex set with linear inequality constraints. Recently, Lalitha and Mehta [156] derived some Lagrange multiplier characterizations of the solution set of an optimization problem involving pseudolinear functions in terms of bifunction h. Very recently, Mishra *et al.* [200] established Lagrange multiplier characterizations for the solution sets of nonsmooth constrained optimization problems by using the properties of locally Lipschitz pseudolinear functions.

In this chapter, we study the Lagrange multiplier characterization for the linear as well as nonlinear constrained optimization problems. Using the properties of different types of differentiable and nondifferentiable pseudolinear functions, studied in Chapter 2, we present the characterizations for the solution sets.

4.2   Definitions and Preliminaries

Let $X \subseteq \mathbb{R}^n$ be a nonempty set. Throughout the chapter, let $x\alpha z := (1 - \alpha)x + \alpha z, \alpha \in [0, 1]$.

The proof of the following proposition from Dinh *et al.* [67] can be deduced from the proofs of Theorems 2.1, 2.2, and 2.3 of Chapter 2.

Proposition 4.1 *Let* $f : \mathbb{R}^n \to \mathbb{R}$ *be pseudolinear on* X *and* $x, z \in X$. *Then, the following statements are equivalent:*

(i) $f(x) = f(z)$;

(ii) $\langle \nabla f(z), x - z \rangle = 0$;

(iii) $\langle \nabla f(x), x - z \rangle = 0$;

(iv) $f(x\alpha z) = f(x), \forall \alpha \in [0, 1]$;

(v) $\langle \nabla f(x\alpha z), x - z \rangle = 0, \forall \alpha \in \,]0, 1]$;

(vi) $f(x\alpha z) = f(z), \forall \alpha \in [0, 1]$.

4.3   Pseudolinear Optimization Problems

In this section, we present Lagrange multiplier characterizations of a differentiable pseudolinear optimization problem with linear inequality constraints. We consider the following constrained pseudolinear optimization problem:

$$(\text{PLP}) \quad \min f(x)$$

$$\text{subject to } x \in D := \{x \in C | a_i^T \leq b_i, i = 1, 2, ..., m\},$$

where $f : X \to \mathbb{R}$ is a differentiable pseudolinear function and X is an open convex set containing $D, a_i \in \mathbb{R}^n, b_i \in \mathbb{R}$ and $C \subseteq \mathbb{R}^n$ is a closed convex subset of $\mathbb{R}^n$. Let $I := \{1, 2, ..., m\}$ and let

$$S := \{x \in D | f(x) \leq f(y), \forall y \in D\} \tag{4.1}$$

denote the solution set of the (PLP). We further assume that $S \neq \phi$. For $z \in D$, let $I(z) := \{i \in I | a_i^T z = b_i\}$.

Remark 4.1 *Let* $x, z \in S$, *then* $f(x) = f(z)$. *Using Proposition 4.1, it follows that* $f(x\alpha z) = f(x)$ *and thus* $x\alpha z \in S$, *for every* $\alpha \in [0, 1]$. *Then, the solution set* S *of (PLP) is a convex subset of* $\mathbb{R}^n$.

Now, we state the following necessary and sufficient optimality conditions from Dinh *et al.* [67].

Proposition 4.2 *Let $z \in D$ and let the (PLP) satisfy some suitable constraint qualification such as, there exists $x_0 \in \mathrm{ri}C$, such that, $a_i x_0 \leq b_i$ for all $i \in I$, where $\mathrm{ri}C$ is the relative interior of the set C. Then, $z \in S$ if and only if there exists a Lagrange multiplier $\lambda = (\lambda_i) \in \mathbb{R}^n$, such that*

$$\left.\begin{array}{c} \nabla f(z) + \displaystyle\sum_{i \in I} \lambda_i a_i + N_c(z) = 0 \\[2mm] \lambda \geq 0, \lambda_i(a_i^T z - b_i) = 0, \forall i \in I, \end{array}\right\} \tag{4.2}$$

where $N_c(z)$ is the normal cone of C at z.

The following theorem from Dinh *et al.* [67] establishes that the active constraints corresponding to a known solution of the (PLP) remain active at all the other solutions of the (PLP).

Theorem 4.1 *Let $z \in S$ and let the (PLP) satisfy the optimality condition (4.2) with a Lagrange multiplier $\lambda \in \mathbb{R}^m$. Then, for $x \in S$, $\displaystyle\sum_{i \in I(z)} \lambda_i(a_i^T x - b_i) = 0$ and $f(.) + \displaystyle\sum_{i \in I(z)} \lambda_i(a_i(.) - b_i)$ is a constant function on S.*

Proof Using (4.2), for all $x \in C$, we have

$$\langle \nabla f(z), x - z \rangle + \sum_{i \in I(z)} \lambda_i a_i^T(x - z) \geq 0. \tag{4.3}$$

Since, f is a pseudolinear function on X, by Theorem 2.2, there exists a function $p : X \times X \to \mathbb{R}_{++}$ such that for $x, z \in X$, we have

$$f(x) = f(z) + p(z, x)\langle \nabla f(z), x - z \rangle. \tag{4.4}$$

Using (4.2), (4.3), and (4.4), we get

$$f(x) - f(z) + p(z, x) \sum_{i \in I(z)} \lambda_i(a_i^T x - b_i) \geq 0, \forall x \in C.$$

Since, for each $x \in S, f(x) = f(z)$, it follows that

$$p(z, x) \sum_{i \in I(z)} \lambda_i(a_i^T x - b_i) \geq 0.$$

Since $p(z, x) > 0$, we get

$$\sum_{i \in I(z)} \lambda_i(a_i^T x - b_i) \geq 0$$

From the feasibility of x, we get

$$\sum_{i \in I(z)} \lambda_i(a_i^T x - b_i) \leq 0.$$

Hence, we have

$$\sum_{i \in I(z)} \lambda_i(a_i^T x - b_i) = 0, \tag{4.5}$$

which implies that $f(.) + \sum_{i \in I(z)} \lambda_i(a_i(.) - b_i)$ is a constant function on S.

Now, we assume that $z \in S$ and $\lambda \in \mathbb{R}^m$ is a Lagrange multiplier corresponding to z. Let $\overline{I}(z) := \{i \in I(z) | \lambda_i > 0\}$. Dinh *et al.* [67] established the following characterizations of the solution set of the (PLP).

Theorem 4.2 *Let $z \in S$ and let the (PLP) satisfy the optimality condition (4.2) with a Lagrange multiplier $\lambda \in \mathbb{R}_m$. Let*

$$\overline{S}_1 := \{x \in C | a_i^T x = b_i, \forall i \in I \backslash \overline{I}(z), \langle \nabla f(x), x - z \rangle = 0\},$$

$$\overline{S}_2 := \{x \in C | a_i^T x = b_i, \forall i \in I \backslash \overline{I}(z), \langle \nabla f(z), x - z \rangle = 0\},$$

$$\overline{S}_3 := \{x \in C | a_i^T x = b_i, \forall i \in I \backslash \overline{I}(z), \langle \nabla f(x \alpha z), x - z \rangle = 0, \forall \alpha \in]0, 1]\}.$$

Then $S = \overline{S}_1 = \overline{S}_2 = \overline{S}_3$.

Proof First, we establish the equality $S = \overline{S}_2$ and the other equalities can be proved similarly. Suppose that $x \in \overline{S}_2$, then $x \in C, a_i^T x \leq b_i$, for all $i \in I$ and we have

$$\langle \nabla f(z), x - z \rangle = 0.$$

Now, using Proposition 4.1, it follows that $x \in D$ and $f(x) = f(z)$, which implies that $x \in S$. Thus, $\overline{S}_2 \subseteq S$.

Conversely, suppose that $x \in S$, then $f(x) = f(z)$. Using Theorem 4.1, we get

$$\sum_{i \in I(z)} \lambda_i(a_i^T x - b_i) = 0.$$

Hence, $a_i^T x = b_i$, for all $i \in I \backslash \overline{I}(z)$. Again, since $f(x) = f(z)$, using Proposition 4.1, we get $\langle \nabla f(z), x - z \rangle = 0$, which gives that $x \in \overline{S}_2$. Thus, $S \subseteq \overline{S}_2$. Hence, $S = \overline{S}_2$.

Corollary 4.1 *Let $z \in S$ and let the (PLP) satisfy the optimality conditions (4.2) with a Lagrange multiplier $\lambda \in \mathbb{R}^m$. Let*

$$\widetilde{S}_1 := \{x \in C | a_i^T x = b_i \forall i \in \overline{I}(z), a_i^T \leq b_i \forall i \in I \backslash \overline{I}(z), \langle \nabla f(x), x - z \rangle \leq 0\},$$

$$\widetilde{S}_2 := \{x \in C | a_i^T x = b_i \forall i \in \overline{I}(z), a_i^T x \leq b_i \forall i \in I \backslash \overline{I}(z), \langle \nabla f(z), x - z \rangle \leq 0\},$$

$$\widetilde{S}_3 := \{x \in C | a_i^T x = b_i \forall i \in \overline{I}(z), a_i^T x \leq b_i \forall i \in I \backslash \overline{I}(z), \langle \nabla f(x \alpha z), x - z \rangle \leq 0,$$
$$\forall \alpha \in]0, 1]\}.$$

Then, $S = \widetilde{S}_1 = \widetilde{S}_2 = \widetilde{S}_3$.

Proof It is evident that $S = \bar{S}_1 \subseteq \tilde{S}_1, \bar{S}_2 \subseteq \tilde{S}_2$ and $\bar{S}_3 \subseteq \tilde{S}_3$. Let us assume that $x \in \bar{S}_1$. Then, $a_i^T = b_i$, for all $i \in \bar{I}(z)$ and $a_i^T \leq b_i$, for all $i \in I \setminus \bar{I}(z)$. Since f is pseudolinear on X, by using Theorem 2.2, we have

$$f(z) - f(x) = p(x, z)\langle \nabla f(x), z - x \rangle \geq 0, \tag{4.6}$$

which implies that $f(x) \leq f(z)$. As $z \in S$, we get $f(x) = f(z)$. This proves that $x \in S$. Again from pseudolinearity of f, we have

$$f(x) - f(z) = p(z, x)\langle \nabla f(z), x - z \rangle$$

and

$$f(z) - f(x\alpha z) = (1 - \alpha)p(x\alpha z, z)\langle \nabla f(x\alpha z), z - x \rangle, \forall \alpha \in \,]0, 1]\,.$$

Using the above equalities, we can establish that $\tilde{S}_2 \subseteq \bar{S}_2$ and $\tilde{S}_3 \subseteq \bar{S}_3$, respectively.

The following theorem by Dinh *et al.* [67] characterizes the solution set of (PLP) independently from the objective function f. This establishes that for the case $C = \mathbb{R}^n$, the solution set of the (PLP) is a polyhedral convex subset of $\mathbb{R}^n$.

Theorem 4.3 *Suppose that for $C = \mathbb{R}^n$ and $z \in S$, the (PLP) satisfies the optimality condition (4.2) with a Lagrange multiplier $\lambda \in \mathbb{R}^m$. Then*

$$S = \{x \in \mathbb{R}^n | a_i^T x = b_i, \forall i \in \bar{I}(z), a_i^T x \leq b_i, \forall i \in I \backslash \bar{I}(z)\},$$

and hence, S is a convex polyhedral subset of $\mathbb{R}^n$.

Proof Using (4.2), we get

$$\langle \nabla f(z), x - z \rangle + \sum_{i \in I(z)} \lambda_i a_i^T (x - z) \geq 0, \forall x \in \mathbb{R}^n. \tag{4.7}$$

Let $x \in \mathbb{R}^n$ satisfy $a_i^T x = b_i$, for all $i \in \bar{I}(z)$ and $a_i^T x \leq b_i$, for all $i \in I \backslash \bar{I}(z)$), then from (4.7), we get

$$\langle \nabla f(z), x - z \rangle = 0.$$

Now, using Theorem 4.2 the required result follows.

Corollary 4.2 *Suppose that $C \in \mathbb{R}^n$ and $z \in S$, the (PLP) satisfies the optimality condition (4.2) with a Lagrange multiplier $\lambda \in \mathbb{R}^m$. Then, the solution set S is bounded if and only if*

$$\{x \in \mathbb{R} | a_i^T x = 0, \forall i \in \bar{I}(z), a_i^T x \leq 0, \forall i \in I \backslash \bar{I}(z)\} = \{0\}.$$

Proof Using Theorem 4.3, we get

$$S^\infty = \{x \in \mathbb{R}^n | a_i^T x = 0, \forall i \in \bar{I}(z), a_i^T x \leq 0, \forall i \in I \backslash \bar{I}(z)\},$$

where S^∞ is the recession cone of S. It is obvious by Theorem 1.9, that the set S is bounded if and only if $S^\infty = \{0\}$.

In the next section, we present the characterization for the solution set of a linear fractional programming problem with linear inequality constraints following Dinh *et al.* [67].

4.4   Linear Fractional Programming Problems

We consider the following linear fractional programming problem with linear inequality constraints:

$$(LFP) \quad \min \left(\frac{c^T x + \alpha}{d^T x + \beta} \right)$$

$$\text{subject to } x \in D := \{x \in C : a_i^T x \le b_i, i = 1, \dots, m\},$$

where $c, d \in \mathbb{R}^n, \alpha, \beta \in \mathbb{R}, a_i \in \mathbb{R}^n, b_i \in \mathbb{R}$ and $C = \{x \in \mathbb{R}^n : x \ge 0\}$. Let $I := \{1, \dots m\}, S$ be the solution set of (LFP), $X := \{x \in \mathbb{R}^n : d^T x + \beta > 0\}$ and let $q^T = \alpha d^T - \beta c^T$. Assume that $S \ne \phi$ and $D \subseteq X$.

It is clear that if we take $f(x) = (c^T x + \alpha)/(d^T x + \beta)$, then, f is pseudolinear on X and (LFP) is a pseudolinear program of the model (PLP). The following theorem by Dinh *et al.* [67] establishes the characterizations of the solution set of the problem (LFP).

Theorem 4.4 *Let $z \in S$ and if the (LFP) satisfies the optimality condition (4.2) with the Lagrange multiplier $\lambda \in \mathbb{R}^m$. Then*

$$S = \{x \in C : a_i^T x = b_i, \forall i \in \bar{I}(z), a_i^T x \le b_i, \forall i \in I \backslash \bar{I}(z), (c^T x)(d^T z)$$

$$-(d^T x)(c^T z) = q^T(x - z)\},$$

$$S = \{x \in C : a_i^T x = b_i, \forall i \in \bar{I}(z), a_i^T x \le b_i, \forall i \in I \backslash \bar{I}(z), (c^T x)(d^T z)$$

$$-(d^T x)(c^T z) \le q^T(x - z)\}.$$

Proof For $x \in X$, we have

$$\nabla f(x) = \frac{1}{(d^T x + \alpha)^2}[(d^T x + \beta)c - (c^T x + \alpha)d].$$

Therefore,

$$\langle \nabla f(x), z - x \rangle = 0 \Leftrightarrow (c^T x)(d^T z) - (d^T x)(c^T z) = q^T(x - z),$$

where $q^T = \alpha d^T - \beta c^T$.

Using Theorem 4.2 and Corollary 4.1, the required result follows.

The following example from Dinh *et al.* [67] illustrates the significance of Theorem 4.4.

Example 4.1 *Consider the following linear fractional programming problem:*

$$(FP1) \quad \min \frac{2 - x}{x - y}$$

subject to $(x, y) \in D := \{x \in C | x + y \leq 4, 3y - x \leq 0\}$.

where $C := \{(x, y) \in \mathbb{R}^2 : x \geq 0, y \geq 0\}$. It is clear that (4,0) is a solution of of (FP1) and the optimality condition (4.2) holds with a Lagrange multiplier $\lambda = (\frac{1}{8}, 0) \in \mathbb{R}^2$. By Theorem 4.4, we get the solution set S, which is given by

$$S = \{(x, y) \in C | x + y - 4 = 0, 3y - x \leq 0\}$$

$$= \{(x, y) \in \mathbb{R}^2 | x + y - 4 = 0, 3 \leq x \leq 4\}.$$

4.5 Vector Linear Fractional Programming Problem

We consider the following vector linear fractional programming problem with linear inequality constraints:

$$(VLFP) \quad \min \ f(x) := (f_1(x), f_2(x), ..., f_p(x))$$

subject to $x \in \hat{D} := \{x \in \mathbb{R}^n | a_j^T x \leq b_j, j \in J := \{1, 2, ..., m\}\}$.

where $a_j \in \mathbb{R}^n, b_i \in \mathbb{R}$ and $f_i(x) := \frac{c_i^T x + \alpha_i}{d^T x + \beta}, c_i, d \in \mathbb{R}^n; i \in I := \{1, 2, ..., p\}$. It is clear that each $f_i, i \in I$ is a differentiable pseudolinear function on open convex set $\hat{D} \subseteq X$, with respect to the same proportional function $\bar{p}(x, y)) = \frac{d^T x + \beta}{d^T y + \beta}, x, y \in X$. Let $J(z) := \{j \in J | a_j^T z = b_j\}$. Let us assume that $\hat{S}$ is the set of all efficient solutions of the (VLFP) and $\hat{S} \neq \phi$.

Theorem 4.5 *For the problem (VLFP), let $z \in \hat{D}$. Then the following statements are equivalent:*

(i) $z \in \hat{S}$, that is, z is an efficient solution of the (VLFP);

(ii) there exist $\lambda_i > 0, i = 1, 2, ..., p$ such that

$$\sum_{i=1}^{p} \lambda_i f_i(x) \geq \sum_{i=1}^{p} \lambda_i f_i(z), \forall x \in \hat{D};$$

(iii) z is a properly efficient solution of the (VLFP).

Proof From the definitions of efficient and properly efficient solutions, the implication $(iii) \Rightarrow (i)$ is obvious. Again, the implication $(ii) \Rightarrow (iii)$ has been proved in Geoffrion [86]. Now, we shall establish the implication $(i) \Rightarrow (ii)$.

Let $z \in \hat{D}$ be an efficient solution of (VLFP). Then by Proposition 3.2 in Chew and Choo [47], there exist $\lambda_i > 0, i = 1, 2, ..., p, r_j \geq 0, j \in J(z)$ such that

$$\sum_{i=1}^{p} \lambda_i \nabla f_i(z) + \sum_{j \in J(z)} r_j a_j = 0.$$

Thus, for all $x \in \hat{D}$, we have

$$\sum_{i=1}^{p} \lambda_i \langle \nabla f_i(z), x - z \rangle + \sum_{j \in J(z)} r_j a_j^T (x - z) = 0. \qquad (4.8)$$

Since, for all $j \in J(z), a_j^T z = b_j$, therefore, using (4.8), we get

$$\sum_{i=1}^{p} \lambda_i \langle \nabla f_i(z), x - z \rangle + \sum_{j \in J(z)} r_j (a_j^T x - b_j) = 0, \forall x \in \hat{D}.$$

Using the fact that $a_j^T x - b_j \leq 0$, for any $x \in \hat{D}$, we have

$$\sum_{i=1}^{p} \lambda_i \langle \nabla f_i(z), x - z \rangle \geq 0. \qquad (4.9)$$

From the pseudolinearity of $f_i, i = 1, 2, ..., p$ with respect to $\bar{p}$, we have

$$f_i(x) - f_i(z) = \bar{p}(z, x) \langle \nabla f_i(z), x - z \rangle, \forall x \in \hat{D}. \qquad (4.10)$$

Since, $\bar{p}(z, x) > 0$ and $\lambda > 0, i = 1, 2, ..., p$, therefore from (4.9) and (4.10), it follows that

$$\sum_{i=1}^{p} \lambda_i f_i(x) \geq \sum_{i=p}^{1} \lambda_i f_i(z), \forall x \in \hat{D},$$

which implies that (ii) holds.

The characterizations of the solution sets of nonsmooth optimization problems employing generalized subdifferentials have always been an interesting research topic. See Burke and Ferris [32], Jeyakumar and Yang [118], and references therein. Recently, Lu and Zhu [171] have established some characterizations for the solution set of nonsmooth locally Lipschitz pseudolinear optimization problems using the Clarke subdifferential. Motivated by Jeyakumar and Yang [119], Lu and Zhu [171], and Dinh *et al.* [67], Mishra *et al.* [200] considered a class of nonsmooth pseudolinear optimization problems with linear inequality constraints and established several Lagrange multiplier characterizations of the solution set under the assumption of locally Lipschitz pseudolinear functions. In the next section, results are from Mishra *et al.* [200].

4.6   Nonsmooth Pseudolinear Optimization Problems with Linear Inequality Constraints

We consider the following constrained nonsmooth pseudolinear optimization problem:

$$(NPLP) \quad \min \ f(x)$$

$$\text{subject to } x \in D := \{x \in C | a_i^T x \le b_i, i = 1, 2, ..., m\},$$

where $f : X \to \mathbb{R}$ is a locally Lipschitz pseudolinear function and X is an open convex set containing D, $a_i \in \mathbb{R}^n$, $b_i \in \mathbb{R}$ and $C \subseteq \mathbb{R}^n$ is a closed convex subset of $\mathbb{R}^n$. Let $I := \{1, 2, ..., m\}$ and let

$$S := \{x \in D | f(x) \le f(y), \forall y \in D\} \tag{4.11}$$

denote the solution set of the (NPLP). We further assume that $S \ne \phi$. For $z \in D$, let $I(z) := \{i \in I | a_i^T z = b_i\}$.

Remark 4.2 *Let $x, z \in S$, then $f(x) = f(z)$. Using Theorem 2.17, it follows that $f(x\alpha z) = f(x)$ and thus $x\alpha z \in S$, for every $\alpha \in [0,1]$. This implies that S is a convex subset of $\mathbb{R}^n$.*

Now, we state the following necessary and sufficient optimality conditions which are nonsmooth versions of the result given in Dinh *et al.* [67] (see Jeyakumar and Wolkowicz [117], Clarke [49]).

Proposition 4.3 *Let $z \in D$ and let the (NPLP) satisfy some suitable constraint qualification such as, there exists $x_0 \in riC$ such that, $a_i x_0 \le b_i$ for all $i \in I$ where riC is the relative interior of the set C. Then, $z \in S$ if and only if there exists a Lagrange multiplier $\lambda = (\lambda_i) \in \mathbb{R}^m$, such that*

$$\left. \begin{array}{c} 0 \in \partial^c f(z) + \sum_{i \in I} \lambda_i a_i + N_c(z), \\[2mm] \lambda_i \ge 0, \lambda_i(a_i^T z - b_i) = 0, \forall i \in I, \end{array} \right\} \tag{4.12}$$

where $N_c(z)$ is the Clarke normal cone of C at z.

The next theorem establishes that the active constraints corresponding to a known solution of the (NPLP) remain active at all the other solutions of the (NPLP).

Theorem 4.6 *Let $z \in S$ and let the (NPLP) satisfy the optimality condition (4.12) with a Lagrange multiplier $\lambda \in \mathbb{R}^m$. Then for each*

$$x \in S, \ \sum_{i \in I(z)} \lambda_i(a_i^T x - b_i) = 0$$

and

$$f(.) + \sum_{i \in I(z)} \lambda_i(a_i(.) - b_i)$$

is a constant function on S.

Proof Using (4.12), there exists $\xi \in \partial^c f(z)$, such that

$$\langle \xi, x - z \rangle + \sum_{i \in I(z)} \lambda_i a_i^T(x - z) \geq 0, \forall x \in C. \tag{4.13}$$

Since, f is locally Lipschitz pseudolinear on X, by Theorem 2.19, there exists a positive real-valued function $p : X \times X \to \mathbb{R}_{++}$ and $\xi' \in \partial^c f(z)$, such that

$$f(x) = f(z) + p(z, x)\langle \xi', x - z \rangle. \tag{4.14}$$

Using Proposition 2.3, there exists $\mu > 0$, such that $\xi' = \mu \xi$. Hence, from (4.14), we get

$$f(x) = f(z) + \mu p(z, x)\langle \xi, x - z \rangle.$$

Using (4.12), (4.13) and (4.14), we get

$$f(x) - f(z) + \mu p(z, x) \sum_{i \in I(z)} \lambda_i(a_i^T x - b_i) \geq 0, \forall x \in C.$$

Since for each $x \in S$, $f(x) = f(z)$, it follows that

$$\mu p(z, x) \sum_{i \in I(z)} \lambda_i(a_i^T x - b_i) \geq 0.$$

Since $p(z, x) > 0$ and $\mu > 0$, we get

$$\sum_{i \in I(z)} \lambda_i(a_i^T x - b_i) \geq 0.$$

From the feasibility of x, we get

$$\sum_{i \in I(z)} \lambda_i(a_i^T x - b_i) \leq 0.$$

Hence, we have

$$\sum_{i \in I(z)} \lambda_i(a_i^T x - b_i) = 0.$$

which implies that

$$f(.) + \sum_{i \in I(z)} \lambda_i(a_i(.) - b_i)$$

is a constant function on S.

Now, we assume that $z \in S$ and $\lambda \in \mathbb{R}^m$ is a Lagrange multiplier corresponding to z. Let $\overline{I}(z) := \{i \in I(z) | \lambda_i > 0\}$. Now, we establish the characterizations of the solution set of the (NPLP).

Theorem 4.7 *Let $z \in S$ and let the (NPLP) satisfy the optimality condition (4.12) with a Lagrange multiplier $\lambda \in \mathbb{R}^m$. Let*

$$\overline{S}_1 := \{x \in C | a_i^T x = b_i, \forall i \in \overline{I}(z), a_i^T x \le b_i, \forall i \in I \backslash \overline{I}(z), \exists \zeta \in \partial^c f(x) :$$

$$\langle \zeta, x - z \rangle = 0\};$$

$$\overline{S}_2 := \{x \in C | a_i^T x = b_i, \forall i \in \overline{I}(z), a_i^T x \le b_i, \forall i \in I \backslash \overline{I}(z), \exists \xi \in \partial^c f(z) :$$

$$\langle \xi, x - z \rangle = 0\};$$

$$\overline{S}_3 := \{x \in C | a_i^T x = b_i, \forall i \in \overline{I}(z), a_i^T x \le b_i, \forall i \in I \backslash \overline{I}(z), \exists \varsigma \in \partial^c f(x \alpha z) :$$

$$\langle \varsigma, x - z \rangle = 0\}.$$

Then $S = \overline{S}_1 = \overline{S}_2 = \overline{S}_3$.

Proof We establish the equality $S = \overline{S}_2$ and the other equalities can be proved similarly. Suppose that $x \in \overline{S}_2$, then $x \in C, a_i^T x \le b_i$, for all $i \in I$ and there exists $\xi \in \partial^c f(z)$, such that $\langle \xi, x - z \rangle = 0$. Now, using Theorem 2.18, it follows that $x \in D$ and $f(x) = f(z)$, which implies that $x \in S$. Thus, $\overline{S}_2 \subseteq S$.

Conversely, suppose that $x \in S$, then $f(x) = f(z)$. Using Theorem 4.6, we get

$$\sum_{i \in I(z)} \lambda_i (a_i^T x - b_i) = 0.$$

Hence, $a_i^T x = b_i$, for all $i \in I \backslash \overline{I}(z)$. Again, since $f(x) = f(z)$, using Theorem 2.18, there exits $\xi \in \partial^c f(z)$ such that $\langle \xi, x - z \rangle = 0$, which gives that $x \in \overline{S}_2$. Thus, $S \subseteq \overline{S}_2$. Hence, $S = \overline{S}_2$.

The following example from Mishra *et al.* [200] illustrates the significance of Proposition 4.3, Theorems 4.6 and 4.7.

Example 4.2 *Consider the following constrained optimization problem:*

$$(COP) \quad \min f(x,y) = \begin{cases} (-x + y + 2), & \text{if } y < x, \\ 2(-x + y + 1), & \text{if } y \ge x, \end{cases}$$

subject to $(x, y) \in D := \{(x, y) \in C | x - y - 2 \le 0, x + y \le 4\},$

where $C := \{(x,y) \in \mathbb{R}^2 | x \ge 0, y \ge 0\}$ is a closed convex set. Evidently, the function f is a locally Lipschitz continuous function on C and nondifferentiable at points $(x,y) \in C$, where $x = y$ (see Figure 4.1).

The Clarke generalized gradient at any point $\overline{z} := (\overline{x}, \overline{y}) \in C$ is given by

$$\partial^c f(\overline{x}, \overline{y}) = \begin{cases} (-1, 1), & \text{if } \overline{y} < \overline{x}, \\ \{(k,t) : k \in [-2,-1], t \in [1,2]\}, & \text{if } \overline{y} = \overline{x}, \\ (-2, 2), & \text{if } \overline{y} > \overline{x}. \end{cases}$$

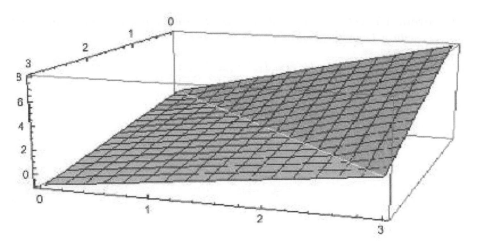

FIGURE 4.1: Objective Function Plot for (COP)

Then, it is easy to check that the function f is pseudoconvex as well as pseudoconcave on C. Hence, the function f is pseudolinear on C. Now, it is obvious that $\bar{z} = (\bar{x}, \bar{y}) = (2, 0)$ is an optimal solution of the (COP) and the optimality condition (4.19) holds with a Lagrange multiplier $\lambda = (1, 0)$ for $(-1, 1) \in \partial^c f(\bar{x}, \bar{y}) = \{-1, 1\}$. Using Theorem 4.7, the solution set S of (COP) is given by

$$S = \{(x, y) \in C : x - y - 2 = 0, x + y \le 4\}$$

$$= \{(\bar{x}, \bar{y}) \in C : x - y - 2 = 0, 0 \le y \le 1\} = \bar{S}_1 = \bar{S}_2 = \bar{S}_3.$$

Clearly, $\bar{z} = (\bar{x}, \bar{y}) = (2, 0) \in S$ and $I(\bar{z}) = \bar{I}(\bar{z}) = \{1\}$. Moreover,

$$f(.) + \sum_{i \in Iz} \lambda_i(a_i(.) - b_i)$$

is a constant function on S.

Corollary 4.3 *Let $z \in S$ and let the (NPLP) satisfy the optimality conditions (4.12) with a Lagrange multiplier $\lambda \in \mathbb{R}^m$. Let*

$$\tilde{S}_1 := \{x \in C | a_i^T x = b_i, \forall i \in \bar{I}(z), a_i^T x \le b_i, \forall i \in I\backslash\bar{I}(z), \exists \zeta \in \partial^c f(x) :$$

$$\langle \zeta, x - z \rangle \le 0\};$$

$$\tilde{S}_2 := \{x \in C | a_i^T x = b_i, \forall i \in \bar{I}(z), a_i^T x \le b_i, \forall i \in I\backslash\bar{I}(z), \exists \xi \in \partial^c f(z) :$$

$$\langle \xi, x - z \rangle \le 0\};$$

$$\tilde{S}_3 := \{x \in C | a_i^T x = b_i, \forall i \in \bar{I}(z), a_i^T x \le b_i, \forall i \in I\backslash\bar{I}(z), \exists \varsigma \in \partial^c f(x\alpha z) :$$

$$\langle \varsigma, x - z \rangle \le 0, \forall \alpha \in]0, 1]\}.$$

Then, $S = \tilde{S}_1 = \tilde{S}_2 = \tilde{S}_3$.

Proof It is clear that $S = \bar{S}_1 \subseteq \tilde{S}_1, \bar{S}_3 \subseteq \tilde{S}_3$, and $\bar{S}_2 \subseteq \tilde{S}_2$. Let us assume that $x \in \tilde{S}_1$. Then, $a_i^T x = b_i$, for all $i \in \bar{I}(z)$ and $a_i^T x \leq b_i$, for all $i \in I \backslash \bar{I}(z)$. Using Theorem 2.19, there exists $\zeta \in \partial^c f(x)$, such that

$$f(z) - f(x) = p(x, z)\langle \zeta, z - x \rangle \geq 0, \tag{4.15}$$

which implies that $f(x) \leq f(z)$. As $z \in S$, we get $f(x) = f(z)$. This proves that $x \in S$. Similarly, by Theorem 2.19, there exist $\xi \in \partial^c f(z)$ and $\varsigma \in \partial^c f(x\alpha z)$, such that

$$f(x) - f(z) = p(z, x)\langle \xi, x - z \rangle,$$

and

$$f(z) - f(x\alpha z) = (1 - \alpha)p(x\alpha z, z)\langle \varsigma, x - z \rangle, \forall \alpha \in \,]0, 1]\,.$$

Using the above equalities, we can establish that $\tilde{S}_2 \subseteq \bar{S}_2$ and $\tilde{S}_3 \subseteq \bar{S}_3$, respectively.

Now, we will show that for the case $C = \mathbb{R}^n$ the solution set of the (NPLP) is a polyhedral convex subset of $\mathbb{R}^n$. We establish the characterization independently from the objective function f.

Theorem 4.8 *Suppose that for $C - \mathbb{R}^n$ and $z \in S$, the (NPLP) satisfies the optimality condition (4.12) with a Lagrange multiplier $\lambda \in \mathbb{R}^m$. Then,*

$$S = \{x \in \mathbb{R}^n | a_i^T x = b_i, \forall i \in \bar{I}(z), a_i^T x \leq b_i, \forall i \in I \backslash \bar{I}(z)\}$$

and hence, S is a convex polyhedral subset of $\mathbb{R}^n$.

Proof Using (4.12), there exists $\xi \subset \partial^c f(z)$, such that

$$\langle \xi, x - z \rangle + \sum_{i \in I(z)} \lambda_i a_i^T (x - z) \geq 0, \forall x \in \mathbb{R}^n \tag{4.16}$$

Let $x \in \mathbb{R}^n$ satisfy $a_i^T x = b_i$, for all $i \in \bar{I}(z)$ and $a_i^T x \leq b_i$, for all $i \in I \backslash \bar{I}(z)$. Then from (4.16), there exists $\xi \in \partial^c f(z)$, such that

$$\langle \xi, x - z \rangle = 0.$$

Now, using Theorem 4.7, the required result follows.

Corollary 4.4 *Suppose that for $C = \mathbb{R}^n$ and $z \in S$, the (NPLP) satisfies the optimality condition (4.12) with a Lagrange multiplier $\lambda \in \mathbb{R}^m$. Then, the solution set S is bounded if and only if*

$$\{x \in \mathbb{R}^n | a_i^T x = 0, \forall i \in \bar{I}(z), a_i^T x \leq 0 \forall i \in I \backslash \bar{I}(z)\} = \{0\}.$$

Proof Using Theorem 4.8, we get

$$S^\infty = \{x \in \mathbb{R}^n | a_i^T x = 0, \forall i \in \bar{I}(z), a_i^T x \leqq 0, \forall i \in I \backslash \bar{I}(z)\},$$

where S^∞ is the recession cone of S. It is obvious by the definition that the set S is bounded if and only if $S^\infty = \{0\}$.

4.7   Nonsmooth Vector Pseudolinear Optimization Problems with Linear Inequality Constraints

In this section, we consider a constrained nonsmooth vector pseudolinear optimization problem and derive the conditions under which an efficient solution becomes a properly efficient solution.

We consider the following vector linear fractional programming problem with linear inequality constraints:

$$(NVOP) \quad \min f(x) := (f_1(x), f_2(x), ..., f_p(x))$$

$$\text{subject to } x \in \hat{D} := \{x \in \mathbb{R}^n | a_j^T x \le b_i, j \in J := \{1, 2, ..., m\}\},$$

where $a_j \in \mathbb{R}^n, b_i \in \mathbb{R}$ and $f_i : X \to \mathbb{R}, i \in I := \{1, 2, .., p\}$ are locally Lipschitz pseudolinear functions on open convex set X, containing $\hat{D}$ with respect to the same proportional function $\bar{p}$.

Let $J(z) := \{j \in J | a_j^T z = b_j\}$. Let us assume that $\hat{S}$ be the set of all efficient solution of the (NVOP) and $\hat{S} \ne \phi$.

In the following theorem, Mishra *et al.* [200] established the conditions under which an efficient solution becomes a properly efficient solution.

Theorem 4.9 *For the problem (NVOP), let $z \in \hat{D}$. Then the following statements are equivalent:*

(i) $z \in \hat{S}$, that is z is an efficient solution of the (NVOP);

(ii) there exists $\lambda_i > 0, i = 1, 2, .., p$ such that

$$\sum_{i=1}^{p} \lambda_i f_i(x) \ge \sum_{i=1}^{p} \lambda_i f_i(z), \forall x \in \hat{D};$$

(iii) z is a properly efficient solution of the (NVOP).

Proof From the definitions of efficient and properly efficient solutions, the implication $(iii) \Rightarrow (i)$ is obvious. Again, the implication $(ii) \Rightarrow (iii)$ has been proved in Geoffrion [86]. Now, we shall establish the implication $(i) \Rightarrow (ii)$.

Let $z \in \hat{D}$ be an efficient solution of (NVOP). Then by Theorem 6.1, there exist $\lambda_i > 0, i = 1, 2, ..., p, r_j \ge 0, j \in J(z)$ and there exists $\xi_i \in \partial^c f_i(z)$, such that

$$\sum_{i=1}^{p} \lambda_i \xi_i + \sum_{j \in J(z)} r_j a_j = 0.$$

Thus, for all $x \in \hat{D}$, there exists $\xi_i \in \partial^c f_i(z)$, such that

$$\sum_{i=1}^{p} \lambda_i \langle \xi_i, x - z \rangle + \sum_{j \in J(z)} r_j a_j^T (x - z) = 0. \tag{4.17}$$

Since, for all $j \in J(z), a_j^T z = b_j$, therefore using (4.17), there exists $\xi_i \in \partial^c f_i(z)$, such that

$$\sum_{i=1}^p \lambda_i \langle \xi_i, x - z \rangle + \sum_{j \in J(z)} r_j \left(a_j^T x - b_j \right) = 0, \forall x \in \hat{D}.$$

Using the fact that $a_j^T x - b_j \le 0$, for any $x \in \hat{D}$, there exists $\xi_i \in \partial^c f_i(z)$, such that

$$\sum_{i=i}^p \lambda_i \langle \xi_i, x - z \rangle \ge 0. \tag{4.18}$$

From the pseudolinearity of $f_i, i = 1, 2, ..., p$ with respect to $\bar{p}$, there exists $\xi_i' \in \partial^c f_i(z)$, such that

$$f_i(x) - f_i(z) = \bar{p}(z, x) \langle \xi_i', x - z \rangle, \forall x \in \hat{D} \tag{4.19}$$

Using Proposition 2.5, there exists $\mu_i > 0, i = 1, 2, ..., p$, such that $\xi_i' = \mu_i \xi_i$. Hence from (4.19), we get

$$f_i(x) - f_i(z) = \bar{p}(z, x) \mu_i \langle \xi_i, x - z \rangle, \forall x \in \hat{D}. \tag{4.20}$$

Since $\bar{p}(z, x) > 0$ and $\lambda_i > 0, i = 1, 2, ..., p$, therefore from (4.18) and (4.20), it follows that

$$\sum_{i=1}^p \lambda_i f_i(x) \ge \sum_{i=1}^p \lambda_i f_i(z), \forall x \in \hat{D},$$

which implies that (ii) holds.

4.8 Necessary Optimality Conditions for a Mathematical Program in Terms of Bifunctions

We consider the following mathematical programming problem:

$$(MPP) \quad \min f(x)$$

$$\text{subject to } g_j \le 0, \ j = 1, ..., m$$

where $f : X \to \mathbb{R}$, $g_j : X \to \mathbb{R}, j \in J = \{1, ..., m\}$ are radially continuous functions defined on a nonempty convex set $X \subseteq \mathbb{R}^n$. Let $S := \{x \in X : g_j(x) \le 0, j \in J\}$ and $J(x) = \{j \in J : g_j(x) = 0\}$ denote the set of all feasible solutions and active constraint index set at $x \in X$, respectively. Let $\bar{S}$ denote the solution set of the problem (MPP). Now, we assume that $\bar{S} \ne \phi$ and bifunctions $h(x; d), h_j(x; d) : X \times \mathbb{R}^n \to \bar{\mathbb{R}}$ associated with functions f

and $g_j, j \in J$, are subadditive functions of d and satisfy (4.21)–(4.22). We also assume that for all $(x, d) \in X \times \mathbb{R}^n$,

$$D_+ f(x; d) \leq h(x; d) \leq D^+ f(x; d)$$

and

$$D_+ g_j(x; d) \leq h_j(x; d) \leq D^+ g_j(x; d), \forall j = 1, ..., m.$$

Lalitha and Mehta [156] derived the following necessary optimality conditions for (MPP), by using Slater's constraint qualification (see Mangasarian [176]).

Theorem 4.10 *Let $\bar{x} \in \bar{S}$ and suppose that for (MPP), the following conditions hold:*

(a) *For each $j \in J, g_j$ is h_j-pseudolinear function with respect to proportional function q_j.*

(b) *Slater's constraint qualification holds, that is there exists some $\hat{x} \in S$ such that $g_j(\hat{x}) < 0$, for all $j \in J(\bar{x})$.*

Then $\forall x \in S$, there exist $\bar{\mu} \in \mathbb{R}^m$, such that

$$\left.\begin{array}{r} h(\bar{x}; x - \bar{x}) + \displaystyle\sum_{j \in J(\bar{x})} \bar{\mu}_j h_j(\bar{x}; x - \bar{x}) \geq 0, \\[2mm] \bar{\mu}^T g(\bar{x}) = 0, \\[2mm] \bar{\mu}_j \geqq 0, j \in J(\bar{x}). \end{array}\right\} \qquad (4.21)$$

Proof Suppose $\bar{x} \in \bar{S}$ is a solution for the problem (MPP). We have to prove that there exist $\bar{\lambda}_i \geq 0, i \in I, \bar{\lambda} \neq 0$ and $\bar{\mu}_j \geqq 0, j \in J(\bar{x})$), such that (4.28) holds. We assert that the following system of inequalities:

$$\left.\begin{array}{r} h(\bar{x}; x - \bar{x}) < 0, \\[1mm] h_j(\bar{x}; x - \bar{x}) \leq 0, j \in J(\bar{x}), \end{array}\right\} \qquad (4.22)$$

has no solution $x \in X$. Suppose to the contrary that $x \in X$, satisfies the above inequalities. Then,

$$\left.\begin{array}{r} D_+ f(\bar{x}; x - \bar{x}) < 0, \\[1mm] h(\bar{x}; x - \bar{x}) \leq 0, j \in J(\bar{x}). \end{array}\right\}$$

The positive homogeneity of h_j implies that there exists a sequence $t_n \to 0+$ and a positive integer m, such that $\forall n \geq m$, we get

$$\left.\begin{array}{r} f(\bar{x} + t_n(x - \bar{x})) < f(\bar{x}), \\[1mm] h_j(\bar{x}, \bar{x} + t_n(x - \bar{x}) - \bar{x}) \leq 0, j \in J(\bar{x}). \end{array}\right\}$$

To arrive at a contradiction, it is sufficient to prove that $\bar{x}+t_p(x-\bar{x})$ is feasible for some $p \geq m$. Using the fact that g_j is h_j-pseudoconcave, it follows that

$$g_j(\bar{x} + t_n(x - \bar{x})) \leq g_j(\bar{x}) = 0, \forall n \geq m.$$

Since, $g_j(\bar{x}) < 0$ for all $j \notin J(\bar{x})$ and g_j is radially continuous on X, there exists a positive integer $p \geq m$, such that

$$g_j(\bar{x} + t_p(x - \bar{x})) < 0, \forall j \notin J(\bar{x}).$$

Hence, $\bar{x} + t_p(x - \bar{x})$ is a feasible solution for (MPP).

Since $h(\bar{x}; x - \bar{x})$ and $h_j(\bar{x}; x - \bar{x}), j \in J$ are sublinear functions of x and hence convex, therefore by Lemma 1.4, there exists nonnegative scalars $\bar{\nu}$ and $\bar{\mu}_j, j \in J(\bar{x})$, not all zero, such that

$$\bar{\nu}h(\bar{x}; x - \bar{x}) + \sum_{j \in J(x)} \bar{\mu}_j h(\bar{x}; x - \bar{x}) \geq 0, \forall x \in X. \qquad (4.23)$$

Now, we shall show that $\nu \neq 0$, otherwise we have

$$\sum_{j \in J(\bar{x})} \bar{\mu}_j h(\bar{x}; x - \bar{x}) \geq 0. \qquad (4.24)$$

By assumption (b) there exists some strictly feasible point $\hat{x}$, that is $g_j(\hat{x}) < 0 = g_j(\bar{x}), \forall j \in J(\bar{x})$. Using the h_j-pseudoconvexity of g_j, we get $h_j(\bar{x}; \hat{x} - \bar{x}) < 0$ for all $j \in J(\bar{x})$. Since $\bar{\mu}_j, j \in J(\bar{x})$ cannot be all zero, so that we must have

$$\sum_{j \in J(\bar{x})} \bar{\mu}_j h_j(\bar{x}; x - \bar{x}) < 0,$$

which contradicts (4.31). Therefore, $\bar{\lambda} \neq 0$. Hence, the result follows by defining $\bar{\mu}_j = 0, j \notin J(\bar{x})$.

Remark 4.3 *If $X = \mathbb{R}^n$ and the function f, g are differentiable pseudolinear functions, then the above necessary optimality conditions (5) reduce to*

$$\left.\begin{array}{r} \nabla f(\bar{x}) + \displaystyle\sum_{j \in J(\bar{x})} \bar{\mu}_j \nabla g(\bar{x}) = 0, \\[2mm] \bar{\mu}^T g(\bar{x}) = 0, \\[1mm] \bar{\mu}_j \geq 0, j \in J(\bar{x}). \end{array}\right\}$$

In the next theorem, Lalitha and Mehta [156] established that the Lagrangian function $L(x, \mu) = f(x) + \mu^T g(x)$ as a function of x remains constant on the solution set $\bar{S}$ for a fixed Lagrange multiplier μ.

For $\bar{x} \in \bar{S}$, let us define

$$M(\bar{x}) := \{\mu \in \mathbb{R}^m_+ : h(\bar{x}; x-\bar{x}) + \sum_{j \in J(\bar{x})} \mu_j h_j(\bar{x}; x-\bar{x}) \geq 0, \forall x \in S, \mu^T g(\bar{x}) = 0\}.$$

For any $\mu \in M(\bar{x})$, let

$$B_\mu(\bar{x}) = \{j \in J(\bar{x}) : \mu_j \neq 0\}$$

and

$$C_\mu(\bar{x}) = I \backslash B_\mu(\bar{x}).$$

Theorem 4.11 *Suppose that for the problem (PMP), $\bar{x} \in \bar{S}$ and all the assumptions of Theorem 4.15 are satisfied. Further, if f is h-convex on X, then for a fixed $\bar{\mu} \in M(\bar{x})$ and each $x \in \bar{S}, \bar{\mu}^T g(\bar{x}) = 0$ and the Lagrangian function $L(.,\bar{\mu})$ is constant on $\bar{S}$.*

Proof Since, $x \in \bar{S}, f(x) = f(\bar{x})$. Hence, by the h-convexity of f on S, we get $h(\bar{x}, x - \bar{x}) \leq 0$. By the optimality condition (4.28), we get

$$\sum_{j \in J(\bar{x})} \mu_j h_j(\bar{x}, x - \bar{x}) \geq 0. \tag{4.25}$$

Therefore, for each $x \in \bar{S}$, we have $g_j(x) \leq 0 = g_j(\bar{x}), \forall j \in J(\bar{x})$. Since, every sublinear functional is subodd, by Theorem 4.14, g_j is h_j-quasiconvex on X. Hence, $h_j(\bar{x}, x - \bar{x}) \leq 0, \forall j \in J(\bar{x})$. Using (4.32), we get that $\bar{\mu}_j h_j(\bar{x}, x - \bar{x}) = 0, \forall j \in J(\bar{x})$. Since, $\mu_j \neq 0$, for $j \in B_\mu(\bar{x})$, it follows that $h_j(\bar{x}, x - \bar{x}) = 0, \forall j \in B_\mu(\bar{x})$. Now, by h_j-pseudoconvexity and h_j-pseudoconcavity of g_j, it follows that

$$g_j(x) = 0 = g_j(\bar{x}), \forall j \in B_\mu(\bar{x}).$$

Moreover, as $\bar{\mu}_j = 0$, for $j \in C_\mu(\bar{x})$, it results that $\bar{\mu}^T g(x) \leq 0$, for each $x \in \bar{S}$. Hence, for each $x \in \bar{S}$, $f(x) + \bar{\mu}^T g(x) = f(\bar{x}) + \bar{\mu}^T g(\bar{x})$, that is, $L(.,\bar{\mu})$ is constant on $\bar{S}$.

4.9　Mathematical Program with h-Convex Objective and h_j-Pseudolinear Constraints

Suppose the function f is h-convex on S and each of the constraint $g_j, j \in J$ is h_j-pseudolinear on S. For $\bar{x} \in \bar{S}$ and $\mu \in M(\bar{x})$, let the cardinality of the set $B_\mu(\bar{x})$ be denoted by k_μ. For $B_\mu(\bar{x}) = \phi$, we take $k_\mu = 0$.

Theorem 4.12 *Let $\bar{x} \in \bar{S}$ and suppose that all the conditions of Theorem 4.15 are satisfied. Further suppose that f is h-convex on S and $\mu \in M(\bar{x})$, then $\bar{S} = \tilde{S} = \hat{S}$, where*

$$\tilde{S} := \{x \in S : h(\bar{x}; x - \bar{x}) = 0, |h(x; \bar{x} - x)| + k_\mu \mu_j h_j(\bar{x}; x - \bar{x}) \leq 0, \forall j \in B_\mu(\bar{x}),$$

$$g_j(x) \leq 0 \forall j \in C_\mu(\bar{x})\},$$

$$\hat{S} := \{x \in S : h(\bar{x}; x - \bar{x}) \leq 0, |h(\bar{x}; x - \bar{x})| + k_\mu h_j(\bar{x}; x - \bar{x}) \leq 0, \forall j \in B_\mu(\bar{z}),$$

$$g_j(x) \leq 0 \forall j \in C_\mu(\bar{x})\}.$$

Proof We first prove the inclusion $\bar{S} \subseteq \tilde{S}$. Let $x \in \bar{S}$, then $f(\bar{x}) = f(\bar{x})$, and by the h-convexity of f, we get

$$h(\bar{x}; x - \bar{x}) \leq 0. \tag{4.26}$$

Since, x is feasible, we get $g_j(x) \leq 0 = g_j(\bar{x}), \forall j \in B_\mu(\bar{x})$. By Theorem 4.14, using the h_j-quasiconvexity of g_j on X, we have $h_j(\bar{x}; x - \bar{x}) \leq 0, \forall j \in B_\mu(\bar{x})$. Hence, we get

$$k_\mu \mu_j h_j(\bar{x}; x - \bar{x}) \leq 0, \forall j \in B_\mu(\bar{x}). \tag{4.27}$$

As $\bar{x}$ is optimal and $\mu \in M(\bar{x})$, therefore, we have

$$h(\bar{x}; x - \bar{x}) + \sum_{j \in J(\bar{x})} \mu_j h_j(\bar{x}; x - \bar{x}) \geq 0.$$

Taking into consideration (4.34), we get

$$h(\bar{x}; x - \bar{x}) \geq 0.$$

Using (4.33), we get

$$h(x; x - \bar{x}) = 0.$$

Interchanging the roles of x and $\bar{x}$ and using the optimality conditions at x it follows that

$$h(x; \bar{x} - x) = 0$$

Therefore, from (4.33)–(4.34) and feasibility of x, it follows that $x \in \tilde{S}$. The inclusion $\tilde{S} \subseteq \hat{S}$ is clearly true. Next, we show that the inclusion $\hat{S} \subseteq S$ holds. Assume that $x \in \hat{S}$. Then, we have

$$k_\mu \mu_j h_j(\bar{x}; x - \bar{x}) \leq 0, \forall j \in B_\mu(\bar{x}),$$

that is,

$$h_j(\bar{x}; x - \bar{x}) \leq 0, \forall j \in B_\mu(\bar{x}).$$

Using the h_j-pseudoconcavity of g_j, it follows that $g_j(x) \leq g_j(\bar{x}) = 0, \forall j \in B_\mu(\bar{x})$. Using the hypothesis $g_j(x) \leq 0, \forall \in C_\mu(\bar{x})$, it follows that x is feasible. Since, $\mu \in M(\bar{x})$, we have

$$h(\bar{x}; x - \bar{x}) + \sum_{j \in J(\bar{x})} \mu_j h_j(\bar{x}; x - \bar{x}) \geq 0.$$

Since, $h(\bar{x}; x - \bar{x}) \leq 0$, it is clear that

$$\sum_{j \in J(\bar{x})} \mu_j h_j(\bar{x}; x - \bar{x}) \geq 0$$

or equivalently

$$\sum_{j \in B_\mu(\bar{x})} \mu_j h_j(\bar{x}; x - \bar{x}) \geq 0.$$

Since,

$$|h(x; \bar{x} - x)| + k_\mu \mu_j h_j(\bar{x}; x - \bar{x}) \leq 0, \forall j \in B_\mu(\bar{x}).$$

Hence, we get

$$h(x; \bar{x} - x) \geq \sum_{j \in B_\mu(\bar{x})} \mu_j h_j(\bar{x}; x - \bar{x}),$$

which implies that $h(x, \bar{x} - x) \geq 0$. By h-convexity of f, it follows that $f(x) \leq f(\bar{x})$. Since, $\bar{x} \in \bar{S}$, we get $f(x) = f(\bar{x})$), that is $x \in \bar{S}$.

As a consequence of Theorem 4.17, we have the following Corollary:

Corollary 4.5 *Suppose that for the problem (MPP), f is a differentiable convex function, the constraints $g_j, j \in J$ are linear that is, $g_j(x) = a_j^T x - b_j, \forall j \in J$ and Slater's constraint qualification holds. Let $\bar{x} \in \bar{S}$ and $\mu \in M(\bar{x})$, then,*

$$\bar{S} = \left\{ x \in S : \langle \nabla f(\bar{x}, x - \bar{x}) \rangle = 0, |\langle \nabla f(\bar{x}, x - \bar{x}) \rangle| + k_\mu \mu_j (a_j^T x - b_j) \leq 0, \right.$$

$$\left. \forall j \in B_\mu(\bar{x}), a_j^T x \leq b_j, \forall j \in C_\mu(\bar{x}) \right\}.$$

Now, for the case $S = \mathbb{R}^n$, we shall establish that the objective function is constant on the solution set of (MPP).

Corollary 4.6 *Suppose that, for the problem (MPP), f is a differentiable convex function, the constraints $g_j, j \in J$ are linear, that is, $g_j(x) = a_j^T - b_j, \forall j \in J$ and Slater's constraint qualification holds. Let $\bar{x} \in \bar{S}$ and $\mu \in M(\bar{x})$, then*

$$\bar{S} \subseteq \left\{ x \in \mathbb{R}^n : \langle \nabla f(\bar{x}), x - \bar{x} \rangle = 0, a_j^T x - b_j = 0, \forall j \in B_\mu(\bar{x}), a_j^T x \leq b_j, \right.$$

$$\left. \forall j \in C_\mu(\bar{x}), \nabla f(x) = \nabla f(\bar{x}) \right\}.$$

Proof Let $x \in \bar{S}$, from Corollary 4.5, we get that x satisfies $\langle \nabla f(\bar{x}), x - \bar{x} \rangle = 0$. Since, f is convex, for each $y \in \mathbb{R}^n$, we have

$$f(y) - f(x) = f(y) - f(\bar{x}) \geq \langle \nabla f(\bar{x}), y - \bar{x} \rangle$$

$$= \langle \nabla f(\bar{x}, y - x) \rangle + \langle \nabla f(\bar{x}, x - \bar{x}) \rangle$$

$$= \langle \nabla f(\bar{x}), y - x \rangle,$$

which implies that $\nabla f(\bar{x})$ is a subgradient of f at x. Since, f is differentiable, the only subgradient of f at x is $\nabla f(x)$, hence, we have $\nabla f(x) = \nabla f(\bar{x})$. The relations $\langle \nabla f(x), x - \bar{x} \rangle = 0$ and $\nabla f(\bar{x}) = \nabla f(x)$ together imply that $\langle \nabla f(x), x - \bar{x} \rangle = 0$. From Remark 4.4, it follows that

$$\langle \nabla f(x), x - \bar{x} \rangle = - \sum_{j \in J(\bar{x})} \mu_j a_j^T (x - \bar{x})$$

$$= - \sum_{j \in J(\bar{x})} \mu_j (a_j^T x - b_j)$$

$$= - \sum_{j \in B_\mu(\bar{x})} \mu_j(a_j^T x - bx_j). \tag{4.28}$$

Hence, we have

$$\sum_{j \in B_\mu(\bar{x})} \mu_j(a_j^T x - b_j) = 0.$$

Since, $\mu_j > 0$ and $a_j^T x - b_j \leq 0, \forall j \in B_\mu(\bar{x})$, it follows that

$$\bar{S} \subseteq \{x \in \mathbb{R}^n : \langle \nabla f(\bar{x}, x - \bar{x}) \rangle = 0, a_j^T x - b_j = 0, \forall j \in B_\mu(\bar{x}), a_j^T x \leq b_j,$$

$$\forall j \in C_\mu(\bar{x}), \nabla f(x) = \nabla f(\bar{x})\}$$

$$\subseteq \{x \in S : \langle \nabla f(\bar{x}, x - \bar{x}) \rangle = 0, |\langle \nabla f(\bar{x}, x - \bar{x}) \rangle| + k_\mu \mu_j(a_j^T x - b_j) \leq 0,$$

$$\forall j \in B_\mu(\bar{x}), a_j^T \leq b_j, \forall j \in C_\mu(\bar{x})\} = \bar{S}.$$

Remark 4.4 *In view of condition (4.35), conditions* $\langle \nabla f(\bar{x}), x - \bar{x} \rangle = 0$ *and* $a_j^T x - b_j = 0, \forall j \in B_\mu(\bar{x})$ *are equivalent. Therefore the solution set* $\bar{S}$ *is actually given by*

$$\bar{S} \subseteq \{x \in R^n : a_j^T x - b_j = 0, \forall j \in B_\mu(\bar{x}), a_j^T \leq b_j,$$

$$\forall j \in C_\mu(\bar{x}), \nabla f(x) = \nabla f(\bar{x})\}.$$

We consider the quadratic program:

$$(QP) \quad \min f(x) = \frac{1}{2} x^T C x + d^T x$$

subject to $a_j^T x \leq b_j, \forall j \in J,$

where C is an $n \times n$ symmetric positive semidefinite matrix, d and $a_j, j \in J$ are vectors in $\mathbb{R}^n$. Assume that the quadratic function f is defined on $\mathbb{R}^n$. Using Remark 4.5, we have the following result from Jeyakumar *et al.* [115].

Corollary 4.7 *Let* $\bar{S}$ *be the solution set of (QP); let* $\bar{x} \in \bar{S}$ *and* $u \in M(\bar{x})$. *Further suppose that Slater's constraint qualification holds. Then,*

$$\bar{S} \subseteq \{x \in \mathbb{R}^n : a_j^T x - b_j = 0, \forall j \in B_\mu(\bar{x}), a_j^T x \leq b_j, \forall j \in C_\mu(\bar{x}), C\bar{x} = Cx\}.$$

The following corollary provides the characterizations for the problem (MPP) in the absence of the constraints, that is, for the problem of minimizing f over the convex domain S. These corollaries are immediate consequences of Theorem 4.17.

Corollary 4.8 *Consider the problem (MPP) with* $J = \phi$. *If* f *is h-convex on* S, *the bifunction* $h(x; d)$ *is subadditive function of d and* $\bar{x} \in \bar{S}$, *then*

$$\bar{S} = \{x \in X : h(\bar{x}; x - \bar{x}) = 0, h(x; \bar{x} - x) = 0\}$$

$$= \{x \in X : h(\bar{x}; x - \bar{x}) \leq 0, h(x; \bar{x} - x) = 0\}.$$

Corollary 4.9 *Consider the problem (MPP) with $J = \phi$. If f is a differentiable convex function and $\bar{x} \in \bar{S}$, then*

$$\bar{S} = \{x \in X : \langle \nabla f(\bar{x}); x - \bar{x} \rangle = 0, \langle \nabla f(\bar{x}), \bar{x} - x \rangle = 0\}$$

$$= \{x \in X : \langle \nabla f(\bar{x}); x - \bar{x} \rangle \leq 0, \langle \nabla f(\bar{x}), \bar{x} - x \rangle = 0\}.$$

Remark 4.5 *If f is a twice continuously differentiable convex function on the open convex set S, then as in the proof of Lemma 1 of Mangasarian [177], $\nabla f(x) = \nabla f(\bar{x})$. Under these conditions the solution set $\bar{S}$ is given by*

$$\bar{S} = \{x \in X : \langle \nabla f(\bar{x}); x - \bar{x} \rangle = 0, \nabla f(x) = \nabla f(\bar{x})\}$$

$$= \{x \in X : \langle \nabla f(\bar{x}); x - \bar{x} \rangle \leq 0, \nabla f(x) = \nabla f(\bar{x})\}.$$

which is precisely Theorem 1 of Mangasarian [177].

The following numerical examples from Lalitha and Mehta [156] illustrate the significance of the above results.

Example 4.3 *Consider the problem:*

$$(P1) \quad \min f(x)$$

$$\text{subject to } g_j(x) \leq 0, j = 1, 2,$$

where $f, g_1, g_2 : \mathbb{R} \to \mathbb{R}$ are functions given by

$$f(x) = \begin{cases} -x, & \text{if } x < 0, \\ 0, & \text{if } 0 \leq x \leq 2, \\ x - 2, & \text{if } x > 0. \end{cases}$$

$$g_1(x) = -x^3 - x,$$

$$g_2(x) = \begin{cases} x - 3, & \text{if } x \geq 3, \\ \frac{x-3}{2}, & \text{if } x < 3. \end{cases}$$

Define

$$h(x, d) = \begin{cases} -d & \text{if } x < 0, \forall d \text{ or } x = 0, \forall d \leq 0, \\ 0, & \text{if } x = 0, d > 0 \text{ or } x = 2, d \leq 0, \\ 0, & \text{if } 0 < x < 2, \forall d, \\ d, & \text{if } x = 2, d > 0 \text{ or } x > 2, \forall d. \end{cases}$$

The set of feasible solutions of the problem (P1) is $\{x \in \mathbb{R} : 0 \leq x \leq 3\}$ and $\bar{x} = 0$ is an optimal solution of the problem (P1). For $\bar{x} = 0$ the set $J(\bar{x}) = \{1\}$ and $M(\bar{x}) = \{(0,0)\}$. Clearly, $k_\mu = 0$ for $\mu = (0,0) \in M(\bar{x})$ as the set $B_\mu(\bar{x}) \neq \phi$. Also, Slater's constraint qualification is satisfied as $g_j(\hat{x}) < 0, \forall j \in J(\bar{x})$, where $\hat{x} = 2$. The solution of (P1) is given by

$$\bar{S} = \{x \in \mathbb{R} : h(0; x) \leq 0, h(x, -x) = 0, g_j(x) \leq 0, j = 1, 2\}$$

$$= \{x \in \mathbb{R} : 0 \leq x \leq 2\}.$$

In the following example, Lalitha and Mehta [156] considered the problem studied by Jeyakumar *et al.* [115] and obtained its solution set using the characterization given in Theorem 4.17.

Example 4.4 *Consider the following problem:*

$$(P2) \quad \min f(x)$$

$$\text{subject to } g_j(x) \le 0, j = 1, 2,$$

where $f, g_1, g_2 : \mathbb{R}^2 \to \mathbb{R}$ *are functions given by*

$$f(x_1, x_2) = (x_1^2 + x_2^2)^{\frac{1}{2}} + x_1 + x_2$$

$$g_1(x_1, x_2) = x_1 + x_2$$

$$g_2(x_1, x_2) = -x_1$$

For $d = (d_1, d_2)$ *and* $x = (x_1, x_2)$ *define*

$$h(x; d) = \begin{cases} (x_1 d_1 + x_2 d_2)/(x_1^2 + x_2^2)^{1/2} + d_1 + d_2, & \text{if } x \ne 0, \forall d, \\ (d_1^2 + d_2^2) + d_1 + d_2, & \text{if } x = 0, \forall d, \end{cases}$$

$$h_1(x, d) = d_1 + d_2,$$

$$h_1(x, d) = -d_1.$$

Then, $f(x_1, x_2) \ge 0$, *for all* (x_1, x_2) *in the feasible set. Clearly,* $\bar{x} = (0, 0) \in \bar{S}$ *and* $J(\bar{x}) = \{1, 2\}$. *For* $\mu(0, 1) \in M(\bar{x})$, *we have* $k_\mu = 1$ *and the solution set is given by*

$$\bar{S} = \{x = (x_1, x_2) \in \mathbb{R}^2 : h(\bar{x}, x - \bar{x}) \le 0, |h(\bar{x}, x - \bar{x})| + h_2(\bar{x}, x - \bar{x}) \le 0,$$

$$g_1(x) \le 0\}$$

$$= \{x = (x_1, x_2) \in \mathbb{R}^2 : (x_1^2 + x_2^2)^{1/2} + x_1 + x_2 \le 0, (x_1^2 + x_2^2)^{1/2} + x_1 + x_2 - x_1 \le 0,$$

$$x_1 + x_2 \le 0\}$$

$$= \{x = (x_1, x_2) \in \mathbb{R}^2 : (x_1^2 + x_2^2)^{1/2} + x_2 \le 0, x_1 + x_2 \le 0\}.$$

Condition $(x_1^2 + x_2^2)^2 + x_2 \le 0$ *is equivalent to* $x_2 \le 0$ *and* $x_1^2 = 0$. *Thus*

$$\bar{S} = \{(x_1, x_2) \in \mathbb{R}^2 : x_1 = 0, x_2 \le 0\}.$$

The following example illustrates the significance of Corollary 4.6.

Example 4.5 *Consider the problem:*

$$(P3) \quad \min f(x)$$

$$\text{subject to } g_j(x) \le 0, j = 1, 2,$$

where $f, g_1, g_2 : \mathbb{R}^2 \to \mathbb{R}$ are functions given by

$$f(x_1, x_2) = x_1^2, g_1(x_1, x_2) = 1 - x_2, g_2(x_1 x_2) = 1 - x_1.$$

Then for $d = (d_1, d_2)$ and $x = (x_1, x_2)$,

$$\nabla f(x)^T d = 2x_1 d_1, \nabla g_1(x)^T d = -d_2, \nabla g_1(x)^T d = -d_1.$$

Clearly, $f(x_1, x_2) \geq 1$, for all (x_1, x_2) in the feasible set. We note that for $\hat{x} = (2, 2) \in \mathbb{R}^2$, Slater's constraint qualification holds. For $\bar{x} = (1, 1) \in \bar{S}$, the set $J(\bar{x}) = \{1, 2\}$ and $M(\bar{x}) = \{(0, 2)\}$. Thus for $\mu = (0, 2)$, we note that $B_\mu(\bar{x}) = \{2\}$ and so $k_\mu = 1$. Using the characterization given in Corollary 4.6, the solution set can be given by

$$\bar{S} = \{x \in \mathbb{R}^2 : \langle \nabla f(\bar{x}), x - \bar{x} \rangle = 0, a_2^T x - b_2 = 0, a_2^T x \leq b_2, \nabla f(x) = \nabla f(\bar{x})\}.$$

4.10   Mathematical Program with h-Pseudolinear Objective and h_j-Pseudolinear Constraints

In this section, characterizations for the solution set for the problem (MPP) are derived under the assumption that the function f is a h-pseudolinear function on X and each constraint g_j is a h_j-pseudolinear function on X.

Theorem 4.13 *Let $\bar{x} \in \bar{S}$ and all the conditions of Theorem 4.6 are satisfied. Further, suppose that f is h-pseudolinear on X and $\mu \in M(\bar{x})$, then,*

$$\bar{S} = \{x \in X : h(\bar{x}; x - \bar{x}) \leq 0, h_j(\bar{x}; x - \bar{x}) \leq 0, \forall j \in B_\mu(\bar{x}), g_j(\bar{x}) \leq 0,$$

$$\forall j \in C_\mu(\bar{x})\}.$$

Proof Let us denote the set on the right hand side by

$$\bar{S} = \{x \in X : h(\bar{x}; x - \bar{x}) \leq 0, h_j(\bar{x}; x - \bar{x}) \leq 0, \forall j \in B_\mu(\bar{x}), g_j(\bar{x}) \leq 0,$$

$$\forall j \in C_\mu(\bar{x})\}.$$

Assume that $x \in \bar{S}$. Since, f is h-pseudoconvex and $h(x; .)$ is sublinear, therefore, f is h-quasiconvex which together with $f(x) = f(\bar{x})$ implies that $h(\bar{x}; x - \bar{x}) \leq 0$. Since, $\mu \in M(\bar{x})$, we have

$$h(\bar{x}; x - \bar{x}) + \sum_{j \in J(\bar{x})} \mu_j h_j(\bar{x}; x - \bar{x}) \geq 0.$$

Therefore, we get $\sum_{j \in J(\bar{x})} \mu_j h_j(\bar{x}; x - \bar{x}) \geq 0$, which is equivalent to

$$\sum_{j \in B_\mu(\bar{x})} \mu_j h_j(\bar{x}; x - \bar{x}) \geq 0. \tag{4.29}$$

Since, x is feasible, we have $g_j(x) \le 0 = g_j(\bar{x}), \forall j \in J(\bar{x})$. Using Theorem 4.14, it follows that g_i is h_j-quasiconvex on X and thus, we have $h_j(\bar{x}; x - \bar{x}) \le 0, \forall j \in B_\mu(\bar{x})$. This implies that

$$\sum_{j \in B_\mu(\bar{x})} \mu_j h_j(\bar{x}; x - \bar{x}) \le 0.$$

Therefore, from (4.36), we get

$$\sum_{j \in B_\mu(\bar{x})} h_j(\bar{x}; x - \bar{x}) = 0.$$

Since, $h_j(\bar{x}; x - \bar{x}) \le 0, \forall j \in B_\mu(\bar{x})$, therefore, we have $h_j(\bar{x}; x - \bar{x}) = 0, \forall j \in B_\mu(\bar{x})$. Hence, $x \in \tilde{S}$.

Conversely, suppose that $x \in \tilde{S}$. As $h_j(\bar{x}; x - \bar{x}) = 0, \forall j \in B_\mu(\bar{x})$ by h_j-pseudoconcavity of g_j we get $g_j(x) \le 0 = g_j(\bar{x}), \forall j \in J(\bar{x})$, which along with the hypothesis $g_j(x) \le 0, \forall j \in C_\mu(\bar{x})$ proves that x is feasible. Since, $\bar{x}$ is optimal therefore by optimality condition (4.28), we have $h(\bar{x}; x - \bar{x}) \ge 0$ as $h_j(\bar{x}; x - \bar{x}), \forall j \in B_\mu(\bar{x})$. Now as $x \in \tilde{S}$, we have $h(\bar{x}; x - \bar{x}) \le 0$ and hence $h(\bar{x}; x - \bar{x}) = 0$. By h-pseudoconvexity and h-pseudoconcavity of f, it follows that $f(x) = f(\bar{x})$. Thus $x \in \tilde{S}$.

Remark 4.6 *If in the problem (MPP), the function f is differentiable pseudolinear function and $I \ne \phi$, then the above theorem reduces to Corollary 3.1 of Jeyakumar and Yang [119]. Again, if in the problem (MPP) when $X = \mathbb{R}^n$, f is differentiable pseudolinear function and the constraints are affine, then, we get Theorem 4.3, given by Dinh et al. [67].*

Lalitha and Mehta [156] illustrated the significance of Theorem 4.18 by the following example.

Example 4.6 *Consider the problem*

$$(P4) \quad \min f(x)$$

$$\text{subject to } g_j(x) \le 0, \ j = 1, 2,$$

where $f, g_1, g_2 : \mathbb{R}^2 \to \mathbb{R}$ are functions given by

$$f(x) = \begin{cases} x_1 + x_2, & \text{if } x_1 + x_2 > 0, \\ x_1 + x_2 - 1, & \text{if } x_1 + x_2 \le 0, \end{cases}$$

$g_1(x) = -x_1 - x_2, g_2(x) = -x_1$, *where* $x = (x_1, x_2)$.
For $d = (d_1, d_2)$, define

$$h(x, d) = \begin{cases} d_1 + d_2, & \text{if } x_1 + x_2 \ne 0, \\ +\infty, & \text{if } x_1 + x_2 = 0, \forall d_1 + d_2 > 0, \\ d_1 + d_2, & \text{if } x_1 + x_2 = 0, \forall d_1 + d_2 \le 0, \end{cases}$$

$$h_1(x, d) = -d_1 - d_2,$$

$$h_2(x; d) = -d_1.$$

It is obvious that the functions h, h_1, and h_2 are sublinear functions of d for each $x \in \mathbb{R}^2$. The set of feasible solutions is

$$\{(x_1, x_2) \in \mathbb{R}^2 : x_1 + x_2 \geq 0, x_1 \geq 0\}.$$

Now, $f(x_1, x_2) \geq 0$, for every feasible point (x_1, x_2) and thus, $\bar{x} = (0,0)$ is an optimal solution of the problem (P4). Also, $J(\bar{x}) = \{1, 2\}$ and there exists $\hat{x} = (1, 1) \in S$, such that

$$g_j(\hat{x}) < 0, \forall j \in J(\bar{x}).$$

Moreover, $M(\bar{x}) = \{(1, 0)\}$ and $B_\mu(\bar{x}) = \{1\}$ for $\mu = (1, 0)$. Using the above characterizations the set of optimal solutions for the problem (P4) is given by

$$\bar{S} = \{(x_1, x_2) \in \mathbb{R}^2 : h(\bar{x}; x - \bar{x}) \leq 0, h_1(\bar{x}; x - \bar{x}) = 0, g_2(x) \leq 0\}$$

$$= \{(x_1, x_2) \in \mathbb{R}^2 : x_1 + x_2 \leq 0, -x_1 - x_2 = 0, -x_1 \leq 0\}$$

$$= \{(x_1, x_2) \in \mathbb{R}^2 : x_1 + x_2 = 0, x_1 \geq 0\}.$$

Chapter 5

Pseudolinear Multiobjective Programming

5.1 Introduction

The multiobjective optimization problems with several conflicting objectives have been an important research topic during the last few decades. These problems arise in several areas of modern research such as in economics, game theory, optimal control, and decision theory. Many authors have developed necessary and sufficient conditions for multiobjective optimization problems. See for example, Kuhn and Tucker [151], Arrow *et al.* [7], Da Cunha and Polak [59], and Tanino and Sawaragi [269]. Singh [255] established necessary conditions for Pareto optimality by using the concept of convergence of a vector at a point and Motzkin's theorem of alternative and obtained several sufficient optimality conditions under the assumptions of pseudoconvexity and quasiconvexity on the functions involved. Chew and Choo [47] established the necessary and sufficient optimality conditions for multiobjective pseudolinear programming problems. Bector *et al.* [20] derived several duality results for the Mond-Weir type dual of a multiobjective pseudolinear programming problem and studied its applications to certain multiobjective fractional programming problems. Recently, various generalizations of the notion of convexity have been introduced to obtain optimality conditions and duality theorems for differentiable multiobjective optimization problems. For further details, we refer to Kaul and Kaur [137], Preda [224], and Hachimi and Aghezzaf [97].

In fractional programming, we study the optimization problems in which objective functions are ratios of two functions. These problems arise in different areas of modern research such as economics (Cambini *et al.* [35]), information theory (Meister and Oettli [183]), engineering design (Tsai [272]), and heat exchange networking (Zamora and Grossmann [295]). Schaible [247] and Bector [18] studied duality in fractional programming for scalar optimization problems. Kaul *et al.* [139] have studied the Mond-Weir type of dual for the multiobjective fractional programming problems involving pseudolinear functions and derived various duality results. Optimality conditions and duality theorems for differentiable multiobjective fractional programming problems in the framework of generalized convex functions have been widely studied by

Chandra *et al.* [40], Egudo [72], Mukherjee and Rao [210], Weir [281], Kuk *et al.* [153], and Liang *et al.* [162, 163]. For more references and further details in this area, we refer to Mishra and Giorgi [189] and Mishra *et al.* [203, 204].

This chapter deals with the optimality conditions and duality theorems for multiobjective programming and multiobjective fractional programming problems involving differentiable pseudolinear functions. Necessary and sufficient optimality conditions for a feasible solution to be an efficient solution for the primal problems under the assumption of differentiable pseudolinear functions have been established. Furthermore, the equivalence between efficiency and proper efficiency, under certain boundedness conditions has been proved. Several duality results such as weak, strong, and converse duality theorems for the corresponding Mond-Weir type dual problems are given.

5.2   Definitions and Preliminaries

We consider the following multiobjective optimization problem:

$$(MP)   \min f(x) := (f_1(x), \dots, f_k(x))$$

$$\text{subject to } g_j(x) \leq 0, \{j = 1, \dots, m\},$$

where $f_i : K \to \mathbb{R}, i \in I := \{1, \dots, k\}; g_j : K \to \mathbb{R}, j \in J := \{1, \dots, m\}$ are differentiable pseudolinear functions on a nonempty open convex set $K \subseteq \mathbb{R}^n$. Let $P = \{x \in K : g_j(x) \leq 0, j \in J\}$ and $J(x) := \{j \in J : g_j(x) = 0\}$ for some $x \in K$, denote the set of all feasible solutions for (MP) and active constraint index set at $x \in K$, respectively.

5.3   Optimality Conditions

The following necessary and sufficient optimality condition for (MP) stated by Bector *et al.* [20] is a variant of Proposition 3.2 of Chew and Choo [47].

Theorem 5.1 *(Necessary and sufficient optimality conditions) Let $\bar{x}$ be a feasible solution of (MP). Let the functions $f_i, i = 1, \dots, k$ be pseudolinear with respect to proportional function p_i and $g_j, j \in J(\bar{x})$ be pseudolinear with respect to proportional functions $q_j, j \in J(\bar{x})$. Then, $\bar{x}$ is an efficient solution for (MP) if and only if there exist $\bar{\lambda}_i \in \mathbb{R}, i = 1, \dots, k, \bar{\mu}_j \in \mathbb{R}, j \in J(\bar{x})$, such* *that*

$$\sum_{i=1}^{k} \bar{\lambda}_i \nabla f_i(\bar{x}) + \sum_{j \in J(\bar{x})} \bar{\mu}_j \nabla g_j(\bar{x}) = 0, \tag{5.1}$$

$$\bar{\lambda}_i > 0, i = 1, \ldots, k,$$

$$\bar{\mu}_j \geq 0, j \in J(\bar{x}).$$

Proof Suppose $\bar{\lambda}_i$ and $\bar{\mu}_j$ exist and satisfy condition (5.1) and $\bar{x}$ is not an efficient solution for (MP). Then, there exists a point $y \in P$, such that

$$f_i(y) \leq f_i(\bar{x}), \forall i = 1, \ldots, k$$

and

$$f_r(y) < f_r(\bar{x}), \text{ for some } r.$$

Using pseudolinearity of $f_i, i = 1, \ldots, k$, with respect to proportional function p_i, we get

$$p_i(\bar{x}, y) \langle \nabla f_i(\bar{x}), y - \bar{x} \rangle \leq 0, \forall i = 1, \ldots, k$$

and

$$p_r(\bar{x}, y) \langle \nabla f_r(\bar{x}), y - \bar{x} \rangle < 0, \text{ for some } r.$$

Since, $p_i(\bar{x}, y) > 0$, for all $i = 1, \ldots, k$, therefore, we get

$$\langle \nabla f_i(\bar{x}), y - \bar{x} \rangle \leq 0, \forall i = 1, \ldots, k$$

and

$$\langle \nabla f_r(\bar{x}), y - \bar{x} \rangle < 0, \text{ for some } r.$$

Therefore, for $\bar{\lambda}_i > 0, i = 1, \ldots, k$, it follows that

$$\sum_{i-1}^{k} \bar{\lambda}_i \langle \nabla f_i(\bar{x}), y - \bar{x} \rangle < 0. \tag{5.2}$$

Again, we have

$$g_j(y) \leq g_j(\bar{x}), \forall j \in J(\bar{x}).$$

Using pseudolinearity of $g_j, j = 1, \ldots, m$, we get

$$q_j(\bar{x}, y) \langle \nabla g_j(\bar{x}), y - \bar{x} \rangle \leq 0, \forall j \in J(\bar{x}).$$

Since, $q_j(\bar{x}, y) > 0$, for all $j \in J(\bar{x})$, therefore, we get

$$\langle \nabla g_j(\bar{x}), y - \bar{x} \rangle \leq 0, \forall j \in J(\bar{x}).$$

Therefore, for $\bar{\mu}_j \geq 0$, for all $j \in J(\bar{x})$, it follows that

$$\sum_{j \in J(\bar{x})} \bar{\mu}_j \langle \nabla g_j(\bar{x}), y - \bar{x} \rangle \leq 0.$$

Using (5.1), we get

$$\sum_{i=1}^{k} \bar{\lambda}_i \langle \nabla f_i(\bar{x}), y - \bar{x} \rangle \geq 0,$$

which is a contradiction to (5.2).

Conversely, suppose $\bar{x}$ be an efficient solution of (MP). For $1 \leq r \leq k$, the following system

$$\left. \begin{array}{r} \langle \nabla g_j(\bar{x}), x - \bar{x} \rangle \leq 0, j \in J(\bar{x}), \\ \langle \nabla f_i(\bar{x}), x - \bar{x} \rangle \leq 0, i = 1, \ldots, r-1, r+1, \ldots, k, \\ \langle \nabla f_r(\bar{x}), x - \bar{x} \rangle < 0, \end{array} \right\} \qquad (5.3)$$

has no solution $x \in P$. For if x is a solution of the system and $y = \bar{x} + t(x - \bar{x})(0 < t \leq 1)$, then for $j \in J(\bar{x})$, we have

$$\langle \nabla g_j(\bar{x}), y - \bar{x} \rangle = t \langle \nabla g_j(\bar{x}), x - \bar{x} \rangle \leq 0,$$

and thus by pseudoconcavity of g_j, we have

$$g_j(y) \leq g_j(\bar{x}) = 0.$$

If $j \notin J(\bar{x})$, then $g_j(\bar{x}) < 0$ and so $g_j(y) < 0$, when t is sufficiently small. Hence, for small t, y is a point of P. Hence, we have

$$f_i(y) - f_i(\bar{x}) = t p_i(\bar{x}, y) \langle \nabla f_i(\bar{x}), x - \bar{x} \rangle \leq 0, i \neq r$$

and

$$f_r(y) - f_r(\bar{x}) < 0, \text{ for some } r,$$

which is a contradiction to our assumption that $\bar{x}$ is an efficient solution for (MP). From which, it follows that the system (5.3) has no solution. Therefore, by Farkas's theorem (Theorem 1.24), there exist $\bar{\lambda}_{r_i} \geq 0$ and $\bar{\mu}_{r_j} \geq 0$, such that

$$-\nabla f_r(\bar{x}) = \sum_{i \neq r} \bar{\lambda}_{r_i} \nabla f_i(\bar{x}) + \sum_{j \in J(x^*)} \bar{\mu}_{r_j} \nabla g_j(\bar{x}). \qquad (5.4)$$

Summing (5.4) over r, we get (5.1) with

$$\bar{\lambda}_i = 1 + \sum_{r \neq i} \bar{\lambda}_{r_i}, \bar{\mu}_j = \sum_{r=1}^{k} \bar{\mu}_{r_j}.$$

The following corollary is a variant of Corollary 3.3 of Chew and Choo [47].

Corollary 5.1 *Let $\bar{x}$ be a feasible solution of the problem (MP). Then the following statements are equivalent:*

(i) $\bar{x}$ is an efficient solution of the problem (MP).

(ii) $\bar{x}$ is an efficient solution of the problem (MLP)

$$\min \ (\nabla f_1(\bar{x})x, \ldots, \nabla f_k(\bar{x})x)$$

$$\text{subject to } x \in K.$$

(iii) There exists positive numbers $\bar{\lambda}_i, i = 1, \ldots, k$, such that $\bar{x}$ minimizes the linear function

$$\sum_{i=1}^{k} \bar{\lambda}_i \nabla f_i(\bar{x})x, \ x \in K. \tag{5.5}$$

Proof (i) $\Rightarrow$ (ii) Suppose $\bar{x}$ is an efficient solution of the problem (MP), then by Theorem 5.1, there exist $\bar{\lambda}_i \in \mathbb{R}, i = 1, \ldots, k$, and $\bar{\mu}_j \in \mathbb{R}, j \in J(\bar{x})$, such that

$$\sum_{i=1}^{k} \bar{\lambda}_i \nabla f_i(\bar{x}) + \sum_{j \in J(\bar{x})} \bar{\mu}_j \nabla g_j(\bar{x}) = 0. \tag{5.6}$$

On the contrary, suppose that $\bar{x}$ is not an efficient solution of the (MLP) problem. Then, there exists some $x \in K$, such that

$$\nabla f_i(\bar{x})x \leq \nabla f_i(\bar{x})\bar{x}, \ \forall i = 1, \ldots, k,$$

$$\nabla f_r(\bar{x})x < \nabla f_r(\bar{x})\bar{x}, \text{ for some } r.$$

Now, we have

$$0 > \sum_{i=1}^{k} \bar{\lambda}_i \langle \nabla f_i(\bar{x}), x - \bar{x} \rangle = - \sum_{j \in J(\bar{x})} \bar{\mu}_j \langle \nabla g_j(\bar{x}), x - \bar{x} \rangle$$

$$= - \left(\sum_{j \in J(\bar{x})} \bar{\mu}_j \left(\frac{g_j(x) - g_j(\bar{x})}{q_j(\bar{x}, x)} \right) \right) \geq 0,$$

which is obviously a contradiction. Hence, $\bar{x}$ must be an efficient solution of the problem (MLP).

(ii) $\Rightarrow$ (iii) Suppose $\bar{x}$ be an efficient solution of the problem (MLP). Then by Theorem 5.1, there exist $\bar{\lambda}_i \in \mathbb{R}, i = 1, \ldots, k, \bar{\mu}_j \in \mathbb{R}, j \in J(\bar{x})$, such that (5.1) holds. Therefore, for any $x \in K$, we get

$$\sum_{i=1}^{k} \bar{\lambda}_i \langle \nabla f_i(\bar{x}), x - \bar{x} \rangle = - \sum_{j \in J(\bar{x})} \bar{\mu}_j \langle \nabla g_j(\bar{x}), x - \bar{x} \rangle$$

$$= - \left(\sum_{j \in J(\bar{x})} \bar{\mu}_j \left(\frac{g_j(x) - g_j(\bar{x})}{q_j(\bar{x}, x)} \right) \right) \geq 0.$$

This shows that $\bar{x}$ minimizes (5.5).

(iii) $\Rightarrow$ (i) Suppose $\bar{x}$ minimizes (5.5) for some positive number $\lambda_1, \ldots, \lambda_k$. If $\bar{x}$ be not an efficient solution for (MP), then, there exists some $x \in K$, such that

$$f_i(x) \leq f_i(\bar{x}), \forall i = 1, \ldots, k,$$

$$f_r(x) < f_r(\bar{x}), \text{ for some } r.$$

Then, by the pseudolinearity of $f_i, i = 1, \ldots, k$ with respect to proportional function p_i, it follows that

$$p_i(\bar{x}, x)\langle \nabla f_i(\bar{x}), x - \bar{x} \rangle \leq 0, \forall i = 1, \ldots, k,$$

$$p_r(\bar{x}, x)\langle \nabla f_r(\bar{x}), x - \bar{x} \rangle \leq 0, \text{ for some } r.$$

Since, $p_i(\bar{x}, x) > 0, i = 1, \ldots, k$, it follows that

$$\langle \nabla f_i(\bar{x}), x - \bar{x} \rangle \leq 0, \forall i = 1, \ldots, k,$$

$$\langle \nabla f_r(\bar{x}), x - \bar{x} \rangle \leq 0, \text{ for some } r.$$

Therefore, for $\bar{\lambda}_i > 0, i = 1, \ldots, k$, it follows that

$$\sum_{i=1}^{k} \bar{\lambda}_i \langle \nabla f_i(\bar{x}), x - \bar{x} \rangle < 0,$$

which is a contradiction to our assumption that $\bar{x}$ minimizes (5.5). Hence, $\bar{x}$ is an efficient solution of (MP).

Definition 5.1 *A feasible point $\bar{x}$ for the problem (MP) is said to satisfy the boundedness condition, if the set*

$$\left\{ \frac{p_i(\bar{x}, x)}{p_j(\bar{x}, x)} | x \in K, f_i(\bar{x}) > f_i(x), f_j(\bar{x}) > f_i(x), 1 \leq i, j \leq k \right\} \tag{5.7}$$

is bounded from above.

The following theorem is a variant of Proposition 3.5 of Chew and Choo [47].

Theorem 5.2 *Every efficient solution of (MP) involving pseudolinear functions, satisfying the boundedness condition is a properly efficient solution of (MP).*

Proof Let $\bar{x}$ be an efficient solution of (MP). Then from Corollary 5.1, there exist positive numbers $\bar{\lambda}_i > 0, i = 1, \ldots, k$, such that $\bar{x}$ minimizes the linear function (5.5). Hence, for any $x \in K$, we have

$$\sum_{i=1}^{k} \bar{\lambda}_i \langle \nabla f_i(\bar{x}), x - \bar{x} \rangle \geq 0. \tag{5.8}$$

Otherwise, we would arrive at a contradiction as in the first part of Theorem 5.1.

Since the set defined by (5.7) is bounded above, therefore, the following set

$$\left\{ (k-1)\frac{\bar{\lambda}_j p_i(\bar{x}, x)}{\bar{\lambda}_i p_j(\bar{x}, x)} \mid x \in K, f_i(\bar{x}) > f_i(x), f_j(\bar{x}) < f_j(x), 1 \leq i, j \leq k \right\} \quad (5.9)$$

is also bounded from above.

Let $M > 0$ be a real number that is an upper bound of the set defined by (5.9). Now, we shall show that $\bar{x}$ is a properly efficient solution of (MP). Assume that there exists r and $x \in K$, such that

$$f_r(x) < f_r(\bar{x}).$$

Then, by the pseudolinearity of f_r, it follows that

$$p_r(\bar{x}, x)\langle \nabla f_r(\bar{x}), x - \bar{x} \rangle < 0.$$

Since $p_r(\bar{x}, x) > 0$, we get

$$\langle \nabla f_r(\bar{x}), x - \bar{x} \rangle < 0. \quad (5.10)$$

Let us define

$$-\bar{\lambda}_s \langle \nabla f_s(\bar{x}), x - \bar{x} \rangle = \max \left\{ -\bar{\lambda}_i \langle \nabla f_i(\bar{x}), x - \bar{x} \rangle \mid \langle \nabla f_i(\bar{x}), x - \bar{x} \rangle > 0 \right\}. \quad (5.11)$$

Using (5.8), (5.10), and (5.11), we get

$$\bar{\lambda}_r \langle \nabla f_r(\bar{x}), x - \bar{x} \rangle \leq (k-1)(-\bar{\lambda}_s \langle \nabla f_s(\bar{x}), x - \bar{x} \rangle).$$

Therefore,

$$(f_r(x) - f_r(\bar{x})) \leq (k-1)\frac{\bar{\lambda}_s p_r(\bar{x}, x)}{\bar{\lambda}_r p_s(\bar{x}, x)}(f_s(\bar{x}) - f_s(x)).$$

Using the definition of M, we get

$$(f_r(x) - f_r(\bar{x})) \leq M(f_s(\bar{x}) - f_s(x)).$$

Hence, $\bar{x}$ is a properly efficient solution of (MP).

Remark 5.1 *We know that every linear fractional function* $f(x) = \frac{ax+b}{cx+d}, a, c, x \in \mathbb{R}^n$ *and* $b, d \in \mathbb{R}$ *is pseudolinear on* $D_f = \{x \in \mathbb{R}^n \mid cx+d > 0\}$ *with respect to the proportional function* $p(x, y) = \frac{cx+d}{cy+d}$, *which is continuous. When K is a compact set, the set (5.7) is bounded. Hence, Theorem 5.2 generalizes the results of Isermann [108] on linear functions and Choo [48] and Tigan [271] on linear fractional functions.*

5.4   Duality for (MP)

Now, in relation to (MP), we consider the following Mond-Weir dual problem:

$$(DP) \quad \max \ f(u) := (f_1(u), \dots, f_k(u))$$

$$\text{subject to} \ \ 0 = \sum_{i=1}^{k} \lambda_i \nabla f_i(u) + \nabla(y^T g(u)), \tag{5.12}$$

$$y^T g(u) \geq 0,$$

$$y \geq 0, \lambda_i > 0, i = 1, \dots, k.$$

Bector *et al.* [20] established the following weak, strong, and converse duality relations between (MP) and (DP).

Theorem 5.3 (Weak duality) *Let $f_i, i = 1, \dots, k$ be pseudolinear with respect to proportional function p_i. If x be a feasible solution for (MP) and (u, λ, y) be a feasible solution for (DP), such that $y^T g$ be pseudolinear with respect to q, then the following cannot hold*

$$f(x) \leq f(u).$$

Proof Suppose the above inequality holds. The pseudolinearity of $f_i, i = 1, \dots, k$ on K with respect to proportional function p_i imply that

$$p_i(x, u)\langle \nabla f_i(u), x - u \rangle \leq 0, \forall i = 1, \dots, k,$$

$$p_r(x, u)\langle \nabla f_r(u), x - u \rangle < 0, \text{ for some } r.$$

Since $p_i(x, u) > 0$, *for all* $i = 1, \dots, k$, it follows that

$$\langle \nabla f_i(u), x - u \rangle \leq 0, \forall i = 1, \dots, k,$$

$$\langle \nabla f_r(u), x - u \rangle < 0, \text{ for some } r.$$

Since $\lambda_i > 0$, *for each* $i = 1, \dots, k$, we get

$$\sum_{i=1}^{k} \lambda_i \langle \nabla f_i(u), x - u \rangle < 0. \tag{5.13}$$

By (5.12) and (5.13), we have

$$\langle \nabla(y^T g(u)), x - u \rangle > 0. \tag{5.14}$$

Since x is feasible for (MP) and (u, λ, y) is feasible for (DP), we have

$$y^T g(x) \leq 0$$

and
$$y^T g(u) \geq 0.$$
Using pseudolinearity of $y^T g$ with respect to q, it follows that
$$q(x, u)\langle \nabla(y^T g(u)), x - u \rangle \leq 0.$$
Again as $q(x, u) > 0$, it follows that
$$\langle \nabla(y^T g(u)), x - u \rangle \leq 0,$$
which is a contradiction to (5.14).

The nonnegative linear combination of pseudolinear functions with respect to different proportional functions may not be pseudolinear. In the following theorem, the conditions of pseudolinearity on the objective functions have been relaxed.

Theorem 5.4 (Weak duality) *Let x be a feasible solution for (NMP) and (u, λ, y) be a feasible solution for (DP), such that $\sum_{i=1}^{k} \lambda_i f_i$ be pseudolinear with respect to p and $y^T g$ be pseudolinear with respect to q. Then, the following cannot hold*
$$f(x) \leq f(u).$$

Proof Suppose the above inequality holds. Since $\lambda_i > 0, i = 1, \ldots, k$, it follows that
$$\sum_{i=1}^{k} \lambda_i f_i(x) < \sum_{i=1}^{k} \lambda_i f_i(u).$$

Using the pseudolinearity of $\sum_{i=1}^{k} \lambda_i f_i$ with respect to p, we get the inequality (5.13) of Theorem 5.3. The rest of the proof of the theorem follows along the lines of Theorem 5.3.

Theorem 5.5 *Let $\bar{x}$ be a feasible solution of (MP) and $(\bar{u}, \bar{\lambda}, \bar{y})$ be a feasible solution of (DP), such that*
$$f(\bar{x}) = f(\bar{u}). \tag{5.15}$$
If for all feasible solutions (u, λ, y) of (DP), $\sum_{i=1}^{k} \lambda_i f_i$ be pseudolinear with respect to p and $y^T g$ be pseudolinear with respect to q, then, $\bar{x}$ is properly efficient solution for (MP) and $(\bar{u}, \bar{\lambda}, \bar{y})$ is properly efficient solution for (DP).

Proof Suppose $\bar{x}$ is not an efficient solution for (DP), then there exists $x \in P$, such that
$$f(x) \leq f(\bar{x}).$$
Using the assumption $f(\bar{x}) = f(u)$, we get a contradiction to Theorem 5.4. Hence, $\bar{x}$ is an efficient solution for (MP). Similarly, it can be shown that $(\bar{u}, \bar{\lambda}, \bar{y})$ is an efficient solution for (DP).

Now suppose that $\bar{x}$ be an efficient solution for (MP). Therefore, for every scalar $M > 0$, there exists $x_0 \in P$ and an index i, such that

$$f_i(\bar{x}) - f_i(x_0) > M(f_j(x_0) - f_j(\bar{x})),$$

for all j satisfying

$$f_j(x_0) > f_j(\bar{x}),$$

whenever

$$f_i(x_0) < f_i(\bar{x}).$$

This means $f_i(\bar{x}) - f_i(x_0)$ can be made arbitrarily large and hence for $\bar{\lambda} > 0$, we get the following inequality

$$\sum_{i=1}^{k} \bar{\lambda}_i(f_i(\bar{x}) - f_i(x_0)) > 0. \tag{5.16}$$

Since, x_0 is feasible for (NMP) and $(\bar{u}, \bar{\lambda}, \bar{y})$ is feasible for (NDP), therefore, we have

$$g(x_0) \leq 0, \tag{5.17}$$

$$\sum_{i=1}^{k} \bar{\lambda}_i \nabla f_i(\bar{u}) + \nabla(\bar{y}^T g(\bar{u})) = 0, \tag{5.18}$$

$$\bar{y}^T g(\bar{u}) \geq 0, \tag{5.19}$$

$$\bar{y} \geq 0, \tag{5.20}$$

$$\bar{\lambda}_i > 0, \; i = 1, \ldots, k. \tag{5.21}$$

From (5.17), (5.19), and (5.20), we get

$$\bar{y}^T g(x_0) \leq \bar{y}^T g(\bar{u}).$$

Since $\bar{y}^T g$ is pseudolinear with respect to q, it follows that

$$q(x_0, \bar{u})\langle \nabla(\bar{y}^T g(u)), x_0 - \bar{u} \rangle \leq 0.$$

Since $q(x_0, \bar{u}) > 0$, we get

$$\langle \nabla(\bar{y}^T g(u)), x_0 - \bar{u} \rangle \leq 0.$$

From (5.18), we get

$$\left\langle \sum_{i=1}^{k} \bar{\lambda}_i \nabla f_i(\bar{u}), x_0 - \bar{u} \right\rangle \geq 0.$$

On using pseudoconvexity of $\sum_{i=1}^{k} \bar{\lambda}_i f_i$, we get

$$\sum_{i=1}^{k} \bar{\lambda}_i f_i(x_0) - \sum_{i=1}^{k} \bar{\lambda}_i f_i(\bar{u}) \geq 0. \tag{5.22}$$

From (5.15) and (5.22), we get

$$\sum_{i=1}^{k} \bar{\lambda}i(f_i(\bar{x}) - f_i(x_0)) \leq 0,$$

which is a contradiction to (5.16). Hence, $\bar{x}$ is a properly efficient solution of (MP).

We now assume that $(\bar{u}, \bar{\lambda}, \bar{y})$ is not a properly efficient solution for (DP). Therefore, for every scalar $M > 0$, there exists a feasible point (u_0, λ_0, y_0) for (DP) and an index i, such that

$$f_i(u_0) - f_i(\bar{u}) > M(f_j(\bar{u}) - f_j(u_0)),$$

for all j, satisfying

$$f_j(u_0) < f_j(\bar{u}),$$

whenever

$$f_i(u_0) > f_i(\bar{u}).$$

This means $f_i(u_0) - f_i(\bar{u})$ can be made arbitrarily large and hence for $\bar{\lambda} > 0$, we get the inequality

$$\sum_{i=1}^{k} \bar{\lambda}_i(f_i(u_0) - f_i(\bar{u})) > 0. \qquad (5.23)$$

Since $\bar{x}$ and $(u, \bar{\lambda}, \bar{y})$ are feasible for (MP) and (DP), respectively, proceeding as in the first part, we get

$$\sum_{i=1}^{k} \bar{\lambda}_i(f_i(u_0) - f_i(\bar{u})) \leq 0,$$

which contradicts (5.23). Hence, $(\bar{u}, \bar{\lambda}, \bar{y})$ is a properly efficient solution for (DP).

Theorem 5.6 *(Strong duality)* *Let $\bar{x}$ be an efficient solution for (MP) and let $f_i, i = 1, \ldots, k$ be pseudolinear with respect to the proportional function p_i and $g_j, j \in J(\bar{x})$ be pseudolinear with respect to q_j. Then there exist $\bar{\lambda} \in \mathbb{R}^k, \bar{y} \in \mathbb{R}^m$, such that $(\bar{x}, \bar{\lambda}, \bar{y})$ be a feasible solution for (DP). Further, if for all feasible (u, λ, y) of (DP), $\sum_{i=1}^{k} \lambda_i f_i$ be pseudolinear with respect to p and $y^T g$ be pseudolinear with respect to q, Then, $(\bar{x}, \bar{\lambda}, \bar{y})$ is a properly efficient solution for (DP).*

Proof The hypothesis of this theorem and Theorem 5.1 imply that there exist $\bar{\lambda} \in \mathbb{R}^k$ and $\bar{y}_j \in \mathbb{R}, j \in J(\bar{x})$, such that

$$\sum_{i=1}^{k} \bar{\lambda}_i \nabla f_i(\bar{x}) + \sum_{j \in J(\bar{x})} \bar{y}_j \nabla g_j(\bar{x}) = 0,$$

$$\bar{y}_j \geq 0, j \in J(\bar{x}),$$

$$\bar{\lambda}_i > 0, i = 1, \ldots, k.$$

Let $\bar{y}_j = 0$, for $j \notin J(\bar{x})$. Then there exist $\bar{\lambda} \in \mathbb{R}^k$ and $\bar{y} \in \mathbb{R}^m$, such that $(\bar{x}, \bar{\lambda}, \bar{y})$ is a feasible solution of (DP). The proper efficiency of $(\bar{x}, \bar{\lambda}, \bar{y})$ for the problem (DP) follows from Theorem 5.5.

Theorem 5.7 *(Converse duality) Let $(\bar{u}, \bar{\lambda}, \bar{y})$ be an efficient solution for (DP) and let for all feasible (u, λ, y) for (DP), $\sum_{i=1}^{k} \lambda_i f_i$ be pseudolinear with respect to proportional function p and $y^T h$ be pseudolinear with respect to q. Moreover, let the $n \times n$ Hessian matrix $\nabla^2(\bar{\lambda}^T f(\bar{u})) + \nabla^2(\bar{\lambda}^T g(\bar{u}))$ be positive or negative definite and $\{\nabla f_i(\bar{u}), i = 1, \ldots, k\}$ be linearly independent. Then, $\bar{u}$ is a properly efficient solution for (MP).*

Proof Since $(\bar{u}, \bar{\lambda}, \bar{y})$ is an efficient solution for (DP), by Fritz John necessary optimality conditions in Weir and Mond [284] there exist $\tau \in \mathbb{R}^k, \nu \in \mathbb{R}^n, p \in \mathbb{R}, s \in \mathbb{R}^m, w \in \mathbb{R}^k$, such that

$$\nabla \tau^T f(\bar{u}) + \nabla \nu^T [\nabla \bar{\lambda}^T f(\bar{u}) + \nabla \bar{y}^T g(\bar{u})] + p \nabla \bar{y}^T g(\bar{u}) = 0, \tag{5.24}$$

$$(\nabla g(\bar{u}))^T \nu + p g(\bar{u}) + s = 0, \tag{5.25}$$

$$(\nabla f(\bar{u}))^T \nu + w = 0, \tag{5.26}$$

$$p \bar{y}^T g(\bar{u}) = 0, \tag{5.27}$$

$$s^T \bar{y} = 0, \tag{5.28}$$

$$w^T \bar{\lambda} = 0, \tag{5.29}$$

$$(\tau, s, p, w) \geqq 0, \tag{5.30}$$

$$(\tau, s, p, w) \neq 0. \tag{5.31}$$

Since, $\bar{\lambda} > 0$, (5.29) implies that $w = 0$. Then, using (5.26), we get

$$\nu^T \nabla f(\bar{u}) = 0. \tag{5.32}$$

Multiplying (5.25) by $\bar{y}$ and using (5.27) and (5.28), we get

$$\nu^T \nabla \bar{y}^T g(\bar{u}) = 0. \tag{5.33}$$

Multiplying (5.24) by ν^T and using (5.32) and (5.33), it follows that

$$\nu^T [\nabla^2 \bar{\lambda}^T f(\bar{u}) + \nabla^2 \bar{\lambda}^T g(\bar{u})] \nu = 0.$$

Since, $\nabla^2(\bar{\lambda}^T f(\bar{u})) + \nabla^2(\bar{\lambda}^T g(\bar{u}))$ is assumed to be positive or negative definite, we get $\nu = 0$. Since, $\nu = 0$, by (5.24) and (5.12), we get

$$\nabla(\tau - p\bar{\lambda})^T f(\bar{u}) = 0.$$

By linear independence of $\{\nabla f_i(\bar{u}), i = 1, \ldots, k\}$, it follows that

$$\tau = p\bar{\lambda}.$$

Since, $\bar{\lambda} > 0, \tau = 0$ implies $p = 0$ and then by (5.25), $s = 0$, giving $(\tau, s, p, w) = 0$, a contradiction to (5.31). Thus $\tau \neq 0$ and $p > 0$. Since $\nu = 0, p > 0$ and $s \geq 0$, (5.25) gives $g(\bar{u}) \leq 0$ and (5.27) gives $\bar{y}^T g(\bar{u}) = 0$. Thus $\bar{u}$ is feasible for (MP). That $\bar{u}$ is an efficient solution for (MP) follows from the assumptions of the theorem and weak duality Theorem 5.4. The proper efficiency of $\bar{u}$ for (MP) follows from Theorem 5.5.

5.5 Applications

Bector *et al.* [20] applied the duality theorems for (DP) to the following multiobjective fractional programming problem:

$$(FP) \quad \min \left(\frac{c_1^T x + \alpha_1}{d^T x + \beta}, \ldots, \frac{c_k^T x + \alpha_k}{d^T x + \beta} \right)$$

$$\text{subject to} \quad Ax \leq b, x \geq 0, \tag{5.34}$$

$$x \in K. \tag{5.35}$$

Here $c_i, d \in \mathbb{R}^n, \alpha_i, \beta \in \mathbb{R}(i = 1, \ldots, k), b \in \mathbb{R}^m$ and $A \in \mathbb{R}^{m \times n}$. Let $K \subseteq \mathbb{R}^n$ be an open convex set and that $d^T x + \beta > 0$, for all $x \in K$. Let the matrix with the columns $c_i(i = 1, \ldots, k)$ be denoted by C. The dual model related with (FP) is given by

$$(FP) \quad \max \left(\frac{c_1^T u + \alpha_1}{d^T u + \beta}, \ldots, \frac{c_k^T u + \alpha_k}{d^T u + \beta} \right)$$

$$\text{subject to} \quad \sum_{i=1}^{k} \lambda_i \left[\nabla \left(\frac{c_1^T x + \alpha_1}{d^T x + \beta} \right) \right] + \sum_{j=1}^{m} \mu_j \nabla \left(b_j - \sum_{l=1}^{n} a_{jl} u_l \right)$$

$$-\nabla \left(\sum_{l=1}^{m} w_l u_i \right) = 0, \tag{5.36}$$

$$\sum_{j=1}^{m} \mu_j \left(b_j - \sum_{i=1}^{n} a_{jl} u_l \right) - \sum_{l=1}^{n} w_l u_l \geq 0, \tag{5.37}$$

$$\lambda > 0, \mu \geq 0, w \geq 0, u \in K. \tag{5.38}$$

Since

$$\nabla\left(\frac{c_i^T x + \alpha_i}{d^T x + \beta}\right) = \frac{1}{d^T x + \beta}\left[c_i - \left(\frac{c_i^T u + \alpha_i}{d^T u + \beta}\right)d\right].$$

Hence, if we take $\sigma_i = \left(\frac{\lambda_i}{d^T u + \beta}\right) > 0$, the model (FD) can be reformulated as

$$(FD) \quad \max \quad \left(\frac{c_1^T u + \alpha_1}{d^T u + \beta}, \dots, \frac{c_k^T u + \alpha_k}{d^T u + \beta}\right)$$

subject to $\displaystyle\sum_{i=1}^{k} \sigma_i\left[c_i - \left(\frac{c_i^T + \alpha_i}{d^T u + \beta}\right)d\right] - \sum_{j=1}^{m} \mu_j a_{jl} - w_l = 0,$ (5.39)

$$(b - Au) - w^T u \geq 0, \tag{5.40}$$

$$\lambda > 0, \mu \geq 0, w \geq 0, u \in K. \tag{5.41}$$

Remark 5.2 *In case the feasible region of (FP) is compact, it is known from Choo [48] that every efficient solution of (FP) is also properly efficient in the sense of Geoffrion [86].*

(Linear vector optimization problem) Let $\alpha_i = 0(i = 1, \dots, k), d = 0, \beta = 1$ and $K = \mathbb{R}^n$. Then (FP) becomes the following linear vector optimization problem:

$$(LP) \quad \min \ (c_1^T x, \dots, c_k^T x)$$

$$\text{subject to } Ax \geq b, \tag{5.42}$$

$$x \geq 0. \tag{5.43}$$

In this case, (FD) reduces to

$$\max \ (c_1^T u, \dots, c_p^T u)$$

$$\text{subject to} \quad \sigma^T C - \mu^T A - w^T = 0, \tag{5.44}$$

$$\mu^T (b - Au) - w^T u \geq 0, \tag{5.45}$$

$$\sigma > 0, \mu \geq 0, w \geq 0. \tag{5.46}$$

Multiplying (5.44) by u in the right and using (5.45), we get

$$\sigma^T C u = \mu^T A u + w^T u \leq \mu^T b. \tag{5.47}$$

In view of (5.47), the dual of (LP) reduces to

$$(LVD) \quad \max \ (c_1^T u, \dots, c_p^T u)$$

$$\text{subject to} \quad \sigma^T C \geq \mu^T A,$$

$$\sigma^T C u \geq \mu^T b,$$

$$\sigma > 0, \mu \geqq 0.$$

Taking $Cu = z \in \mathbb{R}^p$, (LVD) can be rewritten as

$$\max \ z$$

$$\text{subject to} \quad \sigma^T C \geqq \mu^T A,$$

$$\sigma^T z \geqq \mu^T b,$$

$$\sigma > 0, \mu \geqq 0.$$

Remark 5.3 *We note that the (LVD) is the same dual as studied by Schonfeld [248] and is equivalent to the one obtained by Isermann [108] for $b \neq 0$ as shown by Ivanov and Nehse [110].*

5.6 Pseudolinear Multiobjective Fractional Programming

Now, we consider a multiobjective fractional programming problem involving pseudolinear functions and establish the necessary and sufficient optimality conditions for a feasible solution to be an efficient solution for the problem. Furthermore, we establish the equivalence between efficiency and proper efficiency under certain boundedness conditions. We present the weak, strong, and converse duality results for the corresponding Mond-Weir subgradient type dual model derived by Kaul *et al.* [139].

The following lemma, due to Kaul *et al.* [139] establishes that the ratio of two pseudolinear functions is also a pseudolinear function, with respect to a different proportional function.

Lemma 5.1 *Suppose $f : K \to \mathbb{R}$ and $g : K \to \mathbb{R}$ be two pseudolinear functions defined on K with the same proportional function p. Let $f(x) \geq 0$ and $g(x) > 0$ for every $x \in K$. If f and $-g$ are regular on K, then $\frac{f}{g}$ is also pseudolinear with respect to a new proportional function*

$$\bar{p}(x, y) = \frac{g(x)p(x, y)}{g(y)}.$$

Proof Since f and g are pseudolinear with respect to the same proportional function p, it follows that for any x and y in K, we have

$$f(y) = f(x) + p(x, y)\langle \nabla f(x, y - x)\rangle, \tag{5.48}$$

$$g(y) = g(x) + p(x, y)\langle \nabla g(x, y - x)\rangle. \tag{5.49}$$

Now, from (5.48) and (5.49), we have

$$\frac{g(y)}{g(x)}\left(\frac{f(y)}{g(y)} - \frac{f(x)}{g(x)}\right) = \frac{1}{g(y)}(f(y) - f(x)) - \frac{f(x)}{g(x)g(y)}(g(y) - g(x))$$

$$= p(x,y)\left(\frac{g(x)\langle\nabla f(x), y - x\rangle - f(x)\langle\nabla g(x), y - x\rangle}{(g(x))^2}\right)$$

$$= p(x,y)\left\langle\frac{g(x)\nabla f(x) - f(x)\nabla g(x)}{(g(x))^2}, y - x\right\rangle$$

$$= p(x,y)\left\langle\nabla\left(\frac{f(x)}{g(x)}\right), y - x\right\rangle.$$

$$\frac{f(y)}{g(y)} = \frac{f(x)}{g(x)} + \frac{g(x)p(x,y)}{g(y)}\left\langle\nabla\left(\frac{f(x)}{g(x)}\right), y - x\right\rangle,$$

which implies that $\frac{f}{g}$ is pseudolinear with respect to a new proportional function

$$\bar{p}(x,y) = \frac{g(x)p(x,y)}{g(y)}.$$

We consider the following multiobjective fractional programming problem:

$$(MFP) \quad \min\left(\frac{f_1(x)}{g_1(x)}, \frac{f_2(x)}{g_2(x)}, \ldots, \frac{f_k(x)}{g_k(x)}\right)$$

$$\text{subject to } h_j(x) \leq 0, j \in \{1, \ldots, m\},$$

where $f_i : K \to \mathbb{R}, g_i : K \to \mathbb{R}, i \in I := \{1, \ldots, k\}$ and $h_j : K \to \mathbb{R}, j \in J := \{1, \ldots, m\}$ are differentiable functions on a nonempty open convex subset $K \subseteq \mathbb{R}^n$. From now on, we assume that the functions $f(x) \geq 0, g(x) > 0$ for all $x \in \mathbb{R}^n$. Let $P = \{x \in K \mid h_j(x) \leq 0, j \in J\}$ and $J(x) = \{j \in J \mid h_j(x) = 0\}$ for some $x \in K$ denote the set of feasible solutions for (MFP) and active constraints index set at $x \in K$, respectively.

Definition 5.2 (Kaul et al. [139]) *A feasible solution $\bar{x}$ is said to be an efficient (Pareto efficient) solution for (MFP), if there exists no $x \in P$, such that*

$$\frac{f_i(x)}{g_i(x)} \leq \frac{f_i(\bar{x})}{g_i(\bar{x})}, \forall i = 1, \ldots, k$$

and

$$\frac{f_r(x)}{g_r(x)} < \frac{f_r(\bar{x})}{g_r(\bar{x})}, \text{ for some } r.$$

Definition 5.3 (Kaul et al. [139]) *A feasible solution $\bar{x}$ is said to be a properly efficient (properly Pareto efficient) solution for (MFP) if it is efficient and if there exists a scalar $M > 0$, such that for each i,*

$$\frac{f_i(x)}{g_i(x)} - \frac{f_i(\bar{x})}{g_i(\bar{x})} \leq M\left(\frac{f_r(\bar{x})}{g_r(\bar{x})} - \frac{f_r(x)}{g_r(x)}\right),$$

for some r, such that

$$\frac{f_r(x)}{g_r(x)} > \frac{f_r(\bar{x})}{g_r(\bar{x})}, \text{ whenever } x \in P \text{ with } \frac{f_i(x)}{g_i(x)} < \frac{f_i(\bar{x})}{g_i(\bar{x})}.$$

5.7 Optimality Conditions for (MFP)

In this section, we establish the necessary and sufficient conditions for a feasible solution to be an efficient solution for (MFP) and show the equivalence between efficiency and proper efficiency under certain boundedness conditions.

Now, we shall prove the following necessary and sufficient optimality theorem analogous to Proposition 3.2 of Chew and Choo [47] for (MFP).

Theorem 5.8 *Let $\bar{x}$ be a feasible solution for (MFP). Let the functions f_i and $g_i, i = 1, \ldots, k$ be pseudolinear on K with respect to the same proportional function p_i and $h_j, j \in J(\bar{x})$ be pseudolinear on K with respect to proportional function q_j. Then $\bar{x}$ is an efficient solution of (MFP), if and only if there exist $\lambda_i > 0, i = 1, \ldots, k; \mu_j \geq 0, j \in J(\bar{x})$; such that*

$$\sum_{i=1}^{k} \lambda_i \nabla \left(\frac{f_i(\bar{x})}{g_i(\bar{x})} \right) + \sum_{j \in J(\bar{x})} \mu_j h_j(\bar{x}) = 0. \tag{5.50}$$

Proof Suppose λ_i and μ_j exist and satisfy the given conditions and (5.50). Let $\bar{x}$ is not an efficient solution for (MFP). Therefore, there exists a point $y \in P$, such that

$$\frac{f_i(y)}{g_i(y)} \leq \frac{f_i(\bar{x})}{g_i(\bar{x})}, \forall i = 1, \ldots, k$$

and

$$\frac{f_r(y)}{g_r(y)} < \frac{f_r(\bar{x})}{g_r(\bar{x})}, \text{ for some } r.$$

Using pseudolinearity of $\frac{f_i}{g_i}, i = 1, \ldots, k$, we get

$$\bar{p}_i(\bar{x}, y) \left\langle \nabla \left(\frac{f_i(\bar{x})}{g_i(\bar{x})} \right), y - \bar{x} \right\rangle \leq 0, \forall i = 1, \ldots, k$$

and

$$\bar{p}_r(\bar{x}, y) \left\langle \nabla \left(\frac{f_r(\bar{x})}{g_r(\bar{x})} \right), y - \bar{x} \right\rangle < 0, \text{ for some } r,$$

where $\bar{p}_i(\bar{x}, y) = \frac{g_i(\bar{x}) p_i(\bar{x}, y)}{g_i(y)}$.

Since, $\bar{p}_i(\bar{x}, y) > 0, \forall i = 1, \ldots, k$, therefore, we have

$$\left\langle \nabla \left(\frac{f_i(\bar{x})}{g_i(\bar{x})} \right), y - \bar{x} \right\rangle \leq 0, \forall i = 1, \ldots, k$$

and

$$\left\langle \nabla \left(\frac{f_r(\bar{x})}{g_r(\bar{x})} \right), y - \bar{x} \right\rangle < 0, \text{ for some } r.$$

Since, $\lambda_i > 0$, for all $i = 1, \ldots, k$, we get

$$\sum_{i=1}^{k} \lambda_i \left\langle \nabla \left(\frac{f_i(\bar{x})}{g_i(\bar{x})} \right), y - \bar{x} \right\rangle < 0. \qquad (5.51)$$

Again, we have

$$h_j(y) \leq h_j(\bar{x}), \forall j \in J(\bar{x}).$$

Using pseudolinearity of $h_j, j = 1, \ldots, m$, it follows that

$$q_j(\bar{x}, y) \langle \nabla h_j(\bar{x}), y - \bar{x} \rangle \leq 0, \forall j \in J(\bar{x}).$$

Since, $q_j(\bar{x}, y) > 0$, for all $j \in J(\bar{x})$, therefore, we get

$$\langle \nabla h_j(\bar{x}), y - \bar{x} \rangle \leq 0, \forall j \in J(\bar{x}).$$

Since, $\mu_j \geq 0$, for all $j \in J(\bar{x})$, we get

$$\sum_{j \in J(\bar{x})} \mu_j \langle \nabla h_j(\bar{x}), y - \bar{x} \rangle \leq 0.$$

From (5.50), we get

$$\sum_{i=1}^{k} \lambda_i \left\langle \nabla \left(\frac{f_i(\bar{x})}{g_i(\bar{x})} \right), y - \bar{x} \right\rangle \geq 0,$$

which is a contradiction to (5.51).

Conversely, we assume that $\bar{x}$ is an efficient solution for (MFP). Therefore, for $1 \leq r \leq k$, the following system of inequalities

$$\left. \begin{array}{l} \langle \nabla h_j(\bar{x}), x - \bar{x} \rangle \leq 0, \forall j \in J(\bar{x}), \\ \left\langle \nabla \left(\frac{f_i(\bar{x})}{g_i(\bar{x})} \right), x - \bar{x} \right\rangle \leq 0, i = 1, 2, \ldots, r - 1, r + 1, \ldots, k, \\ \left\langle \nabla \left(\frac{f_r(\bar{x})}{g_r(\bar{x})} \right), x - \bar{x} \right\rangle < 0, \end{array} \right\} \qquad (5.52)$$

has no solution $x \in K$. For if x is a solution of the system and $y = \bar{x} + t(x - \bar{x}), 0 \leq t \leq 1$, then for $j \in J(\bar{x})$, using (5.52), we have

$$\langle \nabla h_j(\bar{x}), y - \bar{x} \rangle = t \langle \nabla h_j(\bar{x}), x - \bar{x} \rangle \leq 0.$$

Therefore, by pseudoconvexity of $h_j, j \in J(\bar{x})$, we get

$$h_j(y) \leq h_j(\bar{x}) = 0.$$

If $j \notin J(\bar{x})$, then $h_j(\bar{x}) < 0$ and so $h_j(y) < 0$, when t is sufficiently small. Therefore, for small t, y is a point of P. Using Lemma 5.1 and (5.52), we get

$$\frac{f_i(y)}{g_i(y)} - \frac{f_i(\bar{x})}{g_i(\bar{x})} = t\bar{p}_i(\bar{x}, y) \left\langle \nabla \left(\frac{f_i(\bar{x})}{g_i(\bar{x})} \right), x - \bar{x} \right\rangle \leq 0, \ i \neq r$$

and also,

$$\frac{f_r(y)}{g_r(y)} - \frac{f_r(\bar{x})}{g_r(\bar{x})} < 0, \text{ for some } r.$$

This contradicts the choice of $\bar{x}$. Hence system (5.52) has no solution in the nonempty convex set K. Applying Farkas's theorem (Theorem 1.24), there exist $\lambda_{r_i} \geq 0, \mu_{r_j} \geq 0$, such that

$$-\nabla \left(\frac{f_r(\bar{x})}{g_r(\bar{x})} \right) = \sum_{i \neq r} \lambda_{r_i} \nabla \left(\frac{f_i(\bar{x})}{g_i(\bar{x})} \right) + \sum_{j \in J(x^*)} \mu_{r_j} \nabla h_j(\bar{x}). \tag{5.53}$$

Summing (5.53) over r, we get (5.50) with

$$\lambda_i = 1 + \sum_{r \neq i} \lambda_{r_i}, \ \mu_j = \sum_{r=1}^{k} \mu_{r_j}.$$

This completes the proof.

Corollary 5.2 *Let $\bar{x}$ be a feasible solution of the problem (MFP). Then the following statements are equivalent:*

(i) $\bar{x}$ is an efficient solution of the problem (MFP).

(ii) $\bar{x}$ is an efficient solution of the problem

$$(MLFP) \quad \min \left(\nabla \left(\frac{f_1(\bar{x})}{g_1(\bar{x})} \right) x, \ldots, \nabla \left(\frac{f_k(\bar{x})}{g_k(\bar{x})} \right) x \right)$$

$$\text{subject to} \quad x \in K.$$

(iii) There exists positive numbers $\lambda_1, \ldots, \lambda_k$ such that $\bar{x}$ minimizes the linear function

$$\sum_{i=1}^{k} \lambda_i \nabla \left(\frac{f_i(\bar{x})}{g_i(\bar{x})} \right) x, x \in K. \tag{5.54}$$

Proof (i) $\Rightarrow$ (ii) Suppose $\bar{x}$ is an efficient solution of the problem (MFP), then by Theorem 5.8, there exist $\lambda_i > 0, i = 1, \ldots, k, \mu_j \in \mathbb{R}, j \in J(\bar{x})$, such that

$$\sum_{i=1}^{k} \lambda_i \left(\nabla \frac{f_i(\bar{x})}{g_i(\bar{x})} \right) + \sum_{j \in J(\bar{x})} \mu_j \nabla h_j(\bar{x}) = 0. \tag{5.55}$$

On the contrary, suppose that $\bar{x}$ is not an efficient solution of the (MLFP) problem. Then there exists some $x \in K$, such that

$$\nabla \left(\frac{f_i(\bar{x})}{g_i(\bar{x})} \right) x \leq \nabla \left(\frac{f_i(\bar{x})}{g_i(\bar{x})} \right) \bar{x}, \forall i = 1, \ldots, k,$$

$$\nabla \left(\frac{f_r(\bar{x})}{g_r(\bar{x})} \right) x \leq \nabla \left(\frac{f_r(\bar{x})}{g_r(\bar{x})} \right) \bar{x}, \text{ for some } r.$$

Now, we have

$$0 > \sum_{i=1}^{k} \lambda_i \left\langle \nabla \left(\frac{f_i(\bar{x})}{g_i(\bar{x})} \right) \bar{x}, x - \bar{x} \right\rangle = - \sum_{j \in J(\bar{x})} \mu_j \left\langle \nabla h_j(\bar{x}), x - \bar{x} \right\rangle$$

$$= - \left(\sum_{j \in J(\bar{x})} \mu_j \left(\frac{h_j(x) - h_j(\bar{x})}{q_j(\bar{x}, x)} \right) \right) \geq 0,$$

which is obviously a contradiction. Hence, $\bar{x}$ must be an efficient solution of the problem (MLFP).

(ii) $\Rightarrow$ (iii) Suppose $\bar{x}$ be an efficient solution of the problem (MLFP). Then by Theorem 5.8, there exist $\lambda_i \in \mathbb{R}, i = 1, \ldots, k$ and $\mu_j \in \mathbb{R}, j \in J(\bar{x})$, such that (5.55) holds. Therefore, for any $x \in K$, we get

$$\sum_{i=1}^{k} \lambda_i \left\langle \nabla \left(\frac{f_i(\bar{x})}{g_i(\bar{x})} \right), x - \bar{x} \right\rangle = - \sum_{j \in J(\bar{x})} \mu_j \left\langle \nabla h_j(\bar{x}), x - \bar{x} \right\rangle$$

$$= - \left(\sum_{j \in J(\bar{x})} \mu_j \left(\frac{h_j(x) - h_j(\bar{x})}{q_j(\bar{x}, x)} \right) \right) \geq 0.$$

This shows that $\bar{x}$ minimizes (5.54).

(iii) $\Rightarrow$ (i) Suppose $\bar{x}$ minimizes (5.54) for some positive number $\lambda_1, \ldots, \lambda_k$. If $\bar{x}$ be not an efficient solution for (MFP), then there exists some $x \in K$, such that

$$\frac{f_i(x)}{g_i(\bar{x})} \leq \frac{f_i(\bar{x})}{g_i(\bar{x})}, \forall i = 1, \ldots, k,$$

$$\frac{f_r(x)}{g_r(\bar{x})} \leq \frac{f_r(\bar{x})}{g_r(\bar{x})}, \text{ for some } r.$$

By the pseudolinearity of $\frac{f_i}{g_i}, i = 1, \ldots, k$ with respect to proportional function $\bar{p}_i$, it follows that

$$\bar{p}_i(\bar{x}, x) \left\langle \nabla \left(\frac{f_i(\bar{x})}{g_i(\bar{x})} \right), x - \bar{x} \right\rangle \leq 0, \; \forall i = 1, \ldots, k,$$

$$\bar{p}_r(\bar{x}, x) \left\langle \nabla \left(\frac{f_r(\bar{x})}{g_r(\bar{x})} \right), x - \bar{x} \right\rangle < 0, \text{ for some } r,$$

where $\bar{p}_i(\bar{x}, x) = \frac{g_i(\bar{x})p(\bar{x},x)}{g_i(x)}$.

Since, $\bar{p}_i(\bar{x}, x) > 0, i = 1, \ldots, k$, it follows that

$$\left\langle \nabla \left(\frac{f_i(\bar{x})}{g_i(\bar{x})} \right), x - \bar{x} \right\rangle \leq 0, \; \forall i = 1, \ldots, k,$$

$$\left\langle \nabla \left(\frac{f_r(\bar{x})}{g_r(\bar{x})} \right), x - \bar{x} \right\rangle < 0, \text{ for some } r,$$

Since, $\lambda_i > 0, i = 1, \ldots, k$, it follows that

$$\sum_{i=1}^{k} \lambda_i \left\langle \nabla \left(\frac{f_i(\bar{x})}{g_i(\bar{x})} \right), x - \bar{x} \right\rangle < 0.$$

This is a contradiction to our assumption that $\bar{x}$ minimizes (5.54). Hence, $\bar{x}$ is an efficient solution of (MFP).

Definition 5.4 *A feasible point $\bar{x}$ for the problem (MFP) is said to satisfy the boundedness condition, if the set*

$$\left\{ \frac{\bar{p}_i(\bar{x}, x)}{\bar{p}_j(\bar{x}, x)} \; | \; x \in K, \frac{f_i(\bar{x})}{g_i(\bar{x})} > \frac{f_i(x)}{g_i(x)}, \frac{f_j(\bar{x})}{g_j(\bar{x})} < \frac{f_i(x)}{g_i(x)} \right\} \tag{5.56}$$

is bounded from above, where $\bar{p}_i(\bar{x}, x) = \frac{g_i(\bar{x})p(\bar{x},x)}{g_i(x)}$.

The following theorem extends Proposition 3.5 of Chew and Choo [47] to the multiobjective fractional programming problem (MFP).

Theorem 5.9 *Every efficient solution of (MFP) involving pseudolinear functions, satisfying the boundedness condition is a properly efficient solution of the problem (MFP).*

Proof Let $\bar{x}$ be an efficient solution of (MFP). Then, from Corollary 5.2, it follows that there exist $\lambda_i > 0, i = 1, \ldots, k$, such that $\bar{x}$ minimizes the linear function (5.54). Hence, we must have

$$\sum_{i=1}^{m} \lambda_i \left\langle \nabla \left(\frac{f_i(\bar{x})}{g_i(\bar{x})} \right), x - \bar{x} \right\rangle \geq 0 \tag{5.57}$$

Otherwise, we would arrive at a contradiction as in the first part of Theorem 5.8.

Since the set defined by (5.56) is bounded above, therefore, the following set

$$\left\{ (k-1)\frac{\lambda_j \bar{p}_i(\bar{x}, x)}{\lambda_i \bar{p}_j(\bar{x}, x)} \mid x \in K, \frac{f_i(\bar{x})}{g_i(\bar{x})} > \frac{f_i(x)}{g_i(x)}, \frac{f_j(\bar{x})}{g_j(\bar{x})} < \frac{f_i(x)}{g_i(x)}, 1 \le i, j \le k \right\}$$
$$(5.58)$$

is also bounded from above.

Let $M > 0$ be a real number that is an upper bound of the set defined by (5.58). Now, we shall show that $\bar{x}$ is a properly efficient solution of (MFP). Assume that there exist r and $x \in K$, such that

$$\frac{f_r(x)}{g_r(x)} < \frac{f_r(\bar{x})}{g_r(\bar{x})}.$$

Then, by the pseudolinearity of $\frac{f_r}{g_r}$, we get

$$p_r(\bar{x}, x)\left\langle \nabla \left(\frac{f_r(\bar{x})}{g_r(\bar{x})} \right), x - \bar{x} \right\rangle < 0.$$

Since $p_r(\bar{x}, x) > 0$, we get

$$\left\langle \nabla \left(\frac{f_r(\bar{x})}{g_r(\bar{x})} \right), x - \bar{x} \right\rangle < 0. \qquad (5.59)$$

Let us define

$$-\lambda_r \left\langle \nabla \left(\frac{f_s(\bar{x})}{g_s(\bar{x})} \right), x - \bar{x} \right\rangle =$$

$$\max \left\{ \lambda_i \left\langle \nabla \left(\frac{f_i(\bar{x})}{g_i(\bar{x})} \right), x - \bar{x} \right\rangle : \left\langle \nabla \left(\frac{f_i(\bar{x})}{g_i(\bar{x})} \right), x - \bar{x} \right\rangle > 0 \right\}. \qquad (5.60)$$

Using (5.57), (5.59), and (5.60), we get

$$\lambda_r \left\langle \nabla \left(\frac{f_r(\bar{x})}{g_r(\bar{x})} \right), x - \bar{x} \right\rangle \le (k-1)\left(-\lambda_s \left\langle \nabla \left(\frac{f_s(\bar{x})}{g_s(\bar{x})} \right), x - \bar{x} \right\rangle \right).$$

Therefore,

$$\left(\frac{f_r(x)}{g_r(x)} - \frac{f_r(\bar{x})}{g_r(\bar{x})} \right) \le (k-1)\frac{\lambda_s \bar{p}_r(\bar{x}, x)}{\lambda_r \bar{p}_s(\bar{x}, x)}\left(\frac{f_s(\bar{x})}{g_s(\bar{x})} - \frac{f_s(x)}{g_s(x)} \right).$$

Using the definition of M, we get

$$\left(\frac{f_r(x)}{g_r(x)} - \frac{f_r(\bar{x})}{g_r(\bar{x})} \right) \le M \left(\frac{f_s(\bar{x})}{g_s(\bar{x})} - \frac{f_s(x)}{g_s(x)} \right).$$

Hence, $\bar{x}$ is a properly efficient solution of (MFP).

5.8   Duality for (MFP)

For the multiobjective fractional programming problem (MFP), Kaul *et al.* [139] considered the following Mond-Weir type dual problem

$$(MFD) \quad \max \left(\frac{f_1(u)}{g_1(u)}, \frac{f_2(u)}{g_2(u)}, \ldots, \frac{f_k(u)}{g_k(u)} \right)$$

$$\text{subject to} \quad \sum_{i=1}^{k} \lambda_i \nabla \left(\frac{f_i(u)}{g_i(u)} \right) + \nabla(\mu^T h(u)) = 0 \qquad (5.61)$$

$$\mu^T h(u) \geq 0,$$

$$\mu \geq 0,$$

$$\lambda_i > 0, i = 1, 2, \ldots, k.$$

Theorem 5.10 *(**Weak duality**) Let y be a feasible solution for (MFP) and (u, λ, μ) be a feasible solution for (MFD), such that f_i and $g_i, i = 1, 2, \ldots, k$ are pseudolinear functions with respect to the same proportional function p_i and $\mu^T h$ is pseudolinear with respect to q, then the following cannot hold:*

$$\frac{f(y)}{g(y)} \leq \frac{f(u)}{g(u)}.$$

Proof We assume that the above inequality is satisfied. Using the pseudolinearity of $\frac{f_i}{g_i}$ on K with respect to proportional function $\bar{p}_i$ we get

$$\bar{p}_i(u, y) \left\langle \nabla \left(\frac{f_i(u)}{g_i(u)} \right), y - u \right\rangle \leq 0, \ \forall i = 1, \ldots, k$$

and

$$\bar{p}_r(u, y) \left\langle \nabla \left(\frac{f_r(u)}{g_r(u)} \right), y - u \right\rangle < 0, \text{ for some } r,$$

where $\bar{p}_i(u, y) = \frac{g_i(u)p_i(u,y)}{g_i(y)}$.

Since $\bar{p}_i(u, y) > 0$, for all $i = 1, \ldots, k$, we get

$$\left\langle \nabla \left(\frac{f_i(u)}{g_i(u)} \right), y - u \right\rangle \leq 0, \ \forall i = 1, \ldots$$

and

$$\left\langle \nabla \left(\frac{f_r(u)}{g_r(u)} \right), y - u \right\rangle < 0, \text{ for some } r.$$

Since $\lambda_i > 0$, for all $i = 1, \ldots, k$ we get

$$\sum_{i=1}^{k} \lambda_i \left\langle \nabla \left(\frac{f_i(u)}{g_i(u)} \right), y - u \right\rangle < 0. \tag{5.62}$$

By (5.61) and (5.62), it follows that

$$\langle \nabla(\mu^T g(u)), x - u \rangle > 0. \tag{5.63}$$

As y is feasible for (MFP) and (u, λ, y) is feasible for (MFD), it follows that

$$\mu^T h(y) \leq 0$$

and

$$\mu^T h(u) \geq 0.$$

Using pseudolinearity of $\mu^T h$ with respect to q, we get

$$q(u, y) \langle \nabla(\mu^T g(u)), y - u \rangle \leq 0.$$

Again as $q(u, y) > 0$, it follows that

$$\langle \nabla(y^T g(u)), y - u \rangle \leq 0,$$

which is a contradiction to (5.63).

In the following theorem, we have relaxed the conditions of pseudolinearity on the objective function.

Theorem 5.11 *(Weak duality) Let y be a feasible solution for the problem (MFP) and (u, λ, μ) be a feasible solution for the problem (MFD) involving pseudolinear functions, such that $\sum_{i=1}^{k} \lambda_i \frac{f_i}{g_i}$ is pseudolinear with respect to p and $\mu^T h$ is pseudolinear with respect to the proportional function q, then, the following inequality cannot hold*

$$\frac{f(y)}{g(y)} < \frac{f(u)}{g(u)}.$$

Proof Assume that the above inequality is satisfied. Since $\lambda_i > 0$, for each $i = 1, 2, \ldots, k$, we get

$$\sum_{i=1}^{k} \lambda_i \frac{f_i(y)}{g_i(y)} < \sum_{i=1}^{k} \lambda_i \frac{f_i(u)}{g_i(u)}$$

Using the pseudolinearity of $\sum_{i=1}^{k} \lambda_i \frac{f_i}{g_i}$ inequality (5.62) of Theorem 5.10 is obtained. The rest of the proof is along the lines of the proof of Theorem 5.10.

Theorem 5.12 *Let us assume that $\bar{x}$ is a feasible solution for (MFP) and $(\bar{u}, \bar{\lambda}, \bar{\mu})$ is a feasible solution for (MFD), such that*

$$\frac{f_i(\bar{x})}{g_i(\bar{x})} = \frac{f_i(\bar{u})}{g_i(\bar{u})}, \ \forall i = 1, \ldots, k \tag{5.64}$$

If for all feasible solutions $(\bar{u}, \bar{\lambda}, \bar{\mu})$ of (MFD), $\sum_{i=1}^{k} \lambda_i \frac{f_i}{g_i}$ is pseudolinear with respect to p and $\mu^T h$ is pseudolinear with respect to q, then, $\bar{x}$ is a properly efficient solution for (MFP) and $(\bar{u}, \bar{\lambda}, \bar{\mu})$ is a properly efficient solution for (MFD).

Proof Let us assume that $\bar{x}$ is not an efficient solution of (MFP), then there exists some $y \in K$, such that

$$\frac{f_i(y)}{g_i(y)} \leq \frac{f_i(\bar{u})}{g_i(\bar{u})}, \ i = 1, \ldots, k$$

$$\frac{f_r(y)}{g_r(y)} \leq \frac{f_r(\bar{u})}{g_r(\bar{u})}, \ \text{for some } r.$$

Now, by the given assumption $\frac{f_i(\bar{x})}{g_i(\bar{x})} = \frac{f_i(\bar{u})}{g_i(\bar{u})}$, for all $i = 1, \ldots, k$. Then, we arrive at a contradiction to Theorem 5.11. Hence $\bar{x}$ is an efficient solution for (MFP). Proceeding in the same way we can prove that $(\bar{u}, \bar{\lambda}, \bar{\mu})$ is an efficient solution for (MFD).

Now assume that $\bar{x}$ is an efficient solution for (MFP). Therefore, for every scalar $M > 0$, there exists some $x^* \in P$ and an index i, such that

$$\frac{f_i(\bar{x})}{g_i(\bar{x})} - \frac{f_i(x^*)}{g_i(x^*)} > M \left(\frac{f_r(x^*)}{g_r(x^*)} - \frac{f_r(\bar{x})}{g_r(\bar{x})} \right),$$

for all r satisfying

$$\frac{f_r(x^*)}{g_r(x^*)} > \frac{f_r(\bar{x})}{g_r(\bar{x})},$$

whenever

$$\frac{f_i(x^*)}{g_i(x^*)} < \frac{f_i(\bar{x})}{g_i(\bar{x})}.$$

Therefore, the difference $\left(\frac{f_i(\bar{x})}{g_i(\bar{x})} - \frac{f_i(x^*)}{g_i(x^*)} \right)$ can be made arbitrarily large and hence for $\bar{\lambda} > 0$, we get the following inequality

$$\sum_{i=1}^{k} \bar{\lambda}_i \left(\frac{f_i(\bar{x})}{g_i(\bar{x})} - \frac{f_i(x^*)}{g_i(x^*)} \right) > 0, \tag{5.65}$$

Since x^* is a feasible solution for the problem (MFP) and $(\bar{u}, \bar{\lambda}, \bar{\mu})$ is a feasible solution for the problem (MFD), we get

$$h(x^*) \leq 0, \tag{5.66}$$

$$\sum_{i=1}^{k} \bar{\lambda}_i \nabla \left(\frac{f_i(u)}{g_i(u)} \right) + \nabla(\mu^T h(u)) = 0, \tag{5.67}$$

$$\bar{\mu}^T h(\bar{u}) \geq 0, \tag{5.68}$$

$$\bar{\mu} \geq 0 \tag{5.69}$$

$$\bar{\lambda}_i > 0, i = 1, 2, \ldots, k. \tag{5.70}$$

Using (5.66), (5.68), and (5.69), we get

$$\bar{\mu}^T h(x^*) \leq \bar{\mu}^T h(\bar{u}).$$

Since $\bar{\mu}^T h$ is pseudolinear with respect to q, we get

$$q(\bar{u}, x^*) \left\langle \nabla(\bar{\mu}^T h(\bar{u})), x^* - \bar{u} \right\rangle \leq 0.$$

As $q(\bar{u}, x^*) > 0$, it follows that

$$\left\langle \nabla(\bar{\mu}^T h(\bar{u})), x^* - \bar{u} \right\rangle \leq 0.$$

Using (5.67), we get

$$\left\langle \sum_{i=1}^{k} \bar{\lambda}_i \nabla \left(\frac{f_i(\bar{u})}{g_i(\bar{u})} \right), x^* - \bar{u} \right\rangle \geq 0.$$

Using regularity assumption and the pseudoconvexity of $\sum_{i=1}^{k} \bar{\lambda}_i \frac{f_i}{g_i}$, it follows that

$$\sum_{i=1}^{k} \bar{\lambda}_i \frac{f_i(x^*)}{g_i(x^*)} - \sum_{i=1}^{k} \bar{\lambda}_i \frac{f_i(\bar{u})}{g_i(\bar{u})} \geq 0. \tag{5.71}$$

From (5.64) and (5.71), we get

$$\sum_{i=1}^{k} \bar{\lambda}_i \left(\frac{f_i(\bar{x})}{g_i(\bar{x})} - \frac{f_i(x^*)}{g_i(x^*)} \right) \geq 0,$$

which is a contradiction to (5.65). Hence, $\bar{x}$ is a properly efficient solution for (MFP).

Now, we assume that $(\bar{u}, \bar{\lambda}, \bar{\mu})$ is not a properly efficient solution for (MFD). Therefore, for every scalar $M > 0$, there exist a feasible point (u^*, λ^*, μ^*) for (MFD) and an index i, such that

$$\frac{f_i(u^*)}{g_i(u^*)} - \frac{f_i(\bar{u})}{g_i(\bar{u})} > M \left(\frac{f_r(\bar{u})}{g_r(\bar{u})} - \frac{f_r(u^*)}{g_r(u^*)} \right),$$

for all k, satisfying

$$\frac{f_r(u^*)}{g_r(u^*)} < \frac{f_r(\bar{u})}{g_r(\bar{u})},$$

whenever
$$\frac{f_i(u^*)}{g_i(u^*)} > \frac{f_i(\bar{u})}{g_i(\bar{u})}.$$

Therefore, the difference $\left(\frac{f_i(u^*)}{g_i(u^*)} - \frac{f_i(\bar{u})}{g_i(\bar{u})}\right)$ can be made arbitrarily large and hence for $\bar{\lambda} > 0$, we have

$$\sum_{i=1}^{k} \bar{\lambda}_i \left(\frac{f_i(u^*)}{g_i(u^*)} - \frac{f_i(\bar{u})}{g_i(\bar{u})}\right) > 0. \tag{5.72}$$

Since $\bar{x}$ and $(\bar{u}, \bar{\lambda}, \bar{\mu})$ are feasible solutions for (MFP) and (MFD), respectively, it follows as in the first part of the theorem that

$$\sum_{i=1}^{k} \bar{\lambda}_i \left(\frac{f_i(u^*)}{g_i(u^*)} - \frac{f_i(u)}{g_i(u)}\right) \leq 0,$$

which is a contradiction to (5.72). Thus, $(\bar{u}, \bar{\lambda}, \bar{\mu})$ is a properly efficient solution for (MFD).

Theorem 5.13 (Strong duality) *Let $\bar{x}$ be an efficient solution for the problem (MFP). Then there exist $\bar{\lambda} \in \mathbb{R}^k, \bar{\mu} \in \mathbb{R}^m$, such that $(\bar{x}, \bar{\lambda}, \bar{\mu})$ is a feasible solution for (MFD). Further, let for all feasible solutions (u, λ, μ) for (MFP), $\sum_{i=1}^{k} \lambda_i \frac{f_i}{g_i}$ be pseudolinear with respect to p and $\mu^T h$ be pseudolinear with respect to q. Then $(\bar{x}, \bar{\lambda}, \bar{\mu})$ is a properly efficient solution for (MFD).*

Proof The hypothesis of this theorem and Theorem 5.8 implies that there exist $\bar{\lambda} \in \mathbb{R}^k, \bar{\mu}_j \in \mathbb{R}, j \in J(\bar{x})$, such that

$$\sum_{i=1}^{k} \bar{\lambda}_i \nabla \left(\frac{f_i(\bar{x})}{g_i(\bar{x})}\right) + \sum_{j \in J(\bar{x})} \bar{\mu}_j \nabla h_j(\bar{x}) = 0,$$

$$\bar{\mu}_j \geq 0, j \in J(\bar{x}),$$

$$\bar{\lambda}_i > 0, i = 1, 2, \ldots, k.$$

Let $\bar{\mu}_j = 0$, for $j \notin J(\bar{x})$, then there exist $\bar{\lambda} \in \mathbb{R}^k, \bar{\mu} \in \mathbb{R}^m$, such that $(\bar{x}, \bar{\lambda}, \bar{\mu})$ is a feasible solution of (MFD). The proper efficiency of $(\bar{x}, \bar{\lambda}, \bar{\mu})$ for the problem (MFD) follows from Theorem 5.12.

Theorem 5.14 (Converse duality) *Let $(\bar{u}, \bar{\lambda}, \bar{\mu})$ be an efficient solution for (MFD) and let for all feasible (u, λ, μ) for (MFD), $\sum_{i=1}^{k} \lambda_i \frac{f_i}{g_i}$ be pseudolinear with respect to proportional function p and $\mu^T h$ be pseudolinear with respect to q. Moreover, let the $n \times n$ Hessian matrix $\nabla^2 \left(\bar{\lambda}^T \frac{f(\bar{u})}{g(\bar{u})} + \bar{\mu}^T h(\bar{u})\right)$ be positive or negative definite and $\left\{\nabla \left(\frac{f_i(\bar{u})}{g_i(\bar{u})}\right), i = 1, \ldots, k\right\}$ be linearly independent. Then, $\bar{u}$ is a properly efficient solution for (MFP).*

Proof Since $(\bar{u}, \bar{\lambda}, \bar{\mu})$ is an efficient solution for (MFD), by Fritz John necessary optimality conditions in Weir and Mond [284] there exist $\tau \in \mathbb{R}^k, \nu \in \mathbb{R}^n, p \in \mathbb{R}, s \in \mathbb{R}^m, w \in \mathbb{R}^k$, such that

$$\nabla \tau^T \left(\frac{f_i(\bar{u})}{g_i(\bar{u})} \right) + \nabla \nu^T \left[\nabla \bar{\lambda}^T \frac{f(\bar{u})}{g(\bar{u})} + \nabla \bar{\mu}^T h(\bar{u}) \right] + p \nabla \bar{\mu}^T h(\bar{u}) = 0, \quad (5.73)$$

$$(\nabla h(\bar{u}))^T \nu + p h(\bar{u}) + s = 0, \quad (5.74)$$

$$\left(\frac{\nabla f(\bar{u})}{\nabla g(\bar{u})} \right)^T \nu + w = 0, \quad (5.75)$$

$$p \bar{\mu}^T h(\bar{u}) = 0, \quad (5.76)$$

$$s^T \bar{\mu} = 0, \quad (5.77)$$

$$w^T \bar{\lambda} = 0, \quad (5.78)$$

$$(\tau, s, p, w) \geqq 0, \quad (5.79)$$

$$(\tau, s, p, w) \neq 0. \quad (5.80)$$

Since, $\bar{\lambda} > 0$, (5.78) implies that $w = 0$. Then, using (5.75), we get

$$\nu^T \left(\frac{\nabla f(\bar{u})}{\nabla g(\bar{u})} \right) = 0. \quad (5.81)$$

Multiplying (5.74) by $\bar{y}$ and using (5.76) and (5.77), we get

$$\nu^T \nabla \bar{\mu}^T h(\bar{u}) = 0. \quad (5.82)$$

Multiplying (5.73) by ν^T and using (5.81) and (5.82), it follows that

$$\nu^T \left[\nabla^2 \bar{\lambda}^T \left(\frac{f(\bar{u})}{g(\bar{u})} \right) + \nabla^2 \bar{\mu}^T h(\bar{u}) \right] \nu = 0.$$

Since, $\nabla^2 \left(\bar{\lambda}^T \frac{f(\bar{u})}{g(\bar{u})} + \bar{\mu}^T h(\bar{u}) \right)$ is assumed to be positive or negative definite, we get $\nu = 0$,

Since, $\nu = 0$, by (5.22) and (5.12), we get

$$\nabla (\tau - p \bar{\lambda})^T \frac{f(\bar{u})}{g(\bar{u})} = 0.$$

By linear independence of $\{ \nabla \left(\frac{f_i(\bar{u})}{g_i(\bar{u})} \right), i = 1, \ldots, k \}$, it follows that

$$\tau = p \bar{\lambda}.$$

Since, $\bar{\lambda} > 0, \tau = 0$ implies $p = 0$ and then by (5.74), $s = 0$, giving $(\tau, s, p, w) = 0$, a contradiction to (5.80). Thus $\tau \neq 0$ and $p > 0$. Since $\nu = 0, p > 0$ and $s \geqq 0$, (5.74) gives $h(\bar{u}) \leqq 0$ and (5.76) gives $\bar{\mu}^T h(\bar{u}) = 0$. Thus $\bar{u}$ is feasible for (MFP). That $\bar{u}$ is an efficient solution for (MFP) follows from the assumptions of the theorem and weak duality Theorem 5.11. The proper efficiency of $\bar{u}$ for (MFP) follows from Theorem 5.12.

Chapter 6

Nonsmooth Pseudolinear Multiobjective Programming

6.1 Introduction

Nonsmooth phenomena occur naturally and frequently in optimization theory. Therefore, the study of nondifferentiable mathematical programming by employing generalized directional derivatives and subdifferentials has been a field of intense investigation during the past several decades. Optimality conditions and duality theorems for nonsmooth multiobjective optimization problems have been an interesting research area. Kanniappan [124] obatained Fritz John and Karush-Kuhn-Tucker type necessary optimality conditions for a nondifferentiable convex multiobjective optimization problem by reducing it to a system of scalar minimization problems. Several scholars have developed interesting results by using generalized convex functions and subdifferentials. See for example, Craven [51], Bhatia and Jain [25], Kim and Bae [141], Liu [165], and the references therein. Duality for nonsmooth multiobjective fractional programming problems involving generalized convex functions have been studied by Kim [140], Kuk [152], Kuk *et al.* [153], Stancu-Minasian *et al.* [260], Nobakhtian [216], and Mishra and Upadhyay [197, 198].

This chapter deals with the optimality conditions and duality theorems for nonsmooth multiobjective programming and nonsmooth multiobjective fractional programming problems involving locally Lipschitz pseudolinear functions. Necessary and sufficient optimality conditions for a feasible solution to be an efficient solution for the primal problems under the assumption of locally Lipschitz pseudolinear functions have been established. The equivalence between efficiency and proper efficiency under certain boundedness conditions has been proved. Moreover, weak and strong duality theorems for corresponding Mond-Weir type subgradient dual problems have been established.

We consider the following nonsmooth multiobjective optimization problem:

$$(NMP) \quad \min f(x) := (f_1(x), ..., f_k(x))$$

$$\text{subject to } g_j(x) \leq 0, j = 1, ..., m,$$

where $f_i : K \to \mathbb{R}, i \in I := 1, 2.., k; g_j : K \to \mathbb{R}, j \in J := (1, .., m)$ are locally Lipschitz pseudolinear functions on a nonempty open convex set $K \subseteq \mathbb{R}^n$.

Let $P = \{x \in K : g(x) \leq 0, j \in J\}$ and $J(x) := \{j \in J : g_j(x) = 0\}$ for some $x \in K$, denote the set of all feasible solutions for (NMP) and active constraint index set at $x \in K$, respectively.

6.2 Optimality Conditions

In this section, we derive the necessary and sufficient optimality conditions for a feasible solution to be an efficient solution for (NMP). Under certain boundedness conditions, we establish the equivalence between efficient and properly efficient solutions for (NMP).

Theorem 6.1 (Necessary and sufficient optimality conditions) *Let $\bar{x}$ be a feasible solution of (NMP). Let the functions $f_i, i = 1, ..., k$ be pseudolinear with respect to proportional function p_i and $g_j, j \in J(\bar{x})$ be pseudolinear with respect to proportional function q_j. Then, $\bar{x}$ is an efficient solution for (NMP) if and only if there exist $\bar{\lambda}_i \in \mathbb{R}, i = 1, ...k, \bar{\mu}_j \in \mathbb{R}, j \in J(\bar{x})$ and there exist $\xi_i \in \partial^c f_i(\bar{x})$ and $\zeta_j \in \partial^c g_j(\bar{x})$, such that*

$$\sum_{i=1}^{k} \bar{\lambda}_i \xi_i + \sum_{j \in J(\bar{x})} \bar{\mu}_j \zeta_j = 0, \tag{6.1}$$

$$\bar{\lambda}_i > 0, i = 1, ..., k,$$

$$\bar{\mu}_j \geq 0, j \in J(\bar{x}).$$

Proof Suppose $\bar{\lambda}_i$ and $\bar{\mu}_j$ exist and satisfy condition (6.1). Let $\bar{x}$ is not an efficient solution for (NMP). There exists then a point $y \in P$, such that

$$f_i(y) \leq f_i(\bar{x}), \forall i = 1, ..., k$$

$$f_r(y) < f_r(\bar{x}), \text{ for some } r.$$

Using pseudolinearity of $f_i, i = 1, .., k$, with respect to proportional function p_i, there exists $\xi_i' \in \partial^c f_i(\bar{x})$ such that

$$p_i(\bar{x}, y)\langle \xi_i', y - \bar{x} \rangle \leq 0, \forall i = 1, ..., k$$

and

$$p_r(\bar{x}, y)\langle \xi_r' - \bar{x} \rangle < 0, \text{ for some } r.$$

Since, $p_i(\bar{x}, y) > 0$, for all $i = 1, ..., k$, therefore, there exists $\xi_i' \in \partial^c f_i(\bar{x})$, such that

$$\langle \xi_i', y - \bar{x} \rangle \leq 0, \forall i = 1, ..., k$$

and

$$\langle \xi_r', y - \bar{x} \rangle < 0, \text{ for some } r.$$

By Proposition 2.5, there exist $s_i > 0$ and $\xi_i \in \partial^c f_i(\bar{x})$, such that $\xi_i' = s_i \xi_i, \forall i = 1, ..., k$. Hence, we have

$$s_i \langle \xi_i, y - \bar{x} \rangle \leq 0, \forall i = 1, ..., k$$

and

$$s_r \langle \xi_r, y - \bar{x} \rangle < 0, \text{ for some } r.$$

Since, $s_i > 0$, for all $i = 1, ..., k$, it follows that

$$\langle \xi_i, y - \bar{x} \rangle \leq 0, \forall i = 1, ..., k$$

and

$$\langle \xi_r, y - \bar{x} \rangle < 0, \text{ for some } r.$$

Therefore, for $\bar{\lambda}_i > 0, i = 1, ..., k$, we get

$$\sum_{i=1}^{k} \bar{\lambda}_i \langle \xi_i, y - \bar{x} \rangle \leq 0. \tag{6.2}$$

Again, we have

$$g_j(y) \leq g_j(\bar{x}), \forall j \in J(\bar{x}).$$

Using pseudolinearity of $g_j, j = 1, ..., m$ with respect to proportional function q_j, there exists $\zeta_j' \in \partial^c g_j(\bar{x})$, such that

$$q_j(\bar{x}, y) \langle \zeta_j', y - \bar{x} \rangle \leq 0, \forall j \in J(\bar{x}).$$

Since, $q_j(\bar{x}, y) > 0$, for all $j \in J(\bar{x})$, therefore, we get

$$\langle \zeta_j', y - \bar{x} \rangle \leq 0, \forall j \in J(\bar{x}).$$

By Proposition 2.5, there exist $t_j > 0$ and $\zeta_j \in \partial^c g_j(\bar{x})$, such that $\zeta_j' = t_j \zeta_j$. Hence, we have

$$t_j \langle \zeta_j, y - \bar{x} \rangle \leq 0 \, \forall j \in J(\bar{x}).$$

Since, $t_j > 0$, for all $j \in J(\bar{x})$, we get

$$\langle \zeta_j, y - \bar{x} \rangle \leq 0, \forall j \in J(\bar{x}).$$

Since, $\bar{\mu}_j \geq 0$, for all $j \in J(\bar{x})$, it follows that

$$\sum_{j \in J(\bar{x})} \bar{\mu}_j \langle \zeta_j, y - \bar{x} \rangle \leq 0 \, \nabla j \in J(\bar{x}).$$

Using (6.1), there exists $\xi_i \in \partial^c f_i(\bar{x}), i = 1, ..., k$, such that

$$\sum_{i=1}^{k} \bar{\lambda}_i \langle \xi_i, y - \bar{x} \geq 0,$$

which is a contradiction to (6.2).

Conversely, suppose $\bar{x}$ be an efficient solution of (NMP). For $1 \leq r \leq k$, there exist $\zeta_j \in \partial^c g_j(\bar{x}), \xi_i \in \partial^c f_i(\bar{x})$ and $\xi_r \in \partial^c f_r(\bar{x})$, such that the following system

$$
\left.
\begin{aligned}
\langle \zeta_j, x - \bar{x} \rangle \leq 0, \ \forall j \in J(\bar{x}), \\
\langle \xi_i, x - \bar{x} \rangle \leq 0, \ i = 1, 2, ..., r-1, r+1, .., k, \\
\langle \xi_r, x - \bar{x} \rangle < 0,
\end{aligned}
\right\}
\tag{6.3}
$$

has no solution $x \in P$. For if x is a solution of the system and $y = \bar{x} + t(x - \bar{x}) \ (0 < t \leq 1)$, then for $j \in J(\bar{x})$ there exists $\zeta_j \in \partial^c g_j(\bar{x})$, such that

$$
\langle \zeta_j, y - \bar{x} \rangle = t \langle \zeta_j, x - \bar{x} \rangle \leq 0.
$$

Therefore, by pseudoconcavity of g_j, we get

$$
g_j(y) \leq g_j(\bar{x}) = 0.
$$

If $j \notin J(\bar{x})$, then $g_j(\bar{x}) < 0$ and so $g_j(y) < 0$, when t is sufficiently small. Hence, for small t, y is a point of P. Moreover, there exist $s'_i > 0$, $\xi_i, \xi'_i \in \partial^c f_i(\bar{x})$, $i = 1, ..., k$, such that $\xi'_i = s'_i \xi_i$. Hence, we have

$$
f_i(y) - f_i(\bar{x}) = t p_i(\bar{x}, y) \langle \xi'_i, x - \bar{x} \rangle = t p_i(\bar{x}, y) s'_i \langle \xi_i, x - \bar{x} \rangle \leq 0, \ \ i \neq r
$$

and similarly

$$
f_r(y) - f_r(\bar{x}) < 0,
$$

which is a contradiction to our assumption that $\bar{x}$ is an efficient solution for (NMP). From which, it follows that the system (6.3) has no solution. Therefore, by Farkas's theorem (Theorem 1.24), there exist $\bar{\lambda}_{r_i} \geq 0, \bar{\mu}_{r_j} \geq 0$, $\xi_i \in \partial^c f_i(\bar{x})$ and $\zeta_j \in \partial^c g_j(\bar{x})$, such that

$$
-\xi_r = \sum_{i \neq r} \bar{\lambda}_{r_i} \xi_i + \sum_{j \in J(x^*)} \bar{\mu}_{r_j} \zeta_j.
\tag{6.4}
$$

Summing (6.4) over r, we get (6.1) with

$$
\bar{\lambda}_i = 1 + \sum_{r \neq i} \bar{\lambda}_{r_i}, \ \ \bar{\mu}_j = \sum_{r=1}^{k} \bar{\mu}_{r_j}.
$$

The following definition is a nonsmooth variant of the corresponding definition in Chew and Choo [47].

Definition 6.1 *A feasible point $\bar{x}$ for the problem (NMP) is said to satisfy the boundedness condition, if the set*

$$
\left\{ \frac{p_i(\bar{x}, x)}{p_j(\bar{x}, x)} | x \in K, f_i(\bar{x}) > f_i(x), f_j(\bar{x}) < f_i(x), 1 \leq i, j \leq k \right\}
\tag{6.5}
$$

is bounded from above.

The following theorem extends Proposition 3.5 of Chew and Choo [47] to the nonsmooth multiobjective programming problem (NMP).

Theorem 6.2 *Every efficient solution of (NMP) involving pseudolinear functions, satisfying the boundedness condition is a properly efficient solution of the problem (NMP).*

Proof Let $\bar{x}$ be an efficient solution of (NMP). Then from Theorem 6.1, it follows that there exist $\lambda_i > 0, i = 1, 2, ..., k; \mu_j \geq 0, j \in J(\bar{x}); \xi_i \in \partial^c f_i(\bar{x})$ and $\zeta_j \in \partial^c h_j(\bar{x})$, such that

$$\sum_{i=1}^{k} \lambda_i \xi_i = - \sum_{j \in J(\bar{x})} \mu_j \zeta_j.$$

Therefore, for any feasible x, we have

$$\sum_{i=1}^{k} \lambda_i \langle \xi_i, x - \bar{x} \rangle = - \sum_{j \in J(\bar{x})} \mu_j \langle \zeta_j, x - \bar{x} \rangle.$$

We observe that for any $x \in K$,

$$\sum_{i=1}^{k} \lambda_i \langle \xi_i, x - \bar{x} \rangle \geq 0. \tag{6.6}$$

Otherwise, we would arrive at a contradiction as in the first part of Theorem 6.1.

Since the set defined by (6.5) is bounded above, therefore, the following set

$$\left\{ (k-1) \frac{\lambda_j p_i(\bar{x}, x)}{\lambda_i p_j(\bar{x}, x)} \,|\, x \in K, f_i(\bar{x}) > f_i(x), f_j(\bar{x}) < f_j(x), \ 1 \leq i, j \leq k \right\} \tag{6.7}$$

is also bounded from above.

Let $M > 0$ be a real number that is an upper bound of the set defined by (6.7). Now, we shall show that $\bar{x}$ is a properly efficient solution of (NMP). Assume that there exist r and $x \in K$, such that

$$f_r(x) < f_r(\bar{x}).$$

Then, by the pseudolinearity of f_r, there exists $\xi_r \in \partial^c f_r(\bar{x})$, such that

$$p_r(\bar{x}, x) \langle \xi_r, x - \bar{x} \rangle < 0.$$

Since $p_r(\bar{x}, x) > 0$, we get

$$\langle \xi_r, x - \bar{x} \rangle < 0. \tag{6.8}$$

Let us define

$$\langle -\xi_s, x - \bar{x} \rangle = \max \left\{ \lambda_i \langle \xi_i, x - \bar{x} \rangle \mid \langle \xi_i, x - \bar{x} \rangle > 0 \right\}. \tag{6.9}$$

Using (6.6), (6.8), and (6.9), we get

$$\lambda_r \langle \xi_r, x - \bar{x} \rangle \leq (k - 1) \left(-\lambda_s \langle \xi_s, x - \bar{x} \rangle \right).$$

Therefore,

$$(f_r(x) - f_r(\bar{x})) \leq (k - 1) \frac{\lambda_s p_r(\bar{x}, x)}{\lambda_r p_s(\bar{x}, x)} (f_s(\bar{x}) - f_s(x)).$$

Using the definition of $\bar{x}$, we get

$$(f_r(x) - f_r(\bar{x})) \leq M (f_s(\bar{x}) - f_s(x)).$$

Hence, $\bar{x}$ is a properly efficient solution of (NMP).

6.3   Duality for (NMP)

Now, in relation to (NMP), we consider the following Mond-Weir type subgradient dual problem:

$$(NDP) \quad \max f(u) := (f_1(u), f_2, ..., f_k(u))$$

$$\text{subject to } 0 \in \sum_{i=1}^{k} \lambda_i \partial^c f_i(x) + \partial^c \left(y^T g(u) \right), \tag{6.10}$$

$$y^T g(u) \geq 0,$$

$$y \geq 0, \quad \lambda_i > 0, \quad i = 1, ..., k.$$

Next, we establish weak and strong duality relations between (NMP) and (NDP).

Theorem 6.3 (Weak duality) *Let f_i, $i = 1, ..., k$ be pseudolinear with respect to proportional function p_i. If x is a feasible solution for (NMP) and (u, λ, y) is a feasible solution for (NDP), such that $y^T g$ is pseudolinear with respect to q, then the following cannot hold*

$$f(x) \leq f(u).$$

Proof Suppose the above inequality holds. The pseudolinearity of f_i on K with respect to proportional function p_i, implies that there exist $\xi_i \in \partial^c f_i(u)$, $i = 1, ..., k$, such that

$$p_i(x, u) \langle \xi_i, x - u \rangle \leq 0, \quad \forall i = 1, ..., k,$$

$$p_r(x, u) \langle \xi_r, x - u \rangle < 0, \text{ for some } r.$$

Since $p_i(x, u) > 0$, for all $i = 1, ..., k$, it follows that

$$\langle \xi_i, x - u \rangle \leq 0, \quad \forall i = 1, ..., k,$$

$$\langle \xi_r, x - u \rangle < 0, \text{ for some } r.$$

Since $\lambda_i > 0$, for each $i = 1, ..., k$, we get

$$\sum_{i=1}^{k} \lambda_i \langle \xi_i, x - u \rangle < 0. \tag{6.11}$$

By (6.10) and (6.11), there exists $\zeta \in \partial^c \left(y^T g(u) \right)$, such that

$$\langle \zeta, x - u \rangle > 0. \tag{6.12}$$

Since x is feasible for (NMP) and (u, λ, y) is feasible for (NDP), we have

$$y^T g(x) \leq 0$$

and

$$y^T g(u) \geq 0.$$

Using pseudolinearity of $y^T g$ with respect to q, there exists $\zeta' \in \partial^c \left(y^T g(u) \right)$, such that

$$q(x, u) \langle \zeta', x - u \rangle \leq 0.$$

Again as $q(x, u) > 0$, it follows that

$$\langle \zeta', x - u \rangle \leq 0.$$

By Proposition 2.5, there exist $t > 0$, and $\zeta \in \partial^c \left(y^T g(u) \right)$, such that $\zeta' = t\zeta$. Hence, we get

$$t \langle \zeta, x - u \rangle \leq 0. \tag{6.13}$$

Since, $t > 0$, from (6.13), it follows that

$$\langle \zeta, x - u \rangle \leq 0,$$

which is a contradiction to (6.12).

In the following theorem, we have weakened the conditions of pseudolinearity on the objective functions.

Theorem 6.4 (Weak duality) *If x is a feasible solution for (NMP) and (u, λ, y) is a feasible solution for (NDP), such that $\sum_{i=1}^{k} \lambda_i f_i$ is pseudolinear with respect to p and $y^T g$ is pseudolinear with respect to q. If $f_i, i = 1, ..., k$, is regular near x, then the following cannot hold*

$$f(x) \leq f(u)$$

Proof Suppose the above inequality holds. Since $\lambda_i > 0$, $i = 1, ..., k$, it follows that

$$\sum_{i=1}^{k} \lambda_i f_i(x) < \sum_{i=1}^{k} \lambda_i f_i(u).$$

Using the regularity of $f_i, i = 1, ..., k$ and pseudolinearity of $\sum_{i=1}^{k} \lambda_i f_i$ with respect to p, we get the inequality (6.11) of Theorem 6.3. The rest of the proof of the theorem follows along the lines of Theorem 6.3.

Theorem 6.5 *Let $\bar{x}$ be a feasible solution of (NMP) and $(\bar{u}, \bar{\lambda}, \bar{y})$ be a feasible solution of (NDP), such that*

$$f(\bar{x}) = f(\bar{u}) \tag{6.14}$$

Let $f_i, i = 1, ..., k$ be regular near $\bar{x}$. If for all feasible solutions (u, λ, y) of (NDP), $\sum_{i=1}^{k} \lambda_i f_i$ is pseudolinear with respect to p and $y^T g$ is pseudolinear with respect to q, then $\bar{x}$ is a properly efficient solution for (NMP) and $(\bar{u}, \bar{\lambda}, \bar{y})$ is a properly efficient solution for (NDP).

Proof Suppose $\bar{x}$ is not an efficient solution for (NDP), then there exists $x \in P$, such that

$$f(x) \leq f(\bar{x}).$$

Using the assumption $f(\bar{x}) = f(u)$ we get a contradiction to Theorem 6.4. Hence $\bar{x}$ is an efficient solution for (NMP). Similarly, it can be shown that $(\bar{u}, \bar{\lambda}, \bar{y})$ is an efficient solution for (NDP).

Now, suppose that $\bar{x}$ be an efficient solution for (NMP). Therefore, for every scalar $M > 0$ there exists $x_0 \in P$ and an index i, such that

$$f_i(\bar{x}) - f_i(x_0) > M(f_j(x_0) - f_j(x_0)),$$

for all j satisfying

$$f_j(x_0) > f_j(\bar{x}),$$

whenever

$$f_i(x_0) < f_i(\bar{x}).$$

This means $f_i(\bar{x}) - f_i(x_0)$ can be made arbitrarily large and hence for $\bar{\lambda} > 0$, we get the following inequality

$$\sum_{i=1}^{k} \bar{\lambda}_i (f_i(\bar{x}) - f_i(\bar{x})) > 0 \tag{6.15}$$

Since, x_0 is feasible for (NMP) and $(\bar{u}, \bar{\lambda}, \bar{y})$ is feasible for (NDP), therefore, we have

$$g\left(x_0\right) \leqq 0, \tag{6.16}$$

$$\sum_{i=1}^{k} \bar{\lambda}_i \xi_i + \zeta = 0, \text{ for some } \xi_i \in \partial^c f_i\left(\bar{x}\right) \text{ and } \zeta \in \partial^c \left(\bar{y}^T g\left(\bar{u}\right)\right) \tag{6.17}$$

$$\bar{y}^T g\left(\bar{u}\right) \geq 0, \tag{6.18}$$

$$\bar{y} \geqq 0, \tag{6.19}$$

$$\bar{\lambda}_i > 0, i = 1, 2, ..., k \tag{6.20}$$

From (6.16), (6.18), and (6.19), we get

$$\bar{y}^T g\left(x_0\right) \leq \bar{y}^T g\left(\bar{u}\right).$$

Since $\bar{y}^T g$ is pseudolinear with respect to q, there exist $\zeta' \in \partial^c \left(\bar{y}^T g\left(\bar{u}\right)\right)$, such that

$$q\left(x_0, \bar{u}\right) \langle \zeta', x_0 - \bar{u}\rangle \leq 0.$$

Since $q(x_0, \bar{u}) > 0$, we get

$$\langle \zeta', x_0 - \bar{u}\rangle \leq 0.$$

By Proposition 2.5, there exists some $\mu > 0$, $\zeta \in \partial^c f\left(\bar{u}\right)$, such that $\zeta' = \mu \zeta$. Hence, we get

$$\mu \langle \zeta, x_0 - \bar{u}\rangle \leq 0.$$

Since, $\mu > 0$, it follows that

$$\langle \zeta, x_0 - \bar{u}\rangle \leq 0.$$

From (6.17), we get

$$\left\langle \sum_{i=1}^{k} \bar{\lambda}_i \xi_i, x_0 - \bar{u}\right\rangle \geq 0, \text{ for some } \xi_i \in \partial^c f_i(\bar{u}).$$

On using the regularity assumption and pseudoconvexity of $\sum_{i=1}^{k} \bar{\lambda}_i f_i$, we get

$$\sum_{i=1}^{k} \bar{\lambda}_i f_i\left(x_0\right) - \sum_{i=1}^{k} \bar{\lambda}_i f_i\left(\bar{u}\right) \geq 0. \tag{6.21}$$

On using (6.14) in (6.21), we get

$$\sum_{i=1}^{k} \bar{\lambda}_i \left(f_i\left(\bar{x}\right) - f_i\left(x_0\right)\right) \leq 0,$$

which is a contradiction to (6.15). Hence, $\bar{x}$ is a properly efficient solution of (NMP).

We now suppose that $(\bar{u}, \bar{\lambda}, \bar{y})$ is not a properly efficient solution for (NDP). Therefore, for every scalar $M > 0$ there exists a feasible point (u_0, λ_0, y_0) for (NDP) and an index i, such that

$$f_j(u_0) - f_j \bar{u}) > M(f_j \bar{u} - f_j(u_0)),$$

for all j, satisfying

$$f_j(u_0) < f_j(\bar{u}),$$

whenever

$$f_i(u_0) > f_i(\bar{u}).$$

This means $f_i(u_0) - f_i(\bar{u})$ can be made arbitrarily large and hence for $\bar{\lambda} > 0$, we get the inequality

$$\sum_{i=1}^{k} \bar{\lambda}_i \left(f_i \left(u_0 \right) - f_i \left(\bar{u} \right) \right) > 0. \tag{6.22}$$

Since $\bar{x}$ and $(\bar{u}, \bar{\lambda}, \bar{y})$ are feasible for (NMP) and (NDP), respectively, proceeding as in the first part, we get

$$\sum_{i=1}^{k} \bar{\lambda}_i \left(f_i \left(u_0 \right) - f_i \left(\bar{u} \right) \right) \leq 0,$$

which contradicts (6.22). Hence, $(\bar{u}, \bar{\lambda}, \bar{y})$ is a properly efficient solution for (NDP).

Theorem 6.6 (Strong duality) *Let $x \in P$, be an efficient solution for (NMP) and let $f_i, i = 1, ..., k$ be pseudolinear with respect to the proportional function p_i and $g_j, j \in J(\bar{x})$ be pseudolinear with respect to q_j, then, there exist $\bar{\lambda} \in \mathbb{R}^k, \bar{y} \in \mathbb{R}^m$, such that $(\bar{x}, \bar{\lambda}, \bar{y})$ is a feasible solution for (NDP). Let $f_i, i = 1, ..., k$ and $g_j, j \in J$ be regular functions at $\bar{x}$. Further, if for all feasible (u, λ, y) of (NDP), $\sum_{i=1}^{k} \lambda_i f_i$ is pseudolinear with respect to p and $y^T g$ is pseudolinear with respect to q then $(\bar{x}, \bar{\lambda}, \bar{y})$ is a properly efficient solution for (NDP).*

Proof The hypothesis of this theorem and Theorem 6.1 implies that there exist $\bar{\lambda}_i \in \mathbb{R}^k$ and $\bar{y}_j \in \mathbb{R}$, $j \in J(\bar{x})$, such that

$$\sum_{i=1}^{k} \bar{\lambda}_i \xi_i + \sum_{j \in J(\bar{x})} \bar{y}_j \zeta_j = 0,$$

$$\bar{y}_j \geq 0, j \in J(\bar{x}),$$

$$\bar{(\lambda)}_i > 0, i = 1, 2, ..., k.$$

Let $\bar{y}_j = 0, j \in J(\bar{x})$. Then there exist $\bar{\lambda} \in \mathbb{R}^k$ and $\bar{y} \in \mathbb{R}^m$, such that $(\bar{x}, \bar{\lambda}, \bar{y})$ is a feasible solution of (NDP). The proper efficiency of $(\bar{x}, \bar{\lambda}, \bar{y})$ for the problem (NDP) follows from Theorem 6.5.

6.4 Nonsmooth Pseudolinear Multiobjective Fractional Programming

In the following lemma, we establish that the ratio of two locally Lipschitz pseudolinear functions is also a locally Lipschitz pseudolinear function, with respect to a different proportional function.

Lemma 6.1 *Suppose $f : K \to \mathbb{R}$ and $g : K \to \mathbb{R}$ are two locally Lipschitz pseudolinear functions defined on K with the same proportional function p. Let $f(x) \geq 0$ and $g(x) > 0$, for every $x \in K$. If f and $-g$ are regular on K, then $\frac{f}{g}$ is also pseudolinear with respect to a new proportional function*

$$\bar{p}(x,y) = \frac{g(x)p(x,y)}{g(y)}.$$

Proof Since f and g are pseudolinear with respect to the same proportional function p, it follows that for any x and y in K, there exist $\xi \in \partial^c f(x)$ and $\zeta \in \partial^c g(x)$, such that

$$f(y) = f(x) + p(x,y) \langle \xi, y - x \rangle, \qquad (6.23)$$

$$g(y) = g(x) + p(x,y) \langle \zeta, y - x \rangle. \qquad (6.24)$$

Now, using Proposition 6.2 and Equations (6.23) and (6.24), there exist $\xi \in \partial^c f(x)$ and $\zeta \in \partial^c g(x)$, such that

$$
\begin{aligned}
\frac{g(y)}{g(x)} \left(\frac{f(y)}{g(y)} - \frac{f(x)}{g(x)} \right) &= \frac{1}{g(y)}(f(y) - f(x)) - \frac{f(x)}{g(x)g(y)}(g(y) - g(x)) \\
&= p(x,y) \left(\frac{g(x) \langle \xi, y - x \rangle - f(x) \langle \zeta, y - x \rangle}{(g(x))^2} \right) \\
&= p(x,y) \left\langle \frac{g(x)\xi - f(x)\zeta}{g(x)^2}, y - x \right\rangle \\
&= p(x,y) \langle \varsigma, y - x \rangle, \text{ for some } \varsigma \in \partial^c \left(\frac{f(x)}{g(x)} \right).
\end{aligned}
$$

$$\frac{f(y)}{g(y)} = \frac{f(x)}{g(x)} + \frac{g(x)p(x,y)}{g(y)} \langle \varsigma, y - x \rangle,$$

which implies that $\frac{f}{g}$ is pseudolinear with respect to a new proportional function

$$\bar{p}(x,y) := \frac{g(x)p(x,y)}{g(y)}.$$

We consider the following nonsmooth multiobjective fractional programming problem:

$$(NMFP) \quad \min \left(\frac{f_1(x)}{g_1(x)}, \cdots, \frac{f_k(x)}{g_k(x)} \right)$$

$$\text{subject to } h_j(x) \leq 0, j \in \{1, 2, 3, \cdots, m\},$$

where $f_i : K \to \mathbb{R}$, $g_i : K \to \mathbb{R}, i \in I = \{1, ..., k\}$ and $h_j : K \to \mathbb{R}, j \in J = \{1, ..., m\}$ are locally Lipschitz functions on a nonempty open convex subset $K \subseteq \mathbb{R}^n$. From now on, we assume that the functions f and $-g$ are regular on K and $f(x) \geq 0, g(x) > 0$, for all $x \in \mathbb{R}^n$. Let $P := \{x \in K : h_j(x) \leq 0, j \in J\}$ and $J(x) = \{j \in J : h_j(x) = 0\}$ for some $x \in K$, denote the set of feasible solutions for (NMFP) and active constraints index set at $x \in K$, respectively.

6.5   Optimality Conditions for (NMFP)

In this section, we establish the necessary and sufficient conditions for a feasible solution to be an efficient solution for (NMFP) and show the equivalence between efficiency and proper efficiency under certain boundedness conditions.

Now, we shall prove the following necessary and sufficient optimality theorem analogous to Proposition 3.2 of Chew and Choo [47] for (NMFP).

Theorem 6.7 *Let $\bar{x}$ be a feasible solution for (NMFP). Let the functions f_i and g_i, $i = 1, 2, ..., k$ be pseudolinear on K with respect to the same proportional function p_i and $h_j, j \in J(\bar{x})$ be pseudolinear on K with proportional function q_j. Then, $\bar{x}$ is an efficient solution of (NMFP), if and only if there exist $\lambda_i > 0, i = 1, 2, ..., k$; $\mu_j > 0, j \in J(\bar{x})$; $\xi_i \in \partial^c \left(\frac{f_i(\bar{x})}{g_i(\bar{x})}\right)$ and $\zeta \in \partial^c h_j(\bar{x})$, such that*

$$\sum_{i=1}^{k} \lambda_i \xi_i + \sum_{j \in J(\bar{x})} \mu_j \zeta_j = 0. \tag{6.25}$$

Proof Suppose λ_i and μ_j exist and satisfy the given conditions and (6.25). Let $\bar{x}$ is not an efficient solution for (NMFP). Therefore, there exists a point $y \in P$, such that

$$\frac{f_i(y)}{g_i(y)} < \frac{f_i(\bar{x})}{g_j(\bar{x})}, \forall\, i = 1 \cdots, k$$

and

$$\frac{f_r(y)}{g_r(y)} < \frac{f_r(\bar{y})}{g_r(\bar{y})}, \text{ for some } r.$$

Using pseudolinearity of $\frac{f_i}{g_i}, i = 1, ..., k$, there exists $\xi_i' \in \partial^c \left(\frac{f_i(\bar{x})}{g_i(\bar{x})}\right)$, such that

$$\bar{p}_i(\bar{x}, y) \langle \xi_i', y - \bar{x} \rangle \leq 0, \forall i = 1, \cdots, k$$

and

$$\bar{p}_r(\bar{x}, y) \langle \xi_r', y - \bar{x} \rangle < 0, \text{ for some } r,$$

where
$$\bar{p}_i(\bar{x}, y) = \frac{g_i(\bar{x})p(\bar{x}, y)}{g_i(y)}.$$

Since, $\bar{p}_i(\bar{x}, y) > 0$, for all $i = 1, ..., k$, therefore, we have
$$\langle \xi'_i, y - \bar{x} \rangle \leq 0, \forall i = 1, \cdots, k$$

and
$$\langle \xi'_r, y - \bar{x} \rangle < 0, \text{ for some } r.$$

By Proposition 2.5, there exist $s_i > 0$ and $\xi_i \in \partial^c \left(\frac{f_i(\bar{x})}{g_i(\bar{x})} \right)$, such that $\xi'_i = s_i \xi_i, i = 1, \cdots, k$. Hence, we get
$$s_i \langle \xi_i, y - \bar{x} \rangle \leq 0, \forall i = 1, \cdots, k$$

and
$$s_r \langle \xi_r, y - \bar{x} \rangle < 0, \text{ for some } r.$$

Since, $s_i > 0$, for all $i = 1, ..., k$, it follows that
$$\langle \xi_i, y - \bar{x} \rangle \leq 0, \forall i = 1, \cdots, k$$

and
$$\langle \xi_r, y - \bar{x} \rangle < 0, \text{ for some } r.$$

Since $\lambda_i > 0$, for all $i = 1, \cdots, k$, we get
$$\sum_{i=1}^{k} \lambda_i \langle \xi_i, y - \bar{x} \rangle > 0. \tag{6.26}$$

Again, we have
$$h_j(y) \leq h_j(\bar{x}), \forall \, j \in J(\bar{x}).$$

Using pseudolinearity of $h_j, j = 1, ..., m$, there exists $\zeta'_i \in \partial^c g_j(\bar{x})$, such that
$$q_j(\bar{x}, y) \langle \zeta'_j, y - \bar{x} \rangle \leq 0 \, \forall \, j \in J(\bar{x}).$$

Since $q_j(\bar{x}, y) > 0$, for all $j \in J(\bar{x})$, therefore, we get
$$\langle \zeta'_j, y - \bar{x} \rangle \leq 0, \, \forall \, j \in J(\bar{x}).$$

By Proposition 2.5, there exist $t_j > 0$ and $\zeta_j \in \partial^c h_j(\bar{x})$, such that $\zeta'_j = t_j \zeta_j$. Hence, we have
$$t_j \langle \zeta_j, y - \bar{x} \rangle \leq 0, \, \forall \, j \in J(\bar{x}).$$

Since $t_j > 0$, for all $j \in J(\bar{x})$, we get
$$\langle \zeta_j, y - \bar{x} \rangle \leq 0, \, \forall \, j \in J(\bar{x}).$$

Since, $\mu_j \geq 0$ for all $j \in J(\bar{x})$, it follows that

$$\sum_{j \in J(\bar{x})} \mu_j \langle \zeta_j, y - \bar{x} \rangle \leq 0.$$

Using (6.25), we get

$$\sum_{i=1}^{k} \lambda_i \langle \xi_i, y - \bar{x} \rangle \geq 0,$$

which is a contradiction to (6.26).

Conversely, assume that $\bar{x}$ is an efficient solution for (NMFP). Therefore, for $1 \leq r \leq k$ there exist $\xi_i \in \partial^c \left(\frac{f_i(\bar{x})}{g_i(\bar{x})} \right)$ and $\zeta_j \in \partial^c h_j(\bar{x})$, such that the following system of inequalities

$$\begin{cases} \langle \zeta_j, x - \bar{x} \rangle \leq 0, j \in J(\bar{x}), \\ \langle \xi_i, x - \bar{x} \rangle \leq 0, i = 1, 2, \cdots, r-1, r+1, \cdots, k, \\ \langle \xi_r, x - \bar{x} \rangle < 0, \end{cases} \qquad (6.27)$$

has no solution $x \in P$. For if x is a solution of the system and $y = \bar{x} + t(x - \bar{x}), 0 \leq t \leq 1$. Then, for $j \in J(\bar{x})$. Using (6.27), there exists $\zeta_j \in \partial^c h_j(\bar{x})$, such that

$$\langle \zeta_j, y - \bar{x} \rangle = t \langle \zeta_j, x - \bar{x} \rangle \leq 0.$$

Therefore, by pseudoconvexity of $h_j, j \in J(\bar{x}) = 0$, we get

$$h_j(y) \leq h_j(\bar{x}).$$

If $j \notin J(\bar{x})$, then $h_j(\bar{x}) < 0$ and so $h_j(y) < 0$ when t is sufficiently small. Therefore, for small t, y is a point of P. Using Lemma 6.1 and (6.27), we get

$$\frac{f_i(y)}{g_i(y)} - \frac{f_i(\bar{x})}{g_i(\bar{x})} = t \bar{p}_i(\bar{x}, y) \langle \xi_i, x - \bar{x} \rangle \leq 0, i \neq r$$

and also

$$\frac{f_r(y)}{g_r(y)} - \frac{f_r(\bar{x})}{g_r(\bar{x})} < r, \text{ for some } r.$$

This contradicts the choice of x. Hence system (6.27) has no solution in the nonempty convex set K.

Applying Farkas's theorem (Theorem 1.24), there exist $\bar{\lambda}_{r_i} \geq 0$, $\bar{\mu}_{r_i} \geq 0, \xi_i \in \partial^c \left(\frac{f_i(\bar{x})}{g_i((x))} \right)$, and $\zeta_j \in \partial^c h_j(\bar{x})$, such that

$$-\xi_r = \sum_{i \neq r} \lambda_{r_i} \xi_i + \sum_{j \in J(x^*)} \mu_{r_j} \zeta_j. \qquad (6.28)$$

Summing (6.28) over r, we get (6.23) with

$$\lambda_i = 1 + \sum_{r \neq i} \lambda_{r_i}, \mu_j = \sum_{r=1}^{k} \mu_{r_j}.$$

This completes the proof.

Definition 6.2 *A feasible point $\bar{x}$ for the problem (NMFP) is said to satisfy the boundedness condition, if the set*

$$\left\{ \frac{\bar{p}_i(\bar{x}, x)}{\bar{p}_j(\bar{x}, x)} : x \in K, \frac{f_i(\bar{(x)})}{g_i(\bar{(x)})} > \frac{f_i(x)}{g_i(x)}, \frac{f_j(\bar{(x)})}{g_j(\bar{(x)})} < \frac{f_i(x)}{g_i(x)} \right\} \qquad (6.29)$$

is bounded from above, where $\bar{p}_i(\bar{x}, x) = \frac{g_i(\bar{x})p(\bar{x}, x)}{g_i(x)}$.

The following theorem extends Proposition 3.5 of Chew and Choo [47] to the nonsmooth multiobjective fractional programming problem (NMFP).

Theorem 6.8 *Every efficient solution of (NMFP) involving locally Lipschitz pseudolinear functions, satisfying the boundedness condition is a properly efficient solution of the problem (NMFP).*

Proof Let x be an efficient solution of (NMFP). Then from Theorem 6.6, it follows that there exist $\lambda_i > 0, i = 1, ..., k; \mu_j \geq 0, j \in J(\bar{x}); \xi_i \in \partial^c \left(\frac{f_i(\bar{x})}{g_i(\bar{x})} \right)$, and $\zeta_j \in \partial^c h_j(\bar{x})$, such that

$$\sum_{i=1}^{k} \lambda_i \xi_i = - \sum_{j \in J(\bar{x})} \mu_j \zeta_j.$$

Therefore, for any feasible x, we have

$$\sum_{i=1}^{k} \lambda_i \langle \xi_i, x - \bar{x} \rangle = - \sum_{j \in J(\bar{x})} \mu_j \langle \zeta_j, x - \bar{x} \rangle.$$

We observe that for any $x \in K$,

$$\sum_{i=1}^{k} \lambda_i \langle \xi_i, x - \bar{x} \rangle \geq 0. \qquad (6.30)$$

Otherwise, we would arrive at a contradiction as in the first part of Theorem 6.6. Since the set defined by (6.29) is bounded above, therefore, the following set

$$\left\{ (k-1) \frac{\lambda_j \bar{p}_i(\bar{x}, x)}{\lambda_i \bar{p}_j(\bar{x}, x)}, x \in K, \frac{f_i(\bar{x})}{g_i(\bar{x})} > \frac{f_i(x)}{g_i(x)}, \frac{f_j(\bar{x})}{g_j(\bar{x})} < \frac{f_i(x)}{g_i(x)} \right\}, 1 \leq i, j \leq k \qquad (6.31)$$

is also bounded from above. Let $M > 0$ be a real number that is an upper bound of the set defined by (6.31). Now, we shall show that x is a properly efficient solution of (NMFP). Assume that there exist r and $x \in K$, such that

$$\frac{f_r(x)}{g_r(x)} < \frac{f_r(\bar{x})}{g_r(\bar{x})}.$$

Then, by the pseudolinearity of $\frac{f_r}{g_r}$ there exists $\xi_r \in \partial^c \left(\frac{f_r(\bar{x})}{g_r(\bar{x})} \right)$, such that

$$p_r(\bar{x}, x) \langle \xi_r, x - \bar{x} \rangle < 0.$$

Since $p_r(\bar{x}, x) > 0$, we get

$$\langle \xi_r, x - \bar{x} \rangle < 0. \tag{6.32}$$

Let us define

$$\langle -\xi_s, x - \bar{x} \rangle = \min\{\lambda_i \langle \xi_i, x - \bar{x} \rangle \mid \langle \xi_i, x - \bar{x} \rangle > 0\}. \tag{6.33}$$

Using (6.30), (6.32), and (6.33), we get

$$\lambda_r \langle \xi_r, x - \bar{x} \rangle \leq (k-1)(-\lambda_s \langle \xi_r, x - \bar{x} \rangle).$$

Therefore,

$$\left(\frac{f_r(x)}{g_r(x)} - \frac{f_r(\bar{x})}{g_r(\bar{x})} \right) \leq (k-1) \frac{\lambda_s \bar{p}_r(\bar{x}, x)}{\lambda_r \bar{p}_s(\bar{x}, x)} \left(\frac{f_s(\bar{x})}{g_s(\bar{x})} - \frac{f_s(x)}{g_s(x)} \right)$$

Using the definition of M, we get

$$\left(\frac{f_r(x)}{g_r(x)} - \frac{f_r(\bar{x})}{g_r(\bar{x})} \right) \leq M \left(\frac{f_s(\bar{x})}{g_s(\bar{x})} - \frac{f_s(x)}{g_s(x)} \right).$$

Hence, $\bar{x}$ is a properly efficient solution of (NMFP).

6.6   Duality for (NMFP)

For the nonsmooth multiobjective fractional programming problem (NMFP), we consider the following Mond-Weir subgradient type dual problem

$$(NMFD) \quad \max \left(\frac{f_1(u)}{g_1(u)}, ..., \frac{f_k(u)}{g_k(u)} \right)$$

$$\text{subject to } 0 \in \sum_{i=1}^{k} \lambda_i \partial^c \left(\frac{f_i(u)}{g_i(u)} \right) + \partial^c \left(\mu^T h(u) \right), \tag{6.34}$$

$$\mu^T h(u) \geq 0,$$

$$\mu \geq 0,$$

$$\lambda_i > 0, i = 1, ..., k.$$

Theorem 6.9 (Weak duality) *Let y be a feasible solution for (NMFP) and (u, λ, μ) be a feasible solution for (NMFD), such that f_i and $g_i, i = 1, ..., k$ be pseudolinear functions with respect to the same proportional function p_i and $\mu^T h$ is pseudolinear with respect to q, then the following cannot hold*

$$\frac{f(y)}{g(y)} \leq \frac{f(u)}{g(u)}.$$

Proof We assume that the above inequality is satisfied. Using the pseudo-linearity of $\frac{f_i}{g_i}$ on K with respect to proportional function $\bar{p}_i$, there exist $\xi_i \in \partial^c \left(\frac{f_i(u)}{g_i(u)} \right)$, such that

$$\bar{p}_i(u, y) \langle \xi_i, y - u \rangle \leq 0, \forall i = 1, ..., k$$

and

$$\bar{p}_r(u, y) \langle \xi_r, y - u \rangle < 0, \text{ for some } r.$$

Since $\bar{p}_i(u, y) > 0$, for all $i = 1, ..., k$, we get

$$\langle \xi_i, y - u \rangle \leq 0, \forall i = 1, ..., k$$

and

$$\langle \xi_r, y - u \rangle < 0, \text{ for some } r.$$

Since $\lambda_i > 0$, for each $i = 1, ..., k$, we get

$$\sum_{i=1}^{k} \lambda_i \langle \xi_i, y - u \rangle < 0. \tag{6.35}$$

By (6.34) and (6.35), there exists $\zeta \in \partial^c \left(\mu^T g(u) \right)$, such that

$$\langle \zeta, x - u \rangle > 0. \tag{6.36}$$

As y is feasible for (NMFP) and (u, λ, μ) is feasible for (NMDP), it follows that

$$\mu^T h(y) \leq 0$$

and

$$\mu^T h(u) \geq 0.$$

Using pseudolinearity of $\mu^T h$ with respect to q, there exists $\zeta' \in \partial^c \left(\mu^T h(u) \right)$, such that

$$q(u, y) \langle \zeta', y - u \rangle \leq 0.$$

Again as $q(u, y) > 0$, it follows that

$$\langle \zeta', y - u \rangle \leq 0.$$

By Proposition 2.5, there exists $t > 0$, and $\zeta \in \partial^c \left(\mu^T h\left(u \right) \right)$, such that $\zeta' = t\zeta$. Hence, we get

$$t \left\langle \zeta, y - u \right\rangle \leq 0. \tag{6.37}$$

Since, $t > 0$, from (6.37) it follows that

$$\left\langle \zeta, y - u \right\rangle \leq 0,$$

which is a contradiction to (6.36).

In the following theorem, we have weakened the conditions of pseudolinearity on the objective function.

Theorem 6.10 *(Weak duality)* *If* y *is a feasible solution for the problem (NMFP) and* (u, λ, μ) *is a feasible solution for the problem (NMFD) involving pseudolinear functions, such that* $\sum_{i=1}^{k} \lambda_i \frac{f_i}{g_i}$ *is pseudolinear with respect to p and* $\mu^T h$ *is pseudolinear with respect to q, then the following inequality cannot hold*

$$\frac{f(y)}{g(y)} \leq \frac{f(u)}{g(u)}.$$

Proof Assume that the above inequality is satisfied. Since $p(x, y)$ for each $g(x) > 0$, we get

$$\sum_{i=1}^{k} \lambda_i \frac{f_i\left(y \right)}{g_i\left(y \right)} < \sum_{i=1}^{k} \lambda_i \frac{f_i\left(u \right)}{g_i\left(u \right)}.$$

Using the pseudolinearity of $\sum_{i=1}^{k} \lambda_i \frac{f_i}{g_i}$, inequality (6.35) of Theorem 6.9 is obtained. The rest of the proof is along the lines of the proof of Theorem 6.9.

Theorem 6.11 *Let us assume that* $\bar{x}$ *is a feasible solution for (NMFP) and* $\left(\bar{u}, \bar{\lambda}, \bar{\mu} \right)$ *is a feasible solution for (NMFD), such that*

$$\frac{f_i\left(\bar{x} \right)}{g_i\left(\bar{x} \right)} = \frac{f_i\left(\bar{u} \right)}{g_i\left(\bar{u} \right)}, \forall i = 1, ..., k. \tag{6.38}$$

If for all feasible solutions (u, λ, μ) *of (NMFD),* $\sum_{i=1}^{k} \lambda_i \frac{f_i}{g_i}$ *is pseudolinear with respect to p and* $\mu^T h$ *is pseudolinear with respect to q, then* $\bar{x}$ *is a properly efficient solution for (NMFP) and* $\left(\bar{u}, \bar{\lambda}, \bar{\mu} \right)$ *is a properly efficient solution for (NMFD).*

Proof Let us assume that $\bar{x}$ is not an efficient solution of (NMFP), then there exists some $y \in P$, such that

$$\frac{f_i\left(y \right)}{g_i\left(y \right)} \leq \frac{f_i\left(\bar{u} \right)}{g_i\left(\bar{u} \right)}, \forall i = 1, ..., k,$$

$$\frac{f_r\left(y \right)}{g_r\left(y \right)} < \frac{f_r\left(\bar{u} \right)}{g_r\left(\bar{u} \right)}, \text{ for some } r.$$

Now, by the given assumption $\frac{f_i(\bar{x})}{g_i(\bar{x})} = \frac{f_i(\bar{u})}{g_i(\bar{u})}$, for all $i = 1, ..., k$. Then, we arrive at a contradiction to Theorem 6.10. Hence $\bar{x}$ is an efficient solution for (NMFP). Proceeding in the same way we can prove that $(\bar{u}, \bar{\lambda}, \bar{\mu})$ is an efficient solution for (NMFD).

Now assume that $\bar{x}$ is an efficient solution for (NMFP). Therefore, for every scalar $M > 0$, there exists some $x^* \in P$ and an index i, such that

$$\frac{f_i(\bar{x})}{g_i(\bar{x})} - \frac{f_i(x^*)}{g_i(x^*)} > M\left(\frac{f_r(x^*)}{g_r(x^*)} - \frac{f_r(\bar{x})}{g_r(\bar{x})}\right),$$

for all r, satisfying

$$\frac{f_r(x^*)}{g_r(x^*)} > \frac{f_r(\bar{x})}{g_r(\bar{x})},$$

whenever

$$\frac{f_i(x^*)}{g_i(x^*)} < \frac{f_i(\bar{x})}{g_i(\bar{x})}.$$

Therefore, the difference $\left(\frac{f_i(\bar{x})}{g_i(\bar{x})} - \frac{f_i(x^*)}{g_i(x^*)}\right)$ can be made arbitrarily large and hence for $K \subseteq \mathbb{R}^n$, we get the following inequality

$$\sum_{i=1}^{k} \bar{\lambda}_i \left(\frac{f_i(\bar{x})}{g_i(\bar{x})} - \frac{f_i(x^*)}{g_i(x^*)}\right) > 0. \tag{6.39}$$

Since x^* is a feasible solution for the problem (NMFP) and $(\bar{u}, \bar{\lambda}, \bar{\mu})$ is a feasible solution for the problem (NMFD), there exist $\xi_i \in \partial^c \left(\frac{f_i(\bar{u})}{g_i(\bar{u})}\right)$, $i = 1, ..., k$ and $\zeta \in \partial^c \left(\mu^T h(\bar{u})\right)$, such that

$$h(x^*) \leqq 0, \tag{6.40}$$

$$\sum_{i=1}^{k} \bar{\lambda}_i \xi_i + \zeta = 0, \tag{6.41}$$

$$\bar{\mu}^T h(\bar{u}) \geq 0, \tag{6.42}$$

$$J(x) = \{j \in J : h_j(x) = 0\}, \tag{6.43}$$

$$\bar{\lambda}_i > 0, i = 1, ..., k. \tag{6.44}$$

Using (6.40), (6.42), and (6.43), we get

$$\bar{\mu}^T h(x^*) \leq \bar{\mu}^T h(\bar{u}).$$

Since $\bar{\mu}^T h$ is pseudolinear with respect to q, there exists $\zeta \in \partial^c \left(\bar{\mu}^T h(\bar{u})\right)$, such that

$$q(\bar{u}, x^*) \langle \zeta, (x^* - \bar{u}) \rangle \leq 0.$$

As $q(\bar{u}, x^*) > 0$, it follows that

$$\langle \zeta, (x^* - \bar{u}) \rangle \leq 0.$$

Using (6.41), there exists $\xi_i \in \partial^c \left(\frac{f_i(\bar{u})}{g_i(\bar{u})} \right)$, such that

$$\left\langle \sum_{i=1}^{k} \bar{\lambda}_i \xi_i, (x^* - u) \right\rangle \geq 0.$$

Using regularity assumption and the pseudoconvexity of $\sum_{i=1}^{k} \bar{\lambda}_i \frac{f_i}{g_i}$ with respect to $x \in P$ it follows that

$$\sum_{i=1}^{k} \bar{\lambda}_i \frac{f_i(x^*)}{g_i(x^*)} - \sum_{i=1}^{k} \bar{\lambda}_i \frac{f_i(\bar{u})}{g_i(\bar{u})} \geq 0. \tag{6.45}$$

Using (6.38) in (6.45), we get

$$\sum_{i=1}^{k} \bar{\lambda}_i \left(\frac{f_i(\bar{x})}{g_i(\bar{x})} - \frac{f_i(x^*)}{g_i(x^*)} \right) \geq 0,$$

which is a contradiction to (6.39). Hence, $\bar{x}$ is a properly efficient solution for (NMFP).

Now we assume that $(\bar{u}, \bar{\lambda}, \bar{\mu})$ is not a properly efficient solution for (NMFP). Therefore, for every scalar $M > 0$, there exist a feasible point (u^*, λ^*, μ^*) for (NMFD) and an index i, such that

$$\frac{f_i(u^*)}{g_i(u^*)} - \frac{f_i(\bar{u})}{g_i(\bar{u})} > M \left(\frac{f_r(\bar{u})}{g_r(\bar{u})} - \frac{f_r(u^*)}{g_r(u^*)} \right),$$

for all k, satisfying

$$\frac{f_r(u^*)}{g_r(u^*)} < \frac{f_r(\bar{u})}{g_r(\bar{u})},$$

whenever

$$\frac{f_i(u^*)}{g_i(u^*)} > \frac{f_i(\bar{u})}{g_i(\bar{u})}.$$

Therefore, the difference $\left(\frac{f_i(u^*)}{g_i(u^*)} - \frac{f_i(\bar{u})}{g_i(\bar{u})} \right)$ can be made arbitrarily large and hence for $\bar{\lambda} > 0$, we have

$$\sum_{i=1}^{k} \bar{\lambda}_i \left(\frac{f_i(u^*)}{g_i(u^*)} - \frac{f_i(\bar{u})}{g_i(\bar{u})} \right) > 0. \tag{6.46}$$

Since $\bar{x}$ and $(\bar{u}, \bar{\lambda}, \bar{\mu})$ are feasible solutions for (NMFP) and (NMFD), respectively, it follows as in the first part of the theorem that

$$\sum_{i=1}^{k} \bar{\lambda}_i \left(\frac{f_i(u^*)}{g_i(u^*)} - \frac{f_i(\bar{u})}{g_i(\bar{u})} \right) \leq 0,$$

which is a contradiction to (6.46). Therefore, $(\bar{u}, \bar{\lambda}, \bar{\mu})$ is a properly efficient solution for (NMFD).

Theorem 6.12 *(Strong duality)* *Let $\bar{x}$ be an efficient solution for the problem (NMFP). Then there exist $\bar{\lambda} \in \mathbb{R}^k$ and $\bar{\mu} \in \mathbb{R}^m$, such that $(\bar{x}, \bar{\lambda}, \bar{\mu})$ is a feasible solution for (NMFD). Further, let for all feasible solutions (u, λ, μ) for (NMFP), $\sum_{i=1}^{k} \lambda_i \frac{f_i}{g_i}$ be pseudolinear with respect to p and $\mu^T h$ be pseudolinear with respect to q. If $h_j, j \in J(\bar{x})$ be regular at $\bar{x}$, then $(\bar{x}, \bar{\lambda}, \bar{\mu})$ is a properly efficient solution for (NMFD).*

Proof The hypothesis of this theorem and Theorem 6.7 implies that there exist $\bar{\lambda} \in \mathbb{R}^k, \bar{\mu}_j \in \mathbb{R}, j \in J(\bar{x}), 1 \leq r \leq k$, and $\zeta_j \in \partial^c h_j(\bar{x})$, such that

$$\sum_{i=1}^{k} \bar{\lambda}_i \xi_i + \sum_{j \in J(\bar{x})} \bar{\mu}_j \zeta_j = 0,$$

$$\bar{\mu}_j \geq 0, \ j \in J(\bar{x}),$$

$$x \in K.$$

Let $\bar{\mu}_j$ for $j \in J(\bar{x})$, then there exist $\bar{\lambda} \in \mathbb{R}^k$ such that $(\bar{x}, \bar{\lambda}, \bar{\mu})$ is a feasible solution of (NMFD). The proper efficiency of $(\bar{x}, \bar{\lambda}, \bar{\mu})$ for the problem (NMFD) follows from Theorem 6.11.

Chapter 7

Static Minmax Programming and Pseudolinear Functions

7.1 Introduction

Minmax programming has been an interesting field of active research for a long time. These problems are of pivotal importance in many areas of modern research such as economics (Von Neumann and Morgenstern [277]), game theory (Schroeder [249]), rational Chebyshev approximation (Barrodale [15]), and portfolio selection (Bajona-Xandri and Legaz [12]). Schmitendorf [250] has established the necessary and sufficient optimality conditions for the following minmax problem:

$$(P) \quad \min_{} \ \sup_{y \in Y} \ f(x,y)$$

$$\text{subject to } h(x) \leqq 0,$$

where $f(.,.) : \mathbb{R}^n \times \mathbb{R}^m \to \mathbb{R}$ and $h(.) : \mathbb{R}^n \to \mathbb{R}^p$ are differentiable convex functions and Y is a compact subset of $\mathbb{R}^m$. Tanimoto [268] formulated two dual models for the problem (P) and established duality theorems under the convexity assumption. Chew [46] studied minmax programming problems (P) involving pseudolinear functions. Bector and Bhatia [19] and Weir [282] established sufficient optimality conditions for the problems (P) by relaxing the convexity assumptions to pseudoconvexity and quasiconvexity and employed optimality conditions to formulate several dual models and derived weak and strong duality theorems.

Chandra and Kumar [41] have formulated two modified dual models and established the duality results for the fractional analogue of the problem (P). Liu et al. [166] have derived necessary and sufficient optimality conditions and presented several duality results for fractional minmax programming problems involving pseudoconvex and quasiconvex functions. Optimality conditions and duality theorems for minmax fractional programming problems under several generalized convexity assumptions have been studied by many authors. For details, we refer to Liu and Wu [167, 168], Liang and Shi [164], Yang and Hou [289], and the references cited therein. Mishra [187] has established necessary and sufficient optimality conditions and duality results for pseudolinear and η-pseudolinear minmax fractional programming problems.

In this chapter, we present the necessary and sufficient optimality conditions for static minmax programming and static minmax fractional programming problems involving pseudolinear functions. Furthermore, we present the weak and strong duality theorem established by Chew [46] and Mishra [187] for corresponding Mond-Weir dual models.

We considered the following static minmax programming problem:

$$(MP) \quad \min \ f(x) := \sup_{y \in Y} \varphi(x, y)$$

$$\text{subject to } g_j(x) \leq 0, j \in J = \{1, ..., r\},$$

where $\varphi(.,.) : X \times \mathbb{R}^m \to \mathbb{R}, g_j(.) : X \to \mathbb{R}, j \in J$ are differentiable functions, X is an open convex subset of $\mathbb{R}^n$, and Y is a specified compact subset of $\mathbb{R}^m$. Let the feasible region of the problem (FP) be denoted by $S =: \{x \in X : g_j(x) \leq 0, j \in J\}$. We define the following sets for every $x \in S$:

$$J(x) =: \{j \in J : g_j(x) = 0\},$$

$$Y(x) =: \{y \in Y : \varphi(x, y) = \sup_{z \in Y} \varphi(x, z)\},$$

$$K(x) =: \{(k, \lambda, \bar{y}) : 1 \leq k \leq n+1, \lambda = (\lambda_1, ..., \lambda_k) \in \mathbb{R}_+^k,$$

$$\sum_{i=1}^{k} \lambda_i = 1 \ \text{and} \ \bar{y} = (y_1, ..., y_k), y_i \in Y(x), i = 1, ..., k\}.$$

Definition 7.1 (Mangasarian Fromovitz constraint qualification [176]): *The problem (MP) is said to satisfy the Mangasarian Fromovitz constraint qualification at $\bar{x}$, if*

$$\sum_{j=1}^{r} \bar{\mu}_j \nabla_x h_j(\bar{x}) = 0 \ \text{and} \ \sum_{j=1}^{r} \bar{\mu}_j h_j(\bar{x}) = 0 \Rightarrow \bar{\mu}_j = 0, j = 1, ..., r.$$

It is clear that Mangasarian Fromovitz constraint qualification holds if the active constraints have linearly independent gradients.

Definition 7.2 (Slater's weak constraint qualification) *The problem (MP) is said to satisfy Slater's weak constraint qualification at $\bar{x}$, if h is pseudolinear at $\bar{x}$ and there exists a feasible point $\hat{x}$, such that*

$$h_j(\hat{x}) < 0, j \in J.$$

7.2 Necessary and Sufficient Optimality Conditions for (MP)

Schmitendorf [250] has established the following necessary Fritz John optimality conditions for (MP):

Theorem 7.1 *If $\bar{x}$ minimizes (MP), then there exist positive integer $\bar{k}$, scalar $\bar{\lambda}_i \geq 0, i = 1, ..., \bar{k}$, scalars $\bar{\mu}_j \geq 0, j = 1, ..., m$ and vectors $y^i \in Y(\bar{x}), i = 1, ..., \bar{k}$, such that*

$$\sum_{i=1}^{\bar{k}} \bar{\lambda}_i \nabla_x \varphi(\bar{x}, y^i) + \sum_{j=1}^{r} \bar{\mu}_j \nabla_x g_j(\bar{x}) = 0, \tag{7.1}$$

$$\bar{\mu}_j g_j(\bar{x}) = 0, j \in J, \tag{7.2}$$

$$(\bar{\lambda}_i, i = 1, ..., \bar{k}; \bar{\mu}_j, j \in J) \neq 0. \tag{7.3}$$

Schmitendorf [250] has established the following Karush-Kuhn-Tucker type optimality conditions under the Mangasarian Fromovitz constraint qualification.

Theorem 7.2 *Suppose that $\bar{x}$ minimizes (MP) and the Mangasarian Fromovitz constraint qualification is satisfied at $\bar{x}$, then there exist positive integer $\bar{k}, 1 \leq \bar{k} \leq n + 1$, scalar $\bar{\lambda}_i \geq 0, i = 1, ..., \bar{k}$, not all zero, scalars $\bar{\mu}_j \geq 0, j = 1, ..., m$ and vectors $y^i \in Y(\bar{x}), i = 1, ..., \bar{k}$, such that*

$$\sum_{i=1}^{\bar{k}} \bar{\lambda}_i \nabla_x \varphi(\bar{x}, y^i) + \sum_{j=1}^{r} \bar{\mu}_j \nabla_x h_j(\bar{x}) = 0, \tag{7.4}$$

$$\bar{\mu}_j h_j(\bar{x}) = 0, j \in J. \tag{7.5}$$

Proof If $\bar{x}$ minimizes (MP) and the Mangasarian Fromovitz constraint qualification is satisfied at $\bar{x}$, then the condition

$$\sum_{j=1}^{r} \bar{\mu}_j h_j(\bar{x}) = 0 \text{ and } \sum_{j=1}^{r} \bar{\mu}_j \nabla_x h_j(\bar{x}) = 0 \Rightarrow \bar{\mu}_j = 0, j = 1, ..., r,$$

will imply that $\bar{\lambda} = (\bar{\lambda}_1, ..., \bar{\lambda}_{\bar{k}}) \neq 0$ for if $\bar{\lambda} = 0$, then $\bar{\mu} = 0$ will contradict the Fritz John condition.

Now, we have the following Karush-Kuhn-Tucker type optimality conditions under Slater's weak constraint qualification:

Theorem 7.3 *Suppose that $\bar{x}$ minimizes (MP) and Slater's weak constraint qualification is satisfied at $\bar{x}$, then there exist positive integer $\bar{k}, 1 \leq \bar{k} \leq n+1$, scalar $\bar{\lambda}_i \geq 0, i = 1, ..., \bar{k}$ not all zero, scalars $\bar{\mu}_j \geq 0, j = 1, ..., m$ and vectors $y^i \in Y(\bar{x}), i = 1, ..., \bar{k}$, such that*

$$\sum_{i=1}^{\bar{k}} \bar{\lambda}_i \nabla_x \varphi(\bar{x}, y^i) + \sum_{j=1}^{r} \bar{\mu}_j \nabla_x h_j(\bar{x}) = 0, \tag{7.6}$$

$$\bar{\mu}_j h_j(\bar{x}) = 0, j \in J. \tag{7.7}$$

Proof Since, $\bar{x}$ minimizes (MP), therefore by Fritz John optimality conditions, there exist positive integer $\bar{k}$, scalar $\bar{\lambda}_i \geq 0, i = 1, ..., \bar{k}$, not all zero, scalars $\bar{\mu}_j \geq 0, i = 1, ..., m$ and vectors $y^i \in Y(\bar{x}), i = 1, ..., \bar{k}$, such that

$$\sum_{i=1}^{\bar{k}} \bar{\lambda}_i \nabla_x \varphi(\bar{x}, y^i) + \sum_{j=1}^{r} \bar{\mu}_j \nabla_x h_j(\bar{x}) = 0, \qquad (7.8)$$

$$\bar{\mu}_j h_j(\bar{x}) = 0, j \in J, \qquad (7.9)$$

$$(\bar{\lambda}_i, i = 1, ..., \bar{k}; \bar{\mu}_j, j \in J) \neq 0. \qquad (7.10)$$

Since, Slater's weak constraint qualification is satisfied at $\bar{x}$, therefore, $h_j, j \in J$ is pseudolinear at $\bar{x}$ and there exists a feasible point $\hat{x}$, such that

$$h_j(\hat{x}) < 0, j \in J.$$

Suppose contrary that $\lambda_i = 0$ for all $i = 1, ..., \bar{k}$, then by (7.10), we get $\bar{\mu}_j, j \in J \neq 0$. Also, by (7.8) and (7.9), we have

$$\sum_{j=1}^{r} \bar{\mu}_j \nabla_x h_j(\bar{x}) = 0, \qquad (7.11)$$

$$\bar{\mu}_j h_j(\bar{x}) = 0, j \in J.$$

Therefore, for $j \notin J(\bar{x})$, we have $\mu_j = 0$ and for $j \in J(\bar{x})$, we have

$$h_j(\hat{x}) - h_j(\bar{x}) < 0.$$

Therefore, by pseudolinearity of h_j at $\bar{x}$, there exists a proportional function $q(\bar{x}, \hat{x})$, such that

$$q(\bar{x}, \hat{x}) \langle \nabla_x h_j(\bar{x}), \hat{x} \rangle - \bar{x} \rangle < 0, j \in J(\bar{x}).$$

Since $q(\bar{x}, \hat{x}) > 0$, we get

$$\langle \nabla_x h_j(\bar{x}), \hat{x} - \bar{x} \rangle < 0, j \in J(\bar{x}).$$

Taking into account that $\bar{\mu}_j \neq 0, j \in J$ and $\mu_j = 0, j \notin J(\bar{x})$, it follows that

$$\left\langle \nabla_x \sum_{j=1}^{r} \bar{\mu}_j h_j(\bar{x}), \hat{x} - \bar{x} \right\rangle < 0,$$

which is a contradiction to (7.11). Hence, $(\lambda_i = 0, i = 1, ..., k) \neq 0$.

Now, we establish the following sufficient optimality conditions for (MP):

Theorem 7.4 *Suppose there exist positive integer* $\bar{k}, 1 \leq \bar{k} \leq n+1$, *scalar* $\bar{\lambda}_i \geq 0, i = 1, ..., \bar{k}$, *not all zero, scalars* $\bar{\mu}_j \geq 0, j = 1, ..., m$ *and vectors* $y^i \in Y(\bar{x}), i = 1, ..., \bar{k}$, *such that*

$$\sum_{i=1}^{\bar{k}} \bar{\lambda}_i \nabla_x \varphi(\bar{x}, y^i) + \sum_{j=1}^{r} \bar{\mu}_j \nabla_x h_j(\bar{x}) = 0, \qquad (7.12)$$

$$\bar{\mu}_j h_j(\bar{x}) = 0, j \in J. \qquad (7.13)$$

If $\varphi(., y)$ *is pseudolinear with respect to the proportional function* p *for each* $y \in Y$ *and* $h(.)$ *is pseudolinear with respect to the proportional function* q, *then* $\bar{x}$ *is a minmax solution to (MP).*

Proof Suppose that the condition of the theorem is satisfied but $\bar{x}$ is not a minmax solution to (MP). Then there exists a feasible x, such that

$$\sup_{y \in Y} \varphi(x, y) < \sup_{y \in Y} \varphi(\bar{x}, y).$$

Now,

$$\sup_{y \in Y} \varphi(\bar{x}, y) = \varphi(\bar{x}, y^i), i = 1, ..., \bar{k}.$$

and

$$\varphi(x, y^i) \leq \sup_{y \in Y} \varphi(\bar{x}, y^i), i = 1, \ldots, \bar{k}.$$

Therefore, we have

$$\varphi(x, y^i) < \varphi(\bar{x}, y^i), i = 1, ..., \bar{k}.$$

Hence, by the pseudolinearity of $\varphi(x, y^i)$, it follows that

$$p_i(\bar{x}, x) \left\langle \nabla_x(\varphi(\bar{x}, y^i)), x - \bar{x} \right\rangle < 0, i = 1, ..., \bar{k},$$

Since, $p_i(\bar{x}, x) > 0$, for all $i = 1, ..., \bar{k}$, we get

$$\left\langle \nabla_x(\varphi(\bar{x}, y^i)), x - \bar{x} \right\rangle < 0, i = 1, ..., \bar{k}.$$

Since, $\bar{\lambda} = (\bar{\lambda}_1, ..., \bar{\lambda}_{\bar{k}}) \neq 0$, it follows that

$$\sum_{i=1}^{\bar{k}} \bar{\lambda}_i \left\langle \nabla_x(\varphi(\bar{x}, y^i)), x - \bar{x} \right\rangle < 0, i = 1, ..., \bar{k}. \qquad (7.14)$$

Since, x is feasible for (MP), we have

$$\bar{\mu}_j h_j(x) - \bar{\mu}_j h_j(\bar{x}) \leq 0, j = 1, ..., r.$$

Using pseudolinearity of $h_j(.)$, we get

$$q(\bar{x}, x) \bar{\mu}_j \langle \nabla_x h_j(\bar{x}), x - \bar{x} \rangle \leq 0, j = 1, ..., r.$$

Again as, $q(\bar{x}, x) > 0$, it follows that

$$\sum_{j=1}^{r} \bar{\mu}_j \langle \nabla_x h_j(\bar{x}), x - \bar{x} \rangle \leq 0. \qquad (7.15)$$

Adding (7.14) and (7.15), we arrive at a contradiction to (7.10).

7.3   Duality for (MP)

For the minmax programming problem (MP), we consider the following dual similar to Weir [282]:

$$(MD) \qquad \max t$$

$$\text{subject to } \lambda_i[\varphi(u, w^i) - t] \geq 0, i = 1, ..., k, 1 \leq k \leq n + 1,$$

$$\sum_{j=1}^{r} \mu_j h_j(u) \geq 0, (k, \lambda, w) \in \Omega, (u, \mu) \in \Theta(k, \lambda, w),$$

where Ω denotes the triplet (k, λ, w), where k ranges over the integers $1 \leq k \leq n + 1, \lambda = (\lambda_i, i = 1, ..., k), \lambda_i \geq 0, i = 1, ..., k, \lambda \neq 0, w = (w_i, i = 1, ..., k)$ with $w_i \in Y$, for all $i = 1, ..., k$ and

$$\Theta(k, \lambda, w) :=$$

$$\left\{ (u, \mu) \in \mathbb{R}^n \times \mathbb{R}^m : \sum_{i=1}^{k} \lambda_i \nabla_x \varphi(u, w^i) + \sum_{j=1}^{m} \mu_j \nabla_x h_j(u) = 0, \mu \geq 0 \right\}.$$

Theorem 7.5 (Weak duality) *Let $\varphi(., y)$ be pseudolinear with respect to the proportional function p and let*

$$\sum_{j=1}^{r} \mu_j h_j(.)$$

be pseudolinear with respect to the proportional function q for all feasible x for (MP) and for all feasible $(k, \lambda, w) \in \Omega, (u, \mu) \in \Theta(k, \lambda, w)$ for (MD). Then

$$\inf(MP) \geq \sup(MD).$$

Proof Suppose that there exist x feasible for (MP) and feasible $(k, \lambda, w) \in \Omega, (u, \mu) \in \Theta(k, \lambda, w)$ for (MD) such that

$$\sup_{y \in Y} \varphi(x, y) < t.$$

$$\varphi(x, y) < t, \quad \forall\, y \in Y.$$

Hence,

$$\lambda_i \varphi(x, y) < \lambda_i t, \quad \forall\, i = 1, ..., k,$$

with at least one strict inequality since $\lambda \neq 0$. From the dual constraints, it follows that

$$\lambda_i \varphi(x, y) < \lambda_i \varphi(u, w^i), \quad \forall\, y \in Y \text{ and } i = 1, ..., k,$$

with at least one strict inequality. Therefore

$$\lambda_i \varphi(x, w^i) < \lambda_i \varphi(u, w^i), i = 1, ..., k,$$

By the pseudolinearity of $\varphi(., y)$ with respect to the proportional function p, it follows that

$$p_i(u, x)\lambda_i \left\langle \nabla_x \varphi(u, w^i), x - u \right\rangle \leq 0, i = 1, ..., k,$$

with at least one strict inequality. Since, $p_i(u, x) > 0$, we get

$$\sum_{i=1}^{k} \lambda_i \left\langle \nabla_x \varphi(u, w^i), x - u \right\rangle < 0. \tag{7.16}$$

By the feasibility of x for (MP) and the feasibility of $(k, \lambda, w) \in \Omega, (u, \mu) \in \Theta(k, \lambda, w)$ for (MD), we get

$$\sum_{j=1}^{r} \mu_j h_j(x) - \sum_{j=1}^{r} \mu_j h_j(u) \leq 0.$$

The pseudolinearity of $\sum_{j=1}^{r} \mu_j h_j(.)$ with respect to the proportional function q, implies that

$$q(u, x) \left\langle \sum_{j=1}^{r} \mu_j \nabla x h_j(u), x - u \right\rangle \leq 0.$$

Since, $q(u, x) > 0$, we get

$$\left\langle \sum_{j=1}^{r} \mu_j \nabla_x h_j(u), x - u \right\rangle \leq 0. \tag{7.17}$$

Adding (7.16) and (7.17), we arrive at a contradiction to the equality constraint of the dual problem (MD).

Theorem 7.6 (Strong duality) *Let $\bar{x}$ be optimal for (MP) and let a Slater's weak constraint qualification or Mangasarian Fromovitz constraint qualification be satisfied. Then there exist $(\bar{k}, \bar{\lambda}, \bar{y}) \in \Omega, \bar{\mu}_j \in \mathbb{R}, j \in J, \bar{\mu}_j \geq 0$ with $(\bar{x}, \bar{\mu}) \in \Theta(k, \lambda, w)$ such that $(\bar{k}, \bar{\lambda}, \bar{y})$ and $(\bar{x}, \bar{\mu})$ are feasible for (MD). Moreover, if $\varphi(., y)$ is pseudolinear with respect to the same proportional function p and*

$$\sum_{j=1}^{r} \mu_j h_j(.)$$

is pseudolinear with respect to the proportional function q for all feasible x for (MP) and for all feasible $(k, \lambda, w) \in \Omega, (u, \mu) \in \Theta(k, \lambda, w)$ for (MD), then $(\bar{x}, \bar{\mu}, \bar{k}, \bar{\lambda}, \bar{y})$ is an optimal solution for (MD).

Proof Since, $\bar{x}$ is an optimal solution for (MP) and a suitable constraint qualification is satisfied, then Theorem 7.2 or Theorem 7.3 guarantee the existence
of a positive integer $\bar{k}, 1 \leq \bar{k} \leq n+1$, scalars $\bar{\lambda}_i \geq 0, i = 1, ..., \bar{k}$, not all zero,
scalars $\bar{\mu}_j \geq 0, j \in J$, vectors $y^i \in Y(\bar{x}), i = 1, ..., \bar{k}$, such that

$$\sum_{i=1}^{\bar{k}} \bar{\lambda}_i \nabla_x \varphi(\bar{x}, y^i) + \sum_{j=1}^{r} \bar{\mu}_j \nabla_x h_j(\bar{x}) = 0,$$

$$\bar{\mu}_j h_j(\bar{x}) = 0, j \in J.$$

Thus, denoting $\bar{y} = (y^1, ..., y^{\bar{k}})$ and $\bar{\lambda} = (\bar{\lambda}_1, ..., \bar{\lambda}_{\bar{k}}), (\bar{k}, \bar{\lambda}, \bar{y}) \in \Omega, (\bar{x}, \bar{\mu}) \in$
$\Theta(\bar{k}, \bar{\lambda}, \bar{y})$ and $t = \varphi(\bar{x}, y^i), i = 1, ..., \bar{k}$ are feasible for the dual and the values
of the primal and dual problems are equal. The optimality of $(\bar{x}, \bar{\mu}, \bar{k}, \bar{\lambda}, \bar{y})$ for
(MD) follows by the weak duality theorem.

7.4   Fractional Minmax Programming Problem

Mishra [187] considered the following fractional minmax programming
problem:

$$(FMP) \quad \min \frac{\varphi(x)}{\psi(x)} := \sup_{y \in Y} \left(\frac{\varphi(x,y)}{\psi(x,y)} \right)$$

$$\text{subject to } h_j(x) \leq 0, \quad j \in J := \{1, ..., r\},$$

where $\varphi(.,.) : X \times \mathbb{R}^m \to \mathbb{R}, \psi(.,.) : X \times \mathbb{R}^m \to \mathbb{R}, h_j(.) : X \to \mathbb{R}$ are
differentiable functions, X is an open convex subset of $\mathbb{R}^n$, and Y is a specified
compact subset of $\mathbb{R}^m$. Moreover, suppose that $\psi(x,y) > 0$, for all $x \in X$
and $y \in Y$. Let the feasible region of the problem (FMP) be denoted by
$S =: \{x \in X : h(x) \leq 0\}$. We define the following sets for every $x \in S$:

$$J(x) := \{j \in J : h_j(x) = 0\},$$

$$Y(x) := \left\{ y \in Y : \frac{f(x,y)}{g(x,y)} := \sup_{z \in Y} \left(\frac{f(x,z)}{g(x,z)} \right) \right\},$$

and

$$K(x) := \{(k, \lambda, \bar{y}) : 1 \leq k \leq n+1, \lambda = (\lambda_1, ..., \lambda_k) \in \mathbb{R}_+^k,$$

$$\sum_{i=1}^{k} \lambda_i = 1 \text{ and } \bar{y} = (y_1, ..., y_k), y_i \in Y(x), i = 1, ..., k\}.$$

7.5 Necessary and Sufficient Optimality Conditions for (FMP)

Mishra and Mukherjee [195] established the following necessary Fritz John optimality conditions:

Theorem 7.7 *If $\bar{x}$ minimizes (FMP), then, there exist positive integer $\bar{k}$, scalars $\bar{\lambda}_i \geq 0, i = 1, ..., \bar{k}$, scalars $\bar{\mu}_j \geq 0, j = 1, ..., m$, and vectors $y^i \in Y(\bar{x}), i = 1, ..., \bar{k}$, such that*

$$\sum_{i=1}^{\bar{k}} \bar{\lambda}_i \nabla_x \left(\frac{f(\bar{x}, y^i)}{g(\bar{x}, y^i)} \right) + \sum_{j=1}^{r} \bar{\mu}_j \nabla_x h_j(\bar{x}) = 0, \tag{7.18}$$

$$\bar{\mu}_j h_j(\bar{x}) = 0, j \in J. \tag{7.19}$$

$$(\bar{\lambda}_i, i = 1, ..., \bar{k}, \bar{\mu}_j, j \in J) \neq 0. \tag{7.20}$$

Mishra [187] established the following Karush-Kuhn-Tucker type optimality conditions under Slater's weak constraint qualification:

Theorem 7.8 *Suppose that $\bar{x}$ minimizes (FMP) and Slater's weak constraint qualification is satisfied at $\bar{x}$, then, there exist positive integer $\bar{k}, 1 \leq \bar{k} \leq n+1$, scalars $\bar{\lambda}_i \geq 0, i = 1, ..., \bar{k}$, not all zero, scalars $\bar{\mu}_j \geq 0, j = 1, ..., m$, and vectors $y^i \in Y(\bar{x}), i = 1, ..., \bar{k}$ such that*

$$\sum_{i=1}^{\bar{k}} \bar{\lambda}_i \nabla_x \left(\frac{f(\bar{x}, y^i)}{g(\bar{x}, y^i)} \right) + \sum_{j=1}^{r} \bar{\mu}_j \nabla_x h_j(\bar{x}) = 0, \tag{7.21}$$

$$\bar{\mu}_j h_j(\bar{x}) = 0, j \in J. \tag{7.22}$$

Proof Since, $\bar{x}$ minimizes (FMP), therefore by Fritz John condition there exist positive integer $\bar{k}$, scalars $\bar{\lambda}_i \geq 0, i = 1, ..., \bar{k}$, scalars $\bar{\mu}_j \geq 0, j = 1, ..., m$, and vectors $y^i \in Y(\bar{x}), i = 1, ..., \bar{k}$ such that

$$\sum_{i=1}^{\bar{k}} \bar{\lambda}_i \nabla_x \left(\frac{f(\bar{x}, y^i)}{g(\bar{x}, y^i)} \right) + \sum_{j=1}^{r} \bar{\mu}_j \nabla_x h_j(\bar{x}) = 0, \tag{7.23}$$

$$\bar{\mu}_j h_j(\bar{x}) = 0, j \in J, \tag{7.24}$$

$$(\bar{\lambda}_i, i = 1, ..., \bar{k}, \bar{\mu}_j, j \in J) \neq 0. \tag{7.25}$$

Since, Slater's constraint qualification is satisfied at $\bar{x}$, therefore, $h_j, j \in J$ is pseudolinear at $\bar{x}$ and there exists a feasible point $\hat{x}$, such that

$$h_j(\hat{x}) < 0, j \in J.$$

Suppose to the contrary, that $\lambda_i = 0$, for all $i = 1, ..., \bar{k}$, then by (7.25), we get $(\bar{\mu}_j, j \in J) \neq 0$. Also, by (7.23) and (7.24), we have

$$\sum_{j=1}^{r} \bar{\mu}_j \nabla_x h_j(\bar{x}) = 0, \qquad (7.26)$$

$$\bar{\mu}_j h_j(\bar{x}) = 0, j \in J.$$

Therefore, for $j \notin J(\bar{x})$, we have $\bar{\mu}_j = 0$ and for $j \in J(\bar{x})$, we have

$$h_j(\hat{x}) - h_j(\bar{x}) < 0.$$

Therefore, by pseudolinearity of h_j at $\bar{x}$, there exists a proportional function $p(\bar{x}, \hat{x})$, such that

$$p(\bar{x}, \hat{x})\langle \nabla_x h_j(\bar{x}), x - \bar{x} \rangle < 0, j \in J(\bar{x}).$$

Since, $p(\bar{x}, \hat{x}) > 0$, we get

$$\langle \nabla_x h_j(\bar{x}), x - \bar{x} \rangle < 0, j \in J(\bar{x}).$$

Taking into account that $(\bar{\mu}_j, j \in J) \neq 0$ and $\mu_j = 0, j \notin J(\bar{x})$, it follows that

$$\sum_{j=1}^{r} \bar{\mu}_j \langle \nabla_x h_j(\bar{x}), x - \bar{x} \rangle < 0,$$

which is a contradiction to (7.26). Hence, $(\lambda_i = 0, i = 1, ..., k) \neq 0$.

Remark 7.1 *If $\bar{x}$ minimizes (FMP) and the Mangasarian Fromovitz constraint qualification is satisfied at $\bar{x}$, then the condition*

$$\sum_{j=1}^{r} \bar{\mu}_j \nabla_x h_j(\bar{x}) = 0 \text{ and } \sum_{j=1}^{r} \bar{\mu}_j h_j(\bar{x}) = 0 \Rightarrow \bar{\mu}_j = 0, j = 1, ..., r$$

will imply that $\bar{\lambda} = (\bar{\lambda}_1, ..., \bar{\lambda}_{\bar{k}}) \neq 0$. Otherwise, if $\bar{\lambda} = 0$, then $\bar{\mu} = 0$ will contradict the Fritz John condition. It is clear that Mangasarian Fromovitz constraint qualification holds if the active constraints have linearly independent gradients.

The following sufficient optimality conditions for (FMP) is from Mishra [187]:

Theorem 7.9 *Suppose there exist positive integer $\bar{k}, 1 \leq \bar{k} \leq n + 1$, scalar $\bar{\lambda}_i \geq 0, i = 1, ..., \bar{k}$, not all zero, scalars $\bar{\mu}_j \geq 0, j = 1, ..., m$, and vectors $y^i \in Y(\bar{x}), i = 1, ..., \bar{k}$, such that*

$$\sum_{i=1}^{\bar{k}} \bar{\lambda}_i \nabla_x \left(\frac{\varphi(\bar{x}, y^i)}{\psi(\bar{x}, y^i)} \right) + \sum_{j=1}^{r} \bar{\mu}_j \nabla_x h_j(\bar{x}) = 0, \qquad (7.27)$$

$$\bar{\mu}_j h_j(\bar{x}) = 0, j \in J. \tag{7.28}$$

If $\varphi(., y)$ and $\psi(., y)$ are pseudolinear with respect to proportional function p for every $y \in Y$ and $h(.)$ is pseudolinear with respect to proportional function q, then $\bar{x}$ is a minmax solution to (FMP).

Proof Suppose that the condition of the theorem is satisfied but $\bar{x}$ is not a minmax solution to (FMP). Then there exists a feasible x, such that

$$\sup_{y \in Y} \frac{\varphi(x, y)}{\psi(x, y)} < \sup_{y \in Y} \frac{\varphi(\bar{x}, y)}{\psi(\bar{x}, y)}.$$

Now,

$$\sup_{y \in Y} \frac{\varphi(\bar{x}, y)}{\psi(\bar{x}, y)} = \frac{\varphi(\bar{x}, y^i)}{\psi(\bar{x}, y^i)}, i = 1, ..., \bar{k},$$

and

$$\frac{\varphi(x, y^i)}{\psi(x, y^i)} \leq \sup_{y \in Y} \frac{\varphi(\bar{x}, y^i)}{\psi(\bar{x}, y^i)}, i = 1, ..., \bar{k},$$

Hence, we have

$$\frac{\varphi(x, y^i)}{\psi(x, y^i)} \leq \frac{\varphi(\bar{x}, y^i)}{\psi(\bar{x}, y^i)}, i = 1, ..., \bar{k}.$$

Therefore, by the pseudolinearity of $\frac{\varphi(x, y^i)}{\psi(x, y^i)}$, it follows that

$$\bar{p}_i(\bar{x}, x) \left\langle \nabla_x \left(\frac{\varphi(\bar{x}, y^i)}{\psi(\bar{x}, y^i)} \right), x - \bar{x} \right\rangle < 0, i = 1, ..., \bar{k},$$

where $\bar{p}_i(\bar{x}, x) = p_i(\bar{x}, x) \frac{\varphi(x, y^i)}{\psi(\bar{x}, y^i)}$.

Since, $\bar{p}_i(\bar{x}, x) > 0$, for all $i = 1, ..., \bar{k}$, we get

$$\left\langle \nabla_x \left(\frac{\varphi(\bar{x}, y^i)}{\psi(\bar{x}, y^i)} \right), x - \bar{x} \right\rangle < 0, i = 1, ..., \bar{k}.$$

Since, $\bar{\lambda} = (\bar{\lambda}_1, ..., \bar{\lambda}_{\bar{k}})$, it follows that

$$\sum_{i=1}^{\bar{k}} \bar{\lambda}_i \left\langle \nabla_x \left(\frac{\varphi(\bar{x}, y^i)}{\psi(\bar{x}, y^i)} \right), x - \bar{x} \right\rangle < 0, i = 1, ..., \bar{k}. \tag{7.29}$$

Since, x is feasible for (FMP), we have

$$\bar{\mu}_j h_j(x) - \bar{\mu}_j h_j(\bar{x}) \leq 0, j = 1, ..., r.$$

Using pseudolinearity of $h_j(.)$, we get

$$q(\bar{x}, x)\bar{\mu}_j \langle \nabla_x h_j(\bar{x}), x - \bar{x} \rangle \leq 0, j = 1, ..., r.$$

Again as, $q(\bar{x}, x) > 0$, it follows that

$$\sum_{j=1}^{r} \bar{\mu}_j \langle \nabla_x h_j(\bar{x}), x - \bar{x} \rangle \leq 0. \tag{7.30}$$

Adding (7.29) and (7.30), we arrive at a contradiction to (7.27).

7.6   Duality for (FMP)

For the fractional minmax programming problem (FMP), we consider the following dual similar to Weir [282]:

$$(FMD) \quad \max t$$

$$\text{subject to } \lambda_i \left[\frac{\varphi(u, w^i)}{\psi(u, w^i)} - t \right] \geq 0, i = 1, ..., k, 1 \leq k \leq n + 1,$$

$$\sum_{j=1}^{r} \mu_j h_j(u) \geq 0, (k, \lambda, w) \in \Omega, (u, \mu) \in \Theta(k, \lambda, w),$$

where Ω denotes the triplet (k, λ, w), where k ranges over the integers $1 \leq k \leq n + 1, \lambda = (\lambda_i, i = 1, ..., k), \lambda_i \geq 0, i = 1, ..., k, \lambda \neq 0, w = (w_i, i = 1, ..., k)$ with $w_i \in Y$, for all $i = 1, ..., k$ and

$$\Theta(k, \lambda, w) :=$$

$$\left\{ (u, \mu) \in \mathbb{R}^n \times \mathbb{R}^m : \sum_{i=1}^{k} \lambda_i \nabla_x \left(\frac{\varphi(u, w^i)}{\psi(u, w^i)} \right) + \sum_{j=1}^{m} \mu_j \nabla_x h_i(u) = 0, \mu \geq 0 \right\}.$$

Theorem 7.10 *(Weak duality)* *Let* $\varphi(., y)$ *and* $\psi(., y)$ *be pseudolinear with respect to the proportional function* p *and let*

$$\sum_{j=1}^{r} \mu_j h_j(.)$$

be pseudolinear with respect to the proportional function q, *for all feasible* x *for (FMP) and for all feasible* $(k, \lambda, w) \in \Omega, (u, \mu) \in \Theta(k, \lambda, w)$ *for (FMD). Then*

$$\inf(FMP) \geq \sup(FMD).$$

Proof Suppose that there exist x feasible for (FMP) and feasible $(k, \lambda, w) \in \Omega(u, \mu) \in \Theta(k, \lambda, w)$ for (FMD), such that

$$\sup_{y \in Y} \left(\frac{\varphi(x, y)}{\psi(x, y)} \right) < t.$$

Then

$$\left(\frac{\varphi(x, y)}{\psi(x, y)} \right) < t, \quad \forall \, y \in Y.$$

Hence,

$$\lambda_i \left(\frac{\varphi(x, y)}{\psi(x, y)} \right) < \lambda_i t, \forall \, i = 1, ..., k,$$

with at least one strict inequality, since $\lambda \neq 0$. From the dual constraints, it follows that

$$\lambda_i \left(\frac{\varphi(x,y)}{\psi(x,y)} \right) < \lambda_i \left(\frac{\varphi(u,w^i)}{\psi(u,w)} \right), \quad \forall \, y \in Y \text{ and } i = 1,...,k,$$

with at least one strict inequality. Therefore,

$$\lambda_i \left(\frac{\varphi(x,w^i)}{\psi(x,w^i)} \right) < \lambda_i \left(\frac{\varphi(u,w^i)}{\psi(u,w^i)} \right), \forall \, i = 1,...,k.$$

By the pseudolinearity of $\varphi(.,y)$ and $\psi(.,y)$ with respect to the proportional function p, it follows that

$$p_i(u,x)\lambda_i \left\langle \nabla_x \left(\frac{\varphi(u,w^i)}{\psi(u,w^i)} \right), x - u \right\rangle \leq 0, i = 1,...,k,$$

with at least one strict inequality. Since, $p_i(u,x) > 0$, we get

$$\sum_{i=1}^{k} \lambda_i \left\langle \nabla_x \left(\frac{\varphi(u,w^i)}{\psi(u,w^i)} \right), x - u \right\rangle < 0. \tag{7.31}$$

By the feasibility of x for (FMP) and the feasibility of $(k, \lambda, w) \in \Omega, (u, \mu) \in \Theta(k, \lambda, w)$ for (FMD), we get

$$\sum_{j=1}^{r} \mu_j h_j(x) - \sum_{j=1}^{r} \mu_j h_j(u) \leq 0.$$

The pseudolinearity of $\sum_{j=1}^{r} \mu_j h_j(.)$ with respect to the proportional function q, implies that

$$q(u,x) \left\langle \sum_{j=1}^{r} \mu_j \nabla_x h_j(u), x - u \right\rangle \leq 0.$$

Since, $q(u,x) > 0$, we get

$$\left\langle \sum_{j=1}^{r} \mu_j \nabla_x h_j(u), x - u \right\rangle \leq 0. \tag{7.32}$$

Adding (7.31) and (7.32), we arrive at a contradiction to the equality constraint of the dual problem (FPD).

Theorem 7.11 (Strong duality) *Let $\bar{x}$ be optimal for (FMP) and let a Slater's weak constraint qualification or Mangasarian Fromovitz constraint qualification be satisfied. Then, there exist $(\bar{k}, \bar{\lambda}, \bar{y}) \in \Omega, \bar{\mu}_j \in \mathbb{R}, j \in J, \bar{\mu}_j \geq 0$ with $(\bar{x}, \bar{\mu}) \in \Theta(k, \lambda, w)$ such that $(\bar{k}, \bar{\lambda}, \bar{y})$ and $(\bar{x}, \bar{\mu})$ are feasible for (FMD).*

Moreover, if $\varphi(.,y)$ and $\varphi(.,y)$ are pseudolinear with respect to the same proportional function p and

$$\sum_{j=1}^{r} \mu_j h_j(.)$$

is pseudolinear with respect to the proportional function q, for all feasible x for (FMP) and for all feasible $(k, \lambda, w) \in \Omega, (u, \mu) \in \Theta(k, \lambda, w)$ for (FMD), then, $(\bar{x}, \bar{\mu}, \bar{k}, \bar{\lambda}, \bar{y})$ is an optimal solution for (FMD).

Proof Since, $\bar{x}$ is an optimal solution for (FMP) and a suitable constraint qualification is satisfied, then Theorem 7.7 guarantees the existence of a positive integer $\bar{k}, 1 \leq \bar{k} \leq n + 1$, scalars $\bar{\lambda}_i \geq 0, i = 1, ..., \bar{k}$, not all zero, scalars $\bar{\mu}_j \geq 0, j \in J$, vectors $y^i \in Y(\bar{x}), i = 1, ..., k$, such that

$$\sum_{i=1}^{\bar{k}} \bar{\lambda}_i \nabla_x \left(\frac{f(\bar{x}, y^i)}{g(\bar{x}, y^i)} \right) + \sum_{j=1}^{r} \bar{\mu}_j \nabla_x h_j(\bar{x}) = 0,$$

$$\bar{\mu}_j h_j(\bar{x}) = 0, j \in J.$$

Thus denoting $\bar{y} = (\bar{y}_1, ..., \bar{y}_{\bar{k}})$ and $\bar{\lambda} = (\bar{\lambda}_1, ..., \bar{\lambda}_{\bar{k}}), (\bar{k}, \bar{\lambda}, \bar{y}) \in \Omega, (\bar{x}, \bar{\mu}) \in \Theta(\bar{k}, \bar{\lambda}, \bar{y})$ and $t = \frac{\varphi(\bar{x}, y^i)}{\psi(\bar{x}, y^i)}, i = 1, ..., \bar{k}$ are feasible for the dual and the values of the primal and dual problems are equal. The optimality of $(\bar{x}, \bar{\mu}, \bar{k}, \bar{\lambda}, \bar{y})$ for (FMD) follows by the weak duality theorem. This completes the proof.

7.7   Generalizations

Mishra [187] has shown that the above sufficient optimality conditions and duality results for (FMP) can be proved along the similar lines for a more general class of functions known as η-pseudolinear functions. We assume that X is an open invex subset of $\mathbb{R}^n$ with respect to $\eta : K \times K \to \mathbb{R}^n$, that satisfies condition C. For the definitions and properties of invex sets, condition C, and η-pseudolinear functions, we refer to Chapter 13 of this book.

Theorem 7.12 *Suppose there exists a positive integer $\bar{k}, 1 \leq \bar{k} \leq n + 1$, scalar $\bar{\lambda}_i \geq 0, i = 1, ..., \bar{k}$, not all zero, scalars $\bar{\mu}_j \geq 0, j = 1, ..., m$, and vectors $y^i \in Y(\bar{x}), i = 1, ..., \bar{k}$, such that*

$$\sum_{i=1}^{\bar{k}} \bar{\lambda}_i \nabla_x \left(\frac{\varphi(\bar{x}, y^i)}{\psi(\bar{x}, y^i)} \right) + \sum_{j=1}^{r} \bar{\mu}_j \nabla_x h_j(\bar{x}) = 0, \qquad (7.33)$$

$$\bar{\mu}_j h_j(\bar{x}) = 0, j \in J. \qquad (7.34)$$

If $\varphi(.,y)$ and $\psi(.,y)$ are η-pseudolinear with respect to proportional function p *for every $y \in Y$ and $h(.)$ is η-pseudolinear with respect to proportional function q, then, $\bar{x}$ is a minmax solution to (FMP).*

Theorem 7.13 (Weak duality) *Let $\varphi(.,y)$ and $\psi(.,y)$ be η-pseudolinear with respect to the proportional function p and let*

$$\sum_{j=1}^{r} \mu_j h_j(.)$$

be η-pseudolinear with respect to the proportional function q for all feasible x for (FMP) and for all feasible $(k, \lambda, w) \in \Omega, (u, \mu) \in \Theta(k, \lambda, w)$ for (FMD). Then

$$\inf(FMP) \geq \sup(FMD).$$

Theorem 7.14 (Strong duality) *Let $\bar{x}$ be optimal for (FMP) and let a Slater's weak constraint qualification or Mangasarian Fromovitz constraint qualification be satisfied. Then there exist $(\bar{k}, \bar{\lambda}, \bar{y}) \in \Omega, \bar{\mu}_j \in \mathbb{R}, j \in J, \bar{\mu}_j \geq 0$ with $\bar{x}, \bar{\mu} \in \Theta(k, \lambda, w)$ such that $(\bar{k}, \bar{\lambda}, \bar{y})$ and $\bar{x}, \bar{\mu}$ are feasible for (FMD). Moreover, if $\varphi(.,y)$ and $\psi(.,y)$ are η-pseudolinear with respect to the same proportional function p and*

$$\sum_{j=1}^{r} \mu_j h_j(.)$$

is η-pseudolinear with respect to the proportional function q for all feasible x for (FMP) and for all feasible $(k, \lambda, w) \in \Omega, (u, \mu) \in \Theta(k, \lambda, w)$ for (FMD), then $(\bar{x}, \bar{\mu}, \bar{k}, \bar{\lambda}, \bar{y})$ is an optimal solution for (FMD).

Chapter 8

Nonsmooth Static Pseudolinear Minmax Programming Problems

8.1 Introduction

The nonsmooth minmax programming problems have been the subject of intense investigation during the past few years. Due to their applications in a great variety of optimal decision making situations, these problems have been widely studied. Bhatia and Jain [25] derived sufficient optimality conditions for a general minmax programming problem under nondifferentiable pseudoconvexity assumptions using pseudoconvexity in terms of classical Dini derivatives. Further, Bhatia and Jain [25] introduced a dual in terms of Dini derivatives for a general minmax programming problem and established duality results. Mehra and Bhatia [182] proved optimality conditions and various duality results in the sense of Mond and Weir [207] for a static minmax programming problem in terms of the right derivatives of the functions involved with respect to the same arc. Studniarski and Taha [262] have derived first order necessary optimality conditions for nonsmooth static minmax programming problems. For further details and recent developments about optimality conditions and duality theorems for the nonsmooth minmax programming problems involving generalized convex functions, we refer to Kuk and Tanino [154], Mishra and Shukla [196], Zheng and Cheng [300], Yuan *et al.* [293], Antczak [6], Ho and Lai [107], Yuan and Liu [294], Mishra and Upadhyay [201], and the references cited therein.

In this chapter, we consider classes of nonsmooth minmax and nonsmooth minmax fractional programming problems involving locally Lipschitz pseudolinear functions. We establish some sufficient optimality conditions for the problems under the assumption of locally Lipschitz pseudolinear functions. Moreover, we formulate dual models to the primal problems and derive weak and strong duality theorems. The results of the chapter are an extesnsion of the corresponding results from Chew [46], Weir [282], and Mishra [187].

We consider the following nonsmooth minmax programming problem:

$$(NMP) \quad \min \sup_{y \in Y} \{\psi(x, y)\}$$

$$\text{subject to } h_j(x) \leq 0, \quad j \in J := \{1, ..., r\},$$

where $\psi(.,.) : K \times \mathbb{R}^m \to \mathbb{R}, h_j(.) : K \to \mathbb{R}, j \in J$ are locally Lipschitz functions and K is an open convex subset of $\mathbb{R}^n$ and Y is a specified compact subset of $\mathbb{R}^m$. Let $X = \{x | x \in K, \ h(x) \leq 0\}$ denote the set of all feasible solutions of the problem (NMP).

We define the following sets for every $x \in X$:

$$J(x) := \{j \in J | h_j(x) = 0\},$$

$$Y(x) := \left\{y \in Y | \psi(x, y) = \sup_{z \in Y} \psi(x, z)\right\}$$

and

$$K(x) :=$$

$$\left\{ \begin{array}{l} (k, \lambda, \bar{y}) : 1 \leq k \leq n + 1, \ \lambda = (\lambda_1, \lambda_2, ..., \lambda_k) \in \mathbb{R}_+^k, \ \sum_{i=1}^{k} \lambda_i = 1, \\ \text{and } \bar{y} = (y_1, y_2, ..., y_k), \ y_i \in Y(x), \ i = 1, ..., k \end{array} \right\}.$$

8.2   Necessary and Sufficient Optimality Conditions

In this section, we establish sufficient optimality conditions for (NMP). The following assumptions and constraint qualifications will be needed in the sequel:

Condition (C1) (Studniarski and Taha [262]) Let us assume that

(i) The set Y is compact;

(ii) $\psi(x, y)$ is upper semicontinuous in (x, y);

(iii) $\psi(x, y)$ is locally Lipschitz in x, uniformly for y in Y, (see Shimizu *et al.* [254]);

(iv) $\psi(x, y)$ is regular in x;

(v) The set-valued map $\partial_x^c \psi(x, y)$ is upper semicontinuous in (x, y);

(vi) $h_j, \ j \in J$ are locally Lipschitz functions and regular at local minimizers.

Condition (C2)(Yuan et al. [293]) For each $\mu \in \mathbb{R}^r$ satisfying the conditions:

$$\mu_j = 0, \quad \forall j \in J \backslash J(\bar{x}),$$
$$\mu_j \neq 0, \quad \forall j \in J(\bar{x}),$$

the following implication holds:

$$\bar{z}_j \in \partial^c h_j(\bar{x}), \ \forall j \in J, \ \sum_{j=1}^r \mu_j \bar{z}_j = 0 \ \Rightarrow \ \mu_j = 0, \ j \in J.$$

Now, we recall the following necessary optimality conditions established by Studniarski and Taha [262].

Theorem 8.1 (*Necessary optimality conditions*) *Let $\bar{x}$ be an optimal solution for (NMP) and condition (C1) holds. Then, there exist a positive integer k, scalars $\bar{\lambda}_i \geq 0$, $i = 1, ..., \bar{k}$, not all zero, scalars $\bar{\mu}_j \geq 0$, $j = 1, ..., r$, and vectors $y^i \in Y(\bar{x}), i = 1, ..., \bar{k}$, such that*

$$0 \in \sum_{i=1}^{\bar{k}} \bar{\lambda}_i \partial_x^c \phi(\bar{x}, y^i) + \sum_{j=1}^r \bar{\mu}_j \partial_x^c h_j(\bar{x}), \tag{8.1}$$

$$\bar{\mu}_j h_j(\bar{x}) = 0, \ j = 1, ..., r. \tag{8.2}$$

Furthermore, if α is the number of nonzero $\bar{\lambda}_i$ and if β is the number of nonzero $\bar{\mu}_j, j = 1, ..., r$, then

$$1 \leq \alpha + \beta \leq n + 1. \tag{8.3}$$

Theorem 8.2 (*Sufficient optimality conditions*) *Let $(\bar{x}, \bar{v}, \bar{\mu}, \bar{k}, \bar{\lambda}, \bar{y})$ satisfy conditions (8.1)–(8.4). If $\psi(., y)$ is pseudolinear with respect to the proportional function p, for all $y \in Y$ and if $h_j(.)$, $j \in J$ is pseudolinear with respect to the proportional function q, then $\bar{x}$ is a minmax solution for (NMP).*

Proof Suppose that the condition of the theorem is satisfied. Hence, by Condition (8.1) there exist $\xi_i \in \partial_x^c \psi(\bar{x}, y^i)$, $i = 1, ..., \bar{k}$, and $\zeta_j \in \partial_x^c h_j(\bar{x})$, $j = 1, ..., r$, such that

$$\sum_{i=1}^{\bar{k}} \bar{\lambda}_i \xi_i + \sum_{j=1}^r \bar{\mu}_j \zeta_j = 0. \tag{8.4}$$

Suppose to the contrary that $\bar{x}$ is not a minmax solution to (NMP). Then, there exists a feasible x, such that

$$\sup_{y \in Y} \psi(x, y) < \sup_{y \in Y} \psi(\bar{x}, y).$$

Now,

$$\sup_{y \in Y} \psi(\bar{x}, y) = \psi(\bar{x}, y^i), \ i = 1, ..., \bar{k},$$

and

$$\psi\left(x, y^{i}\right) \leq \sup_{y \in Y} \psi\left(\bar{x}, y^{i}\right), \quad i = 1, ..., \bar{k}.$$

Therefore, by the pseudolinearity of $\psi\left(x, y^{i}\right)$, there exists $\xi_{i}' \in \partial_{x}^{c}\psi\left(\bar{x}, y^{i}\right)$, such that

$$p_{i}\left(\bar{x}, x\right)\left\langle \xi_{i}, x - \bar{x}\right\rangle < 0, \quad i = 1, ..., \bar{k}.$$

Since, $p_{i}\left(\bar{x}, x\right) > 0$, for all $i = 1, ..., \bar{k}$, we get

$$\left\langle \xi_{i}', x - \bar{x}\right\rangle < 0, \quad i = 1, ..., \bar{k}. \tag{8.5}$$

Using Proposition 2.5, there exists $\tau_{i} > 0$, $i = 1, ..., \bar{k}$, such that $\xi_{i}' = \tau_{i}\xi_{i}$. Hence, from (8.5), we get

$$\tau_{i}\left\langle \xi_{i}, x - \bar{x}\right\rangle < 0, \quad i = 1, ..., \bar{k}.$$

Since, $\tau_{i} > 0$, $i = 1, ..., \bar{k}$, and $\bar{\lambda} = \left(\bar{\lambda}_{1}, ..., \bar{\lambda}_{\bar{k}}\right) \neq 0$, there exists $\xi_{i} \in \partial_{x}^{c}\psi\left(\bar{x}, y^{i}\right)$, $i = 1, ..., \bar{k}$, such that

$$\sum_{i=1}^{\bar{k}} \bar{\lambda}_{i}\left\langle \xi_{i}, x - \bar{x}\right\rangle < 0, \quad i = 1, ..., \bar{k}. \tag{8.6}$$

Since, x is feasible for (FMP), we have

$$\bar{\mu}_{j}h_{j}\left(x\right) - \bar{\mu}_{j}h_{j}\left(\bar{x}\right) \leq 0, \quad j = 1, ..., r.$$

Using the pseudolinearity of $h_{j}\left(.\right)$, there exists $\zeta_{j}' \in \partial_{x}^{c}h_{j}\left(\bar{x}\right)$, $j = 1, ..., r$, such that

$$q\left(\bar{x}, x\right)\bar{\mu}_{j}\left\langle \zeta_{j}', x - \bar{x}\right\rangle \leq 0, \quad j = 1, ..., r.$$

Again, as $q\left(\bar{x}, x\right) > 0$, it follows that

$$\bar{\mu}_{j}\left\langle \zeta_{j}', x - \bar{x}\right\rangle \leq 0, \quad j = 1, ..., r. \tag{8.7}$$

Using Proposition 2.5, there exists $\kappa_{j} > 0$, $j = 1, ..., r$, such that $\zeta_{j}' = \kappa_{j}\zeta_{j}$, $j = 1, ..., r$. Hence, from (8.7), we get

$$\bar{\mu}_{j}\kappa_{j}\left\langle \zeta_{j}, x - \bar{x}\right\rangle \leq 0, \quad j = 1, ..., r.$$

Since, $\kappa_{j} > 0$, $\mu_{j} \geq 0$, for all $j = 1, ..., r$, it follows that

$$\sum_{j=1}^{r} \bar{\mu}_{j}\left\langle \zeta_{j}, x - \bar{x}\right\rangle \leq 0. \tag{8.8}$$

Adding (8.6) and (8.8), we arrive at a contradiction to (8.4).

To illustrate the significance of Theorems 8.1 and 8.2, we give the following example:

Example 8.1 *Consider the problem*

$$(P) \qquad \min \ \sup_{y \in Y} \psi\,(x,y)$$

$$\text{subject to } h_j\,(x) \leq 0, \quad j \in J = \{1,2\}\,,$$

where $Y = [-1,1]$ *is a compact set and* $K = \,]-1,5[$ *is an open convex set. Let* $f\,(.,.,.) : K \times \mathbb{R} \to \mathbb{R}, \ g\,(.,.,.) : K \times \mathbb{R} \to \mathbb{R}, \ \phi : K \times \mathbb{R} \to \mathbb{R}$ *and* $h_j\,(.) : K \to \mathbb{R},$ $j \in J = \{1,2\}$ *are functions defined by:*

$$\psi\,(x,y) = \begin{cases} \left(\frac{2x+1}{x+1}\right)y^4, & \text{if } x \geq 0, \\ \left(\frac{3x+1}{x+1}\right)y^4, & \text{if } x < 0, \end{cases}$$

$$h_1\,(x) = \begin{cases} -3x, & \text{if } x \geq 0, \\ -x, & \text{if } x < 0, \end{cases} \quad \text{and } h_2\,(x) = 0.$$

Obviously, $f\,(x,y)$ *is a locally Lipschitz pseudolinear function in* x *for every* $y \in Y$ *with respect to the proportional function* $p\,(x,x') = \frac{x+1}{x'+1}$. *Also,* $h_1\,(.)$ *and* $h_2\,(.)$ *are locally Lipschitz pseudolinear functions on* K.

We have $Y\,(x) = \{y^1, y^2\} = \{-1,1\}$ *and if for* $i = 1,2,$ *we denote by*

$$\psi\,(x,y^i) := \begin{cases} \left(\frac{2x+1}{x+1}\right), & \text{if } x \geq 0, \\ \left(\frac{3x+1}{x+1}\right), & \text{if } x < 0. \end{cases}$$

Now, the set of feasible solutions for the problem (P) is $\{x \in K | 0 \leq x < 5\}$. *The point* $\bar{x} = 0$ *is an optimal solution, as there exist* $\bar{\lambda} = \left(\frac{1}{2},\frac{1}{2}\right), \bar{\mu} = (1,1), \xi_i = 1 \in \partial_x^c \phi\,(\bar{x},y^i) = [1,2]$ *with* $i = 1,2$ *and there exist* $\varsigma_1 = -1 \in \partial^c h_1\,(\bar{x}) = [-3,-1]$ *and* $\varsigma_2 = 0 \subset \partial^c h_2\,(\bar{x}) = \{0\}$, *such that the conditions of Theorems 8.1 and 8.2 are satisfied.*

8.3   Duality for (NMP)

For the minmax programming problem (NMP), we consider the following subgradient type dual similar to Weir [282]:

$$(NMD) \qquad \max v$$

$$\text{subject to } \lambda_i\,[\psi\,(u,y^i) - v] \geq 0, \quad i = 1,...,k, \ 1 \leq k \leq n+1,$$

$$\sum_{j=1}^{r} \mu_j h_j\,(u) \geq 0, \quad (k,\lambda,\bar{y}) \in \Omega, \ (u,\mu) \in \Theta\,(k,\lambda,\bar{y}),$$

where Ω denotes the triplet $(k, \lambda, \bar{y})$, where k ranges over the integers $1 \leq k \leq n+1$, $\lambda = (\lambda_i, i = 1, ..., k)$, $\lambda_i \geq 0$, $i = 1, ..., k$, $\lambda \neq 0$, $\bar{y} = (y^i, i = 1, ..., k)$ with $y_i \in Y$, for all $i = 1, ..., k$ and

$$\Theta(k, \lambda, y) :=$$

$$\left\{ (u, \mu) \in \mathbb{R}^n \times \mathbb{R}^m : \sum_{i=1}^{k} \lambda_i \partial_x^c \phi(u, y^i) + \sum_{j=1}^{m} \mu_j \partial_x^c h_i(u) = 0, \mu \geqq 0 \right\}.$$

Theorem 8.3 (Weak duality) *Let x and $(z, \mu, v, s, \lambda, \bar{y})$ be feasible solutions for (NMP) and (NMD), respectively. Let $\psi(., y^i)$, $i = 1, ..., k$ be pseudolinear with respect to the proportional function p for all $y \in Y$. Let $h_j(.)$, $j = 1, ..., r$ be pseudolinear with respect to the proportional function q, x feasible for (NMP) and for all feasible $(k, \lambda, \bar{y}) \in \Omega$, $(u, \mu) \in \Theta(k, \lambda, \bar{y})$ for (NMD). Then,*

$$\sup_{y \in Y} \psi(x, y) \geq v.$$

Proof Suppose to the contrary that there exist a feasible solution x for (NMP) and a feasible solution $(z, \mu, v, s, \lambda, \bar{y})$ for (NMD), such that

$$\sup_{y \in Y} \psi(x, y) < v.$$

Then,

$$\psi(x, y) < v, \forall y \in Y.$$

Hence,

$$\lambda_i \psi(x, y) < \lambda_i v, \forall i = 1, ..., k,$$

with at least one strict inequality since $\lambda \neq 0$. From the dual constraints, it follows that

$$\lambda_i \psi(x, y^i) < \lambda_i \psi(u, y^i), \forall y \in Y \text{ and } i = 1, ..., k,$$

with at least one strict inequality. Therefore,

$$\lambda_i \psi(x, y^i) < \lambda_i \psi(u, y^i), i = 1, ..., k.$$

By the pseudolinearity of $\psi(., y^i)$, $i = 1, ..., k$, with respect to the proportional function p, there exists $\xi_i' \in \partial_x^c \psi(u, y^i)$, $i = 1, ..., k$, such that

$$p_i(u, x) \lambda_i \langle \xi_i', x - u \rangle \leq 0, i = 1, ..., k,$$

with at least one strict inequality. Since, $p_i(u, x) > 0$, we get

$$\sum_{i=1}^{k} \lambda_i \langle \xi', x - u \rangle < 0. \tag{8.9}$$

Using Proposition 2.5, there exists $\tau_i > 0$, $i = 1, ..., k$, such that $\xi_i' = \tau_i \xi_i$. Hence from (8.9), we get

$$\sum_{i=1}^{k} \lambda_i \langle \xi_i, x - u \rangle < 0. \tag{8.10}$$

By the feasibility of x for (NMP) and the feasibility of $(k, \lambda, \bar{y}) \in \Omega$, $(u, \mu) \in \Theta(k, \lambda, \bar{y})$ for (NMD), we get

$$\sum_{j=1}^{r} \mu_j h_j(x) - \sum_{j=1}^{r} \mu_j h_j(u) \le 0.$$

By the pseudolinearity of $h_j(.)$, $j = 1, ..., r$ with respect to the proportional function q, there exists $\zeta_j' \in \partial_x^c h_j(u)$, $j = 1, ..., r$, such that

$$q(u, x) \sum_{j=1}^{r} \mu_j \langle \zeta_j', x - u \rangle \le 0, j = 1, ..., r.$$

Since, $q(u, x) > 0$, we get

$$\sum_{j=1}^{r} \mu_j \langle \zeta_j', x - u \rangle \le 0, j = 1, ..., r. \tag{8.11}$$

Using Proposition 2.5, there exists $\kappa_j > 0$, $j = 1, ..., r$, such that $\zeta_j' = \kappa_j \zeta_j$, $j = 1, ..., r$. Hence, from (8.11), we get

$$\sum_{j=1}^{r} \mu_j \kappa_j \langle \zeta_j, x - \bar{x} \rangle \le 0, \ j = 1, ..., r.$$

Since, $\kappa_j > 0$, for all $j = 1, ..., r$, it follows that

$$\sum_{j=1}^{r} \mu_j \langle \zeta_j, x - \bar{x} \rangle \le 0. \tag{8.12}$$

Adding (8.10) and (8.11), we arrive at a contradiction to the equality constraint of the dual problem (NMD).

Theorem 8.4 (Strong duality) *Let $\bar{x}$ be optimal for (NMP) and let a suitable (Slater's weak constraint qualification or Mangasarian Fromovitz constraint qualification) be satisfied. Then, there exist $(\bar{k}, \bar{\lambda}, \bar{y}) \in \Omega$, $\bar{\mu}_j \in \mathbb{R}, j \in J$, $\bar{\mu}_j \ge 0$ with $(\bar{x}, \bar{\mu}) \in \Theta(k, \lambda, w)$ such that $(\bar{k}, \bar{\lambda}, \bar{y})$ and $(\bar{x}, \bar{\mu})$ are feasible for (NMD). Moreover, if $\psi(., y)$ is pseudolinear with respect to the proportional function p and $h_j(.)$, $j = 1, ..., r$ is pseudolinear with respect to the proportional function q for all x feasible for (NMP) and for all $(k, \lambda, \bar{y}) \in \Omega$, $(u, \mu) \in \Theta(k, \lambda, \bar{y})$ feasible for (NMD), then, $(\bar{x}, \bar{\mu}, \bar{k}, \bar{\lambda}, \bar{y})$ is an optimal solution for (NMD).*

Proof Since, $\bar{x}$ is an optimal solution for (NMP) and a suitable constraint qualification is satisfied, then Theorem 8.1 guarantees the existence of a positive integer $\bar{k}$, $1 \leq \bar{k} \leq n+1$, scalars $\bar{\lambda}_i \geq 0$, $i = 1, ..., \bar{k}$, not all zero, scalars $\bar{\mu}_j \geq 0$, $j \in J$, vectors $y^i \in Y(\bar{x})$, $i = 1, ..., \bar{k}$, such that

$$0 \in \sum_{i=1}^{\bar{k}} \bar{\lambda}_i \partial_x^c \psi\left(\bar{x}, y^i\right) + \sum_{j=1}^{r} \bar{\mu}_j \partial_x^c h_j\left(\bar{x}\right),$$

$$\bar{\mu}_j h_j\left(\bar{x}\right) = 0, \quad j \in J.$$

Thus, denoting $\bar{y} = \left(y^1, ..., y^{\bar{k}}\right)$ and $\bar{\lambda} = \left(\bar{\lambda}_1, ..., \bar{\lambda}_{\bar{k}}\right)$, $\left(\bar{k}, \bar{\lambda}, \bar{y}\right) \in \Omega$, $(\bar{x}, \bar{\mu}) \in \Theta\left(\bar{k}, \bar{\lambda}, \bar{y}\right)$ and $t = \psi\left(\bar{x}, y^i\right)$, $i = 1, ..., \bar{k}$ are feasible for the dual and the values of the primal and dual problems are equal. The optimality of $\left(\bar{x}, \bar{\mu}, \bar{k}, \bar{\lambda}, \bar{y}\right)$ for (NMD) follows by weak duality Theorem 8.3. This completes the proof.

8.4 Nonsmooth Minmax Fractional Programming

We consider the following nonsmooth minmax fractional programming problem:

$$(NMFP) \quad \min \sup_{y \in Y} \left\{ \psi\left(x, y\right) := \frac{f\left(x, y\right)}{g\left(x, y\right)} \right\}$$

$$\text{subject to} \quad h_j\left(x\right) \leq 0, \quad j \in J := \{1, ..., r\},$$

where $f(.,.): K \times \mathbb{R}^m \to \mathbb{R}, g(.,.): K \times \mathbb{R}^m \to \mathbb{R}$, and $h_j(.): K \to \mathbb{R}$, $j \in J$ are locally Lipschitz functions and K is an open convex subset of $\mathbb{R}^n$ and Y is a specified compact subset of $\mathbb{R}^m$. Let $X = \{x | x \in K,\ h(x) \leq 0\}$ denote the set of all feasible solutions of the problem (NMFP).

We define the following sets for every $x \in X$:

$$J\left(x\right) = \{j \in J | h_j\left(x\right) = 0\},$$

$$Y\left(x\right) = \left\{y \in Y | \psi\left(x, y\right) = \sup_{z \in Y} \psi\left(x, z\right)\right\}$$

and

$$K\left(x\right) :=$$

$$\left\{ \begin{array}{l} (k, \lambda, \bar{y}): 1 \leq k \leq n+1,\ \lambda = (\lambda_1, \lambda_2, ..., \lambda_k) \in \mathbb{R}_+^k,\ \sum_{i=1}^k \lambda_i = 1, \\ \text{and } \bar{y} = (y_1, y_2, ..., y_k),\ y_i \in Y\left(x\right),\ i = 1, ..., k. \end{array} \right\}.$$

8.5 Necessary and Sufficient Optimality Conditions

In this section, we establish sufficient optimality conditions for (NMFP). The following assumptions and constraint qualifications will be needed in the sequel:

Condition (D1) (Yuan *et al.* [293]) Let us assume that

(i) The set Y is compact;

(ii) $\psi(x, y)$ is upper semicontinuous in (x, y);

(iii) $f(x, y)$ and $g(x, y)$ are locally Lipschitz in x, uniformly for y in Y, (see Shimizu *et al.* [254]);

(iv) $f(x, y)$ and $g(x, y)$ are regular in x;

(v) The set-valued map $\partial_x^c \psi(x, y)$ is upper semicontinuous in (x, y);

(vi) h_j, $j \in J$ are locally Lipschitz and regular at local minimizers.

Along the lines of Studniarski and Taha [262], we can establish the following necessary optimality conditions for (NMFP):

Theorem 8.5 *(Necessary optimality conditions)* *Let $\bar{x}$ be an optimal solution for (NMFP) and condition (D1) holds. Then, there exist a positive integer $\bar{s}$, scalars $\bar{\lambda}_i \geq 0$, $i = 1, ..., \bar{k}$, not all zero, scalars $\bar{\mu}_j \geq 0$, $j = 1, ..., r$, and vectors $y^i \in Y(\bar{x})$, $i = 1, ..., \bar{k}$, such that*

$$0 \in \sum_{i=1}^{\bar{k}} \bar{\lambda}_i \partial_x^c \left(\frac{f(\bar{x}, y^i)}{g(\bar{x}, y^i)} \right) + \sum_{j=1}^{r} \bar{\mu}_j \partial_x^c h_j(\bar{x}), \tag{8.13}$$

$$\bar{\mu}_j h_j(x) = 0, \quad j = 1, ..., r. \tag{8.14}$$

Furthermore, if α is the number of nonzero $\bar{\lambda}_i, i = 1, ..., k$ and if β is the number of nonzero $\bar{\mu}_j, j = 1, ..., r$, then

$$1 \leq \alpha + \beta \leq n + 1. \tag{8.15}$$

Theorem 8.6 *(Sufficient optimality conditions)* *Let $(\bar{x}, \bar{v}, \bar{\mu}, \bar{s}, \bar{\lambda}, \bar{y})$ satisfy conditions (8.13)–(8.15). If $f(., y)$ and $g(., y)$ are pseudolinear with respect to the same proportional function p for all $y \in Y$ and if $h_j(.)$, $j \in J$ is pseudolinear with respect to the proportional function q, then $\bar{x}$ is a minmax solution for (NMFP).*

Proof Suppose that the conditions of the theorem are satisfied. Hence, by condition (8.13) there exist $\xi_i \in \partial_x^c \left(\frac{f(\bar{x},y^i)}{g(\bar{x},y^i)} \right), i = 1, ..., \bar{k}$ and $\zeta_j \in \partial_x^c h_j(\bar{x}), j = 1, ..., r$, such that

$$\sum_{i=1}^{\bar{k}} \bar{\lambda}_i \xi_i + \sum_{j=1}^{r} \bar{\mu}_j \zeta_j = 0. \tag{8.16}$$

Suppose to the contrary that $\bar{x}$ is not a minmax solution to (NMFP). Then there exists a feasible x, such that

$$\sup_{y \in Y} \frac{f(x,y)}{g(x,y)} < \sup_{y \in Y} \frac{f(\bar{x},y)}{g(\bar{x},y)}.$$

Now,

$$\sup_{y \in Y} \frac{f(\bar{x},y)}{g(\bar{x},y)} = \frac{f(\bar{x},y^i)}{g(\bar{x},y^i)}, \ i = 1, ..., \bar{k},$$

and

$$\sup_{y \in Y} \frac{f(x,y)}{g(x,y)} \leq \sup_{y \in Y} \frac{f(\bar{x},y^i)}{g(\bar{x},y^i)}, \ i = 1, ..., \bar{k},$$

Therefore, we get

$$\frac{f(x,y^i)}{g(x,y^i)} < \frac{f(\bar{x},y^i)}{g(\bar{x},y^i)}, \ i = 1, ..., \bar{k},$$

Using Lemma 6.1, by the pseudolinearity of $\frac{f(.,y^i)}{g(.,y^i)}$, with respect to $\bar{p}$, there exists $\xi_i \in \partial_x^c \left(\frac{f(\bar{x},y^i)}{g(\bar{x},y^i)}, \right), i = 1, ..., \bar{k}$, such that

$$\bar{p}_i(\bar{x},x) \langle \xi_i', x - \bar{x} \rangle < 0, \ i = 1, ..., \bar{k},$$

where $\bar{p}_i(\bar{x},x) = p_i(\bar{x},x) \frac{g(\bar{x},y^i)}{g(x,y^i)}$.

Since, $\bar{p}_i(\bar{x},x) > 0$, for all $i = 1, ..., \bar{k}$, we get

$$\langle \xi_i', x - \bar{x} \rangle < 0, \ i = 1, ..., \bar{k}. \tag{8.17}$$

Using Proposition 2.5, there exist $\tau_i > 0, i = 1, ..., \bar{k}$, such that $\xi_i' = \tau_i \xi_i$. Hence, from (8.17), we get

$$\tau_i \langle \xi_i, x - \bar{x} \rangle < 0, \ i = 1, ..., \bar{k}.$$

Since, $\tau_i > 0, i = 1, ..., \bar{k}$ and $\bar{\lambda} = (\bar{\lambda}_1, ..., \bar{\lambda}_{\bar{k}}) \neq 0$, there exist $\xi_i \in \partial_x^c \left(\frac{f(\bar{x},y^i)}{g(\bar{x},y^i)} \right), i = 1, ..., \bar{k}$, such that

$$\sum_{i=1}^{\bar{k}} \bar{\lambda}_i \langle \xi_i, x - \bar{x} \rangle < 0, \ i = 1, ..., \bar{k}. \tag{8.18}$$

Since, x is feasible for (NMFP), we have

$$\bar{\mu}_j h_j (x) - \bar{\mu}_j h_j (\bar{x}) \leq 0, \ \ j = 1, ..., r.$$

Using the pseudolinearity of $h_j (.)$, there exist $\zeta'_j \in \partial^c_x h_j (\bar{x})$, $j = 1, ..., r$, such that

$$q (\bar{x}, x) \bar{\mu}_j \langle \zeta'_j, x - \bar{x} \rangle \leq 0, \ j = 1, ..., r.$$

Again as, $q (\bar{x}, x) > 0$, it follows that

$$\bar{\mu}_j \langle \zeta'_j, x - \bar{x} \rangle \leq 0, \ \ j = 1, ..., r. \tag{8.19}$$

Using Proposition 2.5, there exist $\kappa_j > 0$, $j = 1, ..., r$, such that $\zeta'_j = \kappa_j \zeta_j$, $j = 1, ..., r$. Hence, from (8.19), we get

$$\bar{\mu}_j \kappa_j \langle \zeta_j, x - \bar{x} \rangle < 0, \ \ j = 1, ..., r.$$

Since, $\kappa_j > 0$, for all $j = 1, ..., r$, it follows that

$$\sum_{j=1}^{r} \bar{\mu}_j \langle \zeta_j, x - \bar{x} \rangle \leq 0. \tag{8.20}$$

Adding (8.18) and (8.20), we arrive at a contradiction to (8.16). This completes the proof.

8.6   Duality for (NMFP)

For the minmax programming problem (NMFP), we consider the following subgradient type dual similar to Weir [282]:

$$(NMFD) \quad \max v$$

$$\text{subject to } \lambda_i \left[\left(\frac{f (u, y^i)}{g (u, y^i)} \right) - v \right] \geq 0, \ i = 1, ..., k, \ 1 \leq k \leq n+1,$$

$$\sum_{j=1}^{r} \mu_j h_j (u) \geq 0, \ (k, \lambda, \bar{y}) \in \Omega, \ (u, \mu) \in \Theta (k, \lambda, \bar{y}),$$

where Ω denotes the triplet $(k, \lambda, \bar{y})$, where k ranges over the integers $1 \leq k \leq n+1$, $\lambda = (\lambda_i, i = 1, ..., k)$, $\lambda_i \geq 0$, $i = 1, ..., k$, $\lambda \neq 0$, $\bar{y} = (y^i, i = 1, ..., k)$ with $y_i \in Y$, for all $i = 1, ..., k$ and

$$\Theta (k, \lambda, \bar{y}) :=$$

$$\left\{ (u, \mu) \in \mathbb{R}^n \times \mathbb{R}^r : \sum_{i=1}^{k} \lambda_i \partial^c_x \left(\frac{f (u, y^i)}{g (u, y^i)} \right) + \sum_{j=1}^{r} \mu_j \partial^c_x h_i (u) = 0, \mu \geqq 0 \right\}.$$

Theorem 8.7 (Weak duality) *Let x and $(z, \mu, v, s, \lambda, \bar{y})$ be feasible solutions for (NMFP) and (NMFD), respectively. Let $f(.,y)$ and $g(.,y)$ be pseudolinear with respect to the proportional function p for all $y \in Y$. Let $h_j(.)$, $j = 1, ..., r$ be pseudolinear with respect to the proportional function q, feasible x for (NMFP) and for all feasible $(k, \lambda, \bar{y}) \in \Omega$, $(u, \mu) \in \Theta(k, \lambda, \bar{y})$ for (NMFD). Then,*

$$\sup_{y \in Y} \psi(x, y) \geq v.$$

Proof Suppose to the contrary that there exist a feasible solution x for (NMFP) and a feasible solution $(z, \mu, v, s, \lambda, \bar{y})$ for (NMFD), such that

$$\sup_{y \in Y} \psi(x, y) < v.$$

Then,

$$\psi(x, y) < v, \forall y \in Y.$$

Hence,

$$\lambda_i \psi(x, y) < \lambda_i v, \forall i = 1, ..., k,$$

with at least one strict inequality since $\lambda \neq 0$. From the dual constraints, it follows that

$$\lambda_i \psi(x, y) < \lambda_i \psi(u, y^i), \forall y \in Y \text{ and } i = 1, ..., k,$$

with at least one strict inequality. Therefore,

$$\lambda_i \psi(x, y^i) < \lambda_i \psi(u, y^i), i = 1, ..., k.$$

Using Lemma 6.1, by the pseudolinearity of $\frac{f(.,y^i)}{g(.,y^i)}$, with respect to $\bar{p}$, there exists $\xi_i \in \partial_x^c \left(\frac{f(u, y^i)}{g(u, y^i)} \right)$, $i = 1, ..., k$, such that

$$p_i(u, x) \lambda_i \langle \xi_i', x - u \rangle \leq 0, i = 1, ..., k,$$

with at least one strict inequality. Since, $p_i(u, x) > 0$, we get

$$\sum_{i=1}^{k} \lambda_i \langle \xi', x - u \rangle < 0. \tag{8.21}$$

Using Proposition 2.5, there exist $\tau_i > 0$, $i = 1, ..., \bar{k}$, such that $\xi_i' = \tau_i \xi_i$. Hence, from (8.21), we get

$$\sum_{i=1}^{k} \lambda_i \langle \xi_i, x - u \rangle < 0. \tag{8.22}$$

By the feasibility of x for (NMFP) and the feasibility of $(k, \lambda, \bar{y}) \in \Omega$, $(u, \mu) \in \Theta(k, \lambda, \bar{y})$ for (NMFD), we get

$$\sum_{j=1}^{r} \mu_j h_j(x) - \sum_{j=1}^{r} \mu_j h_j(u) \leq 0.$$

By the pseudolinearity of $h_j(.)$, $j = 1, ..., r$ with respect to the proportional function q, there exists $\zeta_j' \in \partial_x^c h_j(u)$, $j = 1, ..., r$, such that

$$q(u, x) \sum_{j=1}^{r} \mu_j \langle \zeta_j', x - u \rangle \leq 0, j = 1, ..., r.$$

Since, $q(u, x) > 0$, we get

$$\sum_{j=1}^{r} \mu_j \langle \zeta_j', x - u \rangle \leq 0, j = 1, ..., r. \tag{8.23}$$

Using Proposition 2.5, there exist $\kappa_j > 0$, $j = 1, ..., r$, such that $\zeta_j' = \kappa_j \zeta_j$, $j = 1, ..., r$. Hence, from (8.23), we get

$$\sum_{j=1}^{r} \mu_j \kappa_j \langle \zeta_j, x - u \rangle \leq 0, \ j = 1, ..., r.$$

Since, $\kappa_j > 0$, for all $j = 1, ..., r$, it follows that

$$\sum_{j=1}^{r} \mu_j \langle \zeta_j, x - u \rangle \leq 0. \tag{8.24}$$

Adding (8.22) and (8.24), we arrive at a contradiction to the equality constraint of the dual problem (NMFD).

Theorem 8.8 (Strong duality) *Let $\bar{x}$ be optimal for (NMFP) and let a suitable (Slater's weak constraint qualification or Mangasarian Fromovitz) constraint qualification be satisfied. Then there exist $(\bar{k}, \bar{\lambda}, \bar{y}) \in \Omega$, $\bar{\mu}_j \in \mathbb{R}, j \in J$, $\bar{\mu}_j \geq 0$ with $(\bar{x}, \bar{\mu}) \in \Theta(k, \lambda, w)$ such that $(\bar{k}, \bar{\lambda}, \bar{y})$ and $(\bar{x}, \bar{\mu})$ are feasible for (NMFD). Moreover, if $f(., y)$ and $g(., y)$ are pseudolinear with respect to the proportional function p and $h_j(.)$, $j = 1, ..., r$ is pseudolinear with respect to the proportional function q, for all feasible x for (NMFP) and for all feasible $(k, \lambda, w) \in \Omega$, $(u, \mu) \in \Theta(k, \lambda, w)$ for (NMFD), then $(\bar{x}, \bar{\mu}, \bar{k}, \bar{\lambda}, \bar{y})$ is an optimal solution for (NMFD).*

Proof Since, $\bar{x}$ is an optimal solution for (NMFP) and a suitable constraint qualification is satisfied, then Theorem 8.5 guarantees the existence of a positive integer $\bar{k}$, $1 \leq \bar{k} \leq n+1$, scalars $\bar{\lambda}_i \geq 0$, $i = 1, ..., \bar{k}$, not all zero, scalars $\bar{\mu}_j \geq 0$, $j \in J$, vectors $y^i \in Y(\bar{x})$, $i = 1, ..., \bar{k}$, such that

$$0 \in \sum_{i=1}^{\bar{k}} \bar{\lambda}_i \partial_x^c \left(\frac{f(\bar{x}, y^i)}{g(\bar{x}, y^i)} \right) + \sum_{j=1}^{r} \bar{\mu}_j \partial_x^c h_j(\bar{x}),$$

$$\bar{\mu}_j h_j (\bar{x}) = 0, \quad j \in J.$$

Thus denoting $\bar{y} = \left(y^1, ..., y^{\bar{k}}\right)$ and $\bar{\lambda} = \left(\bar{\lambda}_1, ..., \bar{\lambda}_{\bar{k}}\right)$, $\left(\bar{k}, \bar{\lambda}, \bar{y}\right) \in \Omega$, $(\bar{x}, \bar{\mu}) \in \Theta \left(\bar{k}, \bar{\lambda}, \bar{y}\right)$ and $t = \psi \left(\bar{x}, y^i\right)$, $i = 1, ..., \bar{k}$ are feasible for the dual and the values of the primal and dual problems are equal. The optimality of $\left(\bar{x}, \bar{\mu}, \bar{k}, \bar{\lambda}, \bar{y}\right)$ for (NMFD) follows by weak duality Theorem 8.7. This completes the proof.

Chapter 9

Optimality and Duality in Terms of Bifunctions

9.1 Introduction

Optimality conditions and duality results for nonlinear multiobjective optimization have been the subject of much interest in the recent past. Ewing [77] introduced semilocally convex functions and applied it to derive sufficient optimality condition for variational and control problems. Elster and Nehse [75] considered a class of convex-like functions and obtained a saddle point optimality condition for mathematical programs involving such functions. Mishra and Mukherjee [195] extended this class of functions to the case of nonsmooth problems and obtained sufficiency and duality results for several problems. Some generalizations of semilocally convex functions and their properties were investigated in Kaul and Kaur [136].

Kaul *et al.* [138] introduced the class of semilocal pseudolinear functions and obtained optimality conditions for a multiobjective optimization problem involving semilocal pseudolinear functions. Aggarwal and Bhatia [1] established the optimality conditions for pseudolinear optimization problems in terms of the Dini derivative. Optimality conditions and duality results using generalized directional derivatives and generalized convex functions have been studied by Preda and Stancu-Minasian [225, 226], Stancu-Minasian [259], Suneja and Gupta [263], and references cited therein. Recently, Lalitha and Mehta [155, 156, 157] introduced the class of h-pseudolinear functions and derived some Lagrange multiplier characterizations of the solution set of an optimization problem involving pseudolinear functions in terms of bifunction h.

In this chapter, we derive necessary and sufficient optimality conditions for a nonsmooth multiobjective optimization problem under the assumption of h-pseudolinear functions. Furthermore, we formulate a Mond-Weir type dual model related to the primal problem and establish weak and strong duality results.

9.2   Definitions and Preliminaries

Let $K \subseteq \mathbb{R}^n$ be a nonempty convex set and $h : K \times \mathbb{R}^n \to \overline{\mathbb{R}}$ be a bifunction. We consider the following nonsmooth multiobjective optimization problem:

$$(NMP) \quad \min \ (f_1(x), ..., f_k(x))$$

$$\text{subject to } g_j(x) \leq 0, \ \ j = 1, ..., m,$$

where $f_i : K \to \mathbb{R}, \ i \in I := \{1, ..., k\} \ ; \ g_j : K \to \mathbb{R}, \ j \in J := \{1, ..., m\}$ are radially continuous functions on a nonempty convex set $K \subseteq \mathbb{R}^n$. Let $S := \{x \in K : g_j(x) \leq 0, j \in J\}$ and $J(x) := \{j \in J : g_j(x) = 0\}$ denote the set of all feasible solutions and active constraint index set at $x \in K$, respectively. Let the bifunctions $h_i(x; d), \ h'_j(x; d) : K \times \mathbb{R}^n \to \overline{\mathbb{R}}$ associated with functions $f_i, \ i \in I$ and $g_j, \ j \in J$, respectively, are finite valued sublinear functions of d on K, such that $f_i, \ i \in I$ and $g_j, \ j \in J$, vanish at $d = 0$. Moreover, we assume that for all $(x, d) \in K \times \mathbb{R}^n$,

$$D_+ f_i(x; d) \leq h_i(x; d) \leq D^+ f_i(x; d), \forall \, i = 1, ..., k$$

and

$$D_+ g_j(x; d) \leq h'_j(x; d) \leq D^+ g_j(x; d), \forall \, j = 1, ..., m.$$

Let $\bar{S}$ denote the set of all efficient solutions of the problem (NMP). Now onwards, we assume that $\bar{S} \neq \psi$.

Related to (NMP), we consider the following scalar minimization problem:

$$(P_r(\bar{x})) \quad \min f_r(x)$$

$$\text{subject to } f_i(x) \leq f_i(\bar{x}) \forall \, i \in I, i \neq r,$$

$$g_j(x) \leq 0, \ \ j \in J,$$

for each $r = 1, ..., k$.

Lemma 9.1 *(Chankong and Haimes [43])* *Any point $\bar{x} \in S$ is an efficient solution for (NMP), if and only if $\bar{x}$ solves the problem $(P_r(\bar{x}))$, for each $r = 1, 2..., k$.*

Definition 9.1 *(Kanniappan [124])* *The problem $(P_r(\bar{x}))$ is said to satisfy the Slater type constraint qualification at $\bar{x} \in K$, if for each $r = 1, 2, ..., k$, there exists some $\hat{x}_r \in S$, such that*

$$g_j(\hat{x}_r) < 0, \forall \, j \in J(\bar{x})$$

and

$$f_i(\hat{x}_r) < f_i(\bar{x}), \forall \, i \in I, i \neq r.$$

9.3 Necessary and Sufficient Optimality Conditions

In this section, we first derive the necessary optimality conditions for (NMP) by using the Slater type constraint qualification and then establish the sufficient optimality conditions for (NMP) in terms of bifunctions.

Theorem 9.1 *Let $\bar{x} \in \bar{S}$ and for (NMP) the following conditions hold:*

1. *For each $i \in I$, f_i is h_i-pseudolinear with respect to proportional function p_i and for each $j \in J$, g_j is h'_j-pseudolinear function with respect to proportional function q_j;*

2. *The problem $(P_r(\bar{x}))$ satisfies the Slater type constraint qualification at $\bar{x} \subset K$, for each $r = 1, ..., k$.*

Then, $\forall\, x \in K$, there exist $\bar{\lambda}_i \in \mathbb{R}$, $i \in I$, $\bar{\mu}_j \in \mathbb{R}$, $j \in J(\bar{x})$, such that

$$\left.\begin{array}{c} \sum_{i=1}^{k} \bar{\lambda}_i h_i(\bar{x}; x - \bar{x}) + \sum_{j \subset J(\bar{x})} \bar{\mu}_j h'_j(\bar{x}; x - \bar{x}) \geq 0, \\ \bar{\mu}^T g(\bar{x}) = 0, \\ \bar{\lambda}_i > 0, \ i \in I, \sum_{i \in I} \bar{\lambda}_i = 1, \\ \bar{\mu}_j \geq 0, \ j \in J(\bar{x}). \end{array}\right\} \qquad (9.1)$$

Proof Suppose $\bar{x} \in \bar{S}$ be an efficient solution for (NMP). We have to prove that there exist $\bar{\lambda}_i > 0$ $i \in I$ and $\bar{\mu}_j \geq 0$ $j \in J(\bar{x})$, such that (9.1) holds. For $1 \leq r \leq k$, we assert that the following system of inequalities

$$\left\{\begin{array}{l} h_i(\bar{x}; x - \bar{x}) \leq 0, \ \forall\, i = 1, ..., r-1, r+1, ..., k \\ h_r(\bar{x}; x - \bar{x}) < 0 \\ h'_j(\bar{x}; x - \bar{x}) \leq 0, \ j \in J(\bar{x}) \end{array}\right. \qquad (9.2)$$

has no solution $x \in K$. Suppose to the contrary that $x \in K$ satisfies the above inequalities. Then,

$$\left\{\begin{array}{l} D_+ f_i(\bar{x}; x - \bar{x}) \leq 0, \ \forall\, i = 1, ..., r-1, r+1, ..., k \\ D_+ f_r(\bar{x}; x - \bar{x}) < 0 \\ h'_j(\bar{x}; x - \bar{x}) \leq 0, \ j \in J(\bar{x}). \end{array}\right.$$

The positive homogeneity of h'_j implies that there exists a sequence $t_n \to 0+$ and a positive integer m such that $\forall\, n \geq m$, we get

$$\left\{\begin{array}{l} f_i(\bar{x} + t_n(x - \bar{x})) \leq f_i(\bar{x}), \forall\, i = 1, ..., r-1, r+1, ..., k \\ f_r(\bar{x} + t_n(x - \bar{x})) < f_r(\bar{x}) \\ h'_j(\bar{x}; \bar{x} + t_n(x - \bar{x}) - \bar{x}) \leq 0, \ \forall j \in J(\bar{x}). \end{array}\right.$$

To arrive at a contradiction, it is sufficient to prove that $\bar{x} + t_p(x - \bar{x})$ is a feasible solution for (NMP) for some $p \geq m$. Using the fact that g_j is h'_j-pseudoconcave, it follows that

$$g_j \left(\bar{x} + t_n \left(x - \bar{x} \right) \right) \leq g_j \left(\bar{x} \right) = 0, \ \forall n \geq m \text{ and } j \in J \left(\bar{x} \right).$$

Also, as $g_j \left(\bar{x} \right) < 0$ for all $j \notin J \left(\bar{x} \right)$ and g_j is radially continuous on K, there exists a positive integer $p \geq m$, such that

$$g_j \left(\bar{x} + t_p \left(x - \bar{x} \right) \right) < 0, \ \forall j \notin J \left(\bar{x} \right).$$

Hence, $\bar{x} + t_p \left(x - \bar{x} \right)$ is a feasible solution for (NMP).

Since $h_i \left(\bar{x}; x - \bar{x} \right)$, $i \in I$ and $h'_j \left(\bar{x}; x - \bar{x} \right), j \in J \left(\bar{x} \right)$ are sublinear functions of x and hence convex, therefore by Lemma 1.3, there exist nonnegative scalars $\bar{\lambda}_i \geq 0$, $i \in I$ and $\bar{\mu}_j \geq 0$, $j \in J \left(\bar{x} \right)$ not all zero, such that

$$\lambda_r h_r \left(\bar{x}; x - \bar{x} \right) + \sum_{\substack{i = 1 \\ i \neq r}}^{k} \lambda_i h_i \left(\bar{x}; x - \bar{x} \right) + \sum_{j \in J(\bar{x})} \mu_j h'_j \left(\bar{x}; x - \bar{x} \right) \geq 0, \forall x \in K.$$

$$(9.3)$$

Now, we shall show that $\lambda_r \neq 0$. If possible let $\lambda_r = 0$, for some r, $1 \leq r \leq k$. By Lemma 9.1, $\bar{x}$ is an optimal solution of $(P_r \left(\bar{x} \right))$. Therefore, by the Slater type constraint qualification there exists some strictly feasible point $\hat{x}_r$, such that

$$f_i \left(\hat{x}_r \right) < f_i \left(\bar{x} \right), \ i = 1, 2, ..., k, \ i \neq r,$$

$$g_j \left(\hat{x}_r \right) < 0 = g_j \left(\bar{x} \right), \forall j \in J \left(\bar{x} \right).$$

Using the h_i-pseudoconvexity of f_i and h'_j-pseudoconvexity of g_j, we get

$$h_i \left(\bar{x}; \hat{x}_r - \bar{x} \right) < 0, i = 1, 2, ..., k, \ i \neq r,$$

$$h'_j \left(\bar{x}; \hat{x}_r - \bar{x} \right) < 0, \forall j \in J \left(\bar{x} \right).$$

Since at least one of the coefficients $\lambda_i, i = 1, 2, ..., k, \ i \neq r$ and $\mu_j, \ j \in J \left(\bar{x} \right)$ is nonzero, therefore, we must have

$$\sum_{i=1}^{k} \lambda_i h_i \left(\bar{x}; \hat{x}_r - \bar{x} \right) + \sum_{j \in J(\bar{x})} \mu_j h'_j \left(\bar{x}; \hat{x}_r - \bar{x} \right) < 0,$$

which contradicts (9.3). Hence, $\lambda_r \neq 0$. Now, dividing (9.3) by λ_r, we get

$$h_r \left(\bar{x}; x - \bar{x} \right) + \sum_{\substack{i = 1 \\ i \neq r}}^{k} \frac{\lambda_i}{\lambda_r} h_i \left(\bar{x}; x - \bar{x} \right) + \sum_{j \in J(\bar{x})} \frac{\mu_j}{\lambda_r} h'_j \left(\bar{x}; x - \bar{x} \right) \geq 0. \quad (9.4)$$

Setting $\bar{\lambda}_r = \dfrac{1}{1+\sum\limits_{\substack{i=1\\i\neq r}}^{k}\frac{\lambda_i}{\lambda_r}}, \bar{\lambda}_i = \dfrac{\frac{\lambda_i}{\lambda_r}}{1+\sum\limits_{\substack{i=1\\i\neq r}}^{k}\frac{\lambda_i}{\lambda_r}}, i \in I, i \neq r,$ and $\bar{\mu}_j =$

$\dfrac{\frac{\mu_j}{\lambda_r}}{1+\sum\limits_{\substack{i=1\\i\neq r}}^{k}\frac{\lambda_i}{\lambda_r}}, j \in J(\bar{x})$ and defining $\bar{\mu}_j = 0$ for $j \notin J(\bar{x})$, the required

result follows.

Theorem 9.2 *(Sufficient optimality conditions) Suppose $\bar{x}$ be a feasible solution for (NMP). Let for each $i \in I$, f_i is h_i-pseudolinear function with respect to proportional function p_i and for each $j \in J$, g_j is h'_j-pseudolinear function with respect to proportional function q_j. If $\forall\, x \in K$, there exist $\bar{\lambda}_i \in \mathbb{R}$, $i \in I$, $\bar{\mu}_j \in \mathbb{R}$, $j \in J(\bar{x})$, such that*

$$\left.\begin{array}{l} \sum_{i=1}^{k} \bar{\lambda}_i h_i\left(\bar{x}; x-\bar{x}\right) + \sum_{j \in J(\bar{x})} \bar{\mu}_j h'_j\left(\bar{x}; x-\bar{x}\right) \geq 0, \\ \bar{\mu}^T g\left(\bar{x}\right) = 0, \\ \bar{\lambda}_i > 0,\ i \in I, \sum_{i\in I} \bar{\lambda}_i = 1, \\ \bar{\mu}_j \geq 0,\ j \in J\left(\bar{x}\right). \end{array}\right\}$$

Then, $\bar{x}$ is an efficient solution for (NMP).

Proof To prove that $\bar{x}$ is an efficient solution for (NMP), suppose to the contrary that there exists $y \in S$, such that

$$f_i\left(y\right) \leq f_i\left(\bar{x}\right), \forall\, i \in I, \tag{9.5}$$

$$f_r\left(y\right) < f_r\left(\bar{x}\right), \text{ for some } r \in I. \tag{9.6}$$

Using the h_i-pseudolinearity of f_i with respect to proportional function p_i, $i \in I$, it follows that

$$f_i\left(y\right) - f_i\left(\bar{x}\right) = p_i\left(\bar{x}, y\right) h_i\left(\bar{x}; y-\bar{x}\right). \tag{9.7}$$

Using the h'_j-pseudolinearity of g_j with respect to proportional function $q_j, j \in J$, it follows that

$$g_j\left(y\right) - g_j\left(\bar{x}\right) = q_j\left(\bar{x}, y\right) h'_j\left(\bar{x}; y-\bar{x}\right). \tag{9.8}$$

Moreover, for $y \in S$ and for any $j \in J\left(\bar{x}\right)$, we have

$$g_j\left(y\right) \leq g_j\left(\bar{x}\right) = 0. \tag{9.9}$$

Thus from (9.5)–(9.9), we have

$$0 \geq \sum_{j \in J(\bar{x})} \frac{\bar{\mu}_j}{q_j\left(\bar{x}, y\right)} \left(g_j\left(y\right) - g_j\left(\bar{x}\right)\right) = \sum_{j \in J(\bar{x})} \bar{\mu}_j h'_j\left(\bar{x}; y-\bar{x}\right)$$

$$\geq - \sum_{i=1}^{k} \bar{\lambda}_i h_i\left(\bar{x}; y - \bar{x}\right)$$

$$= - \sum_{i=1}^{k} \frac{\bar{\lambda}_i}{p_i\left(\bar{x}, y\right)} \left(f_i\left(y\right) - f_i\left(\bar{x}\right)\right) > 0,$$

which is a contradiction. This completes the proof.

Remark 9.1   *(i) If $k = 1$ then, the above necessary optimality conditions (9.1) reduce to that of Theorem 3.1 of Lalitha and Mehta [157].*

(ii) If $K = \mathbb{R}^n$ and the functions f, g are differentiable pseudolinear functions, then, the above necessary optimality conditions (9.1) reduce to that of Chew and Choo [47], i.e.,

$$\sum_{i=1}^{k} \bar{\lambda}_i \nabla f_i(\bar{x}) + \sum_{j \in J(\bar{x})} \bar{\mu}_{ji} g_j(\bar{x}) = 0,$$

$$\bar{\lambda}_i > 0, i \in I,$$

$$\bar{\mu}_j \geq 0, \ j \in J\left(\bar{x}\right).$$

The following example illustrates the significance of Theorems 9.1 and 9.2.

Example 9.1 *Consider the problem:*

$$(P) \quad \min f\left(x\right) := \left(f_1\left(x\right), f_2\left(x\right)\right)$$

$$\text{subject to } g_j\left(x\right) \leq 0, \ j = 1, 2,$$

where f_1, f_2, g_1, $g_2 : K = [-6, 6] \to \mathbb{R}$ are functions defined by:

$$f_1\left(x\right) = \begin{cases} x, & \text{if } \ x < 0, \\ 2x, & \text{if } \ x \geq 0, \end{cases}$$

$$f_2\left(x\right) = \begin{cases} x, & \text{if } \ x < 0, \\ 3x, & \text{if } \ x \geq 0, \end{cases}$$

$$g_1\left(x\right) = -x^3 - x,$$

and

$$g_2\left(x\right) = \begin{cases} x - 5, & \text{if } \ x \geq 5, \\ \frac{x-5}{3}, & \text{if } \ \ x < 5. \end{cases}$$

Define

$$h_1\left(x; d\right) = \begin{cases} d, & \text{if } \ x < 0, \forall d \text{ or } x = 0, d < 0, \\ 2d, & \text{if }, \ x = 0, d \geq 0 \text{ or } x > 0, \forall d, \end{cases}$$

$$h_2\left(x; d\right) = \begin{cases} d, & \text{if } \ x < 0, \forall d \text{ or } x = 0, d < 0, \\ 3d, & \text{if } \ x = 0, d \geq 0 \text{ or } x > 0, \forall d, \end{cases}$$

$$h'_1 (x; d) = - \left(3x^2 + 1\right) d,$$

and

$$h'_2 (x; d) = \begin{cases} d, & \text{if } x > 5, \forall d \text{ or } x = 5, \; d \geq 0, \\ \frac{d}{3}, & \text{if } x = 5, \; d < 0 \text{ or } x < 5, \forall d. \end{cases}$$

It is clear that each of the functions $f_i, i = 1, 2$ is h_i-pseudolinear and each $g_j, j = 1, 2$ is h'_j-pseudolinear function on the convex set $K = [-6, 6]$. The set of feasible solutions of the problem (P) is $\{x \in K | 0 \leq x \leq 5\}$ and $\bar{x} = 0$ is an optimal solution of the problem (P). Obviously, for $\bar{x} = 0$ the set $J(\bar{x}) = \{1\}$ and the problem $(P_r(\bar{x}))$ satisfies the Slater type constraint qualification for $\hat{x} = 2$ and for each $r = 1, 2$ with $g_j(\hat{x}) < 0, j \in J(\bar{x})$. Moreover, let $\bar{\lambda} = \left(\frac{1}{3}, \frac{2}{3}\right)$ and $\bar{\mu} = (1, 0)$ be the Lagrange multipliers. Then the necessary and sufficient optimality conditions are satisfied for $\bar{x} = 0$.

9.4 Duality

For a nonsmooth multiobjective programming problem (NMP), we formulate the following Mond-Weir type dual model (NDP) in terms of bifunctions:

$$(NDP) \quad \max \left(f_1(\bar{u}), ..., f_k(\bar{u})\right)$$

$$\text{subject to } \sum_{i=1}^{k} \lambda_i h_i(\bar{u}; x - \bar{u}) + \sum_{j \in J} \mu_j h'_j(\bar{u}; x - \bar{u}) \geq 0, \quad \forall x \in S,$$

$$\mu^T g(\bar{u}) \geq 0,$$

$$\lambda_i > 0, i \in I, \lambda \neq 0,$$

$$\mu_j \geq 0, j \subset J, \quad \bar{u} \subset K.$$

Theorem 9.3 (Weak duality) *Suppose that $\bar{x}$ is a feasible solution for (NMP) and $(\bar{u}, \lambda, \mu)$ is a feasible solution for (NDP). Let $f_i, i \in I$ be h_i-pseudolinear with respect to proportional function p_i and $g_j, j \in J$ be h'_j-pseudolinear with respect to proportional function q, then, the following cannot hold:*

$$f(\bar{x}) \leq f(\bar{u}).$$

Proof Suppose to the contrary that the above inequality holds. Using Theorem 4.12, the h_i-pseudolinearity of f_i on K with respect to the proportional function p_i implies that

$$p_i(\bar{u}, \bar{x}) h_i(\bar{u}; \bar{x} - \bar{u}) \leq 0, \forall i \in I,$$

$$p_r(\bar{u}, \bar{x}) h_r(\bar{u}; \bar{x} - \bar{u}) < 0, \quad \text{for some } r \in I.$$

Since, $p_i(\bar{u}, \bar{x}) > 0$, $\forall\, i = 1, 2, ..., k$, it follows that

$$h_i(\bar{u}; \bar{x} - \bar{u}) \leq 0, \forall\, i \in I,$$

$$h_r(\bar{u}; \bar{x} - \bar{u}) < 0, \text{ for some } r \in I.$$

Since, $\lambda_i > 0$, $\forall\, i = 1, 2, ..., k$, it follows that

$$\sum_{i=1}^{k} \lambda_i h_i(\bar{u}; \bar{x} - \bar{u}) < 0. \tag{9.10}$$

Since, $\bar{x}$ is feasible for (NMP) and $(\bar{u}, \lambda, \mu)$ is feasible for (NDP), we get

$$\sum_{j \in J} \mu_j g_j(\bar{x}) \leq 0$$

and

$$\sum_{j \in J} \mu_j g_j(\bar{u}) \geq 0.$$

Using the h'_j-pseudolinearity of $g_j, j \in J$ with respect to the proportional function q, we get

$$q(\bar{u}, \bar{x}) \sum_{j \in J} \mu_j h'_j(\bar{u}; \bar{x} - \bar{u}) \leq 0.$$

Again, as $q(\bar{u}, \bar{x}) > 0$, it follows that

$$\sum_{j \in J} \mu_j h'_j(\bar{u}; \bar{x} - \bar{u}) \leq 0. \tag{9.11}$$

From (9.10) and (9.11), we arrive at a contradiction to the first dual constraint of (NDP).

In the following theorem, we have weakened the conditions of pseudolinearity on the objective function, in the above theorem.

Theorem 9.4 (Weak duality) *Suppose that $\bar{x}$ is a feasible solution for (NMP) and $(\bar{u}, \lambda, \mu)$ is a feasible solution for (NDP). Let f_i, $i \in I$ be h_i-pseudolinear with respect to the same proportional function p and g_j, $j \in J$ be h'_j-pseudolinear with respect to the same proportional function q, then, the following cannot hold:*

$$f(\bar{x}) \leq f(\bar{u}).$$

Proof Suppose to the contrary that the above inequality holds. Since, $\lambda_i > 0$, for each $i \in I$, it follows that

$$\sum_{i=1}^{k} \lambda_i f_i(\bar{x}) < \sum_{i=1}^{k} \lambda_i f_i(\bar{u}).$$

Using the h_i-pseudolinearity of f_i, $i \in I$ with respect to the same proportional function p, we get the inequality (9.10) of Theorem 9.3.

The rest of the proof of the theorem is along the lines of the proof of Theorem 9.3.

Theorem 9.5 *Let $\bar{x}$ be a feasible solution for (NMP) and $(\bar{u}, \bar{\lambda}, \bar{\mu})$ be a feasible solution for (NDP), such that*

$$f(\bar{x}) = f(\bar{u}). \tag{9.12}$$

If for all feasible solutions (u, λ, μ) for (NDP) with f_i, $i \in I$ is h_i-pseudolinear with respect to the same proportional function p and g_j, $j \in J$ is h'_j-pseudolinear with respect to the same proportional function q, then, $\bar{x}$ is a properly efficient solution for (NMP) and $(\bar{u}, \bar{\lambda}, \bar{\mu})$ is a properly efficient solution for (NDP).

Proof Suppose that $\bar{x}$ is not an efficient solution for (NMP), then there exists $x \in S$, such that

$$f(x) \leq f(\bar{x}).$$

Using the assumption $f(\bar{x}) = f(\bar{u})$, we get a contradiction to Theorem 9.4. Hence $\bar{x}$ is an efficient solution of (NMP). In the same way we can show that $(\bar{u}, \bar{\lambda}, \bar{\mu})$ is an efficient solution for (NDP).

Suppose that $\bar{x}$ is not a properly efficient solution of (NMP). Then, for every scalar $M > 0$, there exists $x_0 \in S$ and an index i, such that

$$f_i(\bar{x}) - f_i(x_0) > M(f_j(x_0) - f_j(\bar{x})),$$

for all j satisfying

$$f_j(x_0) > f_j(\bar{x}),$$

whenever

$$f_i(x_0) < f_i(\bar{x}).$$

This implies that $f_i(\bar{x}) - f_i(x_0)$ can be made arbitrarily large and hence for $\bar{\lambda} > 0$, we get the following inequality

$$\sum_{i=1}^{k} \bar{\lambda}_i(f_i(\bar{x}) - f_i(x_0)) > 0. \tag{9.13}$$

Since, x_0 is feasible for (NMP) and $(\bar{u}, \bar{\lambda}, \bar{\mu})$ is feasible for (NDP), we have

$$g(x_0) \leqq 0, \tag{9.14}$$

$$\sum_{i=1}^{k} \bar{\lambda}_i h_i(\bar{u}; x_0 - \bar{u}) + \sum_{j \in J} \bar{\mu}_j h'_j(\bar{u}; x_0 - \bar{u}) \geq 0, \tag{9.15}$$

$$\bar{\mu}^T g(\bar{u}) \geq 0, \tag{9.16}$$

$$\bar{\mu} \geqq 0, \tag{9.17}$$

$$\bar{\lambda}_i > 0, \ i \in I. \tag{9.18}$$

From (9.14), (9.16), and (9.17), it follows that

$$\bar{\mu}^T g(x_0) \leq \bar{\mu}^T g(\bar{u}).$$

Since, g_j, $j \in J$ are h'_j-pseudolinear with respect to proportional function q, it follows that

$$q(\bar{u}, x_0) \sum_{j \in J} \bar{\mu}_j h'_j(\bar{u}; x_0 - \bar{u}) \leq 0.$$

Since, $q(\bar{u}, x_0) > 0$, we get

$$\sum_{j \in J} \bar{\mu}_j h'_j(\bar{u}; x_0 - \bar{u}) \leq 0.$$

From (9.15), we get

$$\sum_{i=1}^k \bar{\lambda}_i h_i(\bar{u}; x_0 - \bar{u}) \geq 0.$$

On using h_i-pseudolinearity of f_i, $i \in I$ with respect to proportional function p, we have

$$\sum_{i=1}^k \bar{\lambda}_i f_i(x_0) - \sum_{i=1}^k \bar{\lambda}_i f_i(\bar{u}) \geq 0. \tag{9.19}$$

Using (9.12) and (9.19), we get

$$\sum_{i=1}^k \bar{\lambda}_i (f_i(\bar{x}) - f_i(x_0)) \leq 0,$$

which contradicts (9.13). Hence, $\bar{x}$ is a properly efficient solution for (NMP).

Now, we assume that $(\bar{u}, \bar{\lambda}, \bar{y})$ is not a properly efficient solution for (NDP). Then, for every scalar $M > 0$, there exist a feasible point (u_0, λ_0, μ_0) for (NDP) and an index i, such that

$$f_i(u_0) - f_i(\bar{u}) > M(f_j(\bar{u}) - f_j(u_0)),$$

for all j satisfying

$$f_j(u_0) < f_j(\bar{u}),$$

whenever

$$f_i(u_0) > f_i(\bar{u}).$$

This implies that $f_i(u_0) - f_i(\bar{u})$ can be made arbitrarily large and hence for $\lambda > 0$, we get the following inequality

$$\sum_{i=1}^k \lambda_i (f_i(u_0) - f_i(\bar{u})) > 0. \tag{9.20}$$

Since $\bar{x}$ and $(\bar{u}, \bar{\lambda}, \bar{\mu})$ are feasible for (NMP) and (NDP) respectively, proceeding as in the first part, we get

$$\sum_{i=1}^{k} \bar{\lambda}_i \left(f_i(u_0) - f_i(\bar{u}) \right) \leq 0,$$

which is a contradiction to (9.20).

Hence, $(\bar{u}, \bar{\lambda}, \bar{\mu})$ is a properly efficient solution for (NDP).

Theorem 9.6 (Strong duality) *Suppose that $\bar{x}$ is an efficient solution for (NMP). Let f_i, $i \in I$ be h_i-pseudolinear with respect to the proportional function p_i, and g_j, $j \in J$ be h'_j-pseudolinear with respect to the proportional function q_j, then there exist $\bar{\lambda} \in \mathbb{R}^k$, $\bar{\mu} \in \mathbb{R}^m$, such that $(\bar{x}, \bar{\lambda}, \bar{\mu})$ is a feasible solution for (NDP). Further, if for all feasible (u, λ, μ) of (NDP), each f_i, $i \in I$ is h_i-pseudolinear with respect to the same proportional p and g_j, $j \in J$ is h'_j-pseudolinear with respect to the same proportional function q, then, $(\bar{x}, \bar{\lambda}, \bar{\mu})$ is a properly efficient solution for (NDP).*

Proof Using Theorem 9.1 and the hypothesis of this theorem, it follows that there exist $\bar{\lambda} \in \mathbb{R}^k$, $\bar{\mu}_j \in \mathbb{R}$, $j \in J(\bar{x})$, such that

$$\sum_{i=1}^{k} \bar{\lambda}_i h_i(\bar{x}; x - \bar{x}) + \sum_{j \in J} \bar{\mu}_j h'_j(\bar{x}; x - \bar{x}) \geq 0,$$

$$\bar{\lambda}_i > 0, \ i \in I,$$

$$\bar{\mu}_j \geq 0, \ j \in J(\bar{x}).$$

Let $\bar{\mu}_j = 0$ for $j \notin J(\bar{x})$, then there exist $\bar{\lambda} \in \mathbb{R}^k$, $\bar{\mu} \in \mathbb{R}^m$ such that $(\bar{x}, \bar{\lambda}, \bar{\mu})$ is a feasible solution of (NDP). The proper efficiency of $(\bar{x}, \bar{\lambda}, \bar{\mu})$ for (NDP) follows from Theorem 9.5 and using $\bar{\lambda} > 0$.

Remark 9.2 *If $K = \mathbb{R}^n$ and the functions f, g are differentiable pseudolinear functions, then the above weak and strong duality results reduce to that of Bector et al. [20].*

Chapter 10

Pseudolinear Multiobjective Semi-Infinite Programming

10.1 Introduction

Optimization problems in which infinite many constraints appear in finitely many variables are referred to as semi-infinite programming problems. These problems arise in several areas of modern research such as Chebyshev approximation, robotics, economics, optimal control, and engineering. See for example, Hettich and Still [103], Kaplan and Tichatschke [130], Still [261], Lopez and Still [169], Shapiro [253], and Flaudas and Pardalos [83]. Hettich and Kortanek [102] have presented theoretical results, methodologies, and applications of semi-infinite programming.

Optimality conditions and duality theorems for semi-infinite programming problems have been widely studied. Lopez and Vercher [170] established the characterizations for optimal solutions to the nondifferentiable convex semi-infinite programming problems involving the notion of the Lagrangian saddle point. Borwein [28] derived that under certain regularity conditions, a semi-infinite quasiconvex program possesses finitely constrained subproblems. Jeroslow [113] established necessary and sufficient optimality conditions for convex semi-infinite programming problems. Karney and Morley [134] considered a convex semi-infinite program and established that under certain limiting conditions the optimal value of a perturbed program tends to be the optimal value of the original program. Ruckmann and Shapiro [241] have established first order optimality conditions for generalized semi-infinite programming problems. Shapiro [252] studied Lagrange duality for convex semi-infinite programming problems. Zhang [296, 297] established optimality conditions and Mond-Weir type duality theorems for semi-infinite programming problems involving arcwise and B-arcwise connected functions.

Kanzi and Nobakhtian [127, 128] have introduced several kinds of constraint qualifications and derived necessary and sufficient optimality conditions for nonsmooth semi-infinite programming problems using the Clarke subdifferential. Mishra et al. [191, 192, 193] have studied optimality conditions and Wolfe and Mond-Weir type dual problems for nonsmooth semi-infinite programming problems under several generalized convexity assumptions.

Recently, Kanzi and Nobakhtian [129] have introduced some constraint qualifications and established optimality conditions for nonsmooth semi-infinite multiobjective optimization problems. For recent developments and further details, we refer to Kanzi [125, 126].

In this chapter, we will consider classes of smooth and nonsmooth multiobjective semi-infinite programming problems. Necessary and sufficient optimality conditions for a feasible solution to be an efficient solution are established under the assumption of pseudolinear functions. The extended Farkas's lemma is employed to establish the necessary optimality conditions for smooth semi-infinite programming problems. Suitable examples are given to illustrate the significance of the optimality conditions. Related to the primal problems, we formulate Mond-Weir type dual models to the smooth and nonsmooth primal problems and establish weak, strong, and converse duality theorems.

10.2   Definitions and Preliminaries

Let $X_0 \subseteq \mathbb{R}^n$ be a nonempty set, by $\mathrm{con}(X_0)$ and $\mathrm{cl}(X_0)$, we denote conical convex hull and closure of X_0, respectively.

We consider the following nonlinear multiobjective semi-infinite programming problem:

$$(MSIP) \quad \min f(x) := (f_1(x), ..., f_k(x))$$

$$\text{subject to } g_j(x) \leq 0 \quad j \in J,$$

$$x \in \mathbb{R}^n,$$

where f_i, $i = 1, ..., k$ and g_j, $j \in J$ are differentiable functions from $\mathbb{R}^n$ to $\mathbb{R} \cup \{\infty\}$ and the index set J is arbitrary, not necessarily finite (but nonempty). Let $X := \{x \in \mathbb{R}^n : g_j(x) \leq 0, \forall j \in J\}$ denote the set of all feasible solutions for (MSIP). For some $x \in X$, let us define $J(x) := \{j \in J \mid g_j(x) = 0\}$, the active constraint index set.

For some $1 \leq r \leq k$, we define

$$Z_r(x) := \{\nabla f_i(x), i = 1, ..., r - 1, r + 1, ..., k\} \cup \{\nabla g_j(x), j \in J(x)\}. \quad (10.1)$$

The following lemma is taken from Goberna *et al.* [92].

Lemma 10.1 (Extended homogeneous Farkas's lemma) *The inequality* $a^T x \geq 0$, $a, x \in \mathbb{R}^n$ *is a consequence of the system* $\{a_j^T x \geq 0 \mid j \in J\}$ *if and only if* $a \in \mathrm{cl \, con} \{a_j, j \in J\}$.

Let us recall the following lemma from Rockafellar [238].

Lemma 10.2 *Let $\{C_j, j \in J\}$ be an arbitrary collection of nonempty convex set in $\mathbb{R}^n$ and K be the convex cone generated by the union of the collection. Then, every nonzero vector of K can be expressed as a nonnegative linear combination of n or fewer linearly independent vectors, each belonging to a different C_j.*

10.3 Necessary and Sufficient Optimality Conditions

In this section, we prove the following theorem analogous to Proposition 3.2 in Chew and Choo [47] for (MSIP). It provides the necessary and sufficient conditions under which a feasible solution is an efficient solution for (MSIP).

Theorem 10.1 *Let x^* be a feasible point for (MSIP). Let the functions $f_i, i = 1, \ldots, k$ be pseudolinear with respect to proportional function p_i and $g_j, j \in J(x^*)$ be pseudolinear with respect to proportional function q_j. Then, x^* is an efficient solution of (MSIP), if and only if for each $r, 1 \le r \le k$, $\mathrm{cone}(Z_r(x^*))$ is closed and there exist $\bar{\lambda}_i \in R, i = 1, \ldots, k$ and $\bar{\mu}_j \in R, j \in J(x^*)$, finite number of them not vanishing, such that*

$$\sum_{i=1}^{k} \bar{\lambda}_i \nabla f_i(x^*) + \sum_{j \in J(x^*)} \bar{\mu}_j \nabla g_j(x^*) = 0,$$

$$\bar{\lambda}_i > 0, \ i = 1, \ldots, k, \tag{10.2}$$

$$\bar{\mu}_j \ge 0, \ j \in J(x^*).$$

Proof Suppose to the contrary, that $\bar{\lambda}_i$ and $\bar{\mu}_j$ exist and satisfy the given condition (10.2), yet x^* is not efficient. Then there exists a point $x \in X$ such that

$$f_i(x^*) \ge f_i(x), \ \ \forall i = 1, 2, \ldots, k$$

and

$$f_r(x^*) > f_r(x), \ \text{for some } r.$$

Then,

$$0 \ge \sum_{j \in J(x^*)} \frac{\bar{\mu}_j}{q_j(x^*, x)} (g_j(x) - g_j(x^*)) = \sum_{j \in J(x^*)} \bar{\mu}_j \langle \nabla g_j(x^*), x - x^* \rangle$$

$$= -\sum_{i=1}^{k} \bar{\lambda}_i \langle \nabla f_i(x^*), x - x^* \rangle = -\sum_{i=1}^{k} \frac{\bar{\lambda}_i}{p_i(x^*, x)} (f_i(x) - f_i(x^*)) > 0,$$

which is a contradiction.

Conversely, suppose x^* is an efficient solution for (MSIP). For $1 \leq r \leq k$, we consider the following system of inequalities:

$$\langle \nabla g_j(x^*), x - x^* \rangle \leq 0, \quad j \in J(x^*)$$

$$\langle \nabla f_i(x^*), x - x^* \rangle \leq 0, \quad i = 1, \ldots r - 1, r + 1, \ldots, k \qquad (10.3)$$

$$\langle \nabla f_r(x^*), x - x^* \rangle < 0,$$

having no solution $x \in \mathbb{R}^n$. For if x is a solution of the system and $y := x^* + t(x - x^*)$ $(0 < t \leq 1)$, then for $j \in J(x^*)$, we get

$$\langle \nabla g_j(x^*), y - x^* \rangle = t \langle \nabla g_j(x^*), x - x^* \rangle \leq 0.$$

Thus,

$$g_j(y) \leq g_j(x^*) = 0.$$

If $j \notin J(x^*)$, then $g_j(x^*) < 0$, when t is sufficiently small. Hence for small t, y is a point in X. Moreover, we have

$$f_i(y) - f_i(x^*) = tp_i(x^*, y)\langle \nabla f_i(x^*), x - x^* \rangle \leq 0, \; i \neq r;$$

and

$$f_r(y) - f_r(x^*) < 0.$$

This contradicts the choice of x^*. It follows that the system (10.3) has no solution in the nonempty convex set $\mathbb{R}^n$. In view of Lemma 10.1 and the fact that for each $r, 1 \leq r \leq k$, $\mathrm{con}(Z_r(x^*))$ is closed, it follows that

$$-\nabla f_r(x^*) \in \mathrm{cl} \; \mathrm{con}(Z_r(x^*)) = \mathrm{con}(Z_r(x^*)).$$

Therefore, there exist $\bar{\lambda}_{r_i} \geq 0$, $i = 1, \ldots, k$, and $\bar{\mu}_{r_j} \geq 0$, $j \in J(x^*)$, a finite number of them are not vanishing, such that

$$-\nabla f_r(x^*) = \sum_{i \neq r} \bar{\lambda}_{r_i} \nabla f_i(x^*) + \sum_{j \in J(x^*)} \bar{\mu}_{r_j} \nabla g_j(x^*). \qquad (10.4)$$

Summing (10.4) over r, we get (10.2) with

$$\bar{\lambda}_i = 1 + \sum_{i \neq r} \bar{\lambda}_{r_i}, \quad \bar{\mu}_j = \sum_{r=1}^{k} \bar{\mu}_{r_j}$$

This completes the proof.

To illustrate the significance of the above necessary and sufficient optimality conditions, we give the following example.

Example 10.1 *Consider the following problem:*

$$(P) \quad \min \; f(x) := (f_1(x), f_2(x))$$

$$\text{subject to} \; g_j(x) \leq 0, \quad j \in J = N,$$

$$x \in \mathbb{R},$$

where N is the set of natural numbers, $f : \mathbb{R} \to \mathbb{R}^2$ and $g_j : \mathbb{R} \to \mathbb{R}$ are functions defined as:

$$f_1(x) = x + x^3,$$
$$f_1(x) = exp\ x,$$
$$g_1(x) = -x,$$
$$g_2(x) = -2x,$$
$$g_j(x) = -x - \frac{1}{j},\ j = 3, 4, \ldots$$

It is easy to check that all the functions defined above are differentiable pseudolinear functions on open convex set $\mathbb{R}$. The set of feasible solution of the problem (P) is given by

$$X = \{x \in \mathbb{R} : x \geq 0\}.$$

It is clear that $\bar{x} = 0$ is a feasible solution for (P), $J(\bar{x}) = \{1, 2\}$ and $cone(Z_r(\bar{x}))$ is a closed convex set for each $1 \leq r \leq 2$. The necessary optimality condition is satisfied at $\bar{x}$ as there exist Lagrange multipliers $\bar{\lambda} = (1, 1)$ and $\bar{\mu} = \left(1, \frac{1}{2}, 0, 0, \ldots, 0, \ldots\right)$, such that the conditions in (10.2) are satisfied. Moreover, we observe that $\bar{x} = 0$ is an efficient solution for (P) as there exists no $x \in X$, such that

$$f(x) \leq f(\bar{x}).$$

10.4   Duality for (MSIP)

In this section, we formulate a Mond-Weir type dual problem for (MSIP) and derive several weak, strong, and converse duality theorems.

Related to our problem (MSIP), we consider the following Mond-Weir type dual problem:

$$(MWSID) \quad \max\ f(y) := (f_1(y), \ldots, f_k(y))$$

$$\text{subject to } \sum_{i=1}^{k} \lambda_i \nabla f_i(y) + \sum_{j \in J} \mu_j \nabla g_j(y) = 0, \qquad (A)$$

$$\sum_{j \in J} \mu_j (g_j(y)) \geq 0,$$

$$\lambda_i > 0,\ i = 1, \ldots, k,$$

where $\mu = (\mu_j) \geq 0$, $j \in J$, and $\mu_j \neq 0$ for finitely many j.

Theorem 10.2 *(Weak duality) Let x be a feasible solution for (MSIP) and (y, λ, μ) be a feasible solution for (MWSID). Let $f_i, i = 1, \ldots, k$ be pseudolinear with respect to proportional function p_i and $g_j, \ j \in J$ be pseudolinear with respect to q, then, the following condition cannot hold*

$$f(x) \leq f(y).$$

Proof Let x be a feasible solution for (MSIP) and (y, λ, μ) be a feasible solution for (NMWSID). Then, we have

$$g_j(x) \leq 0, \ \forall j \in J, \tag{10.5}$$

$$\sum_{i=1}^{k} \lambda_i \nabla f_i(y) + \sum_{j \in J} \mu_j (\nabla g_j(y)) = 0, \tag{10.6}$$

$$\sum_{j \in J} \mu_j g_j(y) \geq 0. \tag{10.7}$$

Suppose the above inequalities hold, then, we have

$$f_i(x) \leq f_i(y), \ \forall i = 1, \ldots, k,$$

$$f_r(x) < f_r(y), \ \text{for some } r.$$

Now, by pseudolinearity of f_i with respect to proportional function $p_i, i = 1, \ldots, k$, we have

$$p_i(y, x) \langle \nabla f_i(y), x - y \rangle \leq 0, \ \forall i = 1, \ldots, k,$$

$$p_r(y, x) \langle \nabla f_r(y), x - y \rangle < 0, \ \text{for some } r.$$

Since, $p_i(y, x) > 0$, for all $i = 1, \ldots, k$, we get

$$\langle \nabla f_i(y), x - y \rangle \leq 0, \ \forall i = 1, \ldots, k,$$

$$\langle \nabla f_r(y), x - y \rangle < 0, \ \text{for some } r.$$

Since, $\lambda_i > 0$, for each $i = 1, \ldots, k$, it follows that

$$\sum_{i=1}^{k} \lambda_i \langle \nabla f_i(y), x - y \rangle < 0. \tag{10.8}$$

As x be a feasible solution for (MSIP) and (y, λ, μ) be a feasible solution for (NMWSID), we have

$$\sum_{j \in J} \mu_j g_j(x) \leq 0$$

and

$$\sum_{j \in J} \mu_j g_j(y) \geq 0.$$

Using the pseudolinearity of g_j, $j \in J$ with respect to proportional function q, we get

$$q(y,x) \sum_{j \in J} \langle \mu_j \nabla g_j(y), x - y \rangle \leq 0.$$

Again, as $q(y,x) > 0$, it follows that

$$\sum_{j \in J} \langle \mu_j \nabla g_j(y), x - y \rangle \leq 0. \tag{10.9}$$

Adding (10.8) and (10.9), we arrive at a contradiction to the equality constraint of the dual problem (MWSID).

In the next theorem, we assume that each component of the objective function is pseudolinear with respect to the same proportional function.

Theorem 10.3 (Weak duality) *Let x be a feasible solution for (MSIP) and (y, λ, μ) be a feasible solution for (MWSID). Let $f_i, i = 1, \ldots, k$ be pseudolinear with respect to proportional function p and g_j, $j \in J$ be pseudolinear with respect to proportional function q, then, the following condition cannot hold*

$$f(x) \leq f(y).$$

Proof We suppose to the contrary that the above inequalities hold. Since $\lambda_i > 0$, for each $i = 1, \ldots, k$, we get

$$\sum_{i=1}^{k} \lambda_i f_i(x) \leq \sum_{i=1}^{k} \lambda_i f_i(y).$$

On using the pseudolinearity of $f_i (i = 1, \ldots, k)$, we get the inequality (10.8) of Theorem 10.2. Proceeding along the same lines as in Theorem 10.2, we arrive at a contradiction.

Theorem 10.4 *Let $\bar{x}$ be a feasible solution for (MSIP) and $(\bar{y}, \bar{\lambda}, \bar{\mu})$ be a feasible solution for (MWSID), such that*

$$f(\bar{x}) = f(\bar{y}). \tag{10.10}$$

Let for all feasible (y, λ, μ) of (MWSID), $f_i, i = 1, \ldots, k$ be pseudolinear with respect to proportional function p and g_j, $j \in J$ are pseudolinear with respect to proportional function q, then $\bar{x}$ is properly efficient for (MSIP) and $(\bar{y}, \bar{\lambda}, \bar{\mu})$ is properly efficient for (MWSID).

Proof Suppose to the contrary that $\bar{x}$ is not an efficient solution of (MSIP) then there exists $x \in X$, such that

$$f(x) \leq f(\bar{x}).$$

Using the assumption

$$f(\bar{x}) = f(\bar{y}),$$

we get a contradiction to Theorem 10.3. Hence, $\bar{x}$ is an efficient solution of (MSIP). Similarly, we can prove that $(\bar{y}, \bar{\mu}, \bar{\lambda})$ is an efficient solution of (MWSID).

Now suppose that $\bar{x}$ is not a properly efficient solution of (MSIP). Therefore, for every scalar $M > 0$, there exist $x_0 \in X$ and an index j, such that

$$f_i(\bar{x}) - f_i(x_0) > M(f_r(x_0)) - f_r(\bar{x}),$$

for all r satisfying

$$f_r(x_0) > f_r(\bar{x}),$$

whenever

$$f_i(x_0) < f_i(\bar{x}).$$

This means $f_i(\bar{x}) - f_i(x_0)$ can be made arbitrarily large and hence for $\bar{\lambda} > 0$, we get the following inequality

$$\sum_{i=1}^{k} \bar{\lambda}_i(f_i(\bar{x}) - f_i(x_0)) > 0. \tag{10.11}$$

Now, x_0 is feasible for (MSIP) and $(\bar{y}, \bar{\lambda}, \bar{\mu})$ is feasible for (MWSID), therefore, we have

$$g_j(x_0) \le 0, \ j \in J, \tag{10.12}$$

$$\sum_{i=1}^{k} \bar{\lambda}_i \nabla f_i(\bar{y}) + \sum_{j \in J} \bar{\mu}_j \nabla g_j(\bar{y}) = 0, \tag{10.13}$$

$$\sum_{j \in J} \bar{\mu}_j g_j(\bar{y}) \ge 0, \tag{10.14}$$

$$\bar{\lambda}_i > 0, \ i = 1, \dots, k, \tag{10.15}$$

where $\bar{\mu}_j \ge 0$ and $\bar{\mu}_j \ne 0$, for finitely many $j \in J$.

From (10.12) and (10.14), we get

$$\sum_{j \in J} \bar{\mu}_j g(x_0) \le \sum_{j \in J} \bar{\mu}_j g(\bar{y}).$$

Using the pseudolinearity of $g_j(.), \ j \in J$ with respect to q, it follows that

$$q(\bar{y}, x_0) \langle \sum_{j \in J} \bar{\mu}_j \nabla g_j(\bar{y}), x_0 - \bar{y} \rangle \le 0.$$

Since, $q(\bar{y}, x_0) > 0$, we have

$$\langle \sum_{j \in J} \bar{\mu}_j \nabla g_j(\bar{y}), x_0 - \bar{y} \rangle \le 0.$$

Using (10.13), we get

$$\langle \sum_{i=1}^{k} \bar{\lambda}_i \nabla f_i(\bar{y}), x_0 - \bar{y} \rangle \geq 0.$$

By the pseudolinearity of $f_i, i = 1, \ldots, k$ with respect to proportional function p, it follows that

$$\sum_{i=1}^{k} \bar{\lambda}_i (f_i(x_0) - f_i(\bar{y})) \geq 0. \tag{10.16}$$

On using (10.10) in (10.16), it follows that

$$\sum_{i=1}^{k} \bar{\lambda}_i (f_i(\bar{x}) - f_i(x_0)) \leq 0,$$

which is a contradiction to (10.11). Hence, $\bar{x}$ is a properly efficient solution of (MSIP).

To prove the rest of the theorem, we now suppose that $(\bar{y}, \bar{\lambda}, \bar{\mu})$ is not a properly efficient solution of (MWSID). Therefore, for every scalar $M > 0$, there exist a feasible (y_0, λ_0, μ_0) in (MWSID) and an index i, such that

$$f_i(y_0) - f_i(\bar{y}) > M(f_r(\bar{y}) - f_r(y_0)),$$

for all r satisfying

$$f_r(y_0) < f_r(\bar{y}),$$

whenever

$$f_i(y_0) > f_i(\bar{y}).$$

This means that $f_i(y_0) - f_i(\bar{y})$ can be made arbitrarily large and hence for $\lambda > 0$, we get the following inequality

$$\sum_{i=1}^{k} \bar{\lambda}_i (f_i(y_0) - f_i(\bar{y})) > 0. \tag{10.17}$$

Since, $\bar{x}$ and $(\bar{y}, \bar{\lambda}, \bar{\mu})$ are feasible for (MSIP) and (MWSID), respectively, proceeding on the same lines as in the first part, it follows that

$$\sum_{i=1}^{k} \bar{\lambda}_i (f_i(y_0) - f_i(\bar{y})) \leq 0,$$

which is a contradiction to (10.17). Hence, $(\bar{y}, \bar{\mu}^j, \bar{\lambda}, \bar{\mu})$ is a properly efficient solution for (MWSID).

Theorem 10.5 (Strong duality) *Let $\bar{x}$ be an efficient solution for (MSIP). Let $f_i, \ i = 1, 2, \ldots, k$ be pseudolinear with respect to proportional function p_i*

and $g_j(.)$, $j \in J(\bar{x})$ be pseudolinear with respect to proportional function q_j. such that for each r, $1 \leq r \leq k$, $cone(Z_r(\bar{x}))$ is a closed set, then there exist $\bar{\lambda} \in R^k, \bar{\mu} = (\mu_j), j \in J$, such that $(\bar{x}, \bar{\lambda}, \bar{\mu})$ is feasible for (MWSID). Further, if for all feasible (x, λ, μ) for (MWSID), f_i, $i = 1, 2, \ldots, k$ are pseudolinear with respect to p and $g_j(.)$, $j \in J$ be pseudolinear with respect to q, then, $(\bar{x}, \bar{\lambda}, \bar{\mu})$ is a properly efficient solution for (MWSID).

Proof Using Theorem 10.1 and assumptions of this theorem, it follows that there exist $\bar{\lambda} \in R^k, \bar{\mu}_j \in R, j \in J(\bar{x})$, such that

$$\sum_{i=1}^{k} \bar{\lambda}_i \nabla f_i(\bar{x}) + \sum_{j \in J(\bar{x})} \bar{\mu}_j \nabla g_j(\bar{x}) = 0,$$

$$\bar{\mu}_j \geq 0, j \in J(\bar{x}),$$

$$\bar{\lambda}_i > 0, \ i = 1, 2, \ldots, k.$$

Setting $\bar{\mu}_j \neq 0$, for $j \notin J(\bar{x})$, it follows that there exist $\bar{\lambda} \in R^k, \mu_j, j \in J(\bar{x})$, such that $(\bar{x}, \bar{\lambda}, \bar{\mu})$ is a feasible solution for (MWSID). The proper efficiency of $(\bar{x}, \bar{\lambda}, \bar{\mu})$ for the problem (MWSID) follows from Theorem 10.3.

Theorem 10.6 *(Converse duality)* Let $(\bar{u}, \bar{\lambda}, \bar{y})$ be an efficient solution for (MWSID) and let for all feasible (u, λ, y) for (MWSID), $\sum_{i=1}^{k} \lambda_i f_i$ be pseudolinear with respect to proportional function p and $y^T h$ be pseudolinear with respect to q. Suppose for each $r, 1 \leq r \leq k, cone(Z_r(\bar{u}))$ is a closed set. Moreover, let the $n \times n$ Hessian matrix $\nabla^2(\bar{\lambda}^T f(\bar{u})) + \nabla^2(\bar{y}^T g(\bar{u}))$ be positive or negative definite and $\{\nabla f_i(\bar{u}), i = 1, \ldots, k\}$ be linearly independent. Then, $\bar{u}$ is a properly efficient solution for (MSIP).

Proof Since $(\bar{u}, \bar{\lambda}, \bar{y})$ is an efficient solution for (MWSID), by Theorem 10.1, there exist $\tau \in R^k, \nu \in R^n, p \in R, s \in R^m, w \in R^k$, such that

$$-\nabla \tau^T f(\bar{u}) + \nabla \nu^T [\nabla \bar{\lambda}^T f(\bar{u}) + \nabla \bar{y}^T g(\bar{u})] - p \nabla \bar{y}^T g(\bar{u}) = 0, \tag{10.18}$$

$$(\nabla g(\bar{u}))^T \nu - pg(\bar{u}) - s = 0, \tag{10.19}$$

$$(\nabla g(\bar{u}))^T \nu - w = 0 \tag{10.20}$$

$$p\bar{y}^T g(\bar{u}) = 0 \tag{10.21}$$

$$s^T \bar{y} = 0 \tag{10.22}$$

$$w^T \bar{\lambda} = 0, \tag{10.23}$$

$$\tau \geq 0, \ \nu \text{ unrestricted}, (s, p, w) \geq 0, \tag{10.24}$$

Since, $\bar{\lambda} > 0$, (10.23) implies that $w = 0$. Then, using (10.20), we get

$$\nu^T \nabla f(\bar{u}) = 0. \tag{10.25}$$

Multiplying (10.19) by $\bar{y}$ and using (10.21) and (10.22), we get

$$\nu^T \nabla \left(\bar{y}^T g(\bar{u}) \right) = 0. \tag{10.26}$$

Multiplying (10.18) by ν^T and using (10.25) and (10.26), it follows that

$$\nu^T [\nabla^2 (\bar{\lambda}^T f(\bar{u})) + \nabla^2 (\bar{y}^T g(\bar{u}))]\nu = 0.$$

Since, $\nabla^2(\bar{\lambda}^T f(\bar{u})) + \nabla^2(\bar{y}^T g(\bar{u}))$ is assumed to be positive or negative definite, we get $\nu = 0$.

Since, $\nu = 0$, by (10.18) and equality constraint (A) of (NMWSID), we get

$$\nabla(\tau - p\bar{\lambda})^T f(\bar{u}) = 0.$$

By linear independence of $\{\nabla f_i(\bar{u}), i = 1, \ldots, k\}$, it follows that

$$\tau = p\bar{\lambda}.$$

Since, $\bar{\lambda} > 0, \tau \geq 0$ implies $p > 0$. Thus $\tau \neq 0$ and $p > 0$. Since $p > 0$ and $s \geq 0$, (10.19) gives $g(\bar{u}) \leq 0$ and (10.21) gives $\bar{y}^T g(\bar{u}) = 0$. Thus $\bar{u}$ is feasible for (MSIP). That $\bar{u}$ is an efficient solution for (MSIP) follows from the assumptions of the theorem and weak duality Theorem 10.4. The proper efficiency of $\bar{u}$ for (MSIP) follows from Theorem 10.5.

10.5   Nonsmooth Multiobjective Semi-Infinite Programming

The following definitions form Kanzi and Nobakhtian [129] will be needed in the sequel.

Given a nonempty set $D \subseteq \mathbb{R}^n$, the negative polar cone and the strictly polar cone are defined respectively by

$$D^{\leq} := \{d \in \mathbb{R}^n : \langle x, d \rangle \leq 0, \forall x \in D\}$$

$$D^{<} := \{d \in \mathbb{R}^n : \langle x, d \rangle < 0, \forall x \in D\}.$$

By the bipolar theorem (Hiriart-Urruty and Lemarechal [105]), we have

$$(D^{\leq})^{\leq} = \overline{cone}(D).$$

The conjugate cone

$$\Gamma(D, \bar{x}) := \{d \in \mathbb{R}^n : \exists t_n \downarrow 0, \exists d_n \to d \text{ such that } \bar{x} + t_n d_n \in D, \forall n \in N\}.$$

$$\Gamma(D, \bar{x}) := \left\{ \lim_{n \to \infty} \frac{x_n - \bar{x}}{\lambda_n} : \lambda_n \downarrow 0, \exists d_n \to d \text{ such that } x_n \in D, x_n \to \bar{x} \right\}.$$

We consider the following nonsmooth multiobjective semi-infinite programming problem:

$$(NMSIP) \quad \min f(x) := (f_1(x), \ldots, f_k(x))$$

$$\text{subject to } g_j(x) \le 0, \ j \in J$$

$$x \in \mathbb{R}^n,$$

where $f_i, i = 1, \ldots, k$ and $g_j, j \in J$ are nondifferentiable locally Lipschitz functions from $\mathbb{R}^n$ to $\mathbb{R} \cup \{\infty\}$ and the index set J is arbitrary, not necessarily finite (but nonempty). Let $X := \{x \in \mathbb{R}^n : g_j(x) \le 0, \forall j \in J\}$ denote the set of all feasible solutions for (NMSIP).

For some $x \in X$, let us define

$$J(\bar{x}) := \{j \in J \mid g_j(x) = 0\},$$

$$F(\bar{x}) := \cup_{i=1}^m \partial^c f_i(\bar{x}) \text{ and } G(\bar{x}) := \cup_{j \in J(\bar{x})} \partial^c g_j(\bar{x}).$$

The following definitions will be needed in the sequel:

Definition 10.1 *The regular constraint qualification (RCQ) is said to be satisfied at $\bar{x} \in X$, if*

$$F^<(\bar{x}) \cap G^\le(\bar{x}) \subseteq \Gamma(X, \bar{x}).$$

Definition 10.2 *(Hiriart-Urruty et al. [106]) Let $f : \mathbb{R}^n \to \mathbb{R} \cup \{\infty\}$ be a $C^{1,1}$ function, i.e., the functions whose gradient mapping is locally Lipschitz. Then, the generalized Hessian matrix of f at $\bar{x}$, denoted by $\partial^2 f(\bar{x})$ is the set of matrices defined as the convex hull of the set*

$$\{M : \exists x_i \to \bar{x} \text{ with } f \text{ twice differentiable at } x_i \text{ and } \nabla^2 f(x_i) \to M\}.$$

The space of $n \times n$ is topologized by taking a metrical norm $\| \ \|_M$ on it. By construction itself, $\partial^2 f(\bar{x})$ is a nonempty convex compact subset of symmetric matrices, which reduces to $\{\nabla^2 f(\bar{x})\}$, when ∇f is strictly (or strongly) differentiable at $\bar{x}$ (see, Ortega and Rheinboldt [217]).

10.6   Necessary and Sufficient Optimality Conditions

We recall the following necessary optimality condition from Kanzi and Nobakhtian [129].

Theorem 10.7 (Necessary optimality conditions) *Let $\bar{x}$ be a weakly efficient solution for (NMSIP) and RCQ holds at $\bar{x}$. If cone $(G(\bar{x}))$ is closed, then there exist $\lambda_i \geq 0, i = 1, \ldots, k$ and $\mu_j \geq 0, j \in J(\bar{x})$ with $\mu_j \neq 0$, for finitely many indices, such that*

$$\sum_{i=1}^{k} \lambda_i \partial^c f_i(\bar{x}) + \sum_{j \in J(x^*)} \mu_j \partial^c g_t(\bar{x}) = 0,$$

$$\sum_{i=1}^{m} \bar{\lambda}_i = 1.$$

Theorem 10.8 *Let $\bar{x}$ be a feasible point for (NMSIP). Let the functions $f_i, i = 1, \ldots, k$ be pseudolinear with respect to proportional function p_i and $g_j, j \in J(\bar{x})$ be pseudolinear with respect to proportional function q_j. If there exist $\bar{\lambda}_i \in \mathbb{R}, i = 1, \ldots, k$ and $\bar{\mu}_j \in \mathbb{R}, \ j \in J(\bar{x}), \ \xi_i \in \partial^c f_i(\bar{x})$ and $\zeta_j \in \partial^c g_j(\bar{x})$, such that*

$$\sum_{i=1}^{k} \bar{\lambda}_i \xi_i + \bar{\mu}_j \zeta_j = 0, \bar{\lambda}_i \geq 0, i - 1, \ldots, k, \sum_{i=1}^{k} \bar{\lambda}_i = 1, \bar{\mu}_j \geq 0, j \in J(\bar{x}). \quad (10.27)$$

Then $\bar{x}$ is a weakly efficient solution of (NMSIP).

Proof Suppose $\bar{\lambda}_i$ and $\bar{\mu}_j$ exist and satisfy the conditions in (10.28). Let $\bar{x}$ is not a weakly efficient for (NMSIP). There exists then a point $y \in X$, such that

$$f_i(y) < f_i(\bar{x}), \forall i = 1, \ldots, k.$$

Using the pseudolinearity of $f_i, i = 1, \ldots, k$ with respect to proportional function p_i, there exists $\xi_i' \in \partial^c f_i(\bar{x})$, such that

$$p_i(\bar{x}, y)\langle \xi_i', y - \bar{x} \rangle < 0, \forall i = 1, \ldots, k.$$

Since, $p_i(\bar{x}, y) > 0$, for all $i = 1, \ldots, k$, therefore, there exists $\xi_i' \in \partial^c f_i(\bar{x})$, such that

$$\langle \xi_i', y - \bar{x} \rangle < 0, \forall i = 1, \ldots, k.$$

By Proposition 2.3, there exist $s_i > 0, \xi_i \in \partial^c f_i(\bar{x})$, such that $\xi_i' = s_i \xi_i, i = 1, \ldots, k$, hence, we have

$$s_i \langle \xi_i, y - \bar{x} \rangle < 0, \forall i = 1, \ldots, k.$$

Since, $s_i > 0$, for all $i = 1, \ldots, k$, it follows that

$$\langle \xi_i, y - \bar{x} \rangle < 0, \forall i = 1, \ldots, k.$$

Therefore, for $\bar{\lambda}_i > 0, i = 1, \ldots, k$, we get

$$\sum_{i=1}^{k} \bar{\lambda}_i \langle \xi_i, y - \bar{x} \rangle < 0. \quad (10.28)$$

Again, we have
$$g_j(y) \leq g_j(\bar{x}), \forall j \in J(\bar{x}).$$

Using the pseudolinearity of $g_j, j \in J$, with respect to proportional function q_j, there exists $\zeta'_j \in \partial^c g_j(\bar{x})$, such that

$$q_j(\bar{x}, y)\langle \zeta'_j, y - \bar{x} \rangle \leq 0, \ \forall j \in J(\bar{x}).$$

Since, $q_j(\bar{x}, y) > 0$, for all $j \in J(\bar{x})$, therefore, we get

$$\langle \zeta'_j, y - \bar{x} \rangle \leq 0, \ \forall j \in J(\bar{x}).$$

By Proposition 2.3, there exist $t_j > 0, \zeta_j \in \partial^c g_j(\bar{x}), j \in J(\bar{x})$, such that $\zeta'_j = t_j \zeta_j$. Hence, we have

$$t_j\langle \zeta_j, y - \bar{x} \rangle \leq 0, \ \forall j \in J(\bar{x}).$$

Since, $t_j > 0$, for all $j \in J(\bar{x})$, we get

$$\langle \zeta_j, y - \bar{x} \rangle \leq 0, \ \forall j \in J(\bar{x}).$$

Since, $\bar{\mu}_j \geq 0$, for all $j \in J(\bar{x})$, it follows that

$$\sum_{j \in J(\bar{x})} \bar{\mu}_j \langle \zeta_j, y - \bar{x} \rangle \leq 0.$$

Using (10.28), there exists $\xi_i \in \partial^c f_i(\bar{x})$, such that

$$\sum_{i=1}^{k} \bar{\lambda}_i \langle \bar{\xi}_i, y - \bar{x} \rangle \geq 0,$$

which is a contradiction to (10.29).

The following example illustrates the significance of the necessary and sufficient optimality conditions.

Example 10.2 *Consider the following nonsmooth multiobjective semi-infinite programming problem:*

$$(NP) \quad \min f(x) := (f_1(x), f_2(x))$$

$$\text{subject to } g_j(x) \leq 0, j \in J = N, x \in \mathbb{R},$$

where N is the set of natural numbers, $f : \mathbb{R} \to \mathbb{R}^2$ and $g_j : \mathbb{R} \to \mathbb{R}$ are functions defined as:

$$f_1(x) := 2x + |x|,$$

$$f_2(x) := \exp x,$$

$$g_1(x) := -x,$$

$$g_2(x) := -2x,$$

$$g_j(x) := -x - \frac{1}{j}, \quad j = 3, 4, ...$$

It is easy to check that all the functions defined above are locally Lipschitz pseudolinear functions on open convex set $\mathbb{R}$. *The set of feasible solutions of the problem (P) is given by*

$$X = \{x \in \mathbb{R} : x \geq 0\} \text{ and } J(\bar{x}) = \{1, 2\}.$$

We observe that $\bar{x} = 0$ *is a weakly efficient solution for (P) as there exists no* $x \in X$, *such that*

$$f(x) < f(\bar{x}).$$

Moreover, it is easy to verify that

$$\Gamma(X, \bar{x}) = X, G(\bar{x}) = \bigcup_{j \in J(\bar{x})} \partial^c g_j(\bar{x}) = \{-1, -2\}, \text{cone}(G(\bar{x})) =]-\infty, 0],$$

$$G^{\leq}(\bar{x}) = [0, \infty[, F(\bar{x}) = [1, 3],$$

$$F^<(\bar{x}) =]0, \infty[, F^<(\bar{x}) \bigcap G^{\leq}(\bar{x}) \subseteq \Gamma(X, \bar{x}).$$

We note that the cone $(G(\bar{x}))$ *is closed. Hence, the necessary optimality condition is satisfied at* $\bar{x}$ *as there exist Lagrange multipliers* $\bar{\lambda} = (1, 1)$ *and* $\bar{\mu} = \left(1, \frac{1}{2}, 0, 0, ..., 0, ...\right)$, *such that the conditions in (10.27) are satisfied.*

10.7 Duality for (NMSIP)

In this section, we formulate the Mond-Weir type dual problem (NN-MWSID) and derive several weak and strong duality results. Related to our problem (NMSIP), we consider the following Mond-Weir type dual problem:

$$(NMWSID) \quad \max f(y) = (f_1(y), \ldots, f_k(y))$$

$$\text{subject to } 0 \in \sum_{i=1}^{k} \lambda_i \partial^c f_i(y) + \sum_{j \in J} \mu_j (\partial^c g_j(y)), \quad (B)$$

$$\sum_{j \in J} \mu_j g_j(y) \geq 0,$$

$$\lambda_i \geq 0, i = 1, \ldots, k, \sum_{i=1}^{k} \lambda_i = 1,$$

where $\mu = (\mu_j) \geq 0, j \in J$ and $\mu_j \neq 0$, for finitely many j.

Theorem 10.9 *(**Weak duality**) Let x be feasible for (NMSIP) and (y, λ, μ) be feasible for (NMWSID). Let $f_i, i = 1, \ldots, k$ be pseudolinear with respect to proportional function p_i and $g_j, j \in J$ be pseudolinear with respect to q, then, the following condition cannot hold:*

$$f(x) < f(y).$$

Proof Let x be feasible for (NMSIP) and (y, λ, μ) be feasible for (NMWSID). Then we have

$$g_j(x) \leq 0, \forall j \in J, \tag{10.29}$$

$$0 \in \sum_{i=1}^{k} \lambda_i \partial^c f_i(y) + \sum_{j \in J} \mu_j (\partial^c g_j(y)), \tag{10.30}$$

$$\sum_{j \in J} \mu_j g(y) \geq 0. \tag{10.31}$$

Therefore, there exist $\xi_i \in \partial^c f_i(y)$ and $\zeta_j \in \partial^c g_j(y), j \in J$, such that

$$0 = \sum_{i=1}^{k} \lambda_i \xi_i + \sum_{j \in J} \mu_j \zeta_J. \tag{10.32}$$

Suppose the above inequalities hold, then we have

$$f_i(x) < f_i(y), \forall i = 1, \ldots, k$$

Now, by pseudolinearity of f_i with respect to proportional function $p_i, i = 1, \ldots, k$ and for some $\xi_i \in \partial^c f_i(y)$ and $\xi_r \in \partial^c f_r(y)$, we have

$$p_i(y, x) \langle \xi_i, x - y \rangle < 0, \forall i = 1, \ldots, k.$$

Since, $p_i(y, x) > 0$, for all $i = 1, \ldots, k$, it follows that

$$\langle \xi_i, x - y \rangle < 0, \forall i = 1, \ldots, k$$

Since $\lambda_i > 0$ for each $i = 1, \ldots, k$, therefore for some $\xi_i \in \partial^c f_i(y)$, we get

$$\sum_{i=1}^{k} \lambda_i \langle \xi_i, x - y \rangle < 0. \tag{10.33}$$

As x is feasible for (NMSIP) and (y, λ, μ) is feasible for (NMWSID), it follows that

$$\sum_{j \in J} \mu_j g_j(x) \leq 0$$

and

$$\sum_{j \in J} \mu_j g_j(y) \geq 0.$$

Using the pseudolinearity of $g_j(.), j \in J$ with respect to proportional function q, for some $\zeta_j \in \partial^c g_j(y)$, it follows that

$$q(y,x) \sum_{i \in I} \langle \mu_i \zeta_j, x - y \rangle \leq 0.$$

Again, as $q(y,x) > 0$, it follows that

$$\sum_{j \in J} \langle \mu_j \zeta_j, x - y \rangle \leq 0. \tag{10.34}$$

Adding (10.36) and (10.37), we arrive at a contradiction to the equality constraint of the dual problem (NMWSID).

In the next theorem, we assume that each component of the objective function is pseudolinear with respect to the same proportional function.

Theorem 10.10 *Let x be feasible for (NMSIP) and (y, λ, μ) be feasible for (NMWSID). Let $f_i, i = 1, \ldots, k$ be pseudolinear with respect to proportional function p and $g_j(.), j \in J$ be pseudolinear with respect to proportional function q, then, the following condition cannot hold:*

$$f(x) < f(y).$$

Proof We suppose to the contrary that the above inequalities hold. Since $\lambda_i > 0$ for each $i = 1, \ldots, k$ we obtain that

$$\sum_{i=1}^{k} \lambda_i f_i(x) < \sum_{i=1}^{k} \lambda_i f_i(y).$$

On using the pseudolinearity of $f_i (i = 1, \ldots, k)$ we obtain the inequality (10.36) of Theorem 10.9. Proceeding along the same lines as in Theorem 10.9, we arrive at a contradiction.

Theorem 10.11 *(Strong duality) Let $\bar{x}$ be a weakly efficient solution for (NMSIP) and RCQ holds at $\bar{x}$. Let $cone(G(\bar{x}))$ be closed. Let $f_i, i = 1, \ldots, k$ be pseudolinear with respect to proportional function p_i and $g_j(.), j \in J(\bar{x})$ be pseudolinear with respect to proportional function q_j. Then, there exist $\bar{\lambda} \in R^k, \bar{\mu} = (\mu_j, j \in J), \mu_j \in \mathbb{R}$, such that $(\bar{x}, \bar{\lambda}, \bar{\mu})$ is feasible for (NMWSID). Further, if for all feasible (x, λ, μ) for (NMWSID), $f_i, i = 1, \ldots, k$ be pseudolinear with respect to p and $g_j(.) j \in J$ be pseudolinear with respect to q. Then, $(\bar{x}, \bar{\lambda}, \bar{\mu})$ is a weakly efficient solution for (NMWSID).*

Proof Using Theorem 10.7 and assumptions of this theorem, it follows that there exist $\bar{\lambda} \in R^k, \bar{\mu}_j \in Rj \in J(\bar{x})$, such that

$$\sum_{i=1}^{k} \bar{\lambda}_i \xi_i + \sum_{j \in J(\bar{x})} \bar{\mu}_j \zeta_j = 0,$$

$$\bar{\mu}_j \geqq 0, j \in J(\bar{x}),$$

$$\bar{\lambda}_i \geq 0, i = 1, \ldots, k, \sum_{i=1}^{m} \bar{\lambda}_i = 1.$$

Setting $\bar{\mu}_j = 0$ for $j \notin J(\bar{x})$, it follows that there exist $\bar{\lambda} \in R^k, \mu_j, j \in J(\bar{x})$, such that $(\bar{x}, \bar{\lambda}, \bar{\mu})$ is a feasible solution for (NMWSID). The weak efficiency of $\bar{x}$ for (NMWSID) follows from the weak duality Theorem 10.8.

Theorem 10.12 (*Converse duality*) *Let $(\bar{u}, \bar{\lambda}, \bar{y})$ be a weakly efficient solution for (NMWSID). Let RCQ hold at $\bar{u}$ and $cone(G(\bar{x}))$ is a closed cone. Suppose that for all feasible (u, λ, y) for (NMWSID), $f_i, i = 1, \ldots, k$ be $C^{1,1}$ pseudolinear functions with respect to p and $g_j, j \in J$ are $C^{1,1}$ pseudolinear functions with respect to q. Moreover, let each $n \times n$ Hessian matrix $M \in \partial^2(\bar{\lambda}^T f(\bar{u}) + \bar{y}^T g(\bar{u})$ be positive or negative definite and the following condition holds:*

$$\forall \xi_i \in \partial^c f_i(\bar{x}), i = 1, \ldots, k, \sum_{i=1}^{k} \alpha_i \xi_i = 0 \Rightarrow \alpha_i = 0, i = 1, \ldots, k. \qquad (C)$$

Then $\bar{u}$ is a weakly efficient solution for (NMSIP).

Proof Since $(\bar{u}, \bar{\lambda}, \bar{y})$ is a weakly efficient solution for (NMWSID), by Theorem 10.6, there exist $\tau \in \mathbb{R}^k, \nu \in \mathbb{R}^n, p \in \mathbb{R}, s \in \mathbb{R}^m, w \in \mathbb{R}^k$, such that

$$(\partial^c f(\bar{u}))\tau + [\partial^2(\bar{\lambda}^T f(\bar{u}) + \bar{y}^T g(\bar{u})]\nu + p\partial^c(\bar{y}^T g(\bar{u})) = 0, \qquad (10.35)$$

$$\nu^T(\partial^c g(\bar{u})) - pg(\bar{u}) - s = 0, \qquad (10.36)$$

$$\nu^T(\partial^c f(\bar{u})) - w = 0, \qquad (10.37)$$

$$p\bar{y}^T g(\bar{u}) = 0, \qquad (10.38)$$

$$s^T \bar{y} = 0, \qquad (10.39)$$

$$w^T \bar{\lambda} = 0, \qquad (10.40)$$

$$\tau \geq 0, \nu \text{ unrestricted}, (s, p, w) \geqq 0, \qquad (10.41)$$

Since, $\bar{\lambda} > 0$, (10.43) implies that $w = 0$. Then, using (10.40), we get

$$\nu^T(\partial^c f(\bar{u})) = 0. \qquad (10.42)$$

Multiplying (10.39) by $\bar{y}$ and using (10.41) and (10.42), we get

$$\nu^T \partial(\bar{y}^T g(\bar{u})) = 0. \tag{10.43}$$

$$\tau \geq 0, \nu \text{ unrestricted}, (s, p, w) \geq 0. \tag{10.44}$$

Since, $\bar{\lambda} > 0$, (10.43) implies that $w = 0$. Then, using (10.40), we get

$$\nu^T (\partial^c f(\bar{u})) = 0. \tag{10.45}$$

Multiplying (10.39) by $\bar{y}$ and using (10.41) and (10.42), we get

$$\nu^T \partial(\bar{y}^T g(\bar{u})) = 0. \tag{10.46}$$

Multiplying (10.38) by ν and using (10.45) and (10.48), it follows that

$$\nu^T [\partial^2 \bar{\lambda}^T f(\bar{u}) + \partial^2 \bar{y}^T g(\bar{u})]\nu = 0.$$

Since, each $n \times n$ Hessian matrix $A \in \partial^2(\bar{\lambda}^T f(\bar{u}) + \bar{y}^T g(\bar{u}))$ is assumed to be positive or negative definite, we get $\nu = 0$. Since, $\nu = 0$, by the equality constraint (B) of (NMWSID) and (10.38), we get

$$(\tau - p\bar{\lambda})^T \partial f(\bar{u}) = 0.$$

By condition (C), it follows that

$$\tau = p\bar{\lambda}.$$

Since, $\bar{\lambda}_i \geq 0, i = 1, \ldots, k, \sum_{i=1}^{m} \bar{\lambda}_i = 1, \tau \geq 0$ implies $p > 0$. Thus $\tau \neq 0$ and $p > 0$. Since $p > 0$ and $s \geq 0$, (10.39) gives $g(\bar{u}) \leq 0$ and (10.41) gives $\bar{y}^T g(\bar{u}) = 0$. Thus $\bar{u}$ is feasible for (NMSIP). That $\bar{u}$ is a weakly efficient solution for (NMSIP) follows from the assumptions of the theorem and weak duality Theorem 10.4.

Chapter 11

Vector Variational Inequalities and Vector Pseudolinear Optimization

11.1   Introduction

Variational inequality theory was introduced by Hartman and Stampacchia [100] as a tool for the study of partial differential equations with applications principally drawn from mechanics. A variational inequality is an inequality involving a functional, which has to be solved for all the values of a given variable usually belonging to a convex set. The mathematical theory of variational inequality was initially developed to deal with equilibrium problems, precisely the *Signorini problem* posed by Signorini in 1959 and solved by Fichera [82]. In that model problem, the functional involved was assumed as the first variation of the involved potential energy, therefore this type of inequality has been given the name, the variational inequality.

Variational inequalities are known either in the form introduced by Minty [186] or in the form presented by Stampacchia [258]. Giannessi [94] has introduced its vector valued version in finite dimensional Euclidean spaces. These problems are further generalized in different directions due to their vital applications in several areas of modern research such as traffic equilibrium problems, optimal control problems, engineering, and economic sciences; see Kinderlehrer and Stampacchia [142], Chen and Cheng [45], Yang and Goh [288], Giannessi *et al.* [96], Nagurney [213], and the references therein. Yang [287] has investigated the relations between vector variational inequality and vector pseudolinear optimization problems. Yang *et al.* [292] have investigated the relationship between a Minty vector variational inequality problem and a vector optimization problem using the pseudoconvexity assumption.

Recently, nonsmooth vector variational inequality problems have been used as an efficient tool to study nonsmooth vector optimization problems. In fact, some recent works in vector optimization have shown that optimality conditions for nonsmooth multiobjective optimization problems can be characterized by vector variational inequalities. For the nonsmooth case, Ward and Lee [280] have established the equivalence between weak Pareto optimal solutions of a vector optimization problem and a solution of a vector variational inequality problem using generalized directional derivatives. Lee and Lee [161] have established the relations among a nonsmooth convex vector optimization problem and different kinds of vector variational inequalities with subdifferentials. Recently, Lalitha and Mehta [157] have studied the relationship between vector variational inequality problems and nonsmooth optimization problems using h-pseudolinear functions. For recent developments and updated surveys in this area, we refer to, Ansari and Rezaei [3], Mishra and Laha [194], Mishra and Upadhyay [199], and the references therein.

In this chapter, we discuss the relationship between solutions of Stampacchia type vector variational inequality problems and efficient and properly efficient solutions of the vector optimization problem for both smooth and nonsmooth cases. Moreover, we consider a variational inequality problem in terms of bifunctions and present the characterizations for the solution set using a h-pseudolinear optimization problem.

11.2   Definitions and Preliminaries

Let $S = \mathbb{R}^m_{++}$ be the positive orthant of $\mathbb{R}^m$.

For $x, \ y \in \mathbb{R}^m$, the following ordering for vectors in $\mathbb{R}^m$ will be adopted:

$$x \leq y \ \Leftrightarrow \ y - x \in S \backslash \{0\} \, ; x \nleq y \ \Leftrightarrow \ y - x \notin S \backslash \{0\} \, .$$

Now, we assume that the set $K \subseteq \mathbb{R}^n$ is a nonempty, closed and convex set unless otherwise specified.

We consider the following vector optimization problem:

$$(VOP) \quad \min_{S} f(x) := (f_1(x), ..., f_m(x))^T, \text{ subject to } x \in K, \qquad (11.1)$$

where $\min\limits_{S}$ means vector optimization with respect to closed convex cone S and $f_i : K \to \mathbb{R}, \ i = 1, ..., m$ are differentiable functions. We recall that a point $\bar{x} \in K$ is said to be an efficient (Pareto efficient) solution for (VOP), if there exists no $y \in K$, such that

$$f(y) \leq f(x)$$

or equivalently,

$$f(x) - f(y) \in S \backslash \{0\} \, .$$

We consider the following Stampacchia type vector variational inequality problem:

(SVVIP)   Find $x \in K$, such that

$$\langle \nabla f(x), y - x \rangle = (\langle \nabla f_1(x), y - x \rangle, ..., \langle \nabla f_m(x), y - x \rangle)^T \nleq 0, \quad \forall y \in K.$$
$$(11.2)$$

The following proposition is from Giannessi [94].

Proposition 11.1 *Let $f : K \to \mathbb{R}^m$ be a differentiable convex function. If $\bar{x}$ is a solution of the (SVVIP), then $\bar{x}$ is an efficient solution to the (VOP).*

Proof We assume that $\bar{x}$ solves (SVVIP), but it is not an efficient solution to the (SVOP). Thus there exists some $y \in K$, such that

$$f(y) - f(\bar{x}) \leq 0.$$

Using the convexity of f, we get

$$\langle \nabla f(\bar{x}), y - \bar{x} \rangle \leq 0.$$

Therefore, $\bar{x}$ cannot be a solution to the (SVVIP). This contradiction leads to the result.

The following example from Yang [287] illustrates the fact that (SVVIP) is not a necessary optimality condition for an efficient solution of (VOP).

Example 11.1 *We consider the problem*

$$\min_{S} \ f(x) - (f_1(x), \ f_2(x))^T, \ \text{subject to } x \in [-1, 0],$$

$$\text{where } f_1(x) = x, \ f_2(x) = x^2.$$

Then obviously every $x \in [-1, 0]$ is an efficient solution of (VOP). Let $x = 0$, then for $y = -1$, we have

$$(\langle \nabla f_1(x), y - x \rangle, \ \langle \nabla f_2(x), y - x \rangle)^T = (-1, 0)^T \leq (0, 0)^T.$$

Hence, $x = 0$ is not a solution of (SVVIP).

We consider the following Minty vector variational inequality problem: (MVVIP) Find $x \in K$, such that

$$\langle \nabla f(y), x - y \rangle \not\geq 0, \quad \forall y \in K. \tag{11.3}$$

The following proposition from Giannessi [95] states that for Minty vector variational inequality, the equivalence between (MVVIP) and (VOP) holds.

Proposition 11.2 *Let K be a closed, convex set with a nonempty interior. Let f be a differentiable convex function, on an open set containing K. Then, x is an efficient solution of (VOP) if and only if it is a solution of (MVVIP).*

The following proposition from Yang [287] establishes that (SVVIP) is a necessary optimality condition for a solution to be properly efficient for (VOP).

Proposition 11.3 *If x is a properly efficient solution for (VOP), then, x is a solution of (SVVIP).*

Proof Since x is a properly efficient solution for (VOP), it follows from Geoffrion [86] that there exists $\lambda \in \text{int} \left(\mathbb{R}^p_+ \right)$, such that x solves the following problem:

$$\min \lambda^T f(x), \quad \text{subject to } x \in K.$$

Then, we have

$$\left\langle \nabla \left(\lambda^T f(x) \right), y - x \right\rangle \geq 0, \quad \forall y \in K.$$

Thus, x satisfies

$$\left\langle \nabla f(x), y - x \right\rangle \not\leq 0, \quad \forall y \in K.$$

Therefore, x is a solution of (SVVIP).

11.2.1 Necessary and Sufficient Optimality Conditions

We consider the following vector pseudolinear optimization problem:

$$(VPOP) \quad \min_S f(x) = (f_1(x), ..., f_m(x))^T, \text{ subject to } x \in K, \qquad (11.4)$$

where $\min_S$ means vector optimization with respect to closed convex cone S and the functions f_i, $i = 1, ..., m$ are differentiable pseudolinear functions with respect to the proportional function $p_i(x, y)$, $i = 1, ..., m$, respectively. Therefore,

$$f_i(y) - f_i(x) = p_i(x, y) \left\langle \nabla f_i(x), y - x \right\rangle.$$

Hence, we have

$$f(y) - f(x) = (p_1(x, y) \left\langle \nabla f_1(x), y - x \right\rangle, ..., p_m(x, y) \left\langle \nabla f_m(x), y - x \right\rangle)^T. \tag{11.5}$$

Now, we consider the following generalized Stampacchia type vector variational inequality problem (GSVVIP):
(GSVVIP) Find $x \in K$, such that

$$\left\langle p(x, y) \nabla f(x), y - x \right\rangle \not\leq 0, \quad \forall y \in K. \tag{11.6}$$

Yang [287] established the equivalence between (VPOP) and (GSVVIP).

Theorem 11.1 *The point x is an efficient solution for (VPOP) if and only if x solves (GSVVIP).*

Proof Let x be an efficient solution for (VPOP). Suppose that x does not solve (GSVVIP), then there exists $y \in K$ such that

$$\left\langle p(x, y) \nabla f(x), y - x \right\rangle \leq 0.$$

From which, it follows that

$$(\left\langle p_1(x, y) \nabla f_1(x), y - x \right\rangle, ..., \left\langle p_m(x, y) \nabla f_m(x), y - x \right\rangle)^T \leq 0.$$

By (11.5), it follows that

$$f(y) - f(x) \leq 0,$$

which is a contradiction.

Conversely, suppose that x solves (GSVVIP), if possible, let x be not an efficient solution of (VPOP). Then there exists some $y \in K$, such that

$$f(y) \leq f(x).$$

Hence, by (11.5), we get

$$(\langle p_1(x,y) \nabla f_1(x), y - x \rangle, \ldots, \langle p_m(x,y) \nabla f_m(x), y - x \rangle)^T \leq 0,$$

which implies that

$$\langle p(x,y) \nabla f(x), y - x \rangle \leq 0.$$

Therefore, x cannot be a solution to (GSVVIP).

This contradiction leads to the result.

When the terms $p_i(x,y)$, $i = 1, \ldots, m$, are independent of the index i, and have the same value say $\bar{p}(x,y)$, then we have the following equivalent condition between (VPOP) and (SVVIP).

Theorem 11.2 *The point x is an efficient solution of (VPOP) if and only if x solves (SVVIP).*

Proof Let x be an efficient solution of (VPOP). Suppose that x does not solve (SVVIP). Then, there exists $y \in K$, such that

$$\langle \nabla f(x), y - x \rangle \leq 0,$$

which is equivalent to

$$(\langle \nabla f_1(x), y - x \rangle, \ldots, \langle \nabla f_m(x), y - x \rangle)^T < 0. \tag{11.7}$$

Multiplying with $\bar{p}(x,y) > 0$ to (11.7), we get

$$(\langle \bar{p}(x,y) \nabla f_1(x), y - x \rangle, \ldots, \langle \bar{p}(x,y) \nabla f_m(x), y - x \rangle)^T \leq 0.$$

Therefore, by the above assumption, we get

$$f(y) - f(x) \leq 0.$$

Hence, x cannot be an efficient solution to (VPOP). This contradiction leads to the result.

The necessary part follows from Theorem 11.1.

The following remark shows the existence of such a vector optimization problem.

Remark 11.1 *Consider a vector pseudolinear optimization problem:*

$$\min_S f(x), \text{ subject to } x \in K,$$

where

$$f(x) = (f_1(x),..,f_m(x))^T = \left(\frac{a_1^T x + b_1}{a^T x + b}, ..., \frac{a_m^T x + b_m}{a^T x + b} \right)^T,$$

and $a, b, a_i, b_i \in \mathbb{R}^n$, $i = 1,...,m$. *Then, it is easy to see that*

$$\frac{a_i^T y + b_i}{a^T y + b} - \frac{a_i^T x + b_i}{a^T x + b} = \bar{p}(x, y) \langle \nabla f_i(x), y - x \rangle,$$

where $\bar{p}(x, y) = \frac{a^T x + b}{a^T y + b}$. *Obviously, the term* $\bar{p}(x, y)$ *does not depend on the index* i.

11.3   Nonsmooth Vector Variational Inequality and Vector Optimization Problems

We consider the following nonsmooth vector optimization problem:

$$(NVOP) \quad \min_S f(x) := (f_1(x),...,f_m(x))^T, \text{ subject to } x \in K, \qquad (11.8)$$

where $\min\limits_S$ means vector optimization with respect to closed convex cone S and $f_i : K \to \mathbb{R}$, $i = 1, ..., m$ are nondifferentiable locally Lipschitz functions. The Clarke generalized gradient of f at any point $x \in K$, is the set

$$\partial^c f(x) = \partial^c f_1(x) \times ... \times \partial^c f_m(x).$$

We consider the following nonsmooth Stampacchia type vector variational inequality problem: (NSVVIP) Find $x \in K$, such that there exists $\xi_i \in \partial^c f_i(x)$, $i = 1,...,m$, such that

$$\langle \xi, y - x \rangle = (\langle \xi_1, y - x \rangle, ..., \langle \xi_m, y - x \rangle)^T \not\leq 0, \forall y \in K, \qquad (11.9)$$

The following proposition is from Lee and Lee [161].

Proposition 11.4 *Let* $f : K \to \mathbb{R}^m$ *be a locally Lipschitz and convex function. If* x *is a solution of the (NSVVIP), then* x *is an efficient solution to the (NVOP).*

In general (NSVVIP) is not a necessary optimality condition for an efficient solution of (NVOP). We illustrate this fact by the following example:

Example 11.2 *We consider the following nonsmooth vector optimization problem*

$$(P1) \quad \min_S \ f(x) := (f_1(x), f_2(x))^T,$$

$$\text{subject to } x \in [-1, 0],$$

where

$$f_1(x) = x, \quad f_2(x) = |x|.$$

Then, obviously every $x \in [-1, 0]$ is an efficient solution of (P1). Let $x = 0$, then for $y = -\frac{1}{2}$ there exist $\xi_1 = \{1\} = \partial^c f_1(0) = \{f_1'(0)\}$ and $\xi_2 = \{0\} \in \partial^c f_2(0) = [-1, 1]$, such that

$$(\langle \xi_1, y - x \rangle, \langle \xi_2, y - x \rangle)^T = \left(-\frac{1}{2}, 0 \right)^T \leq (0, 0)^T.$$

Thus, $x = 0$ is not a solution of (NSVVIP).

Lee [159] has shown that (NSVVIP) is a necessary optimality condition for a solution to be properly efficient for the (NVOP).

Proposition 11.5 *If x is a properly efficient solution for (NVOP), then x is a solution of (NSVVIP).*

We note that it is still an interesting problem to look at whether a kind of vector variational inequality can be a sufficient condition for a properly efficient solution of (NVOP). In the next section, we shall establish that a variant of (NSVVIP) is a necessary as well as sufficient optimality condition for (NVOP) under the assumption of locally Lipschitz pseudolinear functions.

11.3.1 Necessary and Sufficient Optimality Conditions

We consider the following nonsmooth vector pseudolinear optimization problem:

$$(NVPOP) \quad \min_S f(x) := (f_1(x), ..., f_m(x))^T, \text{ subject to } x \in K, \quad (11.10)$$

where $\min_S$ means vector optimization with respect to closed convex cone S and the functions $f_i : K \to \mathbb{R}$, $i = 1, ..., m$ are nondifferentiable locally Lipschitz pseudolinear functions with respect to the proportional function p_i, $i = 1, ..., m$, respectively. Then by Theorem 2.24, there exists $\xi_i \in \partial^c f_i(x)$, such that

$$f_i(y) - f_i(x) = p_i(x, y) \langle \xi_i, y - x \rangle, \quad i = 1, ..., m.$$

Therefore, there exist $\xi_i \in \partial^c f_i(x)$, $i = 1, ..., m$, such that

$$f(y) - f(x) = (\langle p_1(x, y) \xi_1, y - x \rangle, ..., \langle p_m(x, y) \xi_m, y - x \rangle)^T. \quad (11.11)$$

Now, we consider the following generalized nonsmooth Stampacchia type vector variational inequality problem (GNSVVIP):

(GNSVVIP) Find $x \in K$, such that for all $y \in K$, there exists $\xi \in \partial^c f(x)$, such that,

$$\langle p(x,y)\xi, y-x \rangle = (\langle p_1(x,y)\xi_1, y-x \rangle, ..., \langle p_m(x,y)\xi_m, y-x \rangle)^T \nleq 0. \tag{11.12}$$

Now, with the help of the following theorem, we shall establish the equivalence between (NVPOP) and (GNSVVIP).

Theorem 11.3 *The point x is an efficient solution of (NVPOP) if and only if x solves (GNSVVIP).*

Proof Let x be an efficient solution of (NVPOP). Suppose that x does not solve (GNSVVIP), then there exists $y \in K$ and $\xi \in \partial^c f(x)$, such that

$$\langle p(x,y)\xi, y-x \rangle \leq 0.$$

From which it follows that

$$(\langle p_1(x,y)\xi_1, y-x \rangle, ..., \langle p_m(x,y)\xi_m, y-x \rangle)^T \leq 0.$$

By (11.1), it follows that
$$f(y) - f(x) \leq 0,$$

which is a contradiction.

Conversely, suppose that x solves (GNSVVIP), if possible let x be not an efficient solution of (NVPOP). Then, there exists some $y \in K$, such that

$$f(y) \leq f(x).$$

Therefore, by pseudolinearity of f_i with respect to proportional function p_i, $i = 1, ..., m$ there exists $\xi_i' \in \partial^c f_i(x)$, such that

$$(\langle p_1(x,y)\xi_1', y-x \rangle, ..., \langle p_m(x,y)\xi_m', y-x \rangle)^T \leq 0.$$

By Proposition 2.5, there exists $\mu_i > 0$, $\xi_i \in \partial^c f_i(x)$, $i = 1, ..., m$, such that $\xi_i' = \mu_i \xi_i$, $i = 1, ..., m$. Hence, we have

$$(\langle p_1(x,y)\mu_1\xi_1, y-x \rangle, ..., \langle p_m(x,y)\mu_m\xi_m, y-x \rangle)^T \leq 0.$$

Since, $\mu_i > 0, i = 1, ..., m$, it follows that there exists $\xi \in \partial^c f(x)$, such that

$$\langle p(x,y)\xi, y-x \rangle \leq 0,$$

therefore x cannot be a solution to (GNSVVIP). This contradiction leads to the result.

If the terms $p_i(x,y)$, $i = 1, 2, .., m$ are independent of the index i, and have the same value say $\bar{p}(x,y)$, then we have the following equivalent condition between (NVPOP) and (NSVVIP).

Theorem 11.4 *The point x is an efficient solution of (NVPOP) if and only if x solves (NSVVIP).*

Proof Let x be an efficient solution of (NVPOP). Suppose that x does not solve (NSVVIP). Then, there exists $y \in K$ and $\xi \in \partial^c f(x)$, such that

$$\langle \xi, y - x \rangle \le 0.$$

From which, it follows that there exists $\xi_i \in \partial^c f_i(x)$, such that

$$(\langle \xi_1, y - x \rangle, ..., \langle \xi_m, y - x \rangle)^T \le 0. \tag{11.13}$$

On multiplying with $\bar{p}(x, y) > 0$ to (11.13), it follows that there exist $\xi_i \in \partial^c f_i(x)$, $i = 1, ..., m$, such that

$$(\langle \bar{p}(x, y) \xi_1, y - x \rangle, ..., \langle \bar{p}(x, y) \xi_m, y - x \rangle)^T \le 0.$$

Therefore, by our assumption, it follows that

$$f(y) - f(x) \le 0.$$

Hence, x cannot be an efficient solution to (NVPOP). This contradiction leads to the result.

The necessary part follows from Theorem 11.3.

The following example illustrates the significance of Theorem 11.3.

Example 11.3 *Consider the following nonsmooth vector optimization problem:*

$\min_S \ f(x) := (f_1(x), f_2(x))^T$, *such that $x \in [1, 3]$, where*

$$f_1(x) = \begin{cases} 16\left(\frac{x+1}{x+2}\right), & \text{if } x > 2, \\ 16\left(\frac{2x-1}{x+2}\right), & \text{if } x \le 2 \end{cases}$$

and

$$f_2(x) = \begin{cases} 16\left(\frac{-2x-1}{x+2}\right), & \text{if } x > 2, \\ 16\left(\frac{-3x+1}{x+2}\right), & \text{if } x \le 2. \end{cases}$$

It is evident that f_1 and f_2 are nonsmooth locally Lipschitz pseudolinear functions on $[1, 3]$ with respect to the same proportional function $\bar{p}(x, y) = \frac{x+2}{y+2}$. Moreover, we have

$$\partial^c f_1(2) = [1, 5], \partial^c f_2(2) = [-7, -3].$$

Then, obviously $x = 2$ is an efficient solution of (P2). Furthermore, for $x = 2$ and for all $y = [1, 3]$, there exist $\xi_1 = 1 \in \partial^c f_1(2)$ and $\xi_2 = -7 \in \partial^c f_2(2)$, such that

$$(\langle \xi_1, y - x \rangle, \langle \xi_2, y - x \rangle)^T = ((y - 2), -7(y - 2))^T \nleq (0, 0)^T.$$

Thus, $x = 2$ is a solution of (NSVVIP).

11.4   Pseudoaffine Variational Inequality Problems

Consider the following optimization problem:

$$(P) \quad \min f(x)$$

$$\text{subject to } x \in K,$$

where $K \subseteq \mathbb{R}^n$ is a convex set, $f : K \to \mathbb{R}$ is an h-pseudolinear function and $h : K \times \mathbb{R}^n \to \bar{\mathbb{R}}$ is a bifunction associated to f such that h is odd in the second argument and satisfies

$$h(x;.) \text{ is positively homogeneous} \tag{11.14}$$

$$h(x;0) = 0, \tag{11.15}$$

$$D_+ f(x;d) \le h(x;d) \le D_+ f(x;d), \ \forall\, x \in K \text{ and } d \in \mathbb{R}^n. \tag{11.16}$$

Let $\bar{S} \ne \phi$ denote the solution set of the problem (P).

Definition 11.1 (Komlosi [146]) *A bifunction $h : K \times \mathbb{R}^n \to \bar{\mathbb{R}}$ is said to be pseudomonotone on K, if for any $x, y \in K$, $x \ne y$, one has*

$$h(x, y - x) > 0 \Rightarrow h(y, x - y) < 0;$$

or equivalently,

$$h(x, y - x) \ge 0 \Rightarrow h(y, x - y) \le 0.$$

The concept of pseudoaffinity for bifunction h may be defined as follows.

Definition 11.2 (Lalitha and Mehta [157]) *A bifunction $h : K \times \mathbb{R}^n \to \overline{\mathbb{R}}$ is said to be pseudoaffine on K, if h and $-h$ are both pseudomonotone on K.*

Lalitha and Mehta [157] have given the following example of pseudoaffine bifunctions, that satisfies conditions (11.14) and (11.15) and is also odd in the second argument.

Example 11.4 *The bifunction $h : \mathbb{R}^2 \times \mathbb{R}^2 \to \bar{\mathbb{R}}$ is defined as*

$$h(x;d) = \begin{cases} e^{x_1} \left(d_1^2 + d_2^2 \right) / d_1 + d_2, & \text{if } d_1 \ne -d_2, \\ 0, & \text{if } d_1 = -d_2, \end{cases}$$

where, $d = (d_1, d_2)$ and $x = (x_1, x_2)$ is pseudoaffine.

Lalitha and Mehta [157] considered the following variational inequality in terms of bifunction h:
(VIP) Find $\bar{x} \in K$, such that

$$h(\bar{x}, x - \bar{x}) \ge 0, \ \forall\, x \in K,$$

where K is a convex subset of $\mathbb{R}^n$ and $h : K \times \mathbb{R}^n \to \overline{\mathbb{R}}$ is a pseudoaffine bifunction satisfying the assumptions (11.14)–(11.15). Let $\bar{K} \neq \emptyset$ denote the solution set of the problem (VIP).

The following necessary and sufficient condition for pseudoaffinity is an extension of the corresponding result for the PPM map by Bianchi and Schaible [27].

Proposition 11.6 *Suppose that h is radially continuous on K in the first argument, odd in the second argument and satisfies conditions (11.14) and (11.15). Then, h is pseudoaffine on K if and only if*

$$h(x, y - x) = 0 \Rightarrow h(y, x - y) = 0, \forall x, y \in K. \tag{11.17}$$

Proof Assume that h is pseudoaffine on K and $x, y \in K$. If $x = y$, then by (11.14) the result is trivially true. Suppose that $x \neq y$, $h(y, x - y) = 0$. By the pseudomonotonicity of h and $-h$ implies that $h(x; y - x) \leq 0$ and $h(y; x - y) \geq 0$. Thus the result follows. Conversely, let h is not a pseudomonotone bifunction. Then, there exist $x, y \in K$, $x \neq y$ such that $h(x; y - x) > 0$ and $h(y; x - y) \geq 0$. Define $g : [0, 1] \to \overline{\mathbb{R}}$ as $g(\lambda) = h(x\lambda y; y - x)$. We have $g(0) > 0$ and $g(1) = h(y; y - x) = -h(y; x - y) \leq 0$. Since, h is radially continuous in the first argument, there exists $\bar{\lambda} \in]0, 1]$ such that $g(\bar{\lambda}) = 0$, that is $h(x\lambda y; y - x) = 0$. Using, the positive homogeneity of h, we have $h(x\lambda y; x\lambda y - x) = 0$. Using the suboddness of h, we have $h(x\lambda y; x - x\lambda y) = 0$. By (11.14), we get $h(x; x\lambda y - x) = 0$. For the positive homogeneity of h, we have $h(x; y - x) = 0$, which is a contradiction. Proceeding as above, we can prove that $-h$ is pseudomonotone on K. This completes the proof.

By the pseudoaffinity property of h, the following proposition follows directly.

Proposition 11.7 *A point $\bar{x} \in K$ is a solution of (VIP), if and only if*

$$h(\bar{x}, x - \bar{x}) \leq 0, \quad \forall x \in K.$$

Theorem 11.5 *(Lalitha and Mehta [157]) Suppose that the bifunction h is radially continuous in the first argument and odd in the second argument. If $\bar{x} \in \bar{K}$, then,*

$$\bar{K} \subseteq \{x \in K : h(x; \bar{x} - x) = 0\}$$

$$= \{x \in K : h(\bar{x}; x - \bar{x}) = 0\}$$

$$= \{x \in K : h(x\lambda\bar{x}; \bar{x} - x) = 0, \ \forall \lambda \in [0, 1]\}.$$

Proof Let $x \in \bar{K}$, then $h(x; y - x) \geq 0$, $\forall y \in K$. For $y = \bar{x}$, we have $h(x; \bar{x} - x) \geq 0$. Since, $\bar{x}$ is a solution of (VIP), by Proposition 11.1, we get $h(x; \bar{x} - x) \leq 0$. Therefore, we have

$$h(x; \bar{x} - x) = 0, \quad \forall x \in K.$$

From which, it follows that

$$\bar{K} \subseteq \{x \in K : h(x; \bar{x} - x) = 0\}.$$

The proof of the rest of the theorem follows from Theorem 3.13.

The following example from Lalitha and Mehta [157] illustrates the fact that if $\bar{x} \in \bar{K}$ and $h(\bar{x}; x - \bar{x}) \geq 0$, for some $x \in K$ does not imply that $x \in \bar{K}$.

Example 11.5 *Let $K = [-1, 1] \times [-1, 1]$ and let $h : K \times \mathbb{R}^2 \to \overline{\mathbb{R}}$ be defined as*

$$h(x; d) = \begin{cases} \left(1 + x_1^2 + x_2^2\right) d_1^3 / d_2^2, & \text{if } d_2 \neq 0, \\ 0, & \text{if } d_2 = 0, \end{cases}$$

where $d = (d_1, d_2)$ and $x = (x_1, x_2)$. It is easy to see that h is a continuous function of x and an odd function of d satisfying (11.4) and (11.5). Moreover, using (11.6), we can easily verify that h is a pseudoaffine bifunction on K. It is clear that $\bar{x} = (-1, 1)$ is a solution of (VIP) and $x = (1, 1)$ satisfies $h(\bar{x}, x - \bar{x}) = 0$ but $x \notin \bar{X}$, because for $y = (0, 1/2)$, we have $h(x, y - x) = -12 < 0$.

Remark 11.2 *(Jeyakumar and Yang [119]) If F is gradient of a pseudolinear function, then,*

$$\bar{K} = \bar{S} = \{x \in K : \langle F(x); \bar{x} - x \rangle = 0\}.$$

The following theorem from Lalitha and Mehta [157] relates the solutions of (P) and (VIP).

Theorem 11.6 *Assume that f is a h-pseudolinear function on a convex set K and h be the bifunction associated with f. If f is odd in the second argument and satisfies (11.14)–(11.16). Then, $\bar{x} \in K$ is a solution of (P) if and only if $\bar{x}$ is a solution of (VIP).*

Proof Suppose that $\bar{x}$ is an optimal solution for (P). Then, for any $y \in K$,

$$f(x) \leq f(x + \lambda(y - x)), \quad \forall \lambda \in \,]0, 1].$$

From which, it follows that

$$D_+ f(x; y - x) \geq 0, \forall y \in K.$$

Using (11.6), we get

$$h(x; y - x) \geq 0, \quad \forall y \in K.$$

Hence, x is a solution of (VIP). To prove the converse, we assume that x is a solution of (VIP). Suppose on the contrary that x is not an optimal solution of (P). Then there exists $y \in K$, such that

$$f(y) < f(x).$$

Then, by h-pseudoconvexity of f, we have

$$h(x; y - x) < 0,$$

which is a contradiction to our assumption that x is a solution of (VIP).

Theorem 11.7 *Suppose that f is h-pseudolinear on K, the bifunction h is odd in the second argument and satisfies (11.14)–(11.16). Let $\bar{x} \in \bar{K}$, then*

$$\bar{K} = \{x \in K : h(x; \bar{x} - x) = 0\}$$
$$= \{x \in K : h(\bar{x}; x - \bar{x}) = 0\}$$
$$= \{x \in K : h(x\lambda\bar{x}; \bar{x} - x) = 0, \ \forall \lambda \in [0,1]\}.$$

Proof The result follows directly from Theorem 11.7 and Theorem 3.13.

Remark 11.3 *The above characterization does not hold if f is only h-pseudoconvex and not h-pseudoconcave on K. For example, if we consider the function $f : \mathbb{R} \to \mathbb{R}$ be defined as*

$$f(x) = \begin{cases} x^2, & \text{if } x \leq 0, \\ 0, & \text{if } 0 < x \leq 1, \\ x + 1, & \text{if } x > 1. \end{cases}$$

Let

$$h(x; d) = D^+ f(x; d) = \begin{cases} 2xd, & \text{if } x \leq 0, \forall d, \\ 0, & \text{if } 0 < x < 1, \forall \text{ or } x = 1, d \leq 0, \\ \infty, & \text{if } x = 1, d > 0, \\ d, & \text{if } x > 1, \forall d. \end{cases}$$

Then, it is easy to see that f is h-pseudoconvex on $\mathbb{R}$ but not h-pseudoconcave on $\mathbb{R}$. For $x = 0$, $y = -1/2$, $f(x) < f(y)$ but $h(x; y - x) = 0$. Now, for $\bar{x} = 1$ and $x = -2$, we note that

$$h(\bar{x}, x - \bar{x}) = 0, \text{ for } x = -2 \notin \bar{S} = [0,1].$$

The following characterization for solution set $\bar{K}$ follows from Theorem 11.7 and Theorem 3.14.

Theorem 11.8 *Suppose that f is h-pseudolinear on K, the bifunction h is odd in the second argument and satisfies (11.14)–(11.16). Let $\bar{x} \in \bar{K}$, then*

$$\bar{X} = \{x \in K : h(x; \bar{x} - x) \geq 0\}$$
$$= \{x \in K : h(\bar{x}; x - \bar{x}) \leq 0\}$$
$$= \{x \in K : h(x\lambda\bar{x}; \bar{x} - x) \geq 0, \ \forall \lambda \in [0,1]\}.$$

The following characterization for solution set $\bar{K}$ follows from Theorem 11.7 and Theorem 3.15.

Theorem 11.9 *Suppose that f is h-pseudolinear on K, the bifunction h is odd in the second argument and satisfies (11.14)–(11.16). Let $\bar{x} \in \bar{K}$, then*

$$\bar{K} = \{x \in K : h(x; \bar{x} - x) = h(\bar{x}; x - \bar{x})\}$$
$$= \{x \in K : h(x; \bar{x} - x) \geq h(\bar{x}; x - \bar{x})\}.$$

Chapter 12

An Extension of Pseudolinear Functions and Variational Inequalities

12.1 Introduction

Generalizations of monotonicity in connection with variational inequality problems have been an interesting research topic during the last few decades. The purpose of these generalizations is to relax the rigid assumption of monotonicity without losing some of the valuable properties of these models. See, for example, Avriel *et al.* [9] and Ansari *et al.* [5].

Differentiable pseudoconvex functions are characterized by pseudomonotone gradients in the sense of Karamardian [132]. Hence differentiable pseudolinear functions are characterized by pseudomonotone gradients for which the negative of the gradient is also pseudomonotone, which are referred to as pseudoaffine maps.

Bianchi and Schaible [27] have established several characterizations for the pseudomonotone map F where $-F$ is also a pseudomonotone and explored their properties in variational inequality problems. In particular, these results extend some earlier results by Jeyakumar and Yang [119], which were derived for optimization problems. Recently, Bianchi *et al.* [26] have extended some earlier results for pseudolinear optimization problems in Jeyakumar and Yang [119] to pseudoaffine maps and pseudoaffine variational inequalities. Bianchi *et al.* [26] have derived the general form of pseudoaffine maps which are defined on the whole space. In this chapter, we present certain characterizations for the PPM-maps and affine PPM-maps. Using the properties of PPM-maps, characterizations for the solution set of PPM variational inequality problems are given. Furthermore, the general form of pseudoaffine maps on the whole space is discussed in detail.

12.2   Definitions and Preliminaries

Let $K \subseteq \mathbb{R}^n$ be a nonempty set.

Definition 12.1 *(PPM-map)* *The map $F : K \to \mathbb{R}^n$ is said to be pseudomonotone (PM) on K if, for every $x, y \in K$, we have*

$$\langle F(x), y - x \rangle \geq 0 \Rightarrow \langle F(y), y - x \rangle \geq 0,$$

or equivalently

$$\langle F(x), y - x \rangle > 0 \Rightarrow \langle F(y), y - x \rangle > 0.$$

A map F is said to be a PPM-map, if F and $-F$ are both pseudomonotone.

Definition 12.2 *(Regular function)* *Let $F : \Omega \subseteq \mathbb{R}^n \to \mathbb{R}^n$ be a continuously differentiable function on open set Ω. Let $K \subseteq \Omega$. The map F is called regular over K, if for every $x \in K$, we have*

$$F(x) = 0 \Rightarrow Det(J_F(x)) \neq 0,$$

where $J_F(x)$ is the Jacobian matrix evaluated at x.

12.3   Characterizations of PPM-Maps

In this section, we present some necessary and sufficient conditions for a map to be a PPM-map, derived by Bianchi and Schaible [27].

Theorem 12.1 *Suppose that $K \subseteq \mathbb{R}^n$ is convex and let $F : K \to \mathbb{R}^n$ is continuous on K. Then, F is a PPM-map on K if and only if, for all $x, y \in K$,*

$$\langle F(x), y - x \rangle = 0 \Rightarrow \langle F(y), y - x \rangle = 0.$$

Proof Suppose that F is a PPM-map. Let us assume that

$$\langle F(x), y - x \rangle = 0.$$

Then, by the pseudomonotonicity of F and $-F$, we have

$$\langle F(y), y - x \rangle \geq 0 \ and \ \langle -F(y), y - x \rangle \geq 0.$$

Hence, we get

$$\langle F(y), y - x \rangle = 0.$$

Conversely, suppose that F is not a pseudomonotone map. Then, there exist two points $x, y \in K$, such that

$$\langle F(x), y - x \rangle > 0 \text{ and } \langle F(y), y - x \rangle \leq 0.$$

Let us define $\varphi : [0, 1] \to \mathbb{R}$, such that

$$\varphi(t) = \langle F(x + t(y - x)), y - x \rangle.$$

Then,

$$\varphi(0) > 0, \varphi(1) \leq 0.$$

Therefore, by continuity there exists $0 < t^* \leq 1$ such that $\varphi(t^*) = 0$, that is,

$$\langle F(x + t^*(y - x)), y - x \rangle = 0.$$

Setting

$$x^* = x + t^*(y - x),$$

we get

$$\langle F(x^*), x^* - x \rangle = 0,$$

From which, it follows that

$$\langle F(x), y - x \rangle = 0,$$

a contradiction. Similarly, the result follows, if $-F$ is not a pseudomonotone map.

Let $M(x, y)$ be the line generated by x and y; that is,

$$M(x, y) =: \{z \in \mathbb{R}^n : z = tx + (1 - t)y, t \in \mathbb{R}\}.$$

Corollary 12.1 *Suppose $K \subseteq \mathbb{R}^n$ is convex, and let $F : K \to \mathbb{R}^n$ be continuous on K. Then, F is a PPM-map on K if and only if, for all $x, y \in K$, one has*

$$\langle F(x), y - x \rangle = 0 \Rightarrow \langle F(z), y - x \rangle = 0, \forall z \in M(x, y) \bigcap K.$$

Proof Suppose F is a PPM-map. Let $x, y \in K$ be such that $\langle F(x), y - x \rangle = 0$ and $z = tx + (1 - t)y$. Since $\langle F(x), z - x \rangle = 0$, from Theorem 12.1, it follows that

$$\langle F(z), z - x \rangle = 0.$$

Therefore, we get

$$\langle F(z), y - x \rangle = 0.$$

The converse is obvious from Theorem 2.1.

The following corollary follows from Theorem 12.1.

Corollary 12.2 *Suppose $K \subseteq \mathbb{R}^n$ is convex, and let $F : K \to \mathbb{R}^n$ be a continuous PPM-map on K. If $F(x^*) = 0$, then $\langle F(y), y - x^* \rangle = 0, \forall y \in K$.*

Remark 12.1 *Let $F = \nabla f$, where f is pseudolinear on an open convex set K. Then by Komlosi [147], the condition*

$$\nabla f\left(x^*\right) = 0 \Rightarrow \nabla f\left(x\right) = 0, \forall\, x \in S.$$

However, if F is a general PPM-map, then this is not true in general. There are PPM-maps which are zero at some point of their domains, but do not vanish anywhere else. For example, consider the linear map $F\left(x\right) = Px$, where P is nonsingular and skew-symmetric, i.e., $P = -P^T$. Obviously, $F\left(0\right) = 0$ and $F\left(x\right) \neq 0$ for $x \neq 0$. To prove that F is a PPM-map in $\mathbb{R}^n$, one can apply Theorem 12.1, taking into account that, for a skew-symmetric matrix,

$$\langle Py, y \rangle = 0, \forall y \in \mathbb{R}^n.$$

The example also shows that, even if F and $-F$ are monotone, rather than just pseudomonotone, F is not identically zero if it is zero at one point. We note that these maps are examples of PPM-maps which are not gradient maps.

The following results from Bianchi and Schaible [27] provide a significant characterization of the zeros of pseudomonotone maps and PPM-maps.

Proposition 12.1 *Suppose that $K \subseteq \mathbb{R}^n$ is open and convex. Let $F : K \to \mathbb{R}^n$ be a continuous pseudomonotone map on K. Then, the set*

$$Z := \{x \in K : F\left(x\right) = 0\}$$

is convex.

Proof Suppose that $z_1, z_2 \in Z$. Then, we show that

$$\lambda z_1 + \left(1 - \lambda\right) z_2 \in Z, \forall\, \lambda \in [0, 1].$$

On the contrary, suppose that there exists $\bar{z} = \bar{\lambda} z_1 + \left(1 - \bar{\lambda}\right) z_2$ with $F\left(\bar{z}\right) \neq 0$. Therefore, by continuity of F, for small negative t, we have

$$\langle F\left(\bar{z} + tF\left(\bar{z}\right)\right), F\left(\bar{z}\right) \rangle > 0.$$

Setting $w = \bar{z} + tF\left(\bar{z}\right)$, it follows that

$$\langle F\left(w\right), \bar{z} - w \rangle > 0.$$

By our assumption, we have

$$\langle F\left(z_1\right), w - z_1 \rangle = 0 \text{ and } \langle F\left(z_2\right), w - z_2 \rangle = 0.$$

By the pseudomonotonicity of F, it follows that

$$\langle F\left(w\right), w - z_1 \rangle \geq 0 \tag{12.1}$$

and

$$\langle F(w), w - z_2 \rangle \geq 0. \tag{12.2}$$

Multiplying the first equality by λ^* and the second one by $1 - \lambda^*$, we get

$$\langle F(w), w - z^* \rangle \geq 0,$$

which is a contradiction. This completes the proof.

In the special case of a PPM-map, we have the following theorem.

Theorem 12.2 *Let $K \subseteq R^n$ be open and convex, and let $F : K \to R^n$ be a continuous PPM-map. The set $Z = \{x \in K : F(x) = 0\}$ contains $M(z_1, z_2) \bigcap K$ for every $z_1, z_2 \in Z$.*

Proof We prove that $z_1, z_2 \in Z$ implies that $z = \lambda z_1 + (1 - \lambda) z_2 \in Z$, for all λ, such that $z \in K$.

On the contrary, suppose that there exists $z^* = \lambda^* z_1 + (1 - \lambda^*) z_2$ with $z^* = \lambda^* z_1 + (1 - \lambda^*) z_2$. Consequently, by continuity of F, we have

$$\langle F(z^* + tF(z^*)), F(z^*) \rangle > 0, \text{ for small } t.$$

Setting $w = z^* + tF(z^*)$, we get

$$\langle F(w), z^* - w \rangle \neq 0.$$

By our assumption, we have

$$\langle F(z_1), w - z_1 \rangle = 0 \text{ and } \langle F(z_2), w - z_2 \rangle = 0.$$

Hence, by Theorem 12.1,

$$\langle F(w), w - z_1 \rangle = 0 \text{ and } \langle F(w), w - z_2 \rangle = 0.$$

Multiplying the first equality by λ^* and the second one by $(1 - \lambda^*)$, we find that

$$\langle F(w), w - z^* \rangle = 0,$$

a contradiction.

We now present some results about differentiable maps. Using the Dini implicit function theorem and the regularity assumption, it follows that the zeros of F are isolated. We formalized the result as follows:

Theorem 12.3 *Let $K \subseteq \mathbb{R}^n$ be open and convex, and let $F : K \to \mathbb{R}^n$ be a continuous differentiable PPM-map, which is regular. Then, F vanishes in at most one point.*

Proof Suppose to the contrary that z_1 and z_2 are zeros of F. Then, all points in $M(z_1, z_2) \bigcap S$ are zeros of F. Thus, the zeros of F are not isolated, a contradiction.

The following proposition for general PPM-maps is needed to prove the next result.

Proposition 12.2 *Suppose $I \subseteq \mathbb{R}$ and let $F : I \to \mathbb{R}$ be a map. Then, F is a PPM-map if and only if F is either positive or negative or zero on I.*

Proof We know that one-dimensional pseudomonotone maps have the strict sign-preserving property on I, that is,

$$F(x) > 0 \Rightarrow F(y) > 0, \forall y \in I, y > x,$$

$$F(x) < 0 \Rightarrow F(y) < 0, \forall y \in I, y < x.$$

Considering that $-F$ is a pseudomonotone map, the conclusion follows.

Theorem 12.4 *Suppose $K \subseteq \mathbb{R}^n$ is open and convex, and let $F : K \to \mathbb{R}^n$ be differentiable on K. Then, F is a PPM-map if and only if, for all $x, y \in K$, we have*

$$\langle F(x), y - x \rangle = 0 \Rightarrow \langle J_F(z)(y - x), y - x \rangle = 0, \quad \forall z \in [x, y].$$

Proof Let $x, y \in K$, be given. Define

$$\psi(t) =: \langle F(x + t(y - x)), y - x \rangle \text{ and } I = \{t \in \mathbb{R} : x + t(y - x) \in K\}.$$

It is known that F is a PPM-map on the open convex set K if and only if, for all $x, y \in K$, ψ is a PPM-map on the open interval I. We have

$$\psi'(t) = \langle J_F(x + t(y - x))(y - x), y - x \rangle.$$

If F is a PPM-map with $\langle F(x), y - x \rangle = 0$, then $\psi(0) = 0$. By Proposition 12.2, we conclude that $\psi(t) = 0$ for all $t \in I$. This means that $\psi'(t) = 0$, for all $t \in I$ and the conclusion follows.

Conversely, suppose that F is not a PPM-map. Then, there exist two points $x, y \in K$, such that

$$\langle F(x), y - x \rangle = 0 \text{ and } \langle F(y), y - x \rangle > (<) 0.$$

Then, $\psi(0) = 0, \psi(1) > (<) 0$, and according to the mean-value theorem there exists $\bar{t} \in]0, 1[$ such that $\psi'(\bar{t}) > (<) 0$, a contradiction.

Remark 12.2 *In general, the weaker condition*

$$\langle F(x), y - x \rangle = 0 \Rightarrow \langle J_F(x)(y - x), y - x \rangle = 0$$

is not sufficient to characterize PPM-maps. For example, the map $F(x) = x^2$ on $\mathbb{R}$.

The following proposition states that the previous implication is sufficient, if the PPM-map is regular.

Proposition 12.3 *Let $K \subseteq \mathbb{R}^n$ be open and convex, and let $F : K \to \mathbb{R}^n$ be continuously differentiable and regular on K. Then, F is a PPM-map if and only if, for all $x, y \in K$, we have*

$$\langle F(x), y - x \rangle = 0 \Rightarrow \langle J_F(x)(y - x), y - x \rangle = 0.$$

Proof The result is an immediate consequence of Proposition 2 in John [123].

Remark 12.3 *In the particular case, where F does not vanish in K, this can also be seen from Corollary 4 in Crouzeix and Ferland [57].*

12.4   Affine PPM-Maps

In this section, we consider the special case, where the map $F : K \to \mathbb{R}^n$ is affine, i.e., $F(x) = Mx + q$, where M is a $n \times n$ matrix and $q \in \mathbb{R}^n$.

From Theorem 5.2 in Karamardian *et al.* [133], we have the following result.

Theorem 12.5 *Suppose $K \subseteq \mathbb{R}^n$ is open and convex, and let $F(x) = Mx + q$. Then, F is a PPM-map if and only if, for all $x, y \in K$,*

$$\langle Mx + q, y - x \rangle = 0 \Rightarrow \langle M(y - x), y - x \rangle = 0.$$

The following proposition is an immediate consequence of Theorem 5.4 in Karamardian *et al.* [133].

Proposition 12.4 *If $F(x) = Mx + q$, is a PPM-map on $\mathbb{R}^n$, then F and $-F$ are monotone, i.e., M and $-M$ are positive semidefinite.*

Now, we recall that M and $-M$ are positive semidefinite, if and only if

$$\langle M(y), y \rangle = 0, \forall y \in \mathbb{R}^n,$$

which is equivalent to M being skew-symmetric. Hence, we have the following corollary.

Corollary 12.3 *The map $F(x) = Mx + q$ is a PPM-map on $\mathbb{R}^n$ if and only if M is skew-symmetric, in which case F and $-F$ are monotone.*

There are PPM affine maps on proper subsets of $\mathbb{R}^n$, which are not associated with a skew-symmetric matrix; for example,

$$F(x) = [0 + x_1 + x_2 + 1]^T \text{ on } K = \mathbb{R}^2_{++}.$$

For an arbitrary open convex set $K \subseteq \mathbb{R}^n$, from Corollary 5.2 in Karamardian *et al.* [133], that the following result holds.

Corollary 12.4 *If $F(x) = Mx + q$ is a PPM-map on an open convex set $K \subseteq \mathbb{R}^n$ such that either F or $-F$ is not monotone, then $F(x) \neq 0$ for all $x \in K$. Hence, the only PPM affine maps which vanish at some point in K are those where M is skew-symmetric.*

12.5 Solution Set of a PPM Variational Inequality Problem

Consider the variational inequality problem:

Find $\bar{x} \in K$ such that $\langle F(\bar{x}), y - \bar{x} \rangle \geq 0, \forall y \in S$.

Let us assume that $\bar{K} \neq \phi$ is the solution set. The following result, which implies the convexity of the solution set, holds true in case of pseudomonotone maps. For PPM-maps, the proof is as follows:

Proposition 12.5 *Suppose that $K \subseteq \mathbb{R}^n$ is convex and let $F : K \to \mathbb{R}^n$ be a PPM-map. A vector $\bar{x} \in \bar{K}$, if and only if*

$$\langle F(\bar{x}), y - \bar{x} \rangle \geq 0, \forall y \in K.$$

Proof Let $\bar{x} \in \bar{K}$, then, we have

$$\langle F(\bar{x}), y - \bar{x} \rangle \geq 0, \forall y \in K,$$

From the pseudomonotonicity of the map F, it follows that

$$\langle F(y), y - \bar{x} \rangle \geq 0, \forall y \in K.$$

Conversely, if $\langle F(y), y - \bar{x} \rangle \geq 0$, for all $\forall y \in S$, then, we have

$$\langle -F(y), \bar{x} - y \rangle \geq 0, \forall y \in K.$$

From the pseudomonotonicity of $-F$, we get

$$\langle -F(\bar{x}), \bar{x} - y \rangle \geq 0, \text{ that is } \langle F(\bar{x}), y - \bar{x} \rangle \geq 0, \forall y \in S.$$

Let $\bar{x} \in \bar{K}$ be given, let us define

$$K(\bar{x}) := \{x \in K : \langle F(\bar{x}), x - \bar{x} \rangle = 0\}.$$

It is easy to prove the following result.

Proposition 12.6 *Suppose that $K \subseteq \mathbb{R}^n$ is convex, and let $F : K \to \mathbb{R}^n$ be a continuous PPM-map. Then, the solution set $\bar{K}$ is a subset of $K(\bar{x})$, for all $\bar{x} \in \bar{K}$. Moreover,*

$$K(\bar{x}) = \{x \in K : \langle F(\bar{x}), x - \bar{x} \rangle = 0\}$$

$$= \{x \in K : \langle F(x), x - \bar{x} \rangle = 0\}$$

$$= \left\{x \in K : \langle F(z), x - x^* \rangle = 0, \ \forall z \in M(x, \bar{x}) \bigcap K\right\}.$$

Proof Let $\bar{x} \in \bar{K}$. Then, we have

$$\langle F(\bar{x}), x - \bar{x} \rangle \geq 0.$$

By Proposition 12.5, we get

$$\langle F(\bar{x}), \bar{x} - x \rangle \geq 0.$$

Thus $\langle F(\bar{x}), x - \bar{x} \rangle = 0$, that is, $x \in K(\bar{x})$. Now, if

$$\langle F(\bar{x}), x - \bar{x} \rangle = 0.$$

Then, from Theorem 12.1, it follows that

$$\langle F(x), x - \bar{x} \rangle = 0.$$

Now, from Corollary 12.1, we get

$$\langle F(z), \bar{x} - x \rangle = 0, \forall z \in M(x, \bar{x}) \bigcap K.$$

Now, choosing $z = \bar{x}$, the reverse can be proved.

Remark 12.4 *If $F = \nabla f$ with f being pseudolinear, then by Theorem 3.1, it is clear that $\bar{K} = K(\bar{x})$. Therefore, $K(\bar{x})$ does not depend on the particular choice of $\bar{x}$. However, these results are not true for general PPM-maps. The following example from Bianchi aand Schaible [27] illustrates the fact. Let $F(x) = [-x_2, x_1]^T$ and $K = \mathbb{R}_+^2$. Then, it is easy to see that*

$$\bar{K} = \{x \in K : x_2 = 0\}, K\left\{[0,0]^T\right\} = K, K\left\{[1,0]^T\right\} = \bar{K}.$$

It is clear that this situation occurs always when a PPM-map vanishes in a point of its domain and the domain does not coincide trivially with the solution set. In fact, if $F(\bar{x}) = 0$, then $\bar{x} \in \bar{K}$ and $K(\bar{x}) = K$. We call a zero of F a trivial solution of the variational inequality. If we use only nontrivial solutions to calculate $K(\bar{x})$, then $\bar{K}$ may well be a proper subset of $K(\bar{x})$.

To see this, consider again the function $F(x) = [-x_2, x_1]^T$, and let $K = \{x \in \mathbb{R}^2 : x_2 \geq 0\}$. In this case, $K = \{x \in \mathbb{R}^2 : x_1 \geq 0, x_2 \geq 0\}$ but $K\left([1,0]^T\right) = \{x \in \mathbb{R}^2 : x_2 = 0\}$.

The following example from Bianchi and Schaible [27] illustrates that, for a nonvanishing PPM-map, the inclusion $\bar{K} \subseteq K(\bar{x})$ may be proper. Consider the skew-symmetric matrix

$$P = \begin{bmatrix} 0 & 1 & 0 & 0 \\ -1 & 0 & 2 & 1 \\ 0 & -2 & 0 & -1 \\ 0 & -1 & 1 & 0 \end{bmatrix}$$

and

$$K = \left\{ x \in \mathbb{R}^4 : x_1 \geq 1, x_2 \leq 0, x_4 \geq 0 \right\}.$$

Since P is nonsingular, P does not vanish in K. The solution set of the variational inequality associated with $F(x) = Px$ in K is given by

$$K = \left\{ x \in \mathbb{R}^4 : x_1 \geq 1, x_2 = 0, 0 \leq x_3 \leq x_1/2, x_4 = 0 \right\}.$$

Since, $\bar{x} = [1, 0, 0, 0]^T \in \bar{K}$, we have

$$K(\bar{x}) = \left\{ x \in \mathbb{R}^4 : x_1 \geq 1, x_2 = 0, x_4 \geq 0 \right\},$$

obviously, $\bar{K} \subseteq K(\bar{x})$ and the inclusion is proper.

For condition $\bar{K} = K(\bar{x})$, for every solution $\bar{x}$, we need some additional assumptions, satisfied by gradient maps. For this purpose, Bianchi and Schaible [27], introduced the following notion of a G-map.

Definition 12.3 *(G-map) Suppose K is a convex subset of $\mathbb{R}^n$ and let $F : K \to \mathbb{R}^n$. The map F is said to be a G-map, if there exists a positive function $k(x, y)$ on $K \times K$ such that, for all $x, y \in S$,*

$$\langle F(x), y - x \rangle = 0 \Rightarrow F(x) = k(x, y) F(y).$$

It is evident that G-maps are PPM-maps, but the converse is not true, in general. To illustrate this fact, we have the following example from Bianchi and Schaible [27]. Consider in $\mathbb{R}^3$ the affine map $F(x) = Px$ with the skew-symmetric matrix

$$P = \begin{bmatrix} 0 & 1 & 0 & 0 \\ -1 & 0 & 2 & 1 \\ 0 & -2 & 0 & -1 \\ 0 & -1 & 1 & 0 \end{bmatrix}.$$

Then, for

$$x^T = [2, 0, 1]^T \text{ and } y^T = [0, -4, 1]^T,$$

we have

$$F(x) = -F(y) \text{ and } \langle F(x), y - x \rangle = 0.$$

Rapcsak [233] and Komlosi [147] have shown that the gradients of (nonconstant) pseudolinear functions are G-maps, where $k(x, y) = \lVert F(x) \rVert / \lVert F(y) \rVert$. However, there are examples of G-maps which are not gradient maps. The following example illustates the fact.

$$F(x) = [0, x_1 + x_2 + 1]^T \text{ in } K \in \mathbb{R}^2_{++}.$$

In this case,

$$k(x, y) = (x_1 + x_2 + 1)/(y_1 + x_2 + 1).$$

We know that if the gradient of a pseudolinear function vanishes in one point of its domain, it vanishes everywhere. For a G-map this property also holds.

Theorem 12.6 *Suppose K is a convex subset of $\mathbb{R}^n$ and let $F : K \to \mathbb{R}^n$ be a continuous G-map. Then, the solution set $\bar{K}$ coincides with $K(\bar{x})$ for every solution $\bar{x} \in \bar{K}$.*

Proof We have to show only that every point x in $K(\bar{x})$ is a solution. Since, $x \in K(\bar{x})$ and F is a G-map, therefore,

$$\langle F(\bar{x}), x - \bar{x} \rangle = 0 \Rightarrow F(\bar{x}) = k(\bar{x}, x) F(x).$$

Therefore, from the inequality

$$\langle F(\bar{x}), y - \bar{x} \rangle \geq 0, \forall y \in K,$$

we get

$$\langle F(x), y - \bar{x} \rangle \geq 0.$$

Therefore,

$$\langle F(x), y - x \rangle = \langle F(x), y - \bar{x} \rangle + \langle F(x), \bar{x} - x \rangle \geq 0.$$

Hence, $x \in \bar{K}$.

We recall from Chapter 2, that the following results hold true for the gradient maps of pseudolinear functions. For G-maps, the result is easy to prove.

Proposition 12.7 (Bianchi and Schaible [27]) *Suppose $K \subseteq \mathbb{R}^n$ is convex and let $F : K \to \mathbb{R}^n$ be a continuous G-map. Then, $\bar{K} = K_1(\bar{x})$, where*

$$K_1(\bar{x}) = \{x \in K : \langle F(\bar{x}), \bar{x} - x \rangle \geq 0\} = \{x \in K : \langle F(x), \bar{x} - x \rangle \geq 0\}$$
$$= \left\{x \in K : \langle F(z), \bar{x} - x \rangle \geq 0, \ \forall z \in M(x, \bar{x}) \bigcap K\right\}.$$

Proof It is known that $\bar{K} = K(\bar{x})$, and that

$$K(\bar{x}) \subseteq \{x \in K : \langle F(\bar{x}), \bar{x} - x \rangle \geq 0\}.$$

Then, we have

$$\{x \in K : \langle F(\bar{x}), \bar{x} - x \rangle \geq 0\} = \{x \in K : \langle F(x), \bar{x} - x \rangle \geq 0\}$$
$$= \left\{x \in K : \langle F(z), \bar{x} - x \rangle \geq 0, \ \forall z \in M(x, x^*) \bigcap K\right\}.$$

We note that there does not exist $x \in K$, such that $\langle F(\bar{x}), \bar{x} - x \rangle > 0$, which is a contradiction to our assumption that $\bar{x} \in \bar{K}$. This completes the proof.

Proposition 12.8 *Suppose $K \subseteq \mathbb{R}^n$ is convex, and let $F : K \to \mathbb{R}^n$ be a continuous G-map. Then, $\bar{K} = K_2(\bar{x})$, where*

$$K_2(\bar{x}) =: \{x \in K : \langle F(\bar{x}), \bar{x} - x \rangle\} = \langle F(x), x - \bar{x} \rangle$$
$$= \{x \in K : \langle F(\bar{x}), \bar{x} - x \rangle\} \geq \langle F(x), x - \bar{x} \rangle.$$

Proof It is clear that $\bar{K} = K(\bar{x})$ and that

$$K(\bar{x}) \subseteq \{x \in K : \langle F(\bar{x}), \bar{x} - x \rangle\} = \langle F(x), x - \bar{x} \rangle.$$

Therefore, to complete the proof, we need to show that any element $\tilde{x}$ of $\{x \in K : \langle F(\bar{x}), \bar{x} - x \rangle\} \geq \langle F(x), x - \bar{x} \rangle$ is a solution. Since $\bar{x} \in \bar{K}, \langle F(\bar{x}), \tilde{x} - \bar{x} \rangle \geq 0$, therefore, by the pseudomonotonicity of F, we get $\langle F(\tilde{x}), \tilde{x} - \bar{x} \rangle \geq 0$. By the definition of $\tilde{x}$, it follows that $\langle F(\bar{x}), \bar{x} - \tilde{x} \rangle \geq 0$. Hence, $\langle F(\bar{x}), \tilde{x} - \bar{x} \rangle = 0$, i.e., $\tilde{x} \in K(\bar{x})$. Now, by Theorem 12.6, this implies $\tilde{x} = \bar{K}$.

In the rest of the chapter, we study pseudomonotone maps T, for which $-T$ is also pseudomonotone, known as pseudoaffine maps. The gradients of pseudolinear functions are a particular case of such maps. Bianchi *et al.* [26] have studied such maps and their properties in detail. We shall present results from Bianchi *et al.* [26]. These results will help to determine the most general form of the pseudoaffine maps.

12.6   General Form of Pseudoaffine Maps Defined on the Whole Space

For the whole space $\mathbb{R}^n$, we state the following result from Thompson and Parke [270].

Theorem 12.7 *A lower semicontinuous function $f : \mathbb{R}^n \to \mathbb{R}$ is quasilinear if and only if it has the form*

$$f(x) = h(\langle u, x \rangle), \tag{12.3}$$

where h is a lower semicontinuous increasing function and $u \in \mathbb{R}^n$.

We know that the class of quasilinear functions is very useful. But, from the above theorem it is clear that there are very few quasilinear functions which are defined on the whole space. However, from Martinez-Legaz [178], we know that a lower semicontinuous function $g : \mathbb{R}^n \to \mathbb{R}$ which is bounded from below is quasiconvex if and only if it is the supremum of differentiable quasilinear functions.

The following corollary follows directly from Theorem 12.7:

Corollary 12.5 *A differentiable function $f : \mathbb{R}^n \to \mathbb{R}$ is pseudolinear if and only if it can be written in the form*

$$f(x) = h(\langle u, x \rangle), \tag{12.4}$$

where $u \in \mathbb{R}^n$ and h is a differentiable function whose derivative is always positive or identically zero.

Proof If f is pseudolinear, then it is quasilinear. Thus by Theorem 12.7, f has the form (12.3) where h is differentiable and increasing, i.e., with nonnegative derivative. If h' is zero at some point, then ∇f vanishes at some point. Therefore, by Remark 2.1, f is constant. Hence, h' is identically zero. Otherwise, h is always positive. The converse is obvious.

We note that if T is pseudoaffine, then for all $x, y \in K$, we have

$$\langle T(x), y - x \rangle = 0 \Leftrightarrow \langle T(y), y - x \rangle = 0. \tag{12.5}$$

Conversely, if the above equivalence holds and T is continuous, then by Theorem 12.1, T is pseudoaffine.

The following is an easy consequence of the above

$$T(0) = 0 \Leftrightarrow \langle T(x), x \rangle = 0, \ \forall x \in \mathbb{R}^n. \tag{12.6}$$

If a map T is defined on the whole space, then for the special case $T = \nabla f$ for some pseudolinear function f, we know that f has the form (12.4), which implies that $T(x) = h'(\langle u, x \rangle) u$, i.e., T is always a positive multiple of a constant vector. Again, when T and $-T$ are monotone, then, we have another example of a pseudoaffine map on the whole space.

The following proposition provides the general form of such a map.

Proposition 12.9 *Suppose $T : \mathbb{R}^n \to \mathbb{R}^n$ is such that T and $-T$ are both monotone. Then there exists a skew-symmetric linear map A and a vector $u \in \mathbb{R}^n$, such that*

$$T(x) = Ax + u, x \in \mathbb{R}^n.$$

Proof Let $u = T(0)$ and define $T' : \mathbb{R}^n \to \mathbb{R}^n$ by $T'(x) = Tx - u$. Then, T' and $-T'$ are monotone. Therefore, for all $x, y \in \mathbb{R}^n$, we have

$$\langle T'(y) - T'(x), y - x \rangle = 0 \tag{12.7}$$

$$T'(0) = 0. \tag{12.8}$$

From (12.7) and (12.8), it follows that

$$\langle T'(x), x \rangle = 0, \ \forall x \in \mathbb{R}^n.$$

From (12.7), we get

$$\langle T'(x), y \rangle = \langle T'(y), x \rangle, \ \forall x, y \in \mathbb{R}^n. \tag{12.9}$$

Therefore, for any $t \in \mathbb{R}, x, y \in \mathbb{R}^n$, using (12.9), we get

$$\langle T'(tx), y \rangle = -\langle tx, T'(y) \rangle = -t \langle x, T'(y) \rangle = t \langle T'(x), y \rangle.$$

Hence,

$$\langle T'(tx) - tT'(x), y \rangle = 0, \ \forall y \in \mathbb{R}^n.$$

From, which it follows that $T'(tx) = tT'(x), \forall x \in \mathbb{R}^n$. Similarly, for all $x, y, z \in \mathbb{R}^n$, and using (12.9), we get

$$\langle T'(x+y), z \rangle = -\langle x+y, T'(z) \rangle = -\langle x, T'(z) \rangle - \langle y, T'(z) \rangle$$

$$= \langle T'x, z \rangle + \langle T'y, z \rangle.$$

Therefore,

$$\langle T'(x+y) - T'(x) - T'(y), z \rangle = 0, \forall z \in \mathbb{R}^n.$$

Then, we get

$$T'(x+y) = T'(x) + T'(y),$$

which implies that T' is linear. Using (12.9), it follows that T' is skew-symmetric. Setting $A = T'$, the result follows.

The following theorem from Bianchi *et al.* [26] characterizes the general form of a pseudoaffine map in terms of a skew-symmetric linear map, defined on the whole space.

Theorem 12.8 *A map $T : \mathbb{R}^n \to \mathbb{R}^n$ is pseudoaffine if and only if there exist a skew-symmetric linear map A, a vector u and a positive function $g : \mathbb{R}^n \to \mathbb{R}$, such that*

$$T(x) = g(x)(Ax + u), \ \forall x \in \mathbb{R}^n. \tag{12.10}$$

Proof If g, A, u is defined as above, then for any $x, y \in \mathbb{R}^n$, we have

$$\langle Ay, y - x \rangle = \langle Ax, y - x \rangle$$

Since A is skew-symmetric, therefore, we have

$$\langle T(x), y - x \rangle \geq 0 \Leftrightarrow \langle Ax + u, y - x \rangle \geq 0 \Leftrightarrow$$

$$\langle Ax, y - x \rangle + \langle u, y - x \rangle \geq 0 \Leftrightarrow \langle Ay, y - x \rangle \langle u, y - x \rangle \geq 0 \Leftrightarrow$$

$$\langle T(y), y - x \rangle \geq 0.$$

The converse will be studied in the subsequent sections. The proofs are along the lines of Bianchi *et al.* [26]. In Section 12.7, Theorem 12.4 is proved for continuous pseudoaffine maps. In Section 12.8, some elementary properties of pseudoaffine maps will be discussed. In Section 12.9, we prove the theorem for $n = 2$. In Section 12.10, we derive some properties of pseudoaffine maps with regard to straight lines. In Section 12.11, we establish the theorem for pseudoaffine maps that vanish in at least one point. In Section 12.12, we prove the theorem for $n = 3$ without any assumption on the number of zeros of the map. In the final section, a theorem will be established for the nonvanishing maps with arbitrary n.

The following terminology and notations are from Bianchi *et al.* [26]. Given $x \in R^n, S \subseteq R^n$, let us denote by

$$x + S := \{x + y : y \in S\},$$

$$\mathbb{R}^{++} :=]0, \infty[,$$

$$\mathbb{R}^{++} S := \{tx : t \in \mathbb{R}^{++}, x \in S\}.$$

For $x_1, ..., x_k \in \mathbb{R}^n$, let $sp(x_1, ..., x_k)$ denote the subspace generated by $\{x_1, ..., x_k\}$. More generally, if $S \subseteq \mathbb{R}^n$, then $sp(S)$ denotes the subspace generated by S. Two vectors $x, y \in \mathbb{R}^n$ will be said to have the same direction if there exists $\alpha > 0$, such that $x = \alpha y$. Given $x, y \in \mathbb{R}^n, x \neq y$, we denote by $l(x, y)$ the straight line generated by x and y, that is

$$l(x, y) = \{z \in \mathbb{R}^n : z = tx + (1 - t) y, t \in \mathbb{R}\}.$$

Definition 12.4 *A map T with values in $\mathbb{R}^n$ will be said to have a constant direction on a set K, if it is identically zero on K, or else it is everywhere nonzero on K and there exists $e \in \mathbb{R}^n$, such that*

$$T(x) = \|T(x)\| e, \forall x \in K.$$

We note that for the case $n = 1$, Theorem 12.8 is trivially true. We know that a pseudoaffine map in $\mathbb{R}$ is a function which is always positive or always negative or identically zero, thus it is sufficient to take in (12.10) $A = 0$ and $u = 1$ or $b = -1$ or $u = 0$.

The propositions of this chapter are obviously true for $n = 1$, unless otherwise specified. For this reason, we assume in all proofs that $n \geq 2$.

12.7 Reduction to the Continuous Case

In this section, we show that to prove Theorem 12.8, without a loss of generality, we may suppose that T is continuous.

The following proposition with the additional assumption that T is continuous has been proved in Theorem 12.2.

Proposition 12.10 *Suppose $K \subseteq \mathbb{R}^n$ is open and convex and let $T : K \to \mathbb{R}^n$ be pseudoaffine. If $z_1, z_2 \in K$ are such that $T(z_1) = T(z_2) = 0$, then $T(z) = 0$ for any $z \in l(z_1, z_2) \bigcap K$.*

Proof For each $v \in \mathbb{R}^n$, let $t \in \mathbb{R}$, be such that $z + tv \in K$. By the pseudoaffine of T, we have

$$\langle T(z_1), z + tv - z_1 \rangle = 0 \Rightarrow \langle T(z + tv), z + tv - z_1 \rangle = 0. \qquad (12.11)$$

Since $z \in l(z_1, z_2)$, there exists $\lambda \in \mathbb{R}$, such that

$$z = \lambda z_1 + (1 - \lambda) z_2.$$

Hence, by (12.11), we get

$$\langle T(z+tv), tv + (1-\lambda)(z_2 - z_1) \rangle = 0. \qquad (12.12)$$

Similarly, using $T(z_2) = 0$, we find $\langle T(z+tv), z + tv - z_2 \rangle = 0$, hence

$$\langle T(z+tv), tv + \lambda(z_1 - z_2) \rangle = 0. \qquad (12.13)$$

From (12.12) and (12.13), it follows that $\langle T(z+tv), tv \rangle = 0$. Therefore, $\langle T(z+tv), z - (z+tv) \rangle = 0$, hence, using again that T is pseudoaffine, we get $\langle T(z), tv \rangle = 0$.

Therefore $\langle T(z), v \rangle = 0$ for all $v \in \mathbb{R}^n$, hence $T(z) = 0$.

The following proposition is a resullt of direct application of induction to Proposition 12.10.

Proposition 12.11 *Suppose $K \subseteq \mathbb{R}^n$ is open and convex and let $T : K \to \mathbb{R}^n$ be pseudoaffine. Let $z_1, ..., z_m \in K$ are such that $T(z_1) = T(z_2) = ... = T(z_m) = 0$. Then, T vanishes on $M \bigcap K$, where M is the affine subspace generated by $z_1, ..., z_m$.*

By Proposition 12.11, it is clear that the set of zeros $V = M \bigcap K$ is closed as a subset of K, thus the set

$$W := K/V = \{x \in K : T(x) \neq 0\}$$

is open.

Lemma 12.1 *Suppose $K \subseteq \mathbb{R}^n$ is open and convex and let $T : K \to \mathbb{R}^n$ be pseudoaffine. Then the map $S : W \to \mathbb{R}^n$ defined by*

$$S(x) := T(x) / \|T(x)\|$$

is continuous.

Proof Assume that $x \in W$. On the contrary, we suppose that there exists a sequence $(x_n)_{n \in N}$ in W, such that $(x_n) \to x$ but $S(x_n) \nrightarrow S(x)$. Since $\|S(x_n)\| = 1$ and the unit sphere is compact in $\mathbb{R}^n$, we suppose that $S(x_n) \to A$, where $A \neq S(x)$ selecting a subsequence if necessary. We note that since $\|A\| = \|S(x)\| = 1$, the vectors $S(x)$ and A do not have the same direction. Therefore, we can choose $\nu \in \mathbb{R}^n$, such that

$$\langle S(x), \nu \rangle < 0 < \langle A, \nu \rangle,$$

choosing, ν small enough so that $w := x + \nu \in K$. Since the map S is also pseudoaffine and $\langle S(x), w - x \rangle < 0$, it follows that

$$\langle S(w), w - x \rangle < 0. \qquad (12.14)$$

Again, $0 < \langle A, \nu \rangle = \langle A, w - x \rangle$ implies that there exists $n_0 \in N$ such that for all $n > n_0$, we have $\langle S(x_n), w - x_n \rangle > 0$. Using again the fact that S is pseudoaffine, it follows that $\langle S(w), w - x_n \rangle > 0$. Taking the limit, we get

$$\langle S(w), w - x \rangle \geq 0,$$

which is a contradiction to (12.14). This completes the proof.

Theorem 12.9 *Suppose $K \subseteq \mathbb{R}^n$ is open and convex and let $T : K \to \mathbb{R}^n$ be pseudoaffine. Then there exists a positive function $g : K \to \mathbb{R}$ and a continuous pseudoaffine map T', such that*

$$T(x) = g(x) T'(x), x \in K.$$

Proof Let V be the set of zeros. If $V = \phi$, then, from Lemma 12.3, the theorem follows. Otherwise, assume that $d(x, V)$ is the distance of $x \in K$ from V. The function $d(., V)$ is continuous, and if $x \to x_0 \in V, d(x, V) \to 0$. Therefore, the map $T' : K \to \mathbb{R}^n$ defined by

$$T'(x) = \begin{cases} d(x, V) S(x), & \text{if } x \in W, \\ 0, & \text{if } x \in V, \end{cases}$$

is continuous. It is obvious that, $T(x) = g(x) T'(x)$, where $g : K \to \mathbb{R}$ is a positive function defined by

$$g(x) = \begin{cases} \frac{\|T(x)\|}{d(x,V)}, & \text{if } x \in W, \\ 1, & \text{if } x \in V. \end{cases}$$

Since, $T'(x) = \frac{1}{g(x)} T(x), x \in K$, T' is pseudoaffine.

From Theorem 12.9, it follows that in order to prove Theorem 12.8, without a loss of generality, we may suppose that T is continuous.

12.8 Elementary Properties of Pseudoaffine Maps Defined on the Whole Space

The following proposition from Bianchi *et al.* [26] explores the relation (12.10).

Proposition 12.12 *Suppose that $T : \mathbb{R}^n \to \mathbb{R}^n$ is pseudoaffine. If T has the form (12.10), where $g : \mathbb{R}^n \to \mathbb{R}$ is a positive function, $u \in \mathbb{R}^n$ and A a linear map, then A is skew-symmetric. If in addition T is continuous, then g is continuous on the set*

$$\{x \in \mathbb{R} : T(x) \neq 0\}.$$

Proof Let us take $T'(x) = T(x)/g(x)$. Then T' is pseudoaffine. Now, we prove that for all $x \in \mathbb{R}^n, \langle Ax, x \rangle = 0$. Assume that for some $x \in \mathbb{R}^n, \langle Ax, x \rangle > 0$. Then, for all sufficiently large t and for any $y \in \mathbb{R}^n$, the expression

$$\langle T'(tx), y - tx \rangle = \langle tAx + u, y - tx \rangle = -t^2 \langle Ax, x \rangle + t [\langle Ax, y \rangle - \langle u, x \rangle] + \langle u, y \rangle$$

will be negative. Since T' is pseudoaffine, and for a sufficiently large t, it follows that

$$\langle T'(y), y - tx \rangle = \langle T'(y), y \rangle - t \langle T'(y), x \rangle,$$

will also be negative. Therefore, $\langle A(y) + u, x \rangle$ is nonnegative for all y. For $y = tx, t \in \mathbb{R}$, it follows that $\langle Atx + u, x \rangle \geq 0$. Letting $t \to -\infty$ and taking into account $\langle Ax, x \rangle < 0$, we get a contradiction. Similarly, $\langle Ax, x \rangle < 0$ leads to a contradiction. Thus, A is skew-symmetric.

If T is continuous, then at any $x \in \mathbb{R}^n$, such that $T(x) \neq 0$, it results that

$$g(x) = \|T(x)\| / \|Ax + u\|,$$

which is continuous at x.

Bianchi *et al.* [26] have remarked that when T is continuous, it does not follow that g is continuous on $\mathbb{R}^n$. As an example, define T on $\mathbb{R}^2$ by (12.10), where $u = 0, A(x_1, x_2) = (x_2, -x_1)$ and $g(x_1, x_2) = 1/\sqrt{\|(x_1, x_2)\|}$, whenever $(x_1, x_2) \neq (0, 0)$ and $g(0, 0) = 1$. Then it is easily seen that T is continuous while g is discontinuous at 0.

The following result will be very useful in the sequel:

Proposition 12.13 *Suppose that V is a subspace in $\mathbb{R}^n$, P_V is the orthogonal projection on V, $x_0 \in \mathbb{R}^n, M = V + x_0$, an affine subspace and $T : \mathbb{R}^n \to \mathbb{R}^n$, a pseudoaffine map. Then,*

(a) the translation $T_1(x) = x - x_0$ is a pseudoaffine map on $\mathbb{R}^n$;

(b) the orthogonal projection $P_V T$ is a pseudoaffine map on M.

Proof Part (a) is obviously true. To prove part (b), we note that

$$\langle P_V T(x), y - x \rangle = \langle T(x), P_V(y - x) \rangle = \langle T(x), y - x \rangle, \forall x, y \in M.$$

This completes the proof.

From Proposition 12.11, it is clear that the set of zeros of a pseudoaffine map defined on the whole space, is either empty or an affine subspace. If T vanishes on a hyperplane M, then it vanishes on the whole space.

The following lemma by Bianchi *et al.* [26] will be used often to prove the results:

Lemma 12.2 *Suppose M is a subspace in $\mathbb{R}^n$ and $z, y \in \mathbb{R}^n$ are such that for all $x \in M, \langle z, y - x \rangle = 0$. Then $z \in M^\perp \bigcap y^\perp$.*

Proof Putting $x = 0$ in the given relation, it results that

$$\langle z, y \rangle = 0.$$

Now, we note that $\langle z, x \rangle = 0, \forall x \in M$, which means that $z \in M^{\perp} \bigcap y^{\perp}$. This completes the proof.

Proposition 12.14 *Let $T : \mathbb{R}^n \to \mathbb{R}^n$ be a pseudoaffine map. If T vanishes on a hyperplane M, then it vanishes on $\mathbb{R}^n$.*

Proof Let us consider a translation of T. Without loss of generality, we may suppose that $0 \in M$. For any $z \notin M$ and any $x \in M, \langle T(x), z - x \rangle = 0$, hence by pseudoaffinity in the form (12.5), we deduce that $\langle T(z), z - x \rangle = 0$. By Lemma 12.2, $T(z) \perp M$ and $T(z) \perp z$. Since M is a hyperplane, it follows that $T(z) = 0$.

The following proposition generalizes Proposition 12.14.

Proposition 12.15 *Suppose that $T : \mathbb{R}^n \to \mathbb{R}^n$ is a pseudoaffine map. If T has a constant direction on a hyperplane M, then it has a constant direction on $\mathbb{R}^n$.*

Proof From Theorem 12.9, without a loss of generality we may assume that T is continuous. If T vanishes on M, then the result follows from Proposition 12.14. Otherwise, T has the form $T(x) = \|T(x)\| e$ with $\|e\| = 1$. By the definition of constant direction, we note that T has no zeros on M. Without a loss of generality, we may consider that M is a subspace. Let $z \notin M$. We consider two cases:

Case (1) If e is orthogonal to M. Then, for all $x \in M$, we have

$$\langle T(x), z - x \rangle = \|T(x)\| \langle e, z \rangle.$$

Therefore, the quantity $\langle T(x), z - x \rangle$ has a constant sign. By pseudoaffinity of T, also the quantity $\langle T(z), z - x \rangle$ has a constant sign. Therefore, in particular, for every $x \in M$ and every $t \in \mathbb{R}$, the quantity

$$\langle T(z), z - x \rangle = \langle T(z), z \rangle - t \langle T(z), x \rangle,$$

has a constant sign, which is possible only if

$$\langle T(z), x \rangle = 0.$$

Therefore, $T(z)$ is also orthogonal to M. Hence, $T(z)$ and e are linearly dependent.

Case (2) If e is not orthogonal to M, we consider the subspace $V = \{y \in \mathbb{R}^n : \langle e, y \rangle = 0\}$ and the hyperplane $M_1 = z + V$. Since V is orthogonal to e, M_1 intersects M at an affine subspace $M \bigcap M_1$ with dimension $n - 2$. Let P be the orthogonal projection to V. Since for all $x \in M, T(x) = \|T(x)\| e \perp V$, therefore, the map PT is pseudoaffine on M_1 and vanishes on $M \bigcap M_1$. From

Proposition 12.14, we get that PT vanishes on M_1. Hence, $T(z)$ and e are linearly dependent.

From cases (1) and (2), it follows that for each $z \in \mathbb{R}^n$, $T(z)$ and e are linearly dependent. We note that $T(z)$ is never zero. Indeed, T is never zero on M by assumption. If $T(z) = 0$, for some $z \notin M$, then, by pseudoaffinity of T, we get

$$\langle T(z), z - x \rangle = 0 \Rightarrow \langle T(x), z - x \rangle = 0 \Rightarrow \langle e, z - x \rangle = 0, \forall x \in M.$$

By Lemma 12.2, this would imply that e is orthogonal both to z and M, which is impossible since $e \neq 0$. Since $T(z), z \in \mathbb{R}^n$ is always a multiple of e and is never zero, continuity of T implies that $T(z)$ always has the same direction.

12.9   The Case $n = 2$

In this section, restricting to continuous maps, Bianchi *et al.* [26] have shown that Theorem 12.8 is true, when $n = 2$.

Theorem 12.10 *Let $T : \mathbb{R}^2 \to \mathbb{R}^2$ be a continuous pseudoaffine map. Then there exists a vector $u \in \mathbb{R}^2$, a skew-symmetric linear map A and a positive function $g : \mathbb{R}^2 \to \mathbb{R}^2$, such that*

$$T(x) = g(x)(Ax + u).$$

Proof We consider the following three cases:
Case (1): Let $T(x) \neq 0$, $x \in \mathbb{R}^2$ and

$$P =: \left\{ y \in \mathbb{R}^2 : \langle T(0), y \rangle = 0 \right\}.$$

Then, by pseudoaffinity of T, for all $y \in P$, we get

$$\langle T(y), y \rangle = 0.$$

Hence, $T(y)$ is orthogonal to P. Since the space is two-dimensional and T is never zero, T has a constant direction on P. By Proposition 12.15, T has a constant direction on $\mathbb{R}^2$. It follows that T has the form of (12.10) with

$$A = 0, u = T(0) / \|T(0)\|$$

and

$$g(x) = \|T(x)\|.$$

Case (2): There exists exactly one point $\bar{x}$, such that $T(\bar{x}) = 0$. Taking $T'(x) = T(x + \bar{x})$, then, T' is pseudoaffine and $T'(0) = 0$. By (12.6), for any $x \in \mathbb{R}^2$, $\langle T'(x), x \rangle = 0$. Let $A : \mathbb{R}^2 \to \mathbb{R}^2$ be the map

$$A(a, b) = (b, -a), \quad \forall (a, b) \in \mathbb{R}^2.$$

Then, $(Ax, x) = 0$, $\forall x \in \mathbb{R}^2$. For any $x \in \mathbb{R}^2 \setminus \{0\}$, both Ax and $T'(x)$ are orthogonal to x, hence, they are linearly dependent. Therefore, there exists $g_1(x) \neq 0$, such that $T'(x) = g_1(x) Ax$. Then,

$$g_1(x) = \frac{\langle T'(x), Ax \rangle}{\|Ax\|^2}$$

is continuous and does not vanish on $\mathbb{R}^2 \setminus \{0\}$.

Thus, g_1 has a constant sign. By changing the sign of A, we may assume that g_1 is positive. Now, we have

$$T'(x) = T'(x - x_0) = g_1(x - x_0)(Ax - Ax_0).$$

Setting $g(x) = g_1(x - x_0)$, for $x \neq x_0$, $g(x_0) = 1$ and $u = -Ax_0$, the result follows.

Case (3): There are at least two points $x_1 \neq x_2$ such that $T(x_1) = T(x_2) = 0$. Then, Propositions 12.10 and 12.14 imply that $T(x) = 0$ for any $x \in \mathbb{R}^2$, i.e., T has the form (12.10) with $g = 1, A = 0, u = 0$.

12.10 Line Properties of Pseudoaffine Maps

In this section, we study the properties of pseudoaffine maps along straight lines. The proof follows along the lines of Bianchi *et al.* [26].

Proposition 12.16 *Suppose that $T : \mathbb{R}^n \to \mathbb{R}^n$ is a continuous pseudoaffine map. Assume that $T = 0$ is on a straight line l. Then T has a constant direction on any straight line parallel to l. Moreover, this direction is orthogonal to l.*

Proof Suppose that $0 \in l$, i.e., $l = \{w \in \mathbb{R}^n : w = te, t \in \mathbb{R}\}$ with $e \neq 0$. We consider any straight line $l' = \{w \in \mathbb{R}^n : y = x_0 + te, t \in \mathbb{R}\}$ parallel to l. If $T(y) = 0$ for some $y \in l'$, then by Proposition 12.10, the map T vanishes on the affine subspace generated by l and y, so T is identically zero on l'.

Let us assume that T is never zero on l'. For all $w \in l$ and $x \in \mathbb{R}^n$, the condition

$$\langle T(w), x - w \rangle = 0$$

holds. Now, by the pseudoaffinity of T, it follows that

$$\langle T(x), x - w \rangle = 0.$$

By Lemma 12.2, $T(x) \perp l$ and $T(x) \perp x$. Hence, we have

$$\langle T(x), w \rangle = \langle T(x), x \rangle = 0, \forall x \in \mathbb{R}^n, w \in l. \qquad (12.15)$$

Consider the hyperplane $V = \{z \in \mathbb{R}^n : \langle T(x_0), z \rangle = 0\}$. For any $z \in V$, by pseudoaffinity of T, we have

$$\langle T(x_0), z + x_0 - x_0 \rangle = 0 \Rightarrow \langle T(z + x_0), z \rangle = 0.$$

Using (12.15), for all $y = x_0 + w, w \in l$, we get

$$\langle T(z + x_0), y - (z + x_0) \rangle = 0.$$

Using again the pseudoaffinity of T, it follows that

$$\langle T(y), y - (z + x_0) \rangle = 0.$$

Using (12.15) again, it results that $\langle T(y), z \rangle = 0$. Hence, $T(y) \in V^{\perp}$, i.e., $T(y)$ and $T(x_0)$ are linearly dependent. Since, there is no zero of T on the line l', by continuity $T(x_0)$ and $T(y)$ have the same direction.

Proposition 12.17 *Assume that $x_1 \neq x_2$ are two points in $\mathbb{R}^n$. Let $T : \mathbb{R}^n \to \mathbb{R}^n$ be a continuous pseudoaffine map, such that*

$$T(x_1) = aT(x_2), a \in \mathbb{R}.$$

Let $l = l(x_1, x_2)$. Then, for any $x \in l$, there exists $\lambda(x) \in \mathbb{R}$, such that $T(x) = \lambda(x) T(x_2)$. In particular, if T never vanishes on l, then $\lambda(x) > 0$.

Proof Suppose that $0 \in l$. If $T(x_2) = 0$, we have $T(x_1) = 0$, and thus by Proposition 12.10, T is zero on l. If $T(x_2) \neq 0$, the following two cases may arise:

Case (1) If $\langle T(x_2), x_2 - x_1 \rangle = 0$. Let $V = \{y \in \mathbb{R}^n : \langle T(x_2), y \rangle = 0\}$ and P_V the orthogonal projection on V. Then $P_V T$ is a pseudoaffine map on $M = V + x_1$. Since $x_1, x_2 \in M$ and $P_V T(x_1) = P_V T(x_2) = 0$, by Proposition 12.10, $P_V T$ vanishes on l and thus, $T(x)$ and $T(x_2)$ are linearly dependent for each $x \in l$.

Case (2) If $\langle T(x_2), x_2 - x_1 \rangle \neq 0$. From (12.5), it follows that $a \neq 0$. Suppose to the contrary that there exists $x \in l$ such that $T(x)$ and $T(x_2)$ are linearly independent. Then, $x \neq x_1, x \neq x_2$, and the equations

$$\langle T(x_2), z - x_2 \rangle = 0 \tag{12.16}$$

and

$$\langle T(x), z - x \rangle = 0. \tag{12.17}$$

Let us define two nonparallel hyperplanes through x_2 and x, respectively. Let z satisfy both (12.16) and (12.17). From pseudoaffinity, it follows that

$$\langle T(z), z - x_2 \rangle = 0 \tag{12.18}$$

$$\langle T(z), z - x \rangle = 0.$$

Thus $\langle T(z), x_2 - x \rangle = 0$. Consequently, we have

$$\langle T(z), x_2 - x_1 \rangle = 0.$$

Using (12.18), we infer that $\langle T(z), x_1 - z \rangle = 0$. Again by pseudoaffinity of T, we have

$$\langle T(x_1), x_1 - z \rangle = 0.$$

By our assumption, we get $\langle T(x_2), x_1 - z \rangle = 0$. In view of (12.16), it follows that $\langle T(x_2), x_1 - x_2 \rangle = 0$, which is a contradiction.

Thus, in all cases there exists $\lambda(x) \in \mathbb{R}$ such that $T(x) = \lambda(x) T(x_2), x \in l$. The function λ is obviously continuous; hence, if it does not vanish, it has a constant sign.

If l is the straight line joining two points x_1, x_2 and $T(x_1)$ and $T(x_2)$ are linearly dependent, then there are three possibilities for the various images $T(x), x \in l$. If T has at least two zeros on the line, by Proposition 12.10, it follows that T is identically zero on l. If T has no zeros, then by Proposition 12.17, $T(x), x \in l$ belong to an open half straight line beginning at the origin. If T has exactly one zero, then again Proposition 12.17 guarantees that all $T(x), x \in l$ belong to a straight line through the origin, but some have opposite directions.

The result follows from the following proposition.

Proposition 12.18 *Suppose that $x_1, x_2 \in l$. Let $T : \mathbb{R}^n \to \mathbb{R}^n$ be a continuous pseudoaffine map, such that*

$$T(x_1) = aT(x_2), a \in \mathbb{R}.$$

If T has exactly one zero on l, then there exist $z_1, z_2 \in l$, such that $T(z_1) = \alpha T(z_2) \neq 0$, where $\alpha < 0$.

Proof Assume that T has exactly one zero at x_0 on l. By making a translation, we may suppose that $x_0 = 0$. It is clear that $T(x_2) \neq 0$, otherwise T would be identically zero on l. Let $V = sp(x_2, T(x_2))$ and P be the orthogonal projection on V. Then, PT is pseudoaffine and for any $x \in l, T(x) = \lambda(x) T(x_2) \in V$. Thus, $PT = T$ on l. Therefore, by Theorem 12.10, there exist a linear map A on V, a vector $u \in V$ and a positive function $g : V \to \mathbb{R}$, such that $PT(x) = g(x)(Ax + u)$ on V. Our assumption $T(0) = 0$ implies that $u = 0$. Since $A(-x_2) = -Ax_2$, therefore, $T(x_2)$ and $T(-x_2)$ have opposite directions.

We now consider the cases, when $T(x_1)$ and $T(x_2)$ are linearly independent:

Proposition 12.19 *Suppose $T : \mathbb{R}^n \to \mathbb{R}^n$ is continuous and pseudoaffine and $x_1, x_2 \in \mathbb{R}^n$ are such that $T(x_1), T(x_2)$ are linearly independent. Assume that $l = l(x_1, x_2)$. Then for any $z \in l, T(z) \in sp(T(x_1), T(x_2))$.*

Proof Since $T(x_1), T(x_2)$ are linearly independent, the set

$$M = \{z \in \mathbb{R}^n : \langle T(x_1), z - x_1 \rangle = 0, \langle T(x_2), z - x_2 \rangle = 0\}$$

is an intersection of two nonparallel hyperplanes and, as such, is an $n - 2$ dimensional affine subspace. Therefore, by the pseudoaffinity of T and for all $z \in M$, we have

$$\langle T(z), z - x_1 \rangle = 0 \ and \ \langle T(z), z - x_1 \rangle = 0.$$

Therefore, for all $t \in \mathbb{R}$, we have

$$\langle T(z), z - (tx_1 + (1-t)x_2) \rangle = 0.$$

Again, by the pseudoaffinity of T, we get

$$\langle T(tx_1 + (1-t)x_2), z - (tx_1 + (1-t)x_2) \rangle = 0, \ \forall z \in M.$$

Therefore,

$$\langle T(w), z - w \rangle = 0, \ \forall w \in l, \forall z \in M. \tag{12.19}$$

Fix $z_0 \in M$. If

$$N = (T(x_1))^\perp \bigcap (T(x_2))^\perp.$$

Then, it is obvious that $M = z_0 + N$. From (12.19), we deduce that $\forall z \in N$,

$$\langle T(w), z + z_0 - w \rangle = 0.$$

By Lemma 12.2, we get

$$T(w) \in N^\perp = sp(T(x_1), T(x_2)), \forall w \in l.$$

12.11   The Case Where T Has at Least One Zero

From the previous discussion, we know that the set of zeros of T is either empty or an affine subspace. We shall begin by showing that Theorem 12.8 is true under the additional assumption that T has exactly one zero. We recall the following two basic results from algebraic topology and projective geometry:

Theorem 12.11 (Schauder domain invariance, Dugundji and Granas [70]) *If U is open in a normed space E and $f : U \to E$ is an injective completely continuous field, then $f(U)$ is open.*

Let L be the lattice of subspaces of $\mathbb{R}^n, n \geq 3$. We recall that an automorphism $F : L \to L$ over L is an onto application, such that for $V_1, V_2 \in L$ with $V_1 \subseteq V - 2$, is equivalent to $F(V_1) \subseteq F(V_2)$. The following fundamental theorem is a simplified form of the theorem of projective geometry.

Theorem 12.12 (Fundamental theorem of projective geometry, Baer [11]) *Suppose that $F : L \to L$ is an automorphism of the lattice of subspaces of $\mathbb{R}^n, n \geq 3$. Then, there exists a linear map A on $\mathbb{R}^n$, such that*

$$F(V) = A(V), \forall V \in L.$$

We know that f is a completely continuous field, if the map $F(x) = x - f(x)$ is continuous and it maps bounded sets to relatively compact sets.

Proposition 12.20 *Suppose that T is continuous and pseudoaffine and has a zero only at 0. Then there exists a positive function g and a linear map A such that*

$$T(x) = g(x) Ax \forall\ x \in \mathbb{R}^n.$$

Proof It is known that the theorem is true for $n = 2$, we may assume that $n \geq 3$. For any straight line l through 0 and any nonzero $y \in l$, $T(y)$ is also nonzero by assumption. By Proposition 12.17, for all $x \in l, T(x) \in sp(T(y))$. Therefore the image of a straight line through 0 is contained in a straight line through 0. Hence, if D is the set of all lines through 0, T defines a map $F_1 : D \to D$ such that for all $x \neq 0, T(x) \in F_1(sp(x))$. It is clear that this map is 1-1. To prove this, we first show that:

$$T(sp(x, y)) \subseteq sp(T(x), T(y)),\ \forall x, y \in \mathbb{R}^n. \tag{12.20}$$

We note that, each $z \in sp(x, y)$ can be written as $z = ax + by$ with $a, b \in R$. By Propositions 12.17 and 12.19, $T(z) \in (sp(T(ax), T(by)))$. Since, we know that $T(ax) \in sp(T(x))$ and $T(ay) \in sp(T(y))$, therefore, $T(z) \in sp(T(x), T(y))$, and (12.20) follows.

First, we show that F_1 is 1-1. On the contrary, we suppose that F_1 is not 1-1. Then there exist x, y linearly independent such that $T(x), T(y)$ are linearly dependent. From (12.20) it follows that $T(sp(x, y)) \subseteq sp(T(x))$. By our assumption on zeros, $T(sp(x, y) \setminus \{0\})$ does not contain 0. In addition, by Proposition 12.18, it contains two points z and z' such that $T(z), T(z')$ have opposite directions. Thus, $T(sp(x, y) \setminus \{0\})$ cannot be connected, while $sp(x, y) \setminus \{0\}$ is connected. This is not possible, since T is continuous. Hence F_1 is 1-1.

Now, we show that F_1 is an onto map. On the contrary, suppose that F_1 is not onto. We note that the set D is a differentiable manifold. In fact D is the real projective $n - 1$ manifold $\mathbb{R}P^{n-1}$. Since D is compact and connected and $F_1(D)$ is continuous, the image $F_1(D)$ is also compact and connected. Thus, if F_1 is not onto, then there exists $l \in D$, such that $F_1(l)$ is a boundary point of the range of F_1. By considering local coordinate systems (U, φ) and (V, ψ) around l and $F_1(l)$, we get the coordinate representation $\tilde{F}_1 \varphi F_1 \psi^{-1}$ of F_1 from the open bounded set $\varphi(U)$ to the open bounded set $\psi(V)$ in $\mathbb{R}^{n-1}$. We note that the image $\tilde{F}_1(U)$ is not open in $\mathbb{R}^n$, since $\psi(F_1(l))$ is a boundary point of $\tilde{F}_1(U)$. However, it is obvious that $\tilde{F}_1$ is a completely continuous field, which is a contradiction to Theorem 12.11.

Let us assume that L be the lattice of subspaces of $\mathbb{R}^n$. We define a map $F : L \to L$ by

$$F(V) = sp(T(V)), V \in L.$$

From the definition of F and relation (12.20), we note that

$$F(sp(x_1, ...x_k)) = sp(T(x_1), ...T(x_k)), \ \forall x_1, ...x_k \in \mathbb{R}^n. \qquad (12.21)$$

Now, we show that the map F is a lattice automorphism.

First we show that it preserves the dimension of subspaces. In fact, for any $V \in L$, let k_1, k_2 be the dimensions of $V, F(V)$. From (12.21) it follows that if $x_1, ..., x_{k_1}$ span V, then $T(x_1), T(x_2), ..., T(x_{k_1})$ span $F(V)$. Hence, $k_2 \leq k_1$. Assume that D_1, D_2 be the sets of lines through the origin in $V F(V)$, respectively. Then D_1, D_2 are differentiable manifolds with dimensions $k_1 - 1, k_2 - 1$, respectively. The map $F_1 : D_1 \to D_2$ is continuous and 1-1. It is well known that this implies $k_2 - 1 \geq k_1 - 1$. Hence, V and $F(V)$ have the same dimension.

Next, we show that F is an onto map. If $U \in L$, let $\{y_1, ...y_k\}$ be a basis of U. Since F_1 is an onto map, there exist $x_1, ...x_k \in \mathbb{R}^n$ such that $F(sp(x_i)) = sp(y_i), i = 1, ...k$. Then from (12.21), we get

$$F(sp(x_1, x_2, ...x_k)) = sp(T(x_1), T(x_2), ...T(x_k)) = U.$$

Finally, we show that F preserves the lattice-theoretic union, i.e., the sum of subspaces. For $V_1, V_2 \in L$, we may find a basis of $V_1 + V_2$ in the form $K_1 \bigcup K_2 \bigcup K_3$, such that $K_1 \bigcup K_3$ is a basis of V_1, $K_2 \bigcup K_3$ is a basis of V_2 and K_3 is a basis of $V_1 \bigcap V_2$ (K_3 is empty if $V_1 \bigcap V_2 = \{0\}$). Then,

$$F\left(V_1 \bigcup V_2\right) = sp(T(K_1), T(K_2), T(K_3))$$

$$= sp(T(K_1), T(K_3)) + sp(T(K_2), T(K_3)) = F(V_1) + F(V_2).$$

Clearly, for $V_1, V_2 \in L, V_1 \subseteq V_2 \Rightarrow F(V_1) \subseteq F(V_2)$. Now, we show that the converse is true. If $F(V_1) \subseteq F(V_2)$, then $F(V_1 + V_2) = F(V_1) + F(V_2) = F(V_2)$. It follows that $V_2, F(V_2), F(V_1 + V_2), V_1 + V_2$ all have the same dimension. Hence, $V_2 = V_1 + V_2$, which implies that $V_1 \subseteq V_2$. Thus $V_1 \subseteq V_2$ is equivalent to $F(V_1) \subseteq F(V_2)$.

Thus, we have shown that the map F is an onto automorphism over the lattice of subspaces of $\mathbb{R}^n, n \geq 3$, which imply that F is a lattice automorphism. By Theorem 12.12, there exists a linear map A such that for all $V \in L, F(V) = A(V)$. In particular, for any $x \neq 0$, it follows that

$$T(x) \in sp(T(x)) = F(sp(x)) = A(sp(x)).$$

Hence, there exists a number $g(x) \neq 0$ such that $T(x) = g(x)Ax$. The function g is obviously continuous on $\mathbb{R}^n \setminus \{0\}$, thus it has a constant sign. By changing the sign, we may suppose that g is positive. We set $g(0) = 1$ and the result follows.

Theorem 12.13 *Assume that T has at least one zero. Then, there exist a positive function g, a skew-symmetric map A, and $u \in \mathbb{R}^n$, such that*

$$T(x) = g(x)(Ax + u), x \in \mathbb{R}^n.$$

Proof In view of Proposition 12.12, it is sufficient to show that T has the form (12.10) with g positive and A linear. Choose x_0, such that $T(x_0) = 0$ and set

$$T_1(x) = T(x + x_0), x \in \mathbb{R}^n.$$

Then, T_1 is pseudoaffine, and by Proposition 12.11, its set of zeros is a subspace V. We proceed by induction on the dimension of V to show that T_1 has the form (12.10) with $u - 0$. By Proposition 12.20, this is true if the dimension of V is 0. We suppose that it is true if the dimension is $k - 1$ and proceed to show that it is also true if the dimension is k. Choose $e \in V \setminus \{0\}$.

We first show that the subspace $\{e\}^\perp$ is invariant under T_1. Indeed, if $x \in \{e\}^\perp$, then $\langle T_1(e), x - e \rangle = 0$, hence, $\langle T_1(x), x - e \rangle = 0$. By Lemma 12.2, $T_1(x) \in \{e\}^\perp$. The set of zeros of the restriction of T_1 on $\{e\}^\perp$ has dimension $k - 1$. Since the theorem is true in this case, there exists a linear map A on $\{e\}^\perp$ and a nonnegative function g on $\{e\}^\perp$, such that

$$T_1(x) = g(x)Ax, x \in \{e\}^\perp. \tag{12.22}$$

Suppose that P is the orthogonal projection on $\{e\}^\perp$. For any $x \in \mathbb{R}^n$, x and Px are on a straight line parallel to e, and by Proposition 12.16, we know that $T_1(x)$ and $T_1 P(x)$ have the same direction, i.e., there exists a positive function $g_1 : R^n \to \mathbb{R}$ such that

$$T_1(x) = g_1(x)T_1 P(x), x \in R^n.$$

Combining with (12.22) we get

$$\forall x \in \mathbb{R}^n, T_1(x) = g_1(x)T_1P(x) = g_1(x)g(Px)AP(x).$$

It follows that

$$\forall x \in \mathbb{R}^n, T_1(x) = g_1(x - x_0)g(P(x - x_0)) = (APx - APx_0).$$

Hence, T has the form (12.10).

12.12   The Case $n = 3$

In this section we present the proof for Theorem 12.8, for $n = 3$ established by Bianchi *et al.* [26]. Since, we know that the theorem is true if T has at least one zero, we may suppose that T has no zeros.

Proposition 12.21 *Suppose* $T : \mathbb{R}^3 \to \mathbb{R}^3$ *is a continuous pseudoaffine map that has no zeros and let* T *have a constant direction on a straight line* $l = \{y \in R^n : y = te, t \in R\}$. *Let* $x' \neq 0$ *be such that,*

$$\langle T(0), x' \rangle = 0, \langle e, x' \rangle = 0.$$

Then T *has a constant direction on* $l' = x' + l$.

Proof We consider the following two cases:
Case (1) If $\langle T(0), e \rangle = 0$. Then, for any $x' + se \in l'$ and any $te \in l, \langle T(0), x' + se - te \rangle = 0$. Since $T(te)$ and $T(0)$ are linearly dependent, we deduce that $\langle T(te), x' + se - te \rangle = 0$. By the pseudoaffinity of T, we get

$$\langle T(x' + se), x' + se - te \rangle = 0, \forall t \in \mathbb{R}.$$

Hence, $\langle T(x' + se), x' \rangle = 0$ and $\langle T(x' + se), e \rangle = 0$. Since the space is three-dimensional, $T(0)$ and $T(x' + se)$ are linearly dependent, for each $s \in \mathbb{R}$. Since, T has no zeros, it has a constant direction on l'.

Case (2) If $\langle T(0), e \rangle \neq 0$. Suppose to the contrary that $T(x')$ and $T(x' + se)$ are linearly independent for some $s \in R \backslash \{0\}$. Thus we can find $z \in \mathbb{R}^3$ such that $\langle T(x'), z - x' \rangle = 0$ and $\langle T(x' + se), z - (x' + se) \rangle = 0$. By the pseudoaffinity of T, it follows that

$$\langle T(z), z - x' \rangle = 0 \text{ and } \langle T(z), z - (x' + se) \rangle = 0.$$

Thus,

$$\langle T(z), e \rangle = 0.$$

It follows that $z \notin l$ since $z \in l$ would imply by assumption that $T(z)$ and $T(0)$ have the same direction, while $\langle T(0), e \rangle \neq 0$. We now consider a plane V containing l and z and let P_V the orthogonal projection on V. The map $P_V T$ is pseudoaffine with a constant direction on l and thus by Proposition 12.15 with a constant direction on V. But $\langle P_V T(z), e \rangle = \langle T(z), e \rangle = 0$ and $\langle P_V T(0), e \rangle = \langle T(0), e \rangle \neq 0$, are a contradiction. Hence $T(x')$ and $T(x' + se)$ are linearly dependent for all $s \in \mathbb{R}$, and as before this implies that T has a constant direction on l'.

Employing Borsuk's antipodal theorem (see Dugundji and Granas [70]), Bianchi *et al.* [26] have established the following characteristics of a continuous pseudoaffine map with no zeros.

Proposition 12.22 *Suppose that* $T : \mathbb{R}^3 \to \mathbb{R}^3$ *is a continuous pseudoaffine map with no zeros. There exists a straight line* $l = \{y \in \mathbb{R}^n : y = \bar{x} + te. t \in R\}$ *such that* $T(y) = \|T(y)\| e$, *for all* $y \in l$.

Proof By Borsuk's antipodal theorem, there exists $w \in \mathbb{R}^n$ with $\|w\| = 1$ such that $T(-w) = \lambda T(w)$, $\lambda > 0$. Hence, by Proposition 12.17, there exists $v \neq 0$ such that

$$T(y) = \|T(y)\| v, \forall y \in l' = \{tw, t \in \mathbb{R}\}.$$

We consider the subspace $M_1 = \{x \in \mathbb{R}^3 : \langle x, w \rangle = 0\}$ and the orthogonal projection P_1 on M_1. If for some $x_1 \in M_1$, we have $P_1 T(x_1) = 0$, then $T(x_1) = \beta w$, for some $\beta \in \mathbb{R}$ and $\langle Tx_1, x_1 \rangle = 0$. By the pseudoaffinity of T, we get $\langle T(0), x_1 \rangle = 0$. Using Proposition 12.20, it follows that T has a constant direction on the line $l = x_1 + l'$. Therefore, for all $y \in l'$, we have $T(y) = \|T(y)\| \sigma w$, where $\sigma = 1$ or -1. Setting, $e = \sigma w$, the desired result follows.

Let us assume that $P_1 T(x) \neq 0$, for each $x \in M_1$. Then, by case (1) of Theorem 12.10, the pseudomonotone map $P_1 T$ has a constant direction on M_1. Consider any $y \in M_1$ with $\langle y, T(0) \rangle \neq 0$. We note that $T(0)$ is not orthogonal, as by our assumption $P_1 T(0) \neq 0$. Therefore, there exists $\alpha > 0$ such that

$$P_1 T(y) = \alpha P_1 T(0).$$

Now, we consider the two-dimensional subspace M_2 containing y and w and let P_2 be the orthogonal projection on M_2. Since, $l' \subseteq M_2$, by Proposition 12.15, the pseudoaffine map $P_2 T$ has a constant direction on M_2. Hence, there exists $\beta > 0$, such that

$$P_2 T(y) = \beta P_2 T(0).$$

Assume that $\{e_1, e_2, e_3\}$ is an orthonormal basis with $e_1 = w_1, e_2 = \frac{y}{\|y\|}$ and let (a, b, c) and (a_1, b_1, c_1) be the coordinates of $T(0)$ and $T(y)$, respectively with respect to this basis. Then $(0, b_1, c_1) = \alpha (0, b, c)$ and $(a_1, b_1, 0) = \beta (a, b, 0)$ and $y \neq 0$.

Therefore, we have $\beta = \alpha$ and $T(y) = \alpha T(0)$, which means that $T(y)$ and $T(0)$ has the same direction. This holds for all $y \in M_1$ such that $\langle y, T(0) \rangle \neq 0$ and by continuity it holds for all $y \in M_1$. Applying again Proposition 12.15, it follows that T has a constant direction on $\mathbb{R}^3$ and choosing $x_0 = 0$ and $e = \frac{T(0)}{\|T(0)\|}$. This completes the proof.

Theorem 12.14 *Suppose* $T : \mathbb{R}^3 \to \mathbb{R}^3$ *is a continuous pseudoaffine map. Then there exists a vector* $u \in \mathbb{R}^3$, *a skew-symmetric linear map* A, *and a positive function* $g : \mathbb{R}^3 \to \mathbb{R}$, *such that*

$$T(x) = g(x)(Ax + u).$$

Proof Suppose that T has no zeros. Let l be the straight line, whose existence is asserted in Proposition 12.21. Let us assume that $x_0 \neq 0$. We consider an orthonormal basis $\{e_1, e_2, e_3\}$, such that $e_3 = e$ and let x_1, x_2, x_3 be the coordinates of $x \in \mathbb{R}^3$ in this basis. Let $M_i =: \{x \in \mathbb{R}^3 : \langle x, e_i \rangle = 0\}$ and P_i be the orthonormal projection on M_i. Then $P_3 T$ is pseudoaffine on M_3 and $P_3 T(0) = 0$ on M_3. Then, T has a constant direction on M_3. Hence, by Proposition 12.17, it has a constant direction on $\mathbb{R}^3$. Hence, (12.10) holds with $A = 0$ and $u = e_3$. Otherwise, by case (2) of Theorem 12.10, there exists a function with a constant sign $g : \mathbb{R}^2 \to \mathbb{R}$ is continuous on $\mathbb{R}^2/\{0\}$, such that

$$P_3 T(x_1, x_2, 0) = g(x_1, x_2) A(x_1, x_2, 0),$$

where
$$A\left(x_1, x_2, x_3\right) = \left(x_2, -x_1, 0\right). \tag{12.23}$$

Define the map $T' : \mathbb{R}^3 \to \mathbb{R}^3$, by

$$T'\left(x_1, x_2, x_3\right) = \frac{T\left(x_1, x_2, 0\right)}{g\left(x_1, x_2\right)}.$$

For any $x = \left(x_1, x_2, x_3\right) \in \mathbb{R}^3$, by Proposition 12.20, T has a constant direction on the line $x + l$. Hence,

$$T\left(x_1, x_2, x_3\right) = \frac{\left\|T\left(x_1, x_2, x_3\right)\right\|}{\left\|T\left(x_1, x_2, 0\right)\right\|} T\left(x_1, x_2, 0\right)$$

$$= \frac{\left\|T\left(x_1, x_2, x_3\right)\right\|}{\left\|T\left(x_1, x_2, 0\right)\right\|} g\left(x_1, x_2\right) T'\left(x_1, x_2, x_3\right). \tag{12.24}$$

We note that, T and T' always have the same direction or opposite direction. T' is pseudoaffine. In addition, it is continuous on $\mathbb{R}^3/l$ and $P_3 T'\left(x\right) = Ax$, for all $x \in M_3$. Hence,

$$T'\left(x\right) = \left(x_2 - x_1, f\left(x_1, x_2\right)\right), \quad \forall x \in \left(x_1, x_2, 0\right) \in M_3, \tag{12.25}$$

where $f : \mathbb{R}^2 \to \mathbb{R}$ is continuous on $\mathbb{R}^2/\{0\}$.

Now, take any $x_2 \neq 0, x_1 \neq 0$ and consider the affine subspace $M_2 + x_2 e_3$. Then, T' is constant on the line $l + x_2 e_2$, which belongs to M_2. Therefore, by Proposition 3.5, $P_2 T'$ has a constant direction on $M_2 + x_2 e_2$. Thus, the vectors

$$P_2 T'\left(x_1, x_2, 0\right) = \left(x_2, 0, f\left(x_1, x_2\right)\right) \ \ and \ \ P_2 T'\left(0, x_2, 0\right) = \left(x_2, 0, f\left(0, x_2\right)\right)$$

have the same direction. It follows that $f\left(x_1, x_2\right) = f\left(0, x_2\right)$. Similarly, we can prove that

$$f\left(x_1, x_2\right) = f\left(x_1, 0\right).$$

Since, x_1 and x_2 are arbitrary, for any $x_1, x_2, x_1', x_2' \in \mathbb{R}\backslash\{0\}$, it follows that

$$f\left(x_1, x_2\right) = f\left(x_1, 0\right) = f\left(x_1, x_2'\right) = f\left(0, x_2'\right) = f\left(x_1', x_2'\right).$$

Continuity implies that f is constant on $\mathbb{R}^2 \backslash \{0\}$. Therefore, combining (12.23) and (12.25), we note that T' has the form

$$T'x = Ax + \lambda e_3, \ x \in M_3 \backslash \{0\}.$$

Setting $g\left(x_1, x_2, x_3\right) = \frac{\|T(x_1, x_2, x_3)\|}{\|T(x_1, x_2, 0)\|} g\left(x_1, x_2\right)$. From (12.24) and for all $x \in \mathbb{R}^3 \backslash l$, we get

$$T\left(x\right) = g_1\left(x\right)\left(Ax + \lambda e_3\right).$$

By changing the sign of A and λ, if necessary, we can take g_1 to be positive. Since, T has no zeros, $Ax + \lambda e_3$ is never zero on $\mathbb{R}^3$ and its norm has a positive infimum. Thus,

$$g_1(x) = \frac{\|T(x)\|}{\|Ax + e_3\|}, x \in \mathbb{R}^3 \backslash l$$

has a positive constant extension g on $\mathbb{R}^3$. By continuity,

$$T(x) = g(x)(Ax + e_3), \quad x \in \mathbb{R}^3.$$

This completes the proof.

12.13   The Case Where T Has No Zeros

In this section, we present the proof for Theorem 12.8 derived by Bianchi *et al.* [26] for maps that have no zeros. This is the only case left unproved so far. First, we obtain some more information on the image of a straight line through T (see also Proposition 12.19).

Proposition 12.23 *Suppose that T is continuous and pseudoaffine. Suppose, further that $x \neq y \in \mathbb{R}^n$ and $l = l(x, y)$. If $T(x), T(y)$ are linearly independent, then $R^{++}T(l)$ is an open half-space in the plane* $\mathrm{sp}\,(T(x), T(y))$.

Proof By Proposition 12.19, we know that $T(l) \subseteq \mathrm{sp}\,(T(x), T(y))$. Without loss of generality, we suppose that $x = 0$. Assume that V is the subspace $\mathrm{sp}\,(y, T(0), T(y))$ and P is the orthogonal projection on V. Then PT is pseudoaffine on V and $PT = T$ on $sp(y)$. Thus, by Theorem 12.14, PT has the form

$$PT(z) = g(z)(Az + u), z \in V$$

with g positive and $A : V \to V$ is linear. Consequently, we get

$$PT(ty) = T(ty) = g(ty)(tAy + u).$$

Hence, we have $T(l) = \{g(ty)(tAy + u) : t \in R\}$. Since, $T(0)$ and $T(y)$ are linearly independent, u and Ay are also linearly independent. Then,

$$\mathbb{R}^{++}T(l) = \{tsAy + su : t \in \mathbb{R}, s \in \mathbb{R}^{++}\}$$

is obviously a half-space.

Bianchi *et al.* [26] have investigated the sets where T has a constant direction. The following proposition generalizes both Propositions 12.11 and 12.17:

Proposition 12.24 *Suppose that* $T : \mathbb{R}^n \to \mathbb{R}^n$ *is a continuous pseudoaffine map and* $z_1, z_2, ..., z_m \in R^n$ *is such that* $T(z_i) \in sp(v)$, $i = 1, ...m$, *for some* $v \in \mathbb{R}^n$. *Then* $T(z) \in sp(v)$ *for all* z *on the affine subspace* M *generated by* $z_1, ..., z_m$. *In particular, if* T *has no zeros, then* T *has a constant direction on* M.

Proof We proceed by induction. If $m = 2$, the result follows by Proposition 12.17. Suppose that it is true for $m - 1$ and let $z = \sum_{i=1}^{m} a_i z_i$ with $\sum_{i=1}^{m} a_i = 1, a_i \neq 0$ and $T(z_i) \in sp(v)$, $i = 1, 2...m$. At least one of the $a_i's$ is different from 1. If, say, $a_1 \neq 1$, then $z = a_1 z_1 + (1 - a_1) z_1'$, where $z_1' = \sum_{i=2}^{m} \frac{a_i}{1-a_1} z_i$ belongs to the affine subspace generated by $z_2, ...z_m$. By assumption, $T\left(z_1'\right) \in sp(v)$. The proposition follows by again applying Proposition 12.17.

The following proposition generalizes Proposition 12.20:

Proposition 12.25 *Let* T *be continuous and pseudoaffine with no zeros. If* T *has a constant direction on a straight line* l, *then it has a constant direction on any straight line* l' *parallel to* l.

Proof We may assume with no loss of generality that l contains the origin. Let $T(w)$ have the same direction with a vector u for all $w \in l$. Consider the two-dimensional subspace V generated by l, l' and let $x_0 \in V \backslash l$. If x is any point on V not belonging to the line $l + x_0$, then the straight line through x_0 and x intersects l at some point y. By Propositions 12.17 and 12.19, $T(x) \in sp(T(x_0), T(y)) = sp(T(x_0), u)$. By continuity, it follows that the same holds for any $x \in V$.

Remark 12.5 *Note that if* $T(x')$ *has the same direction with* u, *for some* $x' \in V \backslash l$, *then, by Proposition 12.23,* T *has a constant direction on* V. *Thus,* T *has a constant direction on* l'. *Hence, we may assume that* $T(x_0)$ *has a direction different from* u, *for all* $x' \in V' \backslash l'$. *Now consider any straight line* l_1 *in* V *intersecting* l' *at* z. *Then* $T(l)$ *and* $T(l')$ *also intersect at* $T(z)$, *and a fortiori* $R^{++}T(l_1)$ *and* $R^{++}T(l')$ *intersect. Since* l_1, *intersects* l *by assumption* T *has not a constant direction on* l_1, *thus, by Proposition 12.22,* $R^{++}T(l_1)$ *is an open half-space in the two-dimensional space* $sp(T(x_0), u)$. *If* $R^{++}T(l')$ *is also an open half-space, then there exists another common point of* $R^{++}T(l_1)$ *and* $R^{++}T(l')$; *linearly independent from* $T(z)$. *Thus, there exist two points* $z_1 \in l_1$, *and* $z' \in l'$ *such that* $T(z)$, *and* $T(z')$, *have the same direction. Let* l_2 *be the straight line joining* z_1, z'. *Then* T *has a constant direction on* l_2. *But* l_2 *intersects* l. *Thus* T *has the direction* u *on* l_2. *This contradicts our assumption on the direction of* T *outside* l. *Thus,* $R^{++}T(l')$ *is not an open half-space. Hence by Proposition 12.22, for any* $w, w' \in l', T(w), T(w')$ *are linearly dependent. Since by our assumption* T *has no zeros,* T *has a constant direction on* l'.

Proposition 12.26 *Suppose* T *is pseudoaffine and continuous with no zeros and let* M *be the set of all* $x \in \mathbb{R}^n$, *such that* $T(x)$ *has the same direction* as $T(0)$. Then M is a nontrivial subspace. In addition, if we set $N = M^{\perp}$, then for any distinct $x, y \in N$, $T(x), T(y)$ are linearly independent. Finally, if $x, y,$ and 0 are all distinct, then $T(0) \notin sp(T(x), T(y))$.

Proof By Proposition 12.23, the set M is a subspace. Proceeding as in the beginning of the proof of Proposition 12.21, we deduce that M contains at least one line through the origin. We show that $T(x), T(y)$ are linearly independent. Indeed, if they are linearly dependent then by Proposition 12.17, T has a constant direction on the straight line $l \subseteq N$ through x, y. Then by Proposition 12.24, T has a constant direction on the straight line l' through 0 which is parallel to l. Since $l' \subseteq N$ this means that M and N intersect at l', a contradiction.

Now suppose that $x, y \neq 0$ and $T(0) \in sp(T(x), T(y))$. For any $z \in sp(x, y)$, let $a, b \in \mathbb{R}$ be such that $z = ax + by$. Applying Proposition 5.4 we find $T \ T(2ax) \in sp(T(0), T(x)) \subseteq sp(T(x), T(y))$, and likewise $T(2by) \subset sp(T(x), T(y))$. Since $z = \frac{1}{2}2ax + \frac{1}{2}2by$, we also find $T(z) \in sp(T(2ax), T(2by)) \subseteq sp(T(x), T(y))$. Hence the restriction of T on $sp(x, y)$ is a continuous map of $sp(x, y)$ into the two-dimensional subspace $sp(T(x), T(y))$. By the first part of the proof, this map sends distinct points to linearly independent points. If we consider any simple continuous closed curve C in $sp(x, y)$, then its image through T will be a simple continuous closed curve in the two-dimensional space $sp(T(x), T(y))$. It is easy to see that at least two points of the latter curve are linearly dependent, which is a contradiction.

Now, we present the proof for the necessary part of Theorem 12.8.

The Necessary part of Theorem 12.8: According to Theorem 2.4 we may suppose that T is continuous. According to Theorems 12.10, 12.13, 12.14, and the remark at the end of Section 12.5, we may suppose that $n > 3$ and T has no zeros. Let M and N be as in Proposition 12.25. For any straight line $l \subseteq N$ through 0 we know by Propositions 12.19 and 12.25 that $T(l)$ is included in a subspace of dimension 2: Let V_l be the subspace generated by l and $T(l)$ and let P_l be the orthogonal projection on V_l: Then V_l is two- or three-dimensional, $P_l T$ is pseudoaffine on V_l and $P_l T = T$ on l. Thus, we know that there exist a positive function $g_l : V_l \to \mathbb{R}$, a linear map $A_l : V_l \to V_l$, and a vector u_l such that $\forall z \in V_l, P_l T z = g_l(z)(A_l z + u_l)$. We can choose g_l so that $g_l(0) = 1$. In particular, on l it follows that

$$T(x) = g_l(z)(A_l z + u_l), \quad \forall x \in l. \tag{12.26}$$

For $x = 0$ we get $T(0) = u_l$, so u_l does not depend on l and we shall denote it by u. Now define the positive function $g : N \to \mathbb{R}$ by $g(0) = 1$ and $g(x) = g_{sp(x)}x, \forall x \in N \setminus \{0\}$. Then (12.26) gives

$$T(x) = gx\left(A_{sp(x)}x + u\right), \forall x \in N,$$

where we put $A_{sp(0)}0 = 0$. If we further define the map $T' : N \to \mathbb{R}^n$ by $T'(x) = T(x)/g(x)$, then T' is pseudoaffine. Finally, if we define the map

$A : N \to \mathbb{R}^n$ by $Ax = A_{sp(x)}x$, for all $x \in N$, we deduce

$$T'(x) = Ax + u. \forall x \in N, \tag{12.27}$$

Now, we show that map A is linear. Note that for all $t \in \mathbb{R}, x \in N$, we have

$$A(tx) = A_{sp(x)}(tx) = tA_{sp(x)}x = tAx. \tag{12.28}$$

Now, we need to show that for any $x, y \in N, A(x + y) = Ax + Ay$. This is evident if x, y are linearly dependent, because then they belong to the same straight line through the origin. Thus, we suppose that they are linearly independent. First we show that for any linearly independent $x, y \in N$ and any $t \in \mathbb{R}, A(tx + (1 - t)y)$ belongs to the straight line joining $T(x), T(y)$. By Proposition 5.19 and the definition of T', there exist $a, b \in R$ (depending on x, y, t), such that

$$T'(tx + (1 - t)y) = aT'(x) + bT'(y).$$

Therefore,

$$A(tx + (1 - t)y) + u = a'Ax + b'Ay + (a' + b')u. \tag{12.29}$$

The above relation is true for all x, y, t. If we take $2x, 2y$ instead of x, y and the same t we deduce that there exist $a', b' \in R$ such that

$$A(t2x + (1 - t)2y) + u = a'A2x + b'A2y + (a' + b')u.$$

Taking into account (12.28), it follows that

$$2A(tx + (1 - t)y) + u = 2a'Ax + 2b'Ay + (a' + b')u \tag{12.30}$$

From (12.29) and (12.30), we get

$$(a - a')Ax + (b - b')Ay = \left(-a - b + \frac{a' + b' + 1}{2}\right)u. \tag{12.31}$$

Now, we prove that $u \notin sp(Ax, Ay)$ and that Ax, Ay are linearly independent. Indeed, suppose first that $u = a''Ax + b''Ay$. Then a'', b'' cannot both be zero, since $u = T(0) \neq 0$. From (12.27) we deduce that

$$u = a''T'x + b''T'y - (a'' + b'')u. \tag{12.32}$$

Note that $a'' + b'' \neq -1$, otherwise, (12.32) would imply that $T'(x)$ and $T'(y)$ are linearly dependent, and this is excluded by Proposition 12.25. Solving (12.32) with respect to u implies that $u \in sp(T'x, T'y)$; thus $u \in sp(Tx, Ty)$. But this contradicts Proposition 12.25. Hence, $u \notin sp(Ax, Ay)$. Writing $Ax = T'(x) - u$ and $Ay = T'(x) - u$ shows easily that Ax, Ay are linearly independent. Then (12.31) implies that $a = a', b = b', a + b = 1$. From

(12.29), it follows that $A\left(tx + (1 - t)y\right) = aAx + bAy$ where $a + b = 1$. Thus, if $w = \frac{1}{2}\left(x + y\right)$, then $\exists \alpha, \beta \in R$ with $\alpha + \beta = 1$ and

$$Aw = \alpha Ax + \beta Ay. \tag{12.33}$$

For any $0 < t < 1, w = t\left(\frac{x}{2t}\right) + (1 - t)\left(\frac{y}{2(t-1)}\right)$ is also in the straight line joining $\frac{x}{2t}$ and $\frac{y}{2(t-1)}$. Hence,

$$Aw = \alpha\left(t\right)A\left(\frac{x}{2t}\right) + \beta\left(t\right)A\left(\frac{y}{2\left(t - 1\right)}\right) = \left(\frac{\alpha\left(t\right)}{2t}\right)Ax + \left(\frac{\beta\left(t\right)}{2\left(t - 1\right)}\right)Ay,$$

where $\alpha\left(t\right), \beta\left(t\right)$ are functions of t, such that $\alpha\left(t\right) + \beta\left(t\right) = 1$. Since Ax, Ay are linearly independent, a comparison with (12.33) shows that

$$\alpha\left(t\right) - 2t\alpha, \beta\left(t\right) = 2\left(1 - t\right)\beta.$$

Therefore, $2t\alpha + 2\left(1 - t\right)\beta = 1, 0 < t < 1$ and this implies $\alpha = \beta = \frac{1}{2}$. Since,

$$A\left(x + y\right) = 2A\left(\frac{1}{2}\left(x + y\right)\right) = 2Aw = Ax + Ay$$

Therefore, A is linear.

It follows from (12.27) that

$$T\left(x\right) = g\left(x\right)\left(Ax + u\right), \quad \forall x \in N, \tag{12.34}$$

where $A : N \to \mathbb{R}^n$ is linear and $g : N \to \mathbb{R}$ is positive.

Assume that P is the orthogonal projection on N. Then for any $x \in \mathbb{R}^n$, the straight line l joining x and Px is parallel to a line $l' \subseteq M$. Since T has a constant direction on l', it also has a constant direction on l by Proposition 12.25. Hence $T\left(x\right)$ and $T\left(Px\right)$ have the same direction. Therefore, for any $x \in \mathbb{R}^n$, there exists $g_1\left(x\right) > 0$ such that $T\left(x\right) = y_1\left(x\right)T\left(Px\right)$. Using (12.34), we get

$$T\left(x\right) = g_1\left(x\right)g\left(Px\right)\left(APx + u\right), \quad \forall x \in \mathbb{R}^n,$$

where $AP : \mathbb{R}^n \to \mathbb{R}^n$ is linear and $g_1\left(x\right)g\left(Px\right)$ is positive. Proposition 12.12 completes the proof of the theorem.

Chapter 13

η-Pseudolinear Functions: Characterizations of Solution Sets

13.1 Introduction

For many problems encountered in economics and engineering the notion of convexity no longer suffices. To meet this demand and the aim of extending the validity of sufficiency of the Karush-Kuhn-Tucker conditions for differentiable mathematical programming problems, the notion of invexity was introduced by Hanson [99] and named by Craven [53]. Hanson [99] introduced a new class of functions, which generalizes both the classes of invex and pseudoconvex functions, called η-pseudoconvex functions by Kaul and Kaur [137] and pseudoinvex functions by Craven [52]. Ben-Israel and Mond [21] have characterized that a real-valued function f is invex if and only if every stationary point is a global minimizer and pointed out that the classes of pseudoinvex functions and invex functions coincide. In 1988, Weir and Mond [283] introduced a new class of functions known as preinvex functions. Differentiable preinvex functions are invex with respect to the same vector function η, however, the converse holds true only if the vector function η satisfies condition C due to Mohan and Neogy [205]. Rueda [240] studied the properties of functions f, for which f and $-f$ are both pseudoinvex with respect to same vector functions η. Later, Ansari *et al.* [5] called such functions η-pseudolinear functions and derived several characterizations for the solution set of η-pseudolinear programming problems. Giorgi and Rueda [91] have derived the optimality conditions for a vector optimization problem involving η-pseudolinear functions and established that every efficient solution is properly efficient under certain boundedness conditions. Zhao and Yang [299] have used the linear scalarization method to establish the equivalence between efficiency and proper efficiency of a multiobjective optimization problem using the properties of η-pseudolinear functions.

Recently, Zhao and Tang [298] and Ansari and Rezaei [4] studied the properties of the class of nondifferentiable locally Lipschitz η-pseudolinear functions and derived several characterizations for the solution sets of nondifferentiable η-pseudolinear programming problems using the Clarke subdifferential. Mishra and Upadhyay [197, 198] established the optimality conditions and duality results for nondifferentiable η-pseudolinear programming problems using the Clarke subdifferential. Indeed, every pseudolinear function is a η-pseudolinear function, for $\eta(y,x) = y - x$, but the converse is not necessarily true. Hence, the class of η-pseudolinear functions is a natural generalization of the class of pseudolinear functions.

In Chapter 2, we studied the characterizations of differentiable and non-differentiable locally Lipschitz pseudolinear functions. In Chapter 3, we derived the characterizations of the solution sets of constrained optimization problems involving differentiable and nondifferentiable locally Lipschitz pseudolinear functions. Moreover, in Chapter 5, we established certain optimality conditions and duality results for multiobjective optimization problems involving differentiable pseudolinear functions. The results in this chapter will extend and generalize naturally the results of Chapters 2, 3, and 5, to a more general class of functions as well as to a more general class of optimization problems. More specifically, in this chapter, we present the characterizations for differentiable and nondifferentiable η-pseudolinear functions. We present the characterizations for the solution sets of constrained optimization problems involving differentiable and nondifferentiable η-pseudolinear functions. Furthermore, the optimality conditions and equivalence between efficient and properly efficient solutions of the multiobjective optimization problems involving differentiable η-pseudolinear functions are given.

13.2   Characterization of Differentiable η-Pseudolinear Functions

Let $\mathbb{R}_{++}$ be the positive orthant of $\mathbb{R}$ and let $X \subseteq \mathbb{R}^n$ be a nonempty set equipped with the Euclidean norm $\|.\|$.

Definition 13.1 (Invex set, [283]) *A nonempty set $X \subseteq \mathbb{R}^n$ is said to be an invex set with respect to a vector function $\eta : X \times X \to \mathbb{R}^n$, if*

$$x + \lambda\eta(y,x) \in X, \quad \forall\, \lambda \in [0,1],\ \forall\, x,y \in X.$$

Definition 13.2 (Hanson [99]) *Let $X \subseteq \mathbb{R}^n$ be an open set and $F : X \to \mathbb{R}$ be a differentiable function on X. Then f is said to be an invex function at $x \in X$, with respect to a vector function $\eta : X \times X \to \mathbb{R}$, if*

$$f(y) - f(x) \geq \langle \nabla f(x), \eta(y,x) \rangle, \forall y \in X.$$

The function f is said to be an invex with respect to $\eta : X \times X \to \mathbb{R}$, on X, if f is invex η for each $x \in X$.

Weir and Mond [283] introduced the following function:

Definition 13.3 (Preinvex function) *A function* $f : X \to \mathbb{R}$ *is said to be preinvex on* X, *if there exists a vector function* $\eta : X \times X \to \mathbb{R}^n$, *such that*

$$f(x + \lambda\eta(y, x)) \leq (1 - \lambda)f(x) + \lambda f(y), \quad \forall \lambda \in [0, 1], \ \forall x, y \in X.$$

Differentiable preinvex functions are invex with respect to the same vector function $\eta : X \times X \to \mathbb{R}^n$. However, the converse holds true only if η satisfies the following condition C due to Mohan and Neogy [205].

Condition C Let X be an invex set. The function $\eta : X \times X \to \mathbb{R}^n$ is said to satisfy condition C, if
(i) $\eta(x, x + t\eta(y, x)) = -t\eta(y, x)$; (ii) $\eta(y, x + t\eta(y, x)) = (1 - t)\eta(y, x)$.

Remark 13.1 *Yang et al. [290] have shown that*

$$\eta(x + t\eta(y, x), x) = t\eta(y, x).$$

Example 13.1 (Mohan and Neogy [205]) *Consider the set* $X = [-7, -2] \cup [2, 10]$. *Then set* X *is an invex set with respect to* η *given by*

$$\eta(x, y) = \begin{cases} x - y, & \text{if } x \geq 0, y \geq 0, \\ x - y, & \text{if } x \leq 0, y \leq 0, \\ -7 - y, & \text{if } x \geq 0, y \leq 0, \\ 2 - y, & \text{if } x \leq 0, y \geq 0. \end{cases}$$

It is easy to see that the function η *satisfies condition C.*

Rueda [240] studied the properties of the functions, which are both pseudoinvex and pseudoincave with respect to the same η. Ansari *et al.* [5] studied the properties of such functions in detail and called them η-pseudolinear functions.

Analytically, differentiable pseudoinvex and pseudoincave functions with respect to vector function $\eta : X \times X \to \mathbb{R}^n$ are defined as follows:

Definition 13.4 *Let* $\eta : X \times X \to \mathbb{R}$ *be a differentiable function on an open set* X. *The function* f *is said to be*
(i) pseudoinvex at $x \in X$ *with respect to* $\eta : X \times X \to \mathbb{R}^n$ *(or* η-*pseudoconvex), if for all* $y \in X$,

$$\langle \nabla f(x), \eta(y, x) \rangle \geq 0 \Rightarrow f(y) - f(x);$$

(ii) pseudoincave at $x \in X$ *with respect to* $\eta : X \times X \to \mathbb{R}^n$ *(or* η-*pseudoconcave), if for all* $y \in X$,

$$\langle \nabla f(x), \eta(y, x) \rangle \leq 0 \to f(y) \leq f(x).$$

The function f *is said to be pseudoinvex (pseudoincave) on* X, *if it is pseudoinvex (pseudoincave) at every* $x \in X$.

The function f *is said to be* η-*pseudolinear on* X, *if* f *is both pseudoinvex and pseudoincave on* X *with respect to the same* η. *To be precise; let* $f : X \to \mathbb{R}$

be a differentiable function on an open set $X \subseteq \mathbb{R}^n$, the function f is said to be η-pseudolinear on X, if for all $x, y \in X$, one has

$$\langle \nabla f(x), \eta(y, x) \rangle \geq 0 \rightarrow f(y) \geq f(x)$$

and

$$\langle \nabla f(x), \eta(y, x) \rangle \leq 0 \rightarrow f(y) \leq f(x).$$

It is clear that, every pseudolinear function is η-pseudolinear for $\eta(y, x) = y - x$, but the converse is not true as illustrated by the following example by Ansari *et al.* [5].

Example 13.2 *Let* $X = \left\{ (x_1 - x_2) \in R^2 : x_1 > -1, -\frac{\pi}{2} \leq x_2 \leq \frac{\pi}{2} \right\}$ *and* $\eta :$ $X \times X \rightarrow \mathbb{R}^2$ *be defined as follows:*

$$\eta(y, x) = \left(y_1 - x_1, \frac{\sin y_2 - \sin x_2}{\cos x_2} \right), \; \forall \; x = (x_1, x_2) \; y = (y_1, y_2) \in X.$$

Then, the function $f : X \rightarrow \mathbb{R}$ *is defined by*

$$f(x) = x_1 + \sin x_2, \; \forall \; x = (x_1, x_2) \in X$$

is η*-pseudolinear but not pseudolinear. As there exists* $x = \left(\frac{\pi}{6}, \frac{\pi}{3} \right)$ *and* $y = \left(\frac{\pi}{3}, 0 \right)$*, such that*

$$\langle \nabla f(x), y - x \rangle = 0 \; but \; f(y) < f(x).$$

The following definitions are from Mohan and Neogy [205]:

Definition 13.5 (Prequasiinvex function) *Let* $K \subseteq X$ *be an invex set with respect to* $\eta : K \times K \rightarrow \mathbb{R}^n$*. The function* f *is said to be prequasiinvex with respect to* η *if for any* $x, y \in K$*,*

$$f(x + \lambda \eta(y, x)) \leq \max\{f(x), f(y)\}.$$

Definition 13.6 (Assumption A) *Let* $K \subseteq X$ *be an invex set with respect to* $\eta : K \times K \rightarrow \mathbb{R}^n$*. We say that* f *satisfies Assumption A, if for any* $x, y \in K$*,*

$$f(x + \eta(y, x)) \leq \max\{f(x), f(y)\}.$$

The following results of Ansari *et al.* [5] characterize the class of differentiable η-pseudolinear functions.

Theorem 13.1 *Let* $f : K \rightarrow \mathbb{R}$ *be a differentiable function on an open set* X *and* $K \subseteq X$ *be an invex set with respect to* $\eta : K \times K \rightarrow \mathbb{R}^n$ *that satisfies condition C. Suppose that* f *is an* η*-pseudolinear function on* K*. Then for all* $x, y \in K$*,* $\langle \nabla f(x), \eta(y, x) \rangle = 0$ *if and only if* $f(y) = f(x)$*.*

Proof Suppose f is η-pseudolinear on K. Then, for any $x, y \in K$, we have

$$\langle \nabla f(x), \eta(y, x) \rangle \geq 0 \Rightarrow f(y) \geq f(x)$$

and

$$\langle \nabla f(x), \eta(y, x) \rangle \leq 0 \Rightarrow f(y) \leq f(x).$$

Combining these two inequalities, we have

$$\langle \nabla f(x), \eta(y, x) \rangle = 0 \Rightarrow f(y) = f(x), \; \forall x, y \in X.$$

Now, we prove that $f(y) = f(x)$ implies $\langle \nabla f(x), \eta(y, x) \rangle = 0, \quad \forall x, y \in X.$ For that, we first show that

$$f(x + \lambda \eta(y, x)) = f(x), \forall \lambda \in [0, 1].$$

When $\lambda - 0$ and $\lambda = 1$, it is obvious. Now, we will prove it for $\lambda \in]0, 1[$.
If $f(x + \lambda \eta(y, x)) > f(x)$, then by the pseudoincavity of f, we have

$$\langle \nabla f(z), \eta(x, z) \rangle < 0, \tag{13.1}$$

where $z = x + \lambda \eta(y, x), \lambda \in]0, 1[$.
Now from condition C, we have

$$\eta(x, z) = \eta(x, x + \lambda \eta(y, x)) = -\lambda \eta(y, x)$$

and

$$\eta(y, z) = \eta(y, x + \lambda \eta(y, x)) = (1 - \lambda) \eta(y, x).$$

Hence, we get

$$\eta(x, z) = \frac{-\lambda}{1 - \lambda} \eta(y, z).$$

This yields with (13.1) that

$$\frac{-\lambda}{1 - \lambda} \langle \nabla f(z), \eta(y, z) \rangle < 0,$$

which implies that

$$\langle \nabla f(z), \eta(y, z) \rangle > 0.$$

By pseudoinvexity of f, we have

$$f(y) \geq f(z),$$

which is a contradiction to the fact that

$$f(z) > f(x) = f(y).$$

Similarly, if $f(x + \lambda \eta(y, x)) < f(x)$, then by the pseudoincavity of f we get a contradiction again. Hence, for any $\lambda \in]0, 1[$,

$$f(x + \lambda \eta(y, x)) = f(x).$$

Thus

$$\langle \nabla f(x), \eta(y, x) \rangle = \lim_{\lambda \to 0^+} \frac{f(x + \lambda \eta(y, x)) - f(x)}{\lambda} = 0.$$

The following example from Ansari *et al.* [5] shows that the converse of the theorem does not hold, that is, if for all $x, y \in K, \langle \nabla f(x), \eta(y, x) \rangle = 0$ if and only if $f(x) = f(y)$, then f need not be η-pseudolinear on K.

Example 13.3 *Let $X = K = \mathbb{R}$ and $f : X \to \mathbb{R}, \eta : X \times X \to \mathbb{R}$ be defined as*

$$f(x) = e^x, \quad \eta(y, x) = e^{-y} - e^{-x}.$$

Then $\langle \nabla f(x), \eta(y, x) \rangle = 0 \Leftrightarrow y = x \Leftrightarrow f(x) = f(y)$. But for $x = 2$ and $y = -1$, we have

$$\langle \nabla f(x), \eta(y, x) \rangle = e^2(e^{-1} - e^{-2}) = e - 1 > 0$$

and $f(y) = e < e^2 = f(x)$. Hence, f is not pseudoinvex on X.

Remark 13.2 *If $\eta(y, x) = y - x, \forall x, y \in X$, then, Theorem 13.1 reduces to Theorem 2.1 for pseudolinear functions.*

Theorem 13.2 *Let $f : X \to \mathbb{R}$ be a differentiable function on X and $K \subseteq X$ be an invex set with respect $\eta : X \times X \to \mathbb{R}^n$ that satisfies condition C. Then f is a η-pseudolinear function on K if and only if for all $x, y \in K$, there exists function $p : K \times K \to \mathbb{R}$ such that*

$$p(x, y) > 0 \text{ and } f(y) = f(x) + p(x, y) \langle \nabla f(x), \eta(y, x) \rangle. \qquad (13.2)$$

Proof Let f be pseudolinear on X. We need to construct a function $p : K \times K \to \mathbb{R}$ such that for all $x, y \in X$,

$$p(x, y) > 0 \text{ and } f(y) = f(x) + p(x, y) \langle \nabla f(x), \eta(y, x) \rangle.$$

If $\langle \nabla f(x), \eta(y, x) \rangle = 0$, for any $x, y \in X$, we define $p(x, y) = 1$. From Theorem 13.1, we get $f(y) = f(x)$ and thus, (13.2) holds.

If $\langle \nabla f(x), \eta(y, x) \rangle \neq 0$ for any $x, y \in X$, we define

$$p(x, y) = \frac{f(y) - f(x)}{\langle \nabla f(x), \eta(y, x) \rangle}. \qquad (13.3)$$

Evidently, $p(x, y)$ satisfies (13.2). Now we have to show that $p(x, y) > 0$, for all $x, y \in K$. If $f(y) > f(x)$, then by pseudoincavity of f, we have $\langle \nabla f(x), \eta(y, x) \rangle = 0$. From (13.3), we get $p(x, y) > 0$. Similarly, if $f(y) < f(x)$, then by psuedoinvexity of f, we have

$$\langle \nabla f(x), \eta(y, x) \rangle > 0.$$

From (13.3), we get $p(x, y) > 0$, for all $x, y \in K$.

Conversely, suppose that for any $x, y \in K$ there exists a function $p : K \times K \to \mathbb{R}$, such that (13.2) holds. If for any $x, y \in K, \langle \nabla f(x), \eta(y, x) \rangle \geq 0$. From (13.2), it follows that

$$f(y) - f(x) = p(x, y)\langle \nabla f(x), \eta(y, x) \rangle \geq 0.$$

Hence, f is pseudoinvex on K. Similarly, if $\langle \nabla f(x), \eta(y, x) \rangle \leq 0$, for any $x, y \in K$, from (13.2), it follows that f is pseudoincave on K.

Remark 13.3 *If $\eta(y, x) = y - x, \forall x, y \in X$, then, Theorem 13.2 reduces to Theorem 2.2 for pseudolinear functions.*

Theorem 13.3 *Let $f : X \to \mathbb{R}$ be a differentiable η-pseudolinear function defined on open set $X \subseteq \mathbb{R}^n$. Let $F : \mathbb{R} \to \mathbb{R}$ be differentiable with $F'(\lambda) > 0$ or $F'(\lambda) < 0$ for all $\lambda \in \mathbb{R}$. Then, the composite function $F \circ f$ is also η-pseudolinear on X.*

Proof Let $g(x) = F(f(x))$, for all $x \in X$. We prove the result for $F'(\lambda) > 0$ and for $F'(\lambda) < 0$, the result follows, similarly. We have

$$\langle \nabla g(x), \eta(y, x) \rangle = \langle F'(f(x))\nabla f(x), \eta(y, x) \rangle.$$

Since, $F'(\lambda) > 0$, for all $\lambda \in \mathbb{R}$. Then,

$$\langle \nabla g(x), \eta(y, x) \rangle \geq 0 (\leq 0) \Rightarrow \langle \nabla f(x), \eta(y, x) \rangle \geq 0 (\leq 0).$$

By the η-psuedolinearity of f, it follows that

$$f(y) \geq f(x) \quad (f(y) \leq f(x)).$$

Since, F is strictly increasing, we get

$$g(y) \geq g(x) \quad (y(y) \leq g(x)).$$

Hence, g is η-pseudolinear on X.

Remark 13.4 *If $\eta(y, x) = y - x, \forall x, y \in X$, then, Theorem 13.3 reduces to Theorem 2.5 for pseudolinear functions.*

The following example from Ansari et al. [5] illustrates that Theorem 13.3 is not true if $F'(\lambda) = 0$, for some λ.

Example 13.4 *Consider f, η, and X be defined as in Example 13.2 and let $F(\lambda) = \lambda^3$, $\lambda \in \mathbb{R}$. Then, it is clear that $F'(0) = 0$. For $x = (0, 0)$, $y = \left(0, -\frac{\pi}{3}\right)$, we have $\nabla g(0, 0) = 0$ and hence, $\langle \nabla g(0, 0), \eta(y, x) \rangle = 0$. But $g(y) = f(F(y)) = \sin\left(-\frac{\pi}{3}\right)^3 = \frac{-3\sqrt{(3)}}{8} < 0 = g(x)$. Therefore, g is not η-pseudoconvex and hence not an η-pseudolinear function.*

The following proposition from Yang *et al.* [291] relates the pseudoinvex and prequasiinvex functions.

Proposition 13.1 *Let K be an open invex set with respect to $\eta : K \times K \to \mathbb{R}$, that satisfies condition C. Suppose f is pseudoinvex with respect to η and satisfies Assumption A. Then f is prequasiinvex with respect to same η on K.*

Proof Suppose that f is pseudoinvex with respect to η on K. On the contrary, suppose that f is not prequasiinvex with respect to the same η. Then, there exists $x, y \in K$, such that

$$f(y) \leq f(x)$$

and there exists $\bar{\lambda} \in]0, 1[$, such that

$$f(z) > f(x) \geq f(y), \qquad (13.4)$$

where $z := x + \lambda\eta(y, x)$.

Let $\psi(\lambda) := f(x + \lambda\eta(y, x))$, $\lambda \in [0, 1]$. Since, f is a differentiable function, $\psi(\lambda)$ is a continuous function and hence, attains its maximum. From (13.4) and $z := x + \lambda\eta(y, x) \in K$, we know that

$$\psi(0) = f(x) < f(z),$$

hence, $\lambda = 0$ is not a maximum point. By Assumption A and (13.4), we have

$$f(x + \lambda\eta(y, x)) = \psi(1) \leq \max\{f(x), f(y)\} = f(x) < f(z),$$

therefore, $\lambda = 1$ is not a maximum point.

Hence, there exists $\bar{\lambda} \in (0, 1)$ such that

$$f(\bar{x}) = \max_{\lambda \in [0,1]} f(x + \lambda\eta(y, x)), \bar{x} = x + \lambda\eta(y, x).$$

Therefore,

$$\langle \nabla f(\bar{x}), \eta(y, x) \rangle = 0.$$

From condition C, we have

$$\eta(y, \bar{x}) = (1 - \bar{\lambda}\eta(y, x), \eta(x, \bar{x}) = -\lambda\eta(y, x).$$

Hence,

$$\langle \nabla f(\bar{x}), \eta(y, \bar{x}) \rangle = 0.$$

Since, f is pseudoinvex with respect to η, we get

$$f(y) \geq f(\bar{x}),$$

which is a contradiction to (13.4).

Theorem 13.4 *Let K be an invex set with respect to $\eta : K \times K \to \mathbb{R}$, that satisfies condition C. Suppose f is η-pseudolinear and satisfies Assumption A. Then, for any $x, y \in K$,*

$$\langle \nabla f(x), \eta(y, x) \rangle = 0 \Rightarrow f(z_\lambda) = f(x) = f(y),$$

where $z_\lambda =: x + \lambda \eta(y, x), \lambda \in [0, 1]$.

Proof For any $x, y \in K$, $\lambda \in [0, 1]$, let $z_\lambda =: x + \lambda \eta(y, x)$. Suppose that the condition

$$\langle \nabla f(x), \eta(y, x) \rangle = 0 \tag{13.5}$$

holds. Since, f is pseudoinvex with respect to η from (13.5), we have

$$f(y) \geq f(x). \tag{13.6}$$

By the prequasiinvexity of f with respect to the same η and Proposition 13.1, we get

$$f(z_\lambda) \leq \max \{f(x), f(y)\} = f(y). \tag{13.7}$$

Since, $-f$ is pseudoinvex with respect to η from (13.5), we have

$$f(y) \leq f(x). \tag{13.8}$$

By the prequasiinvexity of $-f$ with respect to the same η and Proposition 13.1, we get

$$f(z_\lambda) \geq \min \{f(x), f(y)\} = f(y). \tag{13.9}$$

From (13.7) and (13.9), we get

$$f(z_\lambda) = f(y).$$

From (13.6) and (13.8), we get

$$f(y) = f(x).$$

Hence, the proof is complete.

Remark 13.5 *For $\eta(y, x) = y - x, \forall x, y \in X$, Theorem 13.3 reduces to Theorem 2.3 for pseudolinear functions.*

We consider the following constrained optimization problem:

$$(P) \quad \min f(x)$$

$$\text{subject to } x \in C,$$

where $f : X \to \mathbb{R}$ and $X \subseteq \mathbb{R}^n$ is an open set and C is an invex subset of X with respect to η. We assume that the solution set

$$\bar{S} = \arg\min_{x \in C} f(x) := \{x \in C : f(x) \leq f(y), \ \forall \ y \in C\}$$

is nonempty.

Theorem 13.5 (Ansari et al. [5]) *If f is a preinvex function on C, then the solution set $\bar{S}$ of problem (P) is an invex set.*

Proof Let $x, \bar{x} \in \bar{S}$. Then, $f(x) \leq f(y)$ and $f(x) \leq f(y)$, for all $y \in C$. Since, f is preinvex, we get

$$
\begin{aligned}
f(x + \lambda\eta(x, \bar{x})) &\leq (1 - \lambda)f(x) + \lambda f(\bar{x}), \quad \forall \lambda \in [0, 1] \\
&\leq (1 - \lambda)f(y) + \lambda f(y), \quad \forall \lambda \in [0, 1] \\
&= f(y), \quad \forall \lambda \in [0, 1].
\end{aligned}
$$

Hence, $x + \lambda\eta(x, \bar{x}) \in \bar{S}$, $\forall \lambda \in [0, 1]$. Hence, $\bar{S}$ is an invex set.

Remark 13.6 *Let $\bar{x} \in \bar{S}$. Then, it is clear that $x \in \bar{S}$ if and only if $f(x) = f(\bar{x})$. Hence, by Theorem 13.2, f is η-pseudolinear on $\bar{S}$ if and only if for each $\bar{x} \in \bar{S}$, we get*

$$
\bar{S} = \{x \in C : \langle \nabla f(\bar{x}), \eta(x, \bar{x}) \rangle = 0\}.
$$

Corollary 13.1 *It is evident from Theorem 13.1, that if $f : C \to R$ is a η-pseudolinear function and η satisfies condition C, then the solution set $\bar{S}$ of the problem (P) is an invex set.*

Proof Suppose that $x, \bar{x} \in \bar{S}$, then, we have $f(x) = f(\bar{x})$. By Theorem 13.1,

$$
f(x) = f(\bar{x}) \Leftrightarrow \langle \nabla f(\bar{x}, \eta(x, \bar{x})) \rangle = 0.
$$

Now, we note that

$$
\langle \nabla f(\bar{x}), \eta(x, x + \lambda\eta(x, \bar{x})) \rangle = (1 - \lambda)\langle \nabla f(\bar{x}, \eta(x, \bar{x})) \rangle = 0, \;\; for \; all \; \lambda \in [0, 1].
$$

Therefore, by Theorem 13.1, we get $f(x + \lambda\eta(x, \bar{x})) = f(x) = f(\bar{x}), \forall \lambda \in [0, 1]$. Hence for each $x, \bar{x} \in \bar{S}$, we have $x + \lambda\eta(x, \bar{x}) \in \bar{S}, \forall \lambda \in [0, 1]$.

13.3   Characterization of Solution Set of (P) Using Differentiable η-Pseudolinear Functions

In this section, we present the characterizations for the solution set of (P) in terms of any of its solution points.

The following theorems present the first order characterizations of the solution set of (P) under the assumption that the function f is differentiable η-pseudolinear functions and satisfies Assumption A.

Theorem 13.6 *Let $\bar{x} \in \bar{S}$. Then, $\bar{S} = \tilde{S} = \hat{S} = S^*$, where*

$$
\tilde{S} := \{x \in C : \langle \nabla f(x), \eta(\bar{x}, x) \rangle = 0\}; \tag{13.10}
$$

$$
\hat{S} := \{x \in C : \langle \nabla f(\bar{x}), \eta(x, \bar{x}) \rangle = 0\}; \tag{13.11}
$$

$$
S^* := \{x \in C : \langle \nabla f(z_\lambda), \eta(\bar{x}, x) \rangle = 0, \forall \lambda \in [0, 1]\}. \tag{13.12}
$$

Proof The point $x \in \bar{S}$ if and only if $f(x) = f(\bar{x})$. By Theorem 13.1, $f(x) = f(\bar{x})$ if and only if

$$\langle \nabla f(x), \eta(\bar{x}, x) \rangle = 0 = \langle \nabla f(\bar{x}), \eta(x, \bar{x}) \rangle.$$

Therefore, $\bar{S} = \tilde{S} = \hat{S}$. To prove that $\bar{S} = S^*$. Now, by the definitions, we note that $\hat{S} = S^* \subseteq \tilde{S} = \bar{S}$.

To prove the converse implication, let $x \in \tilde{S} = \bar{S}$. Then, we have

$$\langle \nabla f(x), \eta(\bar{x}, x) \rangle = 0.$$

Therefore, from Theorem 13.4 and for each $\lambda \in \,]0, 1]$, we have

$$f(z_\lambda) = f(x), \; z_\lambda = x + \lambda \eta(x, \bar{x}).$$

Therefore,

$$\langle \nabla f(x), (x - z_\lambda) \rangle = \lambda \langle \nabla f(x), x - \bar{x} \rangle = 0.$$

From Theorem 13.1, we get

$$\langle \nabla f(z_\lambda), \eta(x, z_\lambda) \rangle = 0, \; \forall \lambda \in \,]0, 1].$$

By condition C, we have

$$\eta(x, z_\lambda) = -\lambda \eta(\bar{x}, x) = 0.$$

Therefore,

$$\langle \nabla f(x\lambda\bar{x}), \eta(\bar{x}, x) \rangle = \frac{-1}{\lambda} \langle \nabla f(z_\lambda), \eta(x, z_\lambda) \rangle = 0, \forall \lambda \in \,]0, 1].$$

Thus, $x \in S^*$ and hence $\bar{S} \subseteq S^*$. If $\lambda = 0, \bar{S} = S^*$ is obvious.

Corollary 13.2 *Let $\bar{x} \in \bar{S}$. Then, $\bar{S} = \tilde{S}_1 = \hat{S}_1 = S_1^*$, where*

$$\tilde{S}_1 := \{x \in C : \langle \nabla f(x), \eta(\bar{x}, x) \rangle \geq 0\};$$
$$\hat{S}_1 := \{x \in C : \langle \nabla f(\bar{x}), \eta(x, \bar{x}) \rangle \geq 0\};$$
$$S_1^* := \{x \in C : \langle \nabla f(z_\lambda), \eta(\bar{x}, x) \rangle \geq 0, \forall \lambda \in [0, 1]\},$$

where, $z_\lambda = x + \lambda \eta(x, \bar{x}), \lambda \in [0, 1]$.

Proof By Theorem 13.1, it is clear that $\bar{S} \subseteq \tilde{S}_1$. Assume that $x \in \tilde{S}_1$, that is,

$$x \in C \text{ and } \langle \nabla f(x), \eta(\bar{x}, x) \rangle \geq 0.$$

Then, by Theorem 2.2, it follows that

$$f(\bar{x}) = f(x) + p(x, \bar{x}) \langle \nabla f(x), \eta(\bar{x}, x) \rangle \geq f(x),$$

which implies that $x \in \bar{S}$. Hence, $\bar{S} = \tilde{S}_1$. Similarly, we can prove that $\bar{S} = \tilde{S}_1$. Now, to prove that $\bar{S} = S_1^*$ we note that

$$\bar{S} = S^* \subset S_1^* \subset \tilde{S}_1 = \bar{S}.$$

This completes the proof.

Theorem 13.7 *Let f be a continuously differentiable η-pseudolinear function and $\bar{x} \in \bar{S}$. Let*

$$S^{\#} := \{x \in X : \langle \nabla f(x), \eta(\bar{x}, x) \rangle = \langle \nabla f(\bar{x}), \eta(x, \bar{x}) \rangle\},$$

$$S_1^* := \{x \in X : \langle \nabla f(x), \eta(\bar{x}, x) \rangle \geq \langle \nabla f(\bar{x}), \eta(x, \bar{x}) \rangle\}.$$

Then, $\bar{S} = S^{\#} = S_1^{\#}$.

Proof We first prove that $\bar{S} \subset S^{\#}$. Let $x \in \bar{S}$, from Theorem 13.1, we have

$$\langle \nabla f(x), \eta(\bar{x}, x) \rangle = 0, \quad \langle \nabla f(\bar{x}), \eta(x, \bar{x}) \rangle = 0.$$

Then, we get
$$\langle \nabla f(x), \eta(\bar{x}, x) \rangle = 0 = \langle \nabla f(\bar{x}), \eta(x, \bar{x}) \rangle. \tag{13.13}$$

Thus, $x \in S^{\#}$.

It is clear that, $S^{\#} \subset S_1^{\#}$.

Now, we prove that $S_1^{\#} \subset \bar{S}$. Let $x \in S_1^*$. Then, we have

$$\langle \nabla f(x), \eta(\bar{x}, x) \rangle \geq \langle \nabla f(\bar{x}), \eta(x, \bar{x}) \rangle. \tag{13.14}$$

Suppose that $x \notin \bar{S}$. Then, $f(x) > f(\bar{x})$. By the pseudoincavity of f, we have

$$\langle \nabla f(\bar{x}), \eta(x, \bar{x}) \rangle > 0.$$

From (13.14), we have
$$\langle \nabla f(x), \eta(\bar{x}, x) \rangle > 0.$$

By Theorem 13.2, it follows that

$$f(\bar{x}) = f(x) + p(x, \bar{x}) \langle \nabla f(\bar{x}), \eta(x, \bar{x}) \rangle > f(x),$$

which is a contradiction. Hence, $x \in \bar{S}$.

Remark 13.7 *It is clear that Theorem 13.6, Corollary 13.1, and Theorem 13.7 are generalizations of Theorem 3.1, Corollary 3.1, and Theorem 3.2, respectively.*

13.4 η-Pseudolinear Functions and Vector Optimization

We consider the following multiobjective optimization problem:

$$(MP) \quad \min f(x) := (f_1(x), \dots, f_k(x))$$

subject to $g_j(x) \le 0 \ \{j = 1, \dots, m\}$,

where $f_i : K \to R, i \in I := \{1, \dots, k\}; g_j : K \to R, j \in J := \{1, \dots, m\}$ are differentiable η-pseudolinear functions on a nonempty open invex set $K \subseteq \mathbb{R}^n$ with respect to $\eta : K \times K \to \mathbb{R}^n$, that satisfies condition C. Let $P := \{x \in K : g_j(x) \le 0, j \in J\}$ and $J(x) := \{j \in J : g_j(x) = 0\}$ for some $x \in K$, denote the set of all feasible solutions for (MP) and active constraint index set at $x \in K$, respectively.

The following necessary and sufficient optimality conditions for (MP) proved by Zhao and Yang [299] are an extension of Proposition 3.2 of Chew and Choo [47] to an η-pseudolinear case.

Theorem 13.8 (Necessary and sufficient optimality conditions) *Suppose $\bar{x}$ is a feasible solution of (MP). Let the functions $f_i, i = 1, \dots, k$ be η-pseudolinear with respect to proportional function p_i and $g_j, j \in J(\bar{x})$ be η-pseudolinear with respect to proportional function q_j. Then, $\bar{x}$ is an efficient solution for (MP) if and only if there exist $\bar{\lambda}_i \in R, i = 1, \dots, k, \bar{\mu}_j \in R, j \in J(\bar{x})$, such that*

$$\sum_{i=1}^{k} \bar{\lambda}_i \nabla f_i(\bar{x}) + \sum_{j \in J(\bar{x})} \bar{\mu}_j \nabla g_j(\bar{x}) = 0, \tag{13.15}$$

$$\bar{\lambda}_i > 0, i = 1, \dots, k,$$

$$\bar{\mu}_j \ge 0, j \in J(\bar{x}).$$

Proof Suppose $\bar{\lambda}_i$ and $\bar{\mu}_j$ exist and satisfy condition (13.15). As $\bar{x}$ is not an efficient solution for (MP). There exists then a point $y \in P$, such that

$$f_i(y) \le f_i(\bar{x}), \forall i = 1, \dots, k$$

and

$$f_r(y) < f_r(\bar{x}), \text{ for some } r.$$

Using η-pseudolinearity of $f_i, i = 1, \dots, k$, with respect to proportional function p_i, we get

$$p_i(\bar{x}, y)\langle \nabla f_i(\bar{x}), \eta(y, \bar{x})\rangle \le 0, \forall i = 1, \dots, k$$

and

$$p_r(\bar{x}, y)\langle \nabla f_r(\bar{x}), \eta(y, \bar{x})\rangle < 0, \text{ for some } r.$$

Since, $p_i(\bar{x}, y) > 0$, for all $i = 1, \dots, k$, therefore, we get

$$\langle \nabla f_i(\bar{x}), \eta(y, \bar{x})\rangle \le 0, \forall i = 1, \dots, k$$

and

$$\langle \nabla f_r(\bar{x}), \eta(y, \bar{x})\rangle < 0, \text{ for some } r.$$

Therefore, for $\bar{\lambda}_i > 0, i = 1, \ldots, k$, it follows that

$$\sum_{i=1}^{k} \bar{\lambda}_i \langle \nabla f_i(\bar{x}), \eta(y, \bar{x}) \rangle < 0. \tag{13.16}$$

From feasibility, we have

$$g_j(y) \leq g_j(\bar{x}), \forall j \in J(\bar{x}).$$

Using η-pseudolinearity of $g_j, j = 1, \ldots, m$, we get

$$q_j(\bar{x}, y) \langle \nabla g_j(\bar{x}), \eta(y, \bar{x}) \rangle \leq 0, \ \forall j \in J(\bar{x}).$$

Since, $q_j(\bar{x}, y) > 0$, for all $j \in J(\bar{x})$ therefore, we get

$$\langle \nabla g_j(\bar{x}), \eta(y, \bar{x}) \rangle \leq 0, \ \forall j \in J(\bar{x}).$$

Therefore, for $\bar{\mu}_j \geq 0$, for all $j \in J(\bar{x})$, it follows that

$$\sum_{j \in J(\bar{x})} \bar{\mu}_j \langle \nabla g_j(\bar{x}), \eta(y, \bar{x}) \rangle \leq 0.$$

Using (13.15), we get

$$\sum_{i=1}^{k} \bar{\lambda}_i \langle \nabla f_i(\bar{x}), \eta(y, \bar{x}) \rangle \geq 0,$$

which is a contradiction to (13.16).

Conversely, suppose $\bar{x}$ be an efficient solution of (MP). For $1 \leq r \leq k$, the following system

$$\left. \begin{array}{l} \langle \nabla g_j(\bar{x}), \eta(x, \bar{x}) \rangle \leq 0, \ j \in J(\bar{x}), \\ \langle \nabla f_i(\bar{x}), \eta(x, \bar{x}) \rangle \leq 0, \ i = 1, 2, \ldots, r-1, r+1, \ldots, k, \\ \langle \nabla f_r(\bar{x}), \eta(x, \bar{x}) \rangle < 0, \end{array} \right\} \tag{13.17}$$

has no solution $x \in P$. For if x is a solution of the system and $y = \bar{x} + t\eta(x, \bar{x})(0 < t \leq 1)$, then by condition C and Remark 13.1, for $j \in J(\bar{x})$, we have

$$\langle \nabla g_j(\bar{x}), \eta(y, \bar{x}) \rangle = t \langle \nabla g_j(\bar{x}), \eta(x, \bar{x}) \rangle \leq 0,$$

and thus by pseudoincavity of g_j, we have

$$g_j(y) \leq g_j(\bar{x}) = 0.$$

If $j \notin J(\bar{x})$, then $g_j(\bar{x}) < 0$ and so $g_j(y) < 0$ when t is sufficiently small. Hence, for small t, y is a point of P. Hence, by condition C and Remark 13.1, we have

$$f_i(y) - f_i(\bar{x}) = p_i(\bar{x}, y) \langle \nabla f_i(\bar{x}), \eta(y, \bar{x}) \rangle, \ i \neq r$$

$$= t p_i(\bar{x}, y) \langle \nabla f_i(\bar{x}), \eta(x, \bar{x}) \rangle \leq 0, \ i \neq r$$

and

$$f_r(y) - f_r(\bar{x}) < 0, \text{ for some } r,$$

which is a contradiction to our assumption that $\bar{x}$ is an efficient solution for (MP). From which, it follows that the system (13.17) has no solution. Therefore, by Farkas's theorem (Theorem 1.24), there exist $\bar{\lambda}_{r_i} \geq 0, \bar{\mu}_{r_j} \geq 0$, such that

$$-\nabla f_r(\bar{x}) = \sum_{i \neq r} \bar{\lambda}_{r_i} \nabla f_i(\bar{x}) + \sum_{j \in J(x^*)} \bar{\mu}_{r_j} \nabla g_j(\bar{x}). \tag{13.18}$$

Summing (13.18) over r, we get (13.17) with

$$\bar{\lambda}_i = 1 + \sum_{r \neq i} \bar{\lambda}_{r_i}, \bar{\mu}_j = \sum_{r=1}^{k} \bar{\mu}_{r_j}.$$

Remark 13.8 *It is clear that Theorem 13.8 is a generalization of Theorem 5.1 in Chapter 5.*

Theorem 13.9 *Let $\bar{x}$ be a feasible solution of the problem (MP). Assume that the functions $f_i, i = 1, \ldots, k$ are η-pseudolinear with respect to proportional function p_i and $g_j, j \in J(\bar{x})$ are η-pseudolinear with respect to proportional function q_j. Then, the following statements are equivalent:*

(i) $\bar{x}$ is an efficient solution of the problem (MP);

(ii) $\bar{x}$ is an efficient solution of the problem

$$(MLP) \quad \min(\nabla f_1(\bar{x})x, \ldots, \nabla f_k(\bar{x})x)$$

$$\text{subject to } x \in K;$$

(iii) there exist positive numbers $\bar{\lambda}_i, i = 1, \ldots, k$, such that $\bar{x}$ minimizes the linear function

$$\sum_{i=1}^{k} \bar{\lambda}_i \nabla f_i(\bar{x})x, x \in K. \tag{13.19}$$

Proof (i)$\Rightarrow$(ii) Suppose $\bar{x}$ is an efficient solution of the problem (MP), then, by Theorem 13.8, there exist $\bar{\lambda}_i \in R, i = 1, \ldots, k, \bar{\mu}_j \in R, j \in J(\bar{x})$, such that

$$\sum_{i=1}^{k} \bar{\lambda}_i \nabla f_i(\bar{x}) + \sum_{j \in J(\bar{x})} \bar{\mu}_{r_j} \nabla g_j(\bar{x}) = 0. \tag{13.20}$$

On the contrary, suppose that $\bar{x}$ is not an efficient solution of the (MLP) problem. Then, there exists some $x \in K$, such that

$$\nabla f_i(\bar{x})x \leq \nabla f_i(\bar{x})\bar{x}, \forall i = 1, \ldots, k,$$

$$\nabla f_r(\bar{x})x < \nabla f_r(\bar{x})\bar{x}, \text{ for some } r.$$

Now, we have

$$0 > \sum_{i=1}^{k} \bar{\lambda}_i \langle \nabla f_i(\bar{x}), \eta(x, \bar{x}) \rangle = - \sum_{j \in J(\bar{x})} \bar{\mu}_j \langle \nabla g_j(\bar{x}), \eta(x, \bar{x}) \rangle$$

$$= - \left(\sum_{j \in J(\bar{x})} \bar{\mu}_j \left(\frac{g_j(x) - g_j(\bar{x})}{q_j(\bar{x}, x)} \right) \right) \geq 0,$$

which is a contradiction. Hence, $\bar{x}$ must be an efficient solution of the problem (MLP).

(ii) $\Rightarrow$ (iii) Suppose $\bar{x}$ is an efficient solution of the problem (MLP). Then by Theorem 13.8, there exist $\bar{\lambda}_i \in R, i = 1, \ldots, k, \bar{\mu}_j \in R, j \in J(\bar{x})$, such that (13.15) holds. Therefore, for any $x \in K$, we get

$$\sum_{i=1}^{k} \bar{\lambda}_i \langle \nabla f_i(\bar{x}), \eta(x, \bar{x}) \rangle = - \sum_{j \in J(\bar{x})} \bar{\mu}_j \langle \nabla g_j(\bar{x}), \eta(x, \bar{x}) \rangle$$

$$= - \left(\sum_{j \in J(\bar{x})} \bar{\mu}_j \left(\frac{g_j(x) - g_j(\bar{x})}{q_j(\bar{x}, x)} \right) \right) \geq 0.$$

This shows that $\bar{x}$ minimizes (13.19).

(iii) $\Rightarrow$ (i) Suppose $\bar{x}$ minimizes (13.19) for some positive number $\lambda_1, \ldots, \lambda_k$. If $\bar{x}$ is not an efficient solution for (MP), then there exists some $x \in K$, such that

$$f_i(x) \leq f_i(\bar{x}), \forall i = 1, \ldots, k,$$

$$f_r(x) < f_r(\bar{x}), \text{ for some } r.$$

Then, by the η-pseudolinearity of $f_i, i = 1, \ldots, k$ with respect to proportional function p_i, it follows that

$$p_i(\bar{x}, x) \langle \nabla f_i(\bar{x}), \eta(x, \bar{x}) \rangle \leq 0, \forall i = 1, \ldots, k,$$

$$p_r(\bar{x}, x) \langle \nabla f_r(\bar{x}), \eta(x, \bar{x}) \rangle \leq 0, \text{ for some } r.$$

Since, $p_i(\bar{x}, x) > 0, i = 1, \ldots \ldots, k$, it follows that

$$\langle \nabla f_i(\bar{x}), \eta(x, \bar{x}) \rangle \leq 0, \ \forall i = 1, \ldots, k,$$

$$\langle \nabla f_r(\bar{x}), \eta(x, \bar{x}) \rangle \leq 0, \text{ for some } r.$$

Therefore, for $\bar{\lambda}_i > 0, i = 1, \ldots, k$, it follows that

$$\sum_{i=1}^{k} \bar{\lambda}_i \langle \nabla f_i(\bar{x}), \eta(x, \bar{x}) \rangle < 0,$$

which is a contradiction to our assumption that minimizes (13.19). Hence, $\bar{x}$ must be an efficient solution of (MP).

Definition 13.7 *A feasible point $\bar{x}$ for the problem (MP) is said to satisfy the boundedness condition, if the set*

$$\left\{ \frac{p_i(\bar{x}, x)}{p_j(\bar{x}, x)} \mid x \in K, f_i(\bar{x}) > f_i(x), f_j(\bar{x}) < f_j(x), 1 \le i, j \le k \right\} \qquad (13.21)$$

is bounded from above.

The following theorem is a variant of Proposition 2 of Giorgi and Rueda [91].

Theorem 13.10 *Every efficient solution of (MP) involving η-pseudolinear functions, satisfying the boundedness condition is a properly efficient solution of (MP).*

Proof Let $\bar{x}$ be an efficient solution of (MP). Then from Theorem 13.9, there exist positive numbers $\bar{\lambda}_i > 0, i = 1, \ldots, k$, such that $\bar{x}$ minimizes the linear function (13.19). Hence, for any $x \in K$, we have

$$\sum_{i=1}^{k} \bar{\lambda}_i \langle \nabla f_i(\bar{x}), x - \bar{x} \rangle \ge 0. \qquad (13.22)$$

Otherwise, we would arrive at a contradiction as in the first part of Theorem 13.8.

Since, the set defined by (13.21) is bounded above, therefore, the following set

$$\left\{ (k-1) \frac{\bar{\lambda}_j p_i(\bar{x}, x)}{\bar{\lambda}_i p_j(\bar{x}, x)} \mid x \in K, f_i(\bar{x}) > f_i(x), f_j(\bar{x}) < f_j(x), 1 \le i, j \le k \right\}$$
$$(13.23)$$

is also bounded from above.

Let $M > 0$ be a real number that is an upper bound of the set defined by (13.23). Now, we shall show that $\bar{x}$ is a properly efficient solution of (MP). Assume that there exist r and $x \in K$, such that

$$f_r(x) < f_r(\bar{x}).$$

Then, by the η-pseudolinearity of f_r, it follows that

$$p_r(\bar{x}, x) \langle \nabla f_r(\bar{x}), \eta(x, \bar{x}) \rangle < 0.$$

Since $p_r(\bar{x}, x) > 0$ we get

$$\langle \nabla f_r(\bar{x}), \eta(x, \bar{x}) \rangle < 0. \qquad (13.24)$$

Let us define

$$-\bar{\lambda}_s \langle \nabla f_s(\bar{x}), \eta(x, \bar{x}) \rangle = \max\{ -\bar{\lambda}_i \langle \nabla f_i(\bar{x}), \eta(x, \bar{x}) \rangle \mid \langle \nabla f_i(\bar{x}), \eta(x, \bar{x}) \rangle > 0 \}.$$
$$(13.25)$$

Using (13.22), (13.24), and (13.25), we get

$$\bar{\lambda}_r \langle \nabla f_r(\bar{x}), \eta(x, \bar{x}) \rangle \leq (k-1)(-\bar{\lambda}_s \langle \nabla f_s(\bar{x}), \eta(x, \bar{x}) \rangle).$$

Therefore,

$$(f_r(x) - f_r(\bar{x})) \leq (k-1) \frac{\bar{\lambda}_s p_r(\bar{x}, x)}{\bar{\lambda}_r p_s(\bar{x}, x)} (f_s(\bar{x}) - f_s(x)).$$

Using the definition of M, we get

$$(f_r(x) - f_r(\bar{x})) \leq M(f_s(\bar{x}) - f_s(x)).$$

Hence, $\bar{x}$ is a properly efficient solution of (MP).

Remark 13.9 *It is clear that Theorems 13.9 and 13.10 are generalizations of Theorem 5.1, Corollary 5.1, and Theorem 5.2.*

13.5 Characterizations of Locally Lipschitz η-Pseudolinear Functions

Zhao and Tang [298] extended the concept of η-pseudolinear functions to nondifferentiable cases using the Clarke subdifferential.

Definition 13.8 (Zhao and Tang [298]) *Let X be an open invex set with respect to $\eta : X \times X \to \mathbb{R}^n$. A locally Lipschitz function $f : X \to \mathbb{R}$ is said to be pseudoinvex with respect to η on X, if for all $x, y \in X$, one has*

$$f(y) < f(x) \Rightarrow \langle \zeta, \eta(y, x) \rangle < 0, \; \forall \zeta \in \partial^c f(x)$$

or equivalently,

$$\exists \zeta \in \partial^c f(x) : \langle \zeta, \eta(y, x) \rangle \geq 0 \Rightarrow f(y) \geq f(x).$$

The function f is said to be pseudoincave with respect to the η on X, if $-f$ is pseudoinvex with respect to the same η on X. The function $f : X \to \mathbb{R}$ is said to be η-pseudolinear on X, if f is both pseudoinvex and pseudoincave with respect to the same η on X. To be precise; let $f : X \to \mathbb{R}$ be a locally Lipschitz function on an open invex set X with respect to $\eta : X \times X \to \mathbb{R}^n$. The function f is said to be η-pseudolinear at $x \in X$, if for all $y \in X$, one has

$$\exists \zeta \in \partial^c f(x) : \langle \zeta, \eta(y, x) \rangle \geq 0 \Rightarrow f(y) \geq f(x);$$

and

$$\exists \zeta \in \partial^c f(x) : \langle \zeta, \eta(y, x) \rangle \leq 0 \Rightarrow f(y) \leq f(x).$$

The function in the following example is taken from Zhao and Tang [298].

Example 13.5 *Let $X = \mathbb{R}$. It is clear that X is an invex set with respect to η, where*

$$\eta(x, y) = \begin{cases} 3(x - y), & \text{if} \quad x \geq 0, y \geq 0, \\ x - y, & \text{if} \quad x < 0, y < 0, \\ 3x - y, & \text{if} \quad x \geq 0, y < 0, \\ x - 3y, & \text{if} \quad x < 0, y \geq 0. \end{cases}$$

Let $f : X \to \mathbb{R}$ be defined as $f(x) = 2x + |x|$. Then, it is easy to see that:

$$\partial^c(x) = \begin{cases} 3, & \text{if} \quad x > 0, \\ [1, 3], & \text{if} \quad x = 0, \\ 1, & \text{if} \quad x < 0. \end{cases}$$

Evidently, f is η-pseudolinear on $\mathbb{R}$.

Definition 13.9 *A map $\eta : K \times K \to \mathbb{R}^n$ is said to be skew if for all $x, y \in K$, we have*

$$\eta(x, y) + \eta(y, x) = 0.$$

The following results of Zhao and Tang [298] characterize the class of locally Lipschitz η-pseudolinear functions.

Theorem 13.11 *Let $f : X \to R$ be a locally Lipschitz function on an open set X and $K \subseteq X$ be an invex set with respect to $\eta : K \times K \to R^n$ that satisfies condition C. Suppose that f is an η-pseudolinear function on K. Then, for all $x, y \in K, f(x) = f(y)$ if and only if there exists $\xi \in \partial^c f(x)$, such that*

$$\langle \xi, \eta(y, x) \rangle = 0.$$

Proof Suppose f is η-pseudolinear on K. Assume that there exists $\xi \in \partial^c f(x)$, such that

$$\langle \xi, \eta(y, x) \rangle = 0.$$

Then, for any $x, y \in K$, pseudoinvexity and pseudoincavity of f with respect to η imply that

$$\langle \xi, \eta(y, x) \rangle \geq 0 \Rightarrow f(y) \geq f(x)$$

and

$$\langle \xi, \eta(y, x) \rangle \leq 0 \Rightarrow f(y) \leq f(x).$$

Combining these two inequalities, we get

$$\langle \xi, \eta(y, x) \rangle = 0 \Rightarrow f(y) = f(x), \ x, y \in K.$$

Conversely, suppose that $f(x) = f(y)$. Now, we prove that there exists $\xi \in \partial^c f(x)$, such that $\langle \xi, \eta(y, x) \rangle = 0$, for all $x, y \in X$. For that, we first show that

$$f(x + \lambda \eta(y, x)) = f(x), \forall \lambda \in [0, 1].$$

When $\lambda = 0$ and $\lambda = 1$, it is obvious. Now, we will prove it for $\lambda \in \,]0, 1[$.

If $f(x + \lambda \eta(y, x)) > f(x)$, then by the pseudoinvexity of f with respect to η, for any $\xi' \in \partial^c f(z_\lambda)$, where $z_\lambda = x + \lambda \eta(y, x)$, we have

$$\langle \xi', \eta(x, z_\lambda) \rangle < 0, \tag{13.26}$$

where $z = x + \lambda \eta(y, x)$.

From condition C, we have

$$\eta(x, z_\lambda) = \eta(x, x + \lambda \eta(y, x)) = -\lambda \eta(y, x)$$

and

$$\eta(y, z_\lambda) = \eta(y, x + \lambda \eta(y, x)) = (1 - \lambda)\eta(y, x).$$

Hence, we get

$$\eta(x, z_\lambda) = \frac{-\lambda}{1 - \lambda}\eta(y, z_\lambda).$$

This yields with (13.26) that

$$\frac{-\lambda}{1 - \lambda}\langle \xi', \eta(y, z_\lambda) \rangle < 0,$$

which implies that

$$\langle \xi', \eta(y, z_\lambda) \rangle > 0.$$

By pseudoinvexity of f, we have

$$f(y) \geq f(z_\lambda),$$

which is a contradiction to the fact that

$$f(z_\lambda) > f(x) = f(y).$$

Similarly, if $f(z_\lambda) < f(x)$, then by the pseudoincavity of f, we get a contradiction again. Hence, for any $\lambda \in \,]0, 1[$,

$$f(z_\lambda) = f(x).$$

Obviously, $[x, z_\lambda] \subseteq K$. By the Lebourg mean value theorem, there exists $\bar{\lambda} \in \,]0, \lambda[, \xi_{\bar\lambda} \in \partial^c f(z_{\bar\lambda})$, such that,

$$0 = f(z_\lambda) - f(x) = \lambda \langle \xi_{\bar\lambda}, \eta(y, x) \rangle,$$

that is,

$$\langle \xi_{\bar\lambda}, \eta(y, x) \rangle = 0.$$

If $\lambda \to 0$, then $\bar\lambda \to 0$ and $z_{\bar\lambda} \to x \in K$. Since $\xi_{\bar\lambda}$ is a bounded sequence and has a convergent subsequence. Without a loss of generality, assume that $\lim_{\bar\lambda \to 0} \xi_{\bar\lambda} = \xi$. By the closure property of the Clarke subdifferential, it follows that $\xi \in \partial^c f(x)$ and $\langle \xi, \eta(y, x) \rangle = 0$.

The following example from Ansari and Rezaei [4] illustrates that the converse of Theorem 13.11 is not true in general, that is, for all $x, y \in K$, there exists $\xi \in \partial^c f(x)$ such that $\langle \xi, \eta(y, x) \rangle = 0$ if and only if $f(x) = f(y)$ but the function f need not be η-pseudolinear on K.

Example 13.6 *Suppose that $K =]-1, 1[\subseteq \mathbb{R}$. Clearly, K is an invex set with respect to $\eta : K \times K \to \mathbb{R}$, defined by*

$$\eta(y, x) = \begin{cases} y - x, & \text{if } y \geq 0, x \geq 0 \text{ or } x \leq 0, y \leq 0, \\ -x, & \text{if } y \leq 0, \ x > 0 \text{ or } y > 0, \ x \leq 0. \end{cases}$$

Clearly, η satisfies condition C. Let $f : K \to \mathbb{R}$ be a function defined by

$$f(x) = \begin{cases} 0, & \text{if } x \in]-1, 0], \\ -x, & \text{if } x \in]0, 1[. \end{cases}$$

The function f is a locally Lipschitz function with a constant 2. Then, one can see that for all $x, y \in K, f(x) = f(y)$ if and only if there exist $\xi \in \partial^c f(x)$, such that $\langle \xi, \eta(y, x) \rangle = 0$.

If $x < 0, y < 0$, then $f(x) = f(y) = 0$ and $0 = \xi \in \partial^c f(x) = \{0\}$, such that

$$\langle \xi, \eta(y, x) \rangle = 0.$$

If $x = 0, y < 0$, then $f(x) = f(y) = 0$ and there exists $0 = \xi \in \partial^c f(x) = [-1, 0]$, such that

$$\langle \xi, \eta(y, x) \rangle = 0.$$

If $x > 0, y = 0$, then $f(x) \in (-1, 0) \neq f(y) = 0$.
If $x > 0, y > 0$, then in the case $f(x) = f(y)$ if and only if $x = y$, we have $\eta(y, x) = 0$ and then

$$\langle \xi, \eta(y, x) \rangle = 0.$$

If $x = 0, y \in]0, 1[$, then $f(0) = 0$ and $f(y) \in (-1, 0)$, then $f(x) \neq f(y)$. Hence, for all $x, y \in K, f(x) = f(y)$ if and only if there exist $\xi \in \partial^c f(x)$, such that $\langle \xi, \eta(y, x) \rangle = 0$.
If $x = 0, y \in]0, 1[$, then $f(0) = 0$ and $f(y) \in]-1, 0[$. Thus for each $\xi \in \partial^c f(0)$, we have

$$\langle \xi, \eta(y, 0) \rangle \in [0, 1],$$

however, $f(y) < f(0)$. Therefore, f is not pseudoinvex and so f is not η-pseudolinear on K.

Theorem 13.12 *Let $f : X \to R$ be a locally Lipschitz function on X and $K \subseteq X$ be an invex set with respect to $\eta : K \times K \to \mathbb{R}^n$ that satisfies condition C. Then, f is a η-pseudolinear function on K if and only if for all $x, y \in K$, there exists a function $p : K \times K \to \mathbb{R}$ and $\xi \in \partial^c f(x)$, such that*

$$p(x, y) > 0 \text{ and } f(y) = f(x) + p(x, y) \langle \xi, \eta(y, x) \rangle. \tag{13.27}$$

Proof Let f be η-pseudolinear on X. We need to construct a function $p : K \times K \to \mathbb{R}$, such that for all $x, y \in K$, there exists $\xi \in \partial^c f(x)$, such that

$$p(x, y) > 0 \text{ and } f(y) = f(x) + p(x, y) \langle \xi, \eta(y, x) \rangle.$$

If $\langle \xi, \eta\,(y,x) \rangle = 0$, for any $x, y \in K$, we define $p(x, y) = 1$. From Theorem 13.11, we get $f(y) = f(x)$ and thus, (13.27) holds.

If $\langle \xi, \eta\,(y, x) \rangle \neq 0$, for any $x, y \in K$, we define

$$p\,(x,y) = \frac{f\,(y) - f\,(x)}{\langle \xi, \eta\,(y, x) \rangle}. \tag{13.28}$$

Evidently, $p(x, y)$ satisfies (13.28). Now, we have to show that $p(x, y) > 0, \forall x, y \in K$. If $f(y) > f(x)$, then by pseudoincavity of f, we have $\langle \xi, \eta\,(y, x) \rangle > 0$. From (13.28), we get $p(x, y) > 0$. Similarly, if $f(y) < f(x)$, then by pseudoinvexity of f we have $\langle \xi, \eta\,(y, x) \rangle < 0$.

From (13.28), we get $p(x, y) > 0, \forall x, y \in K$.

Conversely, suppose that for any $x, y \in K$, there exists a function $p : K \times K \rightarrow \mathbb{R}$ and $\xi \in \partial^c f\,(x)$, such that (13.27) holds. If for any $x, y \in K$, $\langle \xi, \eta\,(y, x) \rangle \geq 0$. From (13.27), it follows that

$$f\,(y) - f\,(x) = p\,(x,y)\,\langle \xi, \eta\,(y,x) \rangle \geq 0$$

and hence, f is pseudoinvex on K. Similarly, if $\langle \xi, \eta\,(y, x) \rangle \leq 0$, for any $x, y \in K$. Then, from (13.27), it follows that f is pseudoincave on K.

Theorem 13.13 *Let $g : K \rightarrow \mathbb{R}$ be a locally Lipschitz η-pseudolinear function defined on open set $K \subseteq \mathbb{R}^n$. Let F be differentiable with $F'\,(\lambda) > 0$ or $F'\,(\lambda) < 0$, for all $\lambda \in \mathbb{R}$. Then the composite function $f = g \circ F$ is also η-pseudolinear on K.*

Proof Let $f(x) = goF(x), \forall x \in K$. We prove the result for $F'\,(\lambda) > 0$ and for $F'\,(\lambda) < 0$, the result follows, similarly. By Proposition 1.36, we have

$$g^\circ\,(x; \eta\,(y, x)) = F'\,(f\,(x))\,g^\circ\,(x; \eta\,(y, x)).$$

Then for all $\xi \in \partial^c g\,(x)$, we have

$$\langle \xi, \eta\,(y, x) \rangle \geq 0\,(\leq 0) \Rightarrow \langle F'\,(x)\,\zeta, \eta\,(y, x) \rangle \geq 0\,(\leq 0)\,, \forall \zeta \in \partial^c f\,(x).$$

Since, $F'\,(\lambda) > 0$, for all $\lambda \in \mathbb{R}$. Then,

$$\langle \xi, \eta\,(y, x) \rangle \geq 0\,(\leq 0) \Rightarrow \langle \zeta, \eta\,(y, x) \rangle \geq 0\,(\leq 0)\,, \forall \zeta \in \partial^c f\,(x).$$

By the η-pseudolinearity of g, it follows that

$$g(y) \geq g(x)\,(g(y) \leq g(x)).$$

Since, F is strictly increasing, we get

$$f(y) \geq f(x)\,(f(y) \leq f(x)).$$

Hence, f is η-pseudolinear on K.

The following proposition from Zhao and Tang [298] relates pseudoinvexity and prequasiinvexity under Assumption A.

Proposition 13.2 *Let K be an invex set with respect to $\eta : K \times K \to \mathbb{R}$. Suppose f is pseudoinvex with respect to η and satisfies Assumption A. Then f is prequasiinvex with respect to the same η on K.*

Proof Suppose that f is pseudoinvex with respect to η on K. On the contrary, suppose that f is not prequasiinvex with respect to the same η. Then there exists $x, y \in K$, such that

$$f(y) \leq f(x)$$

and there exists $\bar{\lambda} \in]0, 1[$, such that

$$f(z) > f(x) \geq f(y),   \tag{13.29}$$

where $z =: x + \lambda \eta(y, x)$.

Let $\psi(\lambda) =: f(x + \lambda \eta(y, x)), \lambda \in [0, 1]$. Since, f is a locally Lipschitz function, $\psi(\lambda)$ is a continuous function and hence attains its maximum. From (13.29) and $z =: x + \lambda \eta(y, x) \in K$, we know that

$$\psi(0) = f(x) < f(z),$$

hence, $\lambda = 0$ is not a maximum point.

By Assumption A and (13.29), we have

$$f(x + \lambda \eta(y, x)) = \psi(1) \leq \max\{f(x), f(y)\} = f(x) < f(z),$$

therefore, $\lambda = 1$ is not a maximum point.

Hence, there exists $\bar{\lambda} \in]0, 1[$, such that

$$f(\bar{x}) = \max_{\lambda \in [0,1]} f(x + \lambda \eta(y, x)), \ \bar{x} = x + \lambda \eta(y, x).$$

Therefore,

$$0 \in \partial^c f(\bar{x}).$$

Let $\xi = 0$, then, we have

$$\langle \xi, \eta(y, \bar{x}) \rangle = 0.$$

Since, f is pseudoinvex with respect to η, we get

$$f(y) \geq f(\bar{x}),$$

which is a contradiction to (13.29).

Theorem 13.14 *Let K be an invex set with respect to $\eta : K \times K \to \mathbb{R}$, that satisfies condition C. Suppose f is locally Lipschitz η-pseudolinear and satisfies Assumption A. Then for any $x, y \in K$, there exists $\xi \in \partial^c f(x)$, such that*

$$\langle \xi, \eta(y, x) \rangle = 0 \Rightarrow f(z_\lambda) = f(x) = f(y),$$

where $z_\lambda =: x + \lambda \eta(y, x), \lambda \in [0, 1]$.

Proof For any $x, y \in K, \lambda \in [0, 1]$, let $z_\lambda =: x + \lambda \eta (y, x)$. Suppose that there exists $\xi \in \partial^c f (x)$, such that

$$\langle \xi, \eta (y, x) \rangle = 0. \qquad (13.30)$$

Since, f is pseudoinvex with respect to η, from (13.30), we have

$$f (y) \geq f (x). \qquad (13.31)$$

By the prequasiinvexity of f with respect to the same η and Proposition 13.1, we get

$$f (z_\lambda) \leq \max \{f (x), f (y)\} = f (y). \qquad (13.32)$$

Since, $-f$ is pseudoinvex with respect to η, from (13.30), we have

$$f (y) \leq f (x). \qquad (13.33)$$

By the prequasiinvexity of $-f$ with respect to the same η and Proposition 13.1, we get

$$f (z_\lambda) \geq \min \{f (x), f (y)\} = f (y). \qquad (13.34)$$

From (13.32) and (13.34), we get

$$f (z_\lambda) = f (y).$$

From (13.31) and (13.33), we get

$$f (y) = f (x).$$

Hence, the proof is complete.

Remark 13.10 *It is clear that Theorems 13.11, 13.12, 13.13, 13.14, and Proposition 13.2 are generalizations of Theorems 13.1, 13.2, 13.3, and Proposition 13.1, respectively, to the nondifferentiable case. In turn, they are generalizations of Theorems 2.1, 2.2., 2.3, and 2.5, respectively, to the nonsmooth η-pseudolinear case.*

13.6 Characterization of Solution Set of (P) Using Locally Lipschitz η-Pseudolinear Functions

We consider the following constrained optimization problem:

$$(P) \quad \min f (x)$$

$$\text{subject to } x \in K,$$

where $f : X \to \mathbb{R}$ and $X \subseteq \mathbb{R}^n$ is an open set and K is an invex subset of X with respect to $\eta : K \times K \to \mathbb{R}$, that satisfies condition C. We assume that the solution set

$$\bar{S} = \arg \min_{x \in K} f(x) := \{x \in K : f(x) \le f(y), \forall y \in K\}$$

is nonempty.

In this section, we present the characterization of the solution set of (P) under the assumption that K is an open invex set with respect to $\eta : K \times K \to \mathbb{R}$, that satisfies condition C and the function $f : X \to \mathbb{R}$ is a locally Lipschitz η-pseudolinear function and satisfies Assumption A on the set K. Let $z_\lambda := x + \lambda \eta(x, x), \; \lambda \in [0, 1]$.

Remark 13.11 *Let $x \in \bar{S}$. Then, it is clear that $x \in \bar{S}$ if and only if $f(x) - f(\bar{x})$. Hence, by Theorem 13.11, f is η-pseudolinear on $\bar{S}$ if and only if for each $\bar{x} \in \bar{S}$, there exists $\xi \in \partial^c f(\bar{x})$, such that $\langle \xi, \eta(x, \bar{x}) \rangle = 0$. Hence, we have*

$$\bar{S} = \{x \in C : \langle \xi, \eta(x, \bar{x}) \rangle = 0\}.$$

Proposition 13.3 *The solution set $\bar{S}$ of the problem (P) is an invex set with respect to η if $f : K \to \mathbb{R}$ is a locally Lipschitz η-pseudolinear function with respect to $\eta : K \times \mathbb{K} \to \mathbb{R}$ that is skew and satisfies condition C.*

Proof Suppose $x, \bar{x} \in \bar{S}$, then we have $f(x) = f(\bar{x})$. By Theorem 13.1, $f(x) = f(\bar{x})$ if and only if there exists $\xi \in \partial^c f(\bar{x})$, such that

$$\langle \xi, \eta(x, \bar{x}) \rangle = 0.$$

Now, by condition C, we note that

$$\langle \xi, \eta(x, x + \lambda \eta(x, \bar{x})) \rangle = (1 - \lambda) \langle \xi, \eta(x, \bar{x}) \rangle = 0, \text{ for all } \lambda \in [0, 1].$$

Therefore, by Theorem 13.11, we get

$$f(x + \lambda \eta(x, \bar{x})) = f(x) = f(\bar{x}), \forall \lambda \in [0, 1],$$

Hence for each $x, \bar{x} \in \bar{S}$, we have

$$x + \lambda \eta(x, \bar{x}) \in \bar{S}, \; \forall \lambda \in [0, 1].$$

Now, we present the characterizations for the solution set of (P) using the Clarke subdifferential derived by Zhao and Tang [298].

Theorem 13.15 *Let $\bar{x} \in \bar{S}$, then $\bar{S} = \tilde{S} = \hat{S} = S^\#$, where*

$$\tilde{S} := \{x \in K : \exists \xi \in \partial^c f(x) : \langle \xi, \eta(\bar{x}, x) \rangle = 0\};$$

$$\hat{S} := \{x \in K : \exists \zeta \in \partial^c f(\bar{x}) : \langle \zeta, \eta(x, \bar{x}) \rangle = 0\};$$

$$S^\# := \{x \in K : \exists \varsigma \in \partial^c f(z_\lambda) : \langle \varsigma, \eta(\bar{x}, x) \rangle = 0, \forall \lambda \in [0, 1]\}.$$

Proof The point $x \in \bar{S}$ if and only if $f(x) = f(\bar{x})$. By Theorem 13.11, $x \in \bar{S}$ if and only if there exists $\xi \in \partial^c f(x)$ such that $\langle \xi, \eta(\bar{x}, x) \rangle = 0$, hence, $\bar{S} = \tilde{S}$. To prove that $\tilde{S} = \hat{S}$, let $x \in \tilde{S}$, then there exists $\xi \in \partial^c f(x)$, such that $\langle \xi, \eta(\bar{x}, x) \rangle = 0$. By Theorem 13.11, we get $f(x) = f(\bar{x})$, which implies that there exists $\zeta \in \partial^c f(\bar{x})$, such that $\langle \zeta, x - \bar{x} \rangle = 0$, that is $\langle \zeta, \bar{x} - x \rangle = 0$ and we get $\tilde{S} \subseteq \hat{S}$. Similarly, we can show that $\hat{S} \subseteq \tilde{S}$. Hence, $\tilde{S} = \hat{S}$.

Now, to show that $\bar{S} = S^{\#}$. Let $x \in \bar{S}$, then $f(x) = f(\bar{x})$, which implies that there exists $\xi \in \partial^c f(x)$, such that

$$\langle \xi, \eta(\bar{x}, x) \rangle = 0.$$

By Theorem 13.13 for every $\lambda \in]0, 1]$, $z_\lambda = x + \lambda \eta(\bar{x}, x)$, we have $f(z_\lambda) = f(x)$. Moreover, from Theorem 13.11, there exists $\varsigma \in \partial^c f(z_\lambda)$, such that

$$\langle \varsigma, \eta(x, z_\lambda) \rangle = 0.$$

Now, by condition C, we have

$$\langle \varsigma, \eta(\bar{x}, x) \rangle = \frac{1}{\lambda} \langle \varsigma, \eta(x, z_\lambda) \rangle = 0, \forall \lambda \in]0, 1].$$

When $\lambda = 0$, $\langle \varsigma, \eta(\bar{x}, x) \rangle = 0$ is obviously true. Therefore, $\langle \varsigma, \eta(\bar{x}, x) \rangle = 0$, $\forall \lambda \in [0, 1]$. Hence, $x \in S^{\#}$ and $\bar{S} \subseteq S^{\#}$.

Conversely, let $x \in S^{\#}$, then for all $\lambda \in]0, 1]$, $z_\lambda = x + \lambda \eta(\bar{x}, x)$, there exists $\varsigma \in \partial^c f(z_\lambda)$, such that $\langle \varsigma, \eta(\bar{x}, x) \rangle = 0$, hence, $\frac{1}{\lambda} \langle \varsigma, \eta(z_\lambda, x) \rangle = 0$, that is

$$\langle \varsigma, \eta(z_\lambda, x) \rangle = 0 \Rightarrow f(x) = f(z_\lambda) \Rightarrow \exists \xi \in \partial^c f(x) \text{ such that } \langle \xi, \eta(z_\lambda, x) \rangle = 0$$

$$\Rightarrow \lambda \langle \xi, \eta(\bar{x}, x) \rangle = 0 \Rightarrow \langle \xi, \eta(\bar{x}, x) \rangle = 0 \Rightarrow f(x) = f(\bar{x}).$$

Therefore, $S^{\#} \subseteq \bar{S}$. If $\lambda = 0$, then $S^{\#} = \tilde{S} \subseteq \bar{S}$. Hence, $S^{\#} \subseteq \bar{S}$ is always true.

Corollary 13.3 *Let* $\bar{x} \in \bar{S}$, *then* $\bar{S} = \tilde{S}_1 = \hat{S}_1 = S_1^{\#}$, *where*

$$\tilde{S}_1 = \{x \in K : \exists \xi \in \partial^c f(x), \langle \xi, \eta(\bar{x}, x) \rangle \geq 0\};$$

$$\hat{S}_1 = \{x \in K : \exists \zeta \in \partial^c f(\bar{x}), \langle \zeta, \langle \xi, \eta(x, \bar{x}) \rangle \rangle \geq 0\};$$

$$S_1^{\#} = \{x \in K : \exists \varsigma \in \partial^c f(z_\lambda), \langle \varsigma, \eta(\bar{x}, x) \rangle \geq 0, \forall \lambda \in [0, 1]\}.$$

Proof By Theorem 13.14, it is clear that $\bar{S} \subseteq \tilde{S}_1$. Assume that $x \in \tilde{S}_1$, that is $x \in K$ and there exists $\xi \in \partial^c f(x)$, such that

$$\langle \xi, \eta(\bar{x}, x) \rangle \geq 0.$$

Then, by Theorem 13.12, there exists $\xi \in \partial^c f(x)$, such that

$$f(\bar{x}) = f(x) + p(x, \bar{x}) \langle \xi, \eta(\bar{x}, x) \rangle \geq f(x),$$

which implies that $x \in \bar{S}$. Hence, $\bar{S} = \tilde{S}_1$. Similarly, we can prove that $\bar{S} = \hat{S}_1$. Now, to prove that $\bar{S} = S_1^*$, we note that

$$\bar{S} = S^* \subset S_1^* \subset \tilde{S}_1 = \bar{S}.$$

This completes the proof.

Theorem 13.16 *Let* $\bar{x} \in \bar{S}$, *then* $\bar{S} = \tilde{S}_2 = \hat{S}_2$, *where*

$$\tilde{S}_2 := \{x \in K : \exists \xi \in \partial^c f(x), \exists \zeta \in \partial^c f(\bar{x}) := \langle \xi, \eta(\bar{x}, x) \rangle = \langle \zeta, \eta(x, \bar{x}) \rangle\};$$

$$\hat{S}_2 := \{x \in K : \exists \xi \in \partial^c f(x), \exists \zeta \in \partial^c f(\bar{x}) : \langle \xi, \eta(\bar{x}, x) \rangle \geq \langle \zeta, \eta(x, \bar{x}) \rangle\}.$$

Proof First, we prove that $\bar{S} \subseteq \tilde{S}_2$. Let $x \in \bar{S}$, then we have $f(x) = f(z)$. It follows from Theorem 13.11, that $\langle \xi, \eta(\bar{x}, x) \rangle = 0$ and $\langle \zeta, \eta(x, \bar{x}) \rangle = 0$. Then, $\langle \zeta, \eta(\bar{x}, x) \rangle = 0 = \langle \xi, \eta(x, \bar{x}) \rangle$. Thus $x \in \tilde{S}_2$.

The inclusion $\tilde{S}_2 \subseteq \hat{S}_2$ is clearly true.

Finally, we prove that $\hat{S}_2 \subseteq \bar{S}$. Suppose that $x \in \hat{S}_2$, then x satisfies

$$\langle \xi, \eta(\bar{x}, x) \rangle \geq \langle \zeta, \eta(x, \bar{x}) \rangle. \tag{13.35}$$

Since, $x, \bar{x} \in \bar{S}$, from Theorem 13.13, there exists $\zeta \in \partial^c f(\bar{x})$, such that

$$\langle \zeta, \eta(x, \bar{x}) \rangle = 0.$$

Using (13.35) there exists $\xi \in \partial^c f(x)$, such that

$$\langle \xi, \eta(\bar{x}, x) \rangle \geq 0.$$

By pseudoinvexity of f, we get $f(\bar{x}) \geq f(x)$, which is a contradiction. Hence, $x \in \bar{S}$.

Remark 13.12 *It is clear that Theorem 13.15, Corollary 13.3, and Theorem 13.16 are generalizations of Theorem 13.6, Corollary 13.1, and Theorem 13.7, respectively, to the nondifferentiable case. In turn, Theorem 13.15, Corollary 13.3, and Theorem 13.16 generalize, Theorem 3.1, Corollary 3.1, and Theorem 3.2, respectively, from a smooth to nonsmooth case as well as from pseudolinear to the η-pseudolinear case.*

Chapter 14

Smooth Pseudolinear Functions on Riemannian Manifolds

14.1 Introduction

The knowledge of the structure of the gradient and the Hessian matrix of pseudolinear functions have their importance, not only in theory but also in algorithms to solve some special functional optimization problems. Rapcsak [233] has given an explicit formulation of the gradient of smooth pseudolinear functions, by employing differential geometric tools.

In linear topological spaces, the notion of convexity relies on the possibility of connecting any two points of the space by the line segment between them. In several real-world applications, it is not always possible to connect the points through line segments. This led to the idea of the generalization of the classical notion of convexity. Udriste [274] and Rapcsak [234] proposed a generalization of the convexity notion by replacing the linear space by a Riemannian manifold, the line segment by a geodesic segment between any two points, and the convex functions by the positiveness of their Riemann curvature. A geodesic on the Riemannian manifold is a curve, that locally minimizes the arc length. In particular, Udriste [274]) proved the classical Karush-Kuhn-Tucker theorem for the problem:

$$\min f(x), \ \text{subject to} \ x \in X,$$

where $f : M \to \mathbb{R}$ is a geodesically convex function, (M, g) is a complete Riemannian manifold, and $X = \{x \in M : h_j(x) \le 0, j = 1, ..., m\}$, where $h_j : M \to \mathbb{R}, j = 1, ..., m$ are geodesically convex functions. Udriste's [274] generalization is based on the fact that many of the properties of convex programs on Euclidean space carry over to the case of a complete Riemannian manifold. Following Udriste [274], several other generalizations of convex sets and functions have been proposed on the Riemannian manifold. Barani and Pouryayevali [13] introduced the notion of a geodesic invex subset of a Riemannian manifold and defined preinvex and invex functions on a geodesic invex set. Mishra and Giorgi [201] developed a vector programming on differentiable manifolds and derived Karush-Kuhn-Tucker type sufficient optimality conditions of Pareto optimality for a vector optimization problem using generalized convex functions. Recently, Barani and Pouryayevali [14] defined the notion

365

of d-invexity and applied them to obtain the optimality conditions for vector optimization problems on Riemannian manifolds.

Pripoae and Pripoae [229] have extended the notion of Riemannian convexity to affine differentiable convexity by replacing the geodesic segments by segments of auto-parallel curves of some arbitrary linear connections and the Riemannian Hessian by the affine differential one. Recently, Pripoae *et al.* [230] have generalized the classical notion of η-pseudolinearity from classical to differential setting and provided several characterizations for η-pseudolinear functions. The variation in this study was that the auto-parallel curves have been replaced by other relevant family of curves like flow of some vector field families. This theory depends only on some differentiable objects on manifolds and is therefore less restrictive than the affine directional one.

In this chapter, we present an explicit formulation of the gradients of smooth pseudolinear functions, studied by Rapcsak [233]. Moreover, we present some results of Pripoae *et al.* [230] on characterizations of η-pseudolinear functions and the solution sets of η-pseudolinear programming problems. These results are extensions of the corresponding results on η-pseudolinear functions from the n-dimensional Euclidean space studied in Chapter 13, to arbitrary differentiable manifold space. For what concerns more direct applications of the topics of the present chapter, we refer the reader to an earlier work of Miglierina [185] for the case of compact manifolds and to Udriste *et al.* [275] for the general case.

14.2   Characterizations of Pseudolinear Functions

We start the section with the definition of differentiable manifolds:

Definition 14.1 (Differentiable manifold) *A differentiable manifold of dimension n is a set M and a family of injective maps $x_\alpha : U_\alpha \to M$, where $U_\alpha \subseteq \mathbb{R}^n$ are open sets in $\mathbb{R}^n$, such that:*

(i) $\bigcup_\alpha x_\alpha (U_\alpha) = M$;

(ii) for any pair α with $x_\alpha (U_\alpha) \bigcap x_\beta (U_\beta) = W \neq \phi$, the sets $x_\alpha^{-1} (W)$ and $x_\beta^{-1} (W)$ are open sets in $\mathbb{R}^n$ and the mapping $x_\beta^{-1} \circ x_\beta : x_\beta^{-1} (W) \to x_\beta (W)$ is differentiable;

(iii) the family $\{U_\alpha, x_\alpha\}_\alpha$ is maximal relative to conditions (i) and (ii), i.e., every other family $\{V_\alpha, y_\alpha\}_\alpha$ satisfies (i) and (ii) is included in $\{U_\alpha, x_\alpha\}_\alpha$.

The pair (U_α, x_α) with $p \in x_\alpha(U_\alpha)$ is called a parameterization (or system of coordinates) of M at p; $x_\alpha(U_\alpha)$ is then called a coordinate neighborhood at p.

We observe that the definition does not directly inform about the topological structure of the differentiable manifolds. However, each differentiable manifold can be made into a topological space using the family of finite intersections of a coordinate neighborhood as a base of the topology.

Definition 14.2 *Let M and N be differentiable manifolds. A differentiable mapping $f : M \to N$ is differentiable at $p \in M$, if given a parameterization $y : V \subseteq \mathbb{R}^n \to N$ at $f(p)$, there exists a parameterization $x : U \subseteq \mathbb{R}^m \to M$ at p, such that $f(x(U)) \subseteq y(V)$ and the mapping*

$$y \circ f \circ x^{-1} : U \subseteq \mathbb{R}^m \to \mathbb{R}^n$$

is differentiable at $x^{-1}(p)$.

Definition 14.3 *(Tangent space) Let M be a differentiable manifold. A differentiable function $\alpha : (-\epsilon, \epsilon) \to M$ is called a curve in M. Suppose that $\alpha(0) = p$ and let Ω be the set of functions $f : M \to \mathbb{R}$ differentiable at p. The tangent vector to the curve α at $t = 0$ is a function $\alpha(0) : \Omega \to \mathbb{R}$ given by $\alpha'(0)f = (\frac{d(f \circ \alpha)}{dt})_{t=0}$, where $f \in \Omega$. A tangent vector to M at p is the tangent vector at $t = 0$ of some curve $\alpha : (-\epsilon, \epsilon) \to \mathbb{R}$ with $\alpha(0) = p$. The set of all tangent vectors to M at p will be denoted by T_pM and will be called tangent space to M at p.*

Definition 14.4 *(Riemannian metric) A Riemannian metric g on a differentiable manifold M is a correspondence which associates to each point $p \in M$ an inner product g_p (that is, a symmetric bilinear, positive-definite form) on the tangent space T_pM such that, if $x : U \subseteq \mathbb{R}^n \to M$ is a system of coordinates around p , then*

$$g_q \left(\frac{\partial}{\partial x_i}(q), \frac{\partial}{\partial x_j}(q) \right) = g_{ij}(x(q)),$$

is a differentiable function for each $q \in U$.

Definition 14.5 *(Riemannian manifold, Do Carmo [68]) Let M be a differentiable manifold and g be a Riemannian metric on M. The pair (M, g) is called the Riemannian manifold.*

The following lemmas are needed to prove the main theorem.

Lemma 14.1 *(Hicks [104]) A Riemannian submanifold immersed in $\mathbb{R}^n$ is Euclidean if and only if its Riemannian curvature tensor vanishes identically.*

Lemma 14.2 *(Hicks [104]) Let a Riemannian submanifold of codimension 1 (hypersurface) be immersed in $\mathbb{R}^n$. Then,*

$$R_{ijkl} = b_{ik}b_{jl} - b_{il}b_{jk}, \ i,j,k,l = 1,...,n-1, \tag{14.1}$$

where R_{ijkl}, $i, j, k, l = 1, ..., n - 1$ are the components of the Riemannian curvature tensor of the hypersurface b_{ij}, $i, j = 1, ..., n - 1$ are the second fundamental quantities of the hypersurface.

The following lemma from Rapcsak [233] provides a characterization for a twice continuously differentiable function satisfying a particular condition.

Lemma 14.3 *Suppose $f(x)$ is a twice continuously differentiable function on an open set $M \subset \mathbb{R}^n$, such that $\nabla f(x) \neq 0$, $x \in M$. Assume that there exists a continuously differentiable function $l(x), \eta_i(f(x)), i = 1, ..., n, x \in M$, satisfying the following conditions:*

$$\frac{\partial f(x)}{\partial x_i} = l(x)\,\eta_i(f(x)), i = 1, ..., n, \quad x \in M. \tag{14.2}$$

Then, the second fundamental form of hypersurface,

$$M_{f(\bar{x})} = \{x : f(x) = f(\bar{x}), x \in M\}, \tag{14.3}$$

vanishes identically for the arbitrary point $\bar{x} \in M$.

Proof Since $\nabla f(x) \neq 0$, $x \in M$, the level set

$$M_{f(\bar{x})} = \{x : f(x) = f(\bar{x}), x \in M\}, \tag{14.4}$$

is a hypersurface in $\mathbb{R}^n$ for arbitrary $x_0 \in M$. We choose a point $\hat{x}$ of the hypersurface $M_{f(\bar{x})}$. Without a loss of generality, we can assume that $\partial f(\hat{x})/\partial x_n \neq 0$. Then, the second fundamental quantities (the elements of the matrix of second fundamental form) are determined in a neighborhood $B(\hat{x})$ of $\hat{x}$ by using the gradient vectors of $f(x)$ and the Hessian matrices of $f(x)$ as follows (see, Rapcsak [231, 232]):

$$b_{ij}(x) = -\frac{1}{|\nabla f(x)|}\frac{\partial^2 f(x)}{\partial x_i \partial x_j} + \frac{1}{|\nabla f(x)|}\frac{\frac{\partial^2 f(x)}{\partial x_i \partial x_n}\frac{\partial f(x)}{\partial x_j}}{\frac{\partial f(x)}{\partial x_n}} + \frac{1}{|\nabla f(x)|}\frac{\frac{\partial^2 f(x)}{\partial x_n \partial x_j}\frac{\partial f(x)}{\partial x_i}}{\frac{\partial f(x)}{\partial x_n}}$$

$$-\frac{1}{|\nabla f(x)|}\frac{\frac{\partial^2 f(x)}{\partial x_n^2}\frac{\partial f(x)}{\partial x_j}\frac{\partial f(x)}{\partial x_i}}{\left(\frac{\partial f(x)}{\partial x_n}\right)^2}, \quad i, j = 1, ..., n-1, \ x \in B(\hat{x}). \tag{14.5}$$

Using the relations (14.2) and (14.5), we get

$$|\nabla f(x)|\left(\frac{\partial f(x)}{\partial x_n}\right)^2 b_{ij}(x) = -\left(\frac{\partial l(x)}{\partial x_j}\eta_i(f(x)) + l(x)\frac{d\eta_i(f(x))}{df}\frac{\partial f(x)}{\partial x_j}\right)$$

$$\left(\frac{\partial l(x)}{\partial x_n}\right)^2 + \left(\frac{\partial l(x)}{\partial x_n}\eta_i(f(x)) + l(x)\frac{d\eta_i(f(x))}{df}\frac{\partial f(x)}{\partial x_n}\right)\frac{\partial f(x)}{\partial x_j}\frac{\partial f(x)}{\partial x_n}$$

$$+\left(\frac{\partial l(x)}{\partial x_j}\eta_n(f(x)) + l(x)\frac{d\eta_n(f(x))}{df}\frac{\partial f(x)}{\partial x_j}\right)\frac{\partial f(x)}{\partial x_i}\frac{\partial f(x)}{\partial x_n}$$

$$-\left(\frac{\partial l(x)}{\partial x_n}\eta_n(f(x)) + l(x)\frac{d\eta_n(f(x))}{df}\frac{\partial f(x)}{\partial x_n}\right)\frac{\partial f(x)}{\partial x_i}\frac{\partial f(x)}{\partial x_j}$$

$$= -\frac{\partial l\left(x\right)}{\partial x_j}\eta_i\left(f\left(x\right)\right)\left(\frac{\partial f\left(x\right)}{\partial x_n}\right)^2 - l\left(x\right)\frac{d\eta_i\left(f\left(x\right)\right)}{df}\frac{\partial f\left(x\right)}{\partial x_j}\left(\frac{\partial f\left(x\right)}{\partial x_n}\right)^2$$

$$+\frac{\partial l\left(x\right)}{\partial x_n}\eta_i\left(f\left(x\right)\right)\frac{\partial f\left(x\right)}{\partial x_j}\frac{\partial f\left(x\right)}{\partial x_n} + l\left(x\right)\frac{d\eta_i\left(f\left(x\right)\right)}{df}\frac{\partial f\left(x\right)}{\partial x_n}\frac{\partial f\left(x\right)}{\partial x_j}\frac{\partial f\left(x\right)}{\partial x_n}$$

$$+\frac{\partial l\left(x\right)}{\partial x_j}\eta_n\left(f\left(x\right)\right)\frac{\partial f\left(x\right)}{\partial x_i}\frac{\partial f\left(x\right)}{\partial x_n} + l\left(x\right)\frac{d\eta_n\left(f\left(x\right)\right)}{df}\frac{\partial f\left(x\right)}{\partial x_j}\frac{\partial f\left(x\right)}{\partial x_i}\frac{\partial f\left(x\right)}{\partial x_n}$$

$$-\frac{\partial l\left(x\right)}{\partial x_n}\eta_n\left(f\left(x\right)\right)\frac{\partial f\left(x\right)}{\partial x_i}\frac{\partial f\left(x\right)}{\partial x_j} - l\left(x\right)\frac{d\eta_n\left(f\left(x\right)\right)}{df}\frac{\partial f\left(x\right)}{\partial x_n}\frac{\partial f\left(x\right)}{\partial x_i}\frac{\partial f\left(x\right)}{\partial x_j}$$

$$= -\frac{\partial l\left(x\right)}{\partial x_j}\eta_i\left(f\left(x\right)\right)\left(\frac{\partial f\left(x\right)}{\partial x_n}\right)^2 + \frac{\partial l\left(x\right)}{\partial x_n}\eta_i\left(f\left(x\right)\right)\frac{\partial f\left(x\right)}{\partial x_j}\frac{\partial f\left(x\right)}{\partial x_n}$$

$$+\frac{\partial l\left(x\right)}{\partial x_j}\eta_n\left(f\left(x\right)\right)\frac{\partial f\left(x\right)}{\partial x_i}\frac{\partial f\left(x\right)}{\partial x_n} - \frac{\partial l\left(x\right)}{\partial x_n}\eta_n\left(f\left(x\right)\right)\frac{\partial f\left(x\right)}{\partial x_i}\frac{\partial f\left(x\right)}{\partial x_j}$$

$$= -\frac{\partial l\left(x\right)}{\partial x_j}\eta_i\left(f\left(x\right)\right)l^2\left(x\right)\left(\eta_n\left(f\left(x\right)\right)\right)^2 + \frac{\partial l\left(x\right)}{\partial x_n}\eta_i\left(f\left(x\right)\right)l^2\left(x\right)\eta_j\left(f\left(x\right)\right)$$

$$\eta_n\left(f\left(x\right)\right) + \frac{\partial l\left(x\right)}{\partial x_j}\eta_n\left(f\left(x\right)\right)l^2\left(x\right)\eta_i\left(f\left(x\right)\right)\eta_n\left(f\left(x\right)\right) - \frac{\partial l\left(x\right)}{\partial x_n}\eta_n\left(f\left(x\right)\right)l^2\left(x\right)$$

$$\eta_i\left(f\left(x\right)\right)\eta_j\left(f\left(x\right)\right) = 0. \tag{14.6}$$

This completes the proof.

The following theorem from Rapcsak [233] provides the conditions for a three times continuously differentiable function to be a pseudolinear function.

Theorem 14.1 *Suppose that $f\left(x\right)$ is a three times continuously differentiable function defined on an open convex set $M \subseteq \mathbb{R}^n$ and assume that $\nabla f\left(x\right) \neq 0$, $x \in M$. Then, $f\left(x\right)$ is pseudolinear if and only if there exists continuously differentiable functions $l\left(x\right), \eta_i\left(f\left(x\right)\right), i = 1, ..., n, \ x \in M$, such that the following conditions are satisfied:*

$$\frac{\partial f\left(x\right)}{\partial x_i} = l\left(x\right)\eta_i\left(f\left(x\right)\right), i = 1, ..., n, \quad x \in M. \tag{14.7}$$

Proof (Necessity) Suppose that the function $f\left(x\right)$ is pseudolinear on M. Let $\bar{x} \in M$ be any point, then by Lemma 14.1, we get

$$M_{f(\bar{x})} = \left\{x : f\left(x\right) = f\left(\bar{x}\right), x \in M\right\}. \tag{14.8}$$

Therefore, by the pseudolinearity of f, we get

$$M_{f(\bar{x})} = \left\{x : \langle \nabla f\left(\bar{x}\right), \left(x - \bar{x}\right)\rangle = 0, \ x \in M\right\}. \tag{14.9}$$

This relation implies that the direction of the vectors $\nabla f\left(x\right), x \in M_{f(\bar{x})}$, coincides with the direction of the vector $\nabla f\left(\bar{x}\right)$, at most the length of the vectors $\nabla f(x), x \in M_{f(\bar{x})}$, can be different. But, this fact is included in the relation (14.7).

(Sufficiency) Suppose that condition (14.7) is satisfied on M. Since $\nabla f(x) \neq 0,\ x \in M$, the level set

$$M_{f(\bar{x})} = \{x : f(x) = f(\bar{x}), x \in M\}, \tag{14.10}$$

is a hypersurface in $\mathbb{R}^n$ for arbitrary $\bar{x} \in M$. By Lemmas 14.1, 14.2, and 14.3, this is a hyperplane in $\mathbb{R}^n$ for arbitrary $x_0 \in M$. Then by Theorem 2.1, $f(x)$ is pseudolinear on M.

This completes the proof.

The following example from Rapcsak [233] justifies the significance of Theorem 14.1.

Example 14.1 (i) *If $f(x) = e^{a^T x + b}, a \neq 0$, is defined on an open convex set $M \subseteq \mathbb{R}^n$, then, $l(x) = 1$ and $\eta(f(x)) = f(x)a$.*

(ii) *If $f(x) = (a^T x + b) / (c^T x + d)$ is defined on an open convex set $M \subseteq \mathbb{R}^n$, where $c^T x + d > 0$ and $\nabla f(x) \neq 0, x \in M$, then, $l(x) = 1/(c^T x + d)$ and $\eta(f(x)) = a - f(x)c$.*

(iii) *If $f(x, y) = x + \sqrt{x^2 + y + 1}, x > 0, y > 0$, then, $l(x, y) = 1/\sqrt{x^2 + y + 1}$ and $\eta^T(f(x, y)) = (f(x, y), \frac{1}{2})$.*

Corollary 14.1 *Let $f(x)$ be a three times continuously differentiable function and pseudolinear on an open convex set $M \subseteq \mathbb{R}^n$, such that $\nabla f(x) \neq 0$, $x \in M$. Then the Hessian matrix is the following*

$$\nabla^2 f(x) = \eta(f(x)) \nabla l(x) + l^2(x) \frac{d\eta(f(x))}{df} \eta^T(f(x)). \tag{14.11}$$

Proof Applying Theorem 14.1, there exists a continuously differentiable function

$$l(x), \eta_i(f(x)), \quad i = 1, ..., n, \quad x \in M, \tag{14.12}$$

such that

$$\nabla f(x) = l(x) \eta^T(f(x)). \tag{14.13}$$

Differentiating (14.13) with respect to x, we get

$$\nabla^2 f(x) = \eta(f(x)) \nabla l(x) + l(x) \frac{d\eta(f(x))}{df} \nabla f(x)$$

$$= \eta(f(x)) \nabla l(x) + l^2(x) \frac{d\eta(f(x))}{df} \eta^T(f(x)). \tag{14.14}$$

This completes the proof.

The following example from Rapcsak [233] justifies the significance of Corollary 14.1.

Example 14.2 If $f(x) = (a^T x + b) / (c^T x + d)$ is defined on an open convex set $M \subseteq \mathbb{R}^n$, where $c^T x + d > 0$ and $\nabla f(x) \neq 0, x \in M$, then

$$\nabla^2 f(x) = \frac{(a - f(x) c) c^T}{(c^T x + d)^2} - \frac{c (a^T - f(x) c^T)}{(c^T x + d)^2} = \frac{-ca^T - ac^T + 2f(x) cc^T}{(c^T x + d)^2}$$
(14.15)

Corollary 14.2 Let $f(x)$ be a three times continuously differentiable function and pseudolinear on an open convex set $M \subseteq \mathbb{R}^n$ such that $\nabla f(x) \neq 0, x \in M$. Then, for arbitrary $\bar{x} \in M$ the continuously differentiable function $l(x), x \in M$ defined in (14.7), satisfies the following system of partial differential equations:

$$\frac{\partial l(x)}{\partial x_i} = c_{ij} \frac{\partial l(x)}{\partial x_j}, \quad \text{for some } j \in \{1, ..., n\}, \quad i = 1, ..., n, \ i \neq j, \ x \in M_{f(\bar{x})},$$
(14.16)

for arbitrary $\bar{x} \in M$.

Proof Applying Theorem 14.1, there exist continuously differentiable functions

$$l(x), \quad \eta_i(f(x)), \quad i = 1, ..., n, \ i \neq j, \quad x \in M_{f(\bar{x})},$$
(14.17)

such that the following conditions are satisfied:

$$\frac{\partial f(x)}{\partial x_i} = l(x) \eta_i(f(x)), \quad i = 1, ..., n, \quad x \in M.$$
(14.18)

Since, $\nabla f(x) \neq 0, x \in M$, we can assume that $\partial f(x_0) / \partial x_j \neq 0$ for some j. Therefore, we have

$$\frac{\partial l(x)}{\partial x_j} \eta_i(f(\bar{x})) = \frac{\partial^2 f(x)}{\partial x_i \partial x_j} = \frac{\partial^2 f(x)}{\partial x_j \partial x_i} = \frac{\partial l(x)}{\partial x_i} \eta_j(f(\bar{x})), \quad x \in M_{f(\bar{x})}.$$
(14.19)

For $c_{ij} = \eta_i(f(\bar{x})) / \eta_j(f(\bar{x})), \ i = 1, ..., n, \ i \neq j$ the statement holds.

Remark 14.1 The function $l(a^T x + b)$, where l is an arbitrary continuously differentiable function of a single variable satisfies the system (14.6).

Corollary 14.3 Suppose that $f(x)$ is a twice continuously differentiable function defined on an open convex set $M \subseteq \mathbb{R}^n$. If there exist continuously differentiable functions $\eta_i(f(x)), \ i = 1, ..., n, \ x \in M$, such that

$$\frac{\partial f(x)}{\partial x_i} = \eta_i(f(x)), \quad i = 1, ..., n, \quad x \in M.$$
(14.20)

Therefore, there exists a function $\eta(f(x)) > 0$ satisfying the following equations:

$$\eta_i(f(x)) = c_i \eta(f(x)), \quad i = 1, ..., n, \quad x \in M,$$
(14.21)

where $c_i, i = 1, ..., n$ are constants.

Proof Since, the function $f(x)$ is twice continuously differentiable, therefore, we have

$$\frac{\partial^2 f(x)}{\partial x_i \partial x_j} = \frac{d\eta_i(f(x))}{df} \frac{\partial f(x)}{\partial x_j} = \frac{d\eta_j(f(x))}{df} \frac{\partial f(x)}{\partial x_i} = \frac{\partial^2 f(x)}{\partial x_j \partial x_i}, \quad i,j = 1,...,n.$$
(14.22)

Since $\eta_i(f(x)) > 0$, $i = 1,...,n$, $x \in M$, the system

$$\frac{\frac{d\eta_i(f(x))}{df}}{\eta_i(f(x))} = \frac{\frac{d\eta_j(f(x))}{df}}{\eta_j(f(x))}, \quad i,j = 1,...,n, \quad x \in M,$$
(14.23)

may be solved to yield

$$\eta_i(f(x)) = c_{ij}\eta_j(f(x)), \quad i,j = 1,...,n, \quad x \in M.$$
(14.24)

Hence, there exists a function $\eta(f(x)) > 0$, such that

$$\eta_i(f(x)) = c_i\eta(f(x)), \quad i = 1,...,n, \quad x \in M.$$
(14.25)

This completes the proof.

14.3   Construction of Pseudolinear Functions

To construct pseudolinear functions, Rapcsak [233] applied the Frobenius theorem (see, Hicks [104]) to this case.

Theorem 14.2 *Suppose that the functions* $l(x), \eta_i(f(x))$, $i = 1,...,n$, $x \in M$, *have continuous derivatives in all arguments on an open set* $M \subseteq \mathbb{R}^n$ *and they satisfy the compatibility conditions*

$$\frac{\partial l(x)}{\partial x_j}\eta_i(f(x)) + l^2(x)\frac{d\eta_i(f(x))}{df}\eta_j(x) = \frac{\partial l(x)}{\partial x_i}\eta_j(f(x))$$

$$+ l^2(x)\frac{d\eta_j(f(x))}{df}\eta_i(x),$$
(14.26)

$i,j = 1,...,n$, $x \in M$, *then a uniquely determined solution of the system*

$$\frac{\partial f(x)}{\partial x_i} = l(x)\eta_i(f(x)), \quad i = 1,...,n, \quad x \in M,$$
(14.27)

exists in a neighborhood of every point of M *as soon as the value of the function* $f(x)$ *is prescribed at some points of the neighborhood.*

From this point of view the following lemma is an important result.

Lemma 14.4 (Komlosi [144]) *Let a function* $f(x)$ *be differentiable on an open convex set* $M \subset \mathbb{R}^n$. *Then* $f(x)$ *is pseudolinear on* M *if and only if it is pseudolinear in a convex neighborhood of every point of* M.

14.4 Characterizations of η-Pseudolinearity on Differentiable Manifolds

In this section, we present several characterizations for η-pseudolinear functions established by Pripoae *et al.* [230]. These results are generalizations of similar results in the classical settings given in Chapters 2, 3, and 13 from n-dimensional Euclidean space $\mathbb{R}^n$ to arbitrary differentiable manifolds.

Let M be an n-dimensional differentiable manifold and TM be the total space of its tangent bundles. Let $\Omega(M)$ be the algebra of differentiable functions on M and $\chi(M)$ be the $\Omega(M)$-module of (differentiable) vector field.

Let $\eta : M \times M \to TM$ be a function such that $\eta(x,y)$ belongs to the tangent space of M at y, for every $x, y \in M$. The function η is called a bundle function on M. In particular, $\eta(y,.)$ may be a vector field in $\chi(M)$, for every $x \in M$. In this case, the function η depends smoothly on the second variable.

The following definition is from Pini [222].

Definition 14.6 *A function $f \in \Omega(M)$ is said to be η-convex (invex), if for all $x, y \in M$, we have*

$$f(y) - f(x) \geq df_x(\eta(y,x)).$$

Now, we recall following definitions from Pripoae *et al.* [230].

Definition 14.7 *A function $f \in \Omega(M)$ is said to be*

(i) η-pseudoconvex (pseudoinvex) on M, if for all $x, y \in M$, we have

$$df_x(\eta(y,x)) \geq 0 \Rightarrow f(y) \geq f(x);$$

(ii) η-pseudoconcave (pseudoincave) on M, if for all $x, y \in M$, we have

$$df_x(\eta(y,x)) \leq 0 \Rightarrow f(y) \leq f(x).$$

The function f is said to be η-pseudolinear on M, if f is both η-pseudoconvex and η-pseudoconcave on M. To be precise; the function $f \in \Omega(M)$ is said to be η-pseudolinear on M, if for all $x, y \in M$, we have

$$df_x(\eta(y,x)) \geq 0 \Rightarrow f(y) \geq f(x)$$

and

$$df_x(\eta(y,x)) \leq 0 \Rightarrow f(y) \leq f(x).$$

We note that when M is an open set in $\mathbb{R}^n$, the classical notions of η-pseudolinear functions studied in Chapter 13 are obtained. In this definition, the classical notion of the gradient is replaced by the exterior differential, which enables us to extend the corresponding results from open sets in $\mathbb{R}^n$ to arbitrary differentiable manifolds.

Definition 14.8 *A family of parametrized curves, indexed on $M \times M$ is said to be η-compatible, if for every $x, y \in M$, $c_{x,y} : [0,1] \to M$ satisfies the properties*

1. $c_{x,y}(0) = x$,

2. $c_{x,y}(1) = y$,

3. $c'_{x,y}(t) = \eta(y, c_{x,y}(t))$, $\forall t \in [0,1]$.

Definition 14.9 *A subset K of M is called η-convex, if there exists a family of η-compatible curves, indexed on $K \times K$ and whose images are contained in K, containing x with y, for every $x, y \in K$.*

The set K is said to be strongly η-convex, if it is η-convex, and if it contains all the families of η-compatible curves containing any of these two points, which may coincide. More specifically, η-compatible loop curves through its points.

Definition 14.10 *Let K be an η-convex subset of M and c be a family of η-compatible curves, whose images are contained in M. A function $f : K \to \mathbb{R}$ is called preinvex with respect to η and c if for every $x, y \in K$ and $t \in [0,1]$, we have*

$$f(c_{x,y}(t)) \leq tf(y) + (1-t)f(x).$$

Definition 14.11 (Condition C) *The function η satisfies condition C with respect to c and f if, for every $x, y \in K$ and $t \in [0,1]$,*

$$df_{c_{x,y}(t)}(\eta(x, c_{x,y}(t))) = -tdf_x(\eta(y,x))$$

and

$$df_{c_{x,y}(t)}(\eta(y, c_{x,y}(t))) = (1-t)df_x(\eta(y,x)).$$

Definition 14.12 *The function η is said to be skew-symmetric with respect to f, if for every $x, y \in K$,*

$$df_x(\eta(x,y)) = -df_y(\eta(y,x)).$$

For convenience, we shall write $\eta(x,y) = -\eta(y,x)$. It is clear that condition C implies the skew-symmetry of η.

Remark 14.2 *Suppose that $\eta : \mathbb{R}^n \times \mathbb{R}^n \to \mathbb{R}^n$ is a family of vector fields defined by $\eta(y,x) = y - x$ (the "affine structure function" on $\mathbb{R}^n$) and $c : \mathbb{R}^n \times \mathbb{R}^n \times [0,1] \to \mathbb{R}^n$, is a family of parametrized curves, defined by $c_{x,y}(t) = x + t\eta(y,x)$. Then c is η-compatible and condition C is reduced to the classical condition C studied by Mohan and Neogy [205].*

Proposition 14.1 *Suppose that M is a differentiable manifold $f \in \Omega(M), \eta : M \times M \to TM$ is a bundle function on M and K an η-convex subset of M, such that η satisfies condition C on M, with respect to f. If f is η-pseudolinear on M, then, for each $x, y \in K$,*

$$\eta(y, x) \in Ker(df_x) \Leftrightarrow f(x) = f(y).$$

Proof Suppose that f is η-pseudolinear on M and let $x, y \in K$, such that

$$\eta(y, x) \in Ker(df_x).$$

From which, it follows that

$$df_x(\eta(y, x)) = 0.$$

By η-pseudoconvexity of f and $-f$, it follows that

$$f(y) \geq f(x)$$

and

$$-f(y) \geq -f(x).$$

Hence, for each $x, y \in K$, we get

$$f(y) = f(x).$$

Conversely, suppose that $x, y \in K$ and $f(y) = f(x)$. Since, K is an η-convex set, there exists a family of η-compatible curves $c_{x,y} : [0, 1] \to M$, whose images are contained in K, such that

$$c_{x,y}(0) = x, c_{x,y}(1) = y, c'_{x,y}(t) = \eta(y, c_{x,y}(t)), \quad \forall t \in [0, 1].$$

Now, we consider, a fixed $t \in]0, 1[$ and denote $z = c_{x,y}(t)$. Now, for every $t \in]0, 1[$, we show that

$$f(z) = f(x).$$

On the contrary, suppose that $f(z) > f(x)$. By η-pseudoconvexity of f, we get

$$df_z(\eta(x, z)) < 0. \tag{14.28}$$

Now, by condition C, we get

$$df_z(\eta(x, z)) = -t df_x(\eta(y, x)) = \frac{t}{1-t} df_x(\eta(y, z)). \tag{14.29}$$

By (14.28) and (14.29), it follows that

$$df_z(\eta(y, z)) > 0.$$

By η-pseudoconvexity of f, we get $f(y) = f(x) > f(z)$, which is a contradiction to our assumption. Hence, condition $f(z) > f(x)$ is not true.

Similarly, by using η-pseudoconvexity of $-f$, we can show that the assumption $f(z) < f(x)$ is also not true. Hence, we must have

$$f(z) = f(x), \ \forall t \in \,]0,1[\,.$$

Now, we calculate that

$$df_x\left(\eta\left(y,x\right)\right) = \lim_{t\to 0+} df_{c(t)}\left(\eta\left(y, c\left(t\right)\right)\right) = \lim_{t\to 0+} df_{c_{x,y}(t)}\left(c'_{x,y}\left(t\right)\right)$$

$$= \lim_{t\to 0+} d\left(f \circ c_{x,y}\left(t\right)\right) = 0.$$

Hence, we have proved that $\eta\left(y,x\right) \in Ker\left(df_x\right).$

Proposition 14.2 *Suppose that M is a differentiable manifold. $f \in \Omega\left(M\right)$, $\eta : M \times M \to TM$ is a bundle function on M and K an η-convex subset of M, such that η satisfies condition C on M, with respect to f. Then f is η-pseudolinear on M, if there exists a positive function $p : K \times K \to \mathbb{R}$, such that for every $x, y \in K$,*

$$p\left(x,y\right) > 0 \ and \ f\left(y\right) = f\left(x\right) + p\left(x,y\right) df_x\left(\eta\left(y,x\right)\right). \tag{14.30}$$

Proof Suppose that $f \in \Omega\left(M\right)$ is an η-pseudolinear function on M and $x, y \in K$. If $df_x\left(\eta\left(y,x\right)\right) = 0$, we define $p\left(x,y\right) = 1$. From Proposition 16.1, we know that $f\left(y\right) = f\left(x\right)$, so (14.30) holds. If $df_x\left(\eta\left(y,x\right)\right) \neq 0$, we define

$$p\left(x,y\right) = \frac{f\left(y\right) - f\left(x\right)}{df_x\left(\eta\left(y,x\right)\right)}. \tag{14.31}$$

From the η-pseudolinearity of f, it follows that $p\left(x,y\right) > 0$.

Conversely, suppose that there exists a function p satisfying (14.31). Let $x, y \in K$ be such that $df_x\left(\eta\left(y,x\right)\right) \geq 0$. It follows that

$$f\left(y\right) - f\left(x\right) = p\left(x,y\right) df_x\left(\eta\left(y,x\right)\right) \geq 0,$$

which implies that $f\left(y\right) \geq f\left(x\right).$ Hence, f is η-pseudoconvex. Similarly, $df_x\left(\eta\left(y,x\right)\right) \leq 0$, implies that f is η-pseudoconcave on K.

Proposition 14.3 *Suppose that M is a differentiable manifold, $\eta : M \times M \to TM$ is a bundle function on M. Let $f \in \Omega\left(M\right)$ be an η-pseudolinear function, K an η-convex subset of M and $F : \mathbb{R} \to \mathbb{R}$ is a differentiable function, regular or having only isolated critical points. Then, $F \circ f$ is also an η-pseudolinear function on K.*

Proof Any function f is η-pseudolinear if and only if $-f$ is also η-pseudolinear. Thus, to prove the proposition it is sufficient to restrict ourselves to the case $F' \geq 0$. Due to a similar argument, it suffices also to prove only the η-pseudoconvexity of $F \circ f$.

Assume that $x, y \in M$, be such that

$$0 \leq d\left(F \circ f\right)_x\left(\eta\left(y,x\right)\right) = F'\left(f\left(x\right)\right) df_x\left(\eta\left(y,x\right)\right).$$

From continuity properties, it follows that $df_x\left(\eta\left(y,x\right)\right) \geq 0$. The η-pseudoconvexity of f and the monotonicity of F ensures that $f\left(y\right) \geq f\left(x\right).$

14.5 η-Pseudolinear Programming Problem on Differentiable Manifolds

In this section, we present some characterizations for the solution sets of a constrained optimization problem involving an η-pseudolinear function on differentiable manifolds studied by Pripoae *et al.* [230].

Let M be a differentiable manifold, $f \in \Omega(M)$, $\eta : M \times M \to TM$, be a bundle function on M. Let K be an η-convex subset of M with respect to a family of curves c. We consider the following constrained minimization problem

$$(P) \quad \min f(x)$$

$$\text{subject to } x \in K.$$

Let S be the solution set of the problem (P) and let it be nonempty.

Theorem 14.3 *Let f be the preinvex on K with respect to η and c, then the solution set S is η-convex.*

Proof Suppose that $x, y \in \bar{S}$. Then, for any $z \in K$, we have $f(x) \leq f(z)$ and $f(y) \leq f(z)$. Since, K is an η-convex subset with respect to the family of η-compatible curves $c_{x,y} : [0,1] \to M$, whose image is contained in K, we have

$$c_{x,y}(0) = x, \ c_{x,y}(1) = y \text{ and } c'_{x,y}(t) = \eta(y, c(t)), \text{ for any } t \in [0,1].$$

From the preinvexity of f, it follows that, for every $t \in [0,1]$ and $z \in K$, we get

$$f(c_{x,y}(t)) \leq tf(y) + (1-t)f(x) \leq tf(z) + (1-t)f(z) = f(z).$$

It follows that $c_{x,y}(t) \in S$, so $\bar{S}$ is η-convex.

Theorem 14.4 *Suppose M is a differentiable manifold, $f \in \Omega(M)$, $\eta : M \times M \to TM$, is a bundle function on M. Let K be an η-convex subset of M. Suppose f is η-pseudolinear on K and η satisfies condition C with respect to f. Let $\bar{x}$ be a fixed element of the solution set S for (P). Then $S = S_1 = S_2 = S_3 = S_4 = S_5 = S_6$, where*

$$S_1 := \{x \in K : df_x(\eta(\bar{x}, x)) = 0\}; \quad S_2 =: \{x \in K : df_{\bar{x}}(\eta(x, \bar{x})) = 0\};$$

$$S_3 := \{x \in K : df_x(\eta(\bar{x}, x)) \geq 0\}; \quad S_4 =: \{x \in K : df_{\bar{x}}(\eta(x, \bar{x})) \geq 0\};$$

$$S_5 := \{x \in K : df_{\bar{x}}(\eta(x, \bar{x})) = df_x(\eta(\bar{x}, x))\};$$

$$S_6 := \{x \in K : df_{\bar{x}}(\eta(x, \bar{x})) \geq df_x(\eta(\bar{x}, x))\}.$$

Proof To prove $S = S_1$, let $x \in M$ belong to S if and only if $f(x) = f(\bar{x})$. By Proposition 14.1, this is possible if and only if $df_x(\eta(\bar{x}, x)) = 0$. Hence, $S = S_1$.

Now, we prove $S = S_2$. Now $x \in S$ is equivalent successively to $f(x) = f(\bar{x})$, to $df_x(\eta(\bar{x}, x)) = 0$ and to $df_{\bar{x}}(\eta(x, \bar{x})) = 0$, due to the fact that $\eta(\bar{x}, x) = -\eta(x, \bar{x})$. Hence, we get $S = S_2$.

Again to prove that $S = S_3$. From $S = S_1$, it is obvious that $S \subseteq S_3$. If $x \in S_3$, then by Proposition 14.2, there exists a function $p : K \times K \to \mathbb{R}$ such that

$$p(x, y) > 0 \text{ and } f(\bar{x}) = f(x) + p(x, \bar{x}) \langle \nabla f(x), \bar{x} - x \rangle \geq f(x).$$

From which it follows that $f(x) = f(\bar{x})$, so $x \in S$. Hence, we have $S = S_3$.

Following the idea of the proof of $S = S_2$ and $S = S_3$, we can prove that $S = S_4$.

Now, we prove $S = S_5 = S_6$. Let $x \in S$, then from the above discussions, we have

$$df_x(\eta(\bar{x}, x)) = 0 \text{ and } df_{\bar{x}}(\eta(x, \bar{x})) = 0.$$

Therefore, we get $x \in S_5$. Hence, $S \subseteq S_5$.

From the definitions of the solution sets the inclusion $S_5 \subseteq S_6$ is obvious. Now, we assume that $x \in S_6$. Then, we have $df_{\bar{x}}(\eta(x, \bar{x})) \geq df_x(\eta(\bar{x}, x))$.

On the contrary, suppose that $x \notin S$. Then, we must have $f(x) > f(\bar{x})$ and $df_{\bar{x}}(\eta(x, \bar{x})) > 0$, by η-pseudoconvexity of $(-f)$. From condition $\eta(x, \bar{x}) = -\eta(\bar{x}, x)$, it follows that

$$0 > df_{\bar{x}}(\eta(x, \bar{x})) \geq df_x(\eta(\bar{x}, x)) \text{ and } df_x(\eta(x, \bar{x})) > 0.$$

By Proposition 14.2, there exists a function $p : K \times K \to \mathbb{R}$, such that $p(x, \bar{x}) > 0$ and

$$f(\bar{x}) = f(x) + p(x, \bar{x}) df_x(\eta(\bar{x}, x)).$$

Then, we have $f(\bar{x}) > f(x)$, which is a contradiction to our assumption. Hence, $x \in S$. Therefore, we have $S_6 \subseteq S$.

From the proof of the above inclusions, we get $S = S_5 = S_6$.

14.6 New Examples of η-Pseudolinear Functions

Pripoae *et al.* [230] have shown that the class of differentiable submersions provides a large class of η-pseudolinear functions.

Theorem 14.5 *Assume that M is an n-dimensional differentiable manifold and $f \in \Omega(M)$, is a differentiable submersion. Then there exists a function $\eta : M \times M \to TM$, such that f is η-pseudolinear.*

Proof Suppose that $p \in M$ and $(x_1, ..., x_n)$ are the local coordinates in a chart $U^{(p)}$ around p. On $U^{(p)}$, we write $df = \frac{\partial f}{\partial x_i} dx_i$. By our assumption, f is a submersion, so there exists $\bar{i} \in \{1, ..., n\}$, such that $\frac{\partial f}{\partial x_{\bar{i}}}(p) \neq 0$. Without the loss of generality, we may assume that $\frac{\partial f}{\partial x_{\bar{i}}}(p) \neq 0$ on $U^{(p)}$.

Let us define $\eta^{(p)} : M \times U^{(p)} \to TM$, such that

$$\eta^{(p)}(y, x) = (f(y) - f(x)) \frac{\partial f}{\partial x^{\bar{i}}}(x) \frac{\partial}{\partial x^{\bar{i}}}|x.$$

Therefore, for every $y \in M$ and $x \in U^{(p)}$, we have

$$df_x\left(\eta^{(p)}(y, x)\right) = \left(\frac{\partial f}{\partial x_{\bar{i}}}(x)\right)^2 (f(y) - f(x)).$$

Now, we consider a locally open covering, subordinated to the covering $\left\{U^{(p)} : p \in M\right\}$ and a differentiable portion of unity associated to it. A standard argument allows us to construct a function $\eta : M \times M \to TM$, such that $df_x(\eta(y, x))$ and $(f(y) - f(x))$ differ modulo a positive function. From this fact, it follows that f and $-f$ are η-pseudoconvex, hence, f is η-pseudolinear.

Remark 14.3 *The converse of Theorem 14.5 is also true. We assume that M is an n-dimensional differentiable manifold, $\eta : M \times M \to TM$ and $f \in \Omega(M)$ is an η-pseudolinear function. Suppose conversely that there exists a point $\bar{x} \in M$ such that $df_{\bar{x}} = 0$. Therefore, for every $x \in M$, $df_{\bar{x}}(\eta(x, \bar{x})) = 0$. From Proposition 16.2, it follows that $f(x) = f(\bar{x})$, for every $x \in M$. So, the η-pseudolinearity of f implies that f is a submersion.*

Remark 14.4 *The connection between submersion and η-pseudolinearity appears to be quite natural due to the following facts:*

1. *Both the notions are obviously covariant.*

2. *A submersion is locally (modulo some diffeomorphism), a projection, hence, linear.*

Chapter 15

On the Pseudolinearity of Quadratic Fractional Functions

15.1 Introduction

The nonlinear programming problems, where the objective function is the ratio of a quadratic function and an affine one is referred to as quadratic fractional programming. These problems have been widely studied from both a theoretical and algorithmic point of view, due to their applications in several areas of modern research such as risk theory, portfolio theory, and location models. For details, we refer to Schaible [247], Barros [16], Bazaraa *et al.* [17], and Cambini and Carosi [33].

Many solution methods have been given for quadratic fractional programming problems, whose feasible region is a polyhedron. In these cases, the generalized convexity of the objective function plays a fundamental role, since it guarantees the global optimality of local optima. Among generalized convex functions, the generalized linear ones are extremely useful since local optima are global for both the maximum and minimum problems. In order to explicitly describe more general function classes, it is natural to investigate the generalization of both convex and concave functions. In the case of pseudolinear functions, several first and second order characterizations have been derived by Kortanek and Evans [149], Chew and Choo [47], and Komlosi [147]. Rapcsak [233] has given an explicit formulation for the gradient of smooth pseudolinear functions. Bianchi and Schaible [27] and Bianchi *et al.* [26] have studied the maps where the map and the map multiplied by −1 are both pseudomonotone. These results have been studied in Chapter 2, Chapter 12, and Chapter 14.

Cambini and Carosi [34] have characterized the generalized linearity of quadratic fractional functions and established necessary and sufficient conditions which can be easily checked by means of the proposed characterization. Cambini and Carosi [34] proved that a quadratic fractional function is quasilinear if and only if it is pseudolinear and shown that a quadratic fractional function is pseudolinear if and only if it can be rewritten as the sum of a linear function and a linear fractional one with a constant numerator. This result allows us to give conditions characterizing the generalized linearity of a larger class of functions given by the sum of a quadratic fractional function and a linear one. Furthermore, Cambini and Carosi [34] have established that both minimization and maximization problems, involving a generalized linear quadratic fractional function, can be simply solved through equivalent linear ones.

Rapcsak [235] has derived a new characterization and simpler proof for the pseudolinearity of quadratic fractional functions, which is a reformulation of the corresponding results by Cambini and Carosi [34]. Using this reformulation, Ujvari [276] has given a shorter and less involved proof of the Cambini and Carosi theorem. Rapcsak and Ujvari [236] have studied the affinity of quadratic fractional functions and presented the characterizations for the gradient of pseudolinear quadratic fractional functions. Recently, Komlosi [148] employed the implicit function theorem to present some characterizations for twice continuously differentiable pseudolinear functions and applied these results for the local analysis of quadratic fractional functions.

15.2 Definitions and Preliminaries

Let $\nu_-(Q)$ and $\nu_+(Q)$ denote the number of the negative and positive eigenvalues of a symmetric matrix $Q \in \mathbb{R}^{n \times n}$. Let $\nu_0(Q)$ be the algebraic multiplicity of the 0 eigenvalue. To avoid trivial cases, we assume $n \geq 2$.

Definition 15.1 (Moore-Penrose pseudoinverse) *Let $Q \in \mathbb{R}^{m \times n}$ be any $m \times n$ matrix. The Moore-Penrose pseudoinverse of Q is the unique matrix $Q^\# \in R^{n \times m}$, such that the following Moore-Penrose equation is satisfied:*

$$QQ^\#Q = Q, Q^\#QQ^\# = Q^\#, (QQ^\#)^T = QQ^\#, (Q^\#Q)^T = Q^\#Q.$$

The following definition is from Bazaraa *et al.* [17].

Definition 15.2 (Semistrict local minima) *Let $f :]a, b[\to \mathbb{R}$ is a real-valued function. Then f is said to attain a semistrict local maximum at a point $\overline{x} \in]a, b[$ if there exists two points $x_1, x_2 \in]a, b[, x_1 < \overline{x} < x_2$, such that*

$$f(\overline{x}) \geq f(x_2 + \lambda(x_1 - x_2)), \forall \lambda \in [0, 1]$$

and

$$f(\overline{x}) > \min\{f(x_1), f(x_2)\}.$$

The following proposition from Crouzeix [56] will be needed in the sequel:

Proposition 15.1 *Suppose that $h \in \mathbb{R}^n, h \neq 0$, and $Q \in \mathbb{R}^{n \times n}$ is a symmetric matrix and let $Q^\#$ be the Moore-Penrose pseudoinverse of Q. Then the implication*

$$h^T \nu = 0 \Rightarrow \nu^T Q \nu \geq 0$$

is satisfied for all $\nu \in \mathbb{R}^n$ if and only if any one of the following conditions hold:

(i) $\nu_-(Q) = 0$,

(ii) $\nu_-(Q) = 1, h \in Q(\mathbb{R}^n)$ and $\nu_-(Q) = 1, h^T Q^{\#} h \leq 0$.

Cambini and Carosi [34] have remarked that the assumption $h \neq 0$ in Proposition 15.1 leads to some technical difficulties in the application of the proposition to particular problems, where the vector h is not necessarily different from zero. The following corollary improves the result by Crouzeix [56] not requiring the vector h to be different from zero.

Corollary 15.1 *Let $h \in \mathbb{R}^n$ and let $Q \in \mathbb{R}^{n \times n}$ be a symmetric matrix. Then the following implication*

$$h^T \nu = 0 \Rightarrow \nu^T Q \nu \geq 0 \qquad (15.1)$$

is satisfied for all $\nu \in \mathbb{R}^n$ if and only if any one of the following conditions holds:

(i) $\nu_-(Q) = 0$,

(ii) $\nu_-(Q) = 1, h \neq 0, h \in Q(\nu^n)$ and $u^T Q u \leq 0, \forall u \in \mathbb{R}^n$, *such that* $Qu = h$.

Proof (Necessary) If $\nu_-(Q) = 0$ then (i) holds. Now we assume that $\nu_-(Q) \neq 0$. On the contrary, suppose that $h = 0$; then,

$$h^T \nu = 0, \forall \nu \in \mathbb{R}^n.$$

Hence, for condition (15.1) $\nu^T Q \nu \geq 0, \forall \nu \in \mathbb{R}^n$, that is, Q is positive semi-definite, which is a contradiction being $\nu_-(Q) \neq 0$. Hence, we must have $h \neq 0$ and (ii) follows from (ii) of Proposition 15.1.

(Sufficient) If $\nu_-(Q) = 0$, then (15.1) is trivially satisfied; if (ii) holds then, we have

$$u^T Q u \leq 0 \, \forall u \in \mathbb{R}^n \text{ such that } Qu = h \Rightarrow h^T Q^{\#} h \leq 0.$$

The results follow from Proposition 15.1.

The following property is a key tool in characterizing the generalized linearity of quadratic fractional functions.

Corollary 15.2 *Let $h \in \mathbb{R}^n$ and let $Q \in \mathbb{R}^{n \times n}, Q \neq 0$ be a symmetric matrix. Then for all $\nu \in \mathbb{R}^n$, the following implication*

$$h^T \nu = 0 \Rightarrow \nu^T Q(\nu) = 0 \qquad (15.2)$$

holds if and only if any one of the following conditions hold:

(i) $\nu_0(Q) = n - 1, h \neq 0$ and $h \in Q(\mathbb{R}^n)$,

(ii) $\nu_-(Q) = \nu_+(Q) = 1, h \neq 0, h \in Q(\mathbb{R}^n)$ *and* $u^T Q u = 0, \forall u \in \mathbb{R}^n$ *such that*

$$Qu = h.$$

Proof From Corollary 15.1, we note that, for all $\nu \in \mathbb{R}^n$, the implication

$$h^T \nu = 0 \Rightarrow \nu^T Q \nu \leq 0$$

is satisfied if and only if one of the following conditions hold:

(a) $\nu_+(Q) = 0$,

(b) $\nu_+(Q) = 1, h \neq 0, h \in Q(\mathbb{R}^n)$ and $u^T Q u \leq 0, \forall u \in \mathbb{R}^n$, such that

$$Qu = h.$$

(Necessary) We note that condition (15.2) holds if and only if

$$\{h^T \nu = 0 \Rightarrow \nu^T Q \nu \geq 0\} \, and \, \{h^T \nu = 0 \Rightarrow \nu^T Q \nu \leq 0\}.$$

This happens if and only if one of conditions (i) and (ii) of Corollary 15.2 holds together with one of conditions (a) and (b). We observe that conditions (a) and (i) of Corollary 15.2 imply $Q = 0$, which is a contradiction. Conditions (a) and (ii) of Corollary 15.1 imply condition (i) and the same happens if (b) and (i) of Corollary 15.1 hold.

If conditions (b) and (ii) of Corollary 15.1 are satisfied then condition (ii) follows immediately. Since all the possible exhaustive cases have been considered, the result is proved.

(Sufficient) If (i) holds and Q is positive semidefinite, then,

$$\nu_+(Q) = 1, \nu_-(Q) = 0, h \neq 0, h \in Q(\mathbb{R}^n) \, and \, u^T Q u \leq 0, \forall u \in \mathbb{R}^n.$$

Thus (i) of Corollary 15.1 and condition (b) hold, hence,

$$h^T \nu = 0 \Rightarrow \nu^T Q \nu \geq 0 \, and \, \nu^T Q \nu \leq 0,$$

Therefore, (15.2) holds. The case Q negative semidefinite can be proved with the same arguments.

If (ii) holds then both conditions (b) and (ii) of Corollary 15.1 are satisfied. Again, we have

$$h^T \nu = 0 \Rightarrow \nu^T Q \nu \geq 0 \text{ and } \nu^T Q \nu \leq 0.$$

Thus, (15.2) holds.

Now, we recall the following results by Diewert *et al.* [65] (Corollary 4.3 pp. 401, and Theorem 10 pp. 407).

Proposition 15.2 *Suppose that* f *is a differentiable function defined on the open convex set* $A \subseteq \mathbb{R}^n$. *Then,*

(i) f is quasiconvex if and only if $\forall x \in A, \forall \nu \in \mathbb{R}^n \setminus \{0\}$, such that $\nabla f(x)^T \nu = 0$. The function $\phi_\nu(t) = f(x + t\nu)$ does not attain a semistrict local maximum at $t = 0$. In addition, if the function f is also continuously differentiable, then

(ii) f is pseudoconvex if and only if for all $x \in A$ and for all $\nu \in \mathbb{R}^n \setminus \{0\}$, such that $\nabla f(x)^T \nu = 0$. The function $\phi_\nu(t) = f(x + t\nu)$ attains a local minimum at $t = 0$.

Now, we recall the following characterization of pseudolinear functions (see for example Avriel *et al.* [9] and Martos [179]).

Proposition 15.3 *Suppose $f : X \to \mathbb{R}$, is a differentiable function on an open convex set $X \subseteq \mathbb{R}^n$. Then, f is pseudolinear if and only if for every $x \in X, \nu \in \mathbb{R}^n, \nu \neq 0$ and for all $t \in \mathbb{R}$, such that $x + t\nu \in X$, the following implication holds:*

$$\nabla f(x)^T \nu = 0 \Rightarrow \phi_\nu(t) = f(x + t\nu) \text{ is constant.}$$

15.3 Generalized Linearity of Quadratic Fractional Functions

In this section, we present the characterizations for the generalized linearity of quadratic fractional functions of the following form:

$$f(x) = \frac{\frac{1}{2} x^T Q x + q^T x + q_0}{b^T x + b_0} \tag{15.3}$$

defined on the set $X := \{x \in \mathbb{R}^n : b^T x + b_0 > 0\}, n \geq 2$, where $Q \neq 0$ is a $n \times n$ symmetric matrix, $q, x, b \in \mathbb{R}^n, b \neq 0$, and $q_0, b_0 \in \mathbb{R}$. We note that Q is symmetric and $Q \neq 0$ if and only if $\nu_0(Q) \leq n - 1$. Moreover, it follows that:

$$\nabla f(x) = \frac{Qx + q - f(x)b}{b^T x + b_0}, \tag{15.4}$$

where $\nabla f(x)$ is expressed as a row vector denotes the gradient of the function f at x.

Remark 15.1 *We note that f in (15.3) is not a constant function. Suppose on the contrary, that f is a constant, that is $f(x) = k$, so that $\nabla f(x) = 0, \forall x \in X$. Consider now an arbitrary $x_1 \in X$ and let $\alpha \in \mathbb{R}$, such that $\alpha \neq 0, \alpha \neq 1$ and $\alpha x_1 \in X$. Then by (15.4), we have $Qx_1 + q - kb = Q\alpha x_1 + q - kb$ and hence, $Qx_1 = \alpha Q x_1$ which implies $Qx_1 = 0$, that is, $Qx = 0$, for all $x \in X$. Since, X is an n-dimensional half-space it must be $Q = 0$, which is a contradiction to the definition of f in (15.3).*

The next theorem from Cambini and Carosi [34] shows that a quadratic fractional function is pseudolinear if and only if it is quasilinear. Moreover, for these classes of functions, Cambini and Carosi [34] derived a new characterization based on the behavior of $\nu^T Q \nu$, when $\nabla f(x)^T \nu = 0$.

Theorem 15.1 *Suppose that function f is defined by (15.3). Then the following conditions are equivalent:*

(i) *f is pseudolinear on X;*

(ii) *f is quasilinear on X;*

(iii) *for all $x \in X$ and for each $\nu \in R^n \backslash \{0\}$, the following implication holds*

$$\nabla f(x)^T \nu = 0 \Rightarrow \nu^T Q \nu = 0;$$

(iv) *for all $x \in X$ if one of the following conditions hold:*

(a) *$\nu_0(Q) = n - 1, \nabla f(x) \neq 0$ and $\nabla f(x) \in Q(\mathbb{R}^n)$,*

(b) *$\nu_-(Q) = \nu_+(Q) = 1, \nabla f(x) \neq 0, \nabla f(x) \in Q(\mathbb{R}^n)$ and $u^T Q u = 0, \forall u \in \mathbb{R}^n$, such that $Qu = \nabla f(x)$.*

Proof From (15.4), we note that,

$$\phi_\nu(t) = f(x + t\nu) = f(x) + \frac{\frac{1}{2} t^2 \nu^T Q \nu + t(b^T x + b_0) \nabla f(x)^T \nu}{b^T x + b_0 + t b^T \nu}.$$

(i)$\Rightarrow$(ii) The proof is trivial.

(ii)$\Rightarrow$(iii) Since f is both quasiconvex and quasiconcave, from Proposition 15.2, it follows that the function $\phi_\nu(t) = f(x) + \frac{t^2 \nu^T Q \nu}{2b^T x + b_0 + t b^T \nu}$ does not attain either a semistrict local maximum or a semistrict local minimum at $t = 0, \forall x \in X, \forall \nu \in \mathbb{R}^n \backslash \{0\}$ such that $\nabla f(x)^T = 0$. This happens only if $\nu^T Q \nu = 0$, being $b^T x + b_0 + t b^T \nu > 0, \forall t \in \mathbb{R}$, such that $x + t\nu \in X$.

(iii)$\Rightarrow$(i) Since $\nabla(x)^T \nu = 0$ implies $\nu^T Q \nu = 0$, it follows that $\phi_\nu(t) = f(x), \forall t \in \mathbb{R}$, such that $x + t\nu \in X$. Hence, the results follow from Proposition 15.3.

(iii)$\Rightarrow$(iv) The result follows directly from Corollary 15.2.

From Chapter 2, we recall that condition $\nabla f(x) \neq 0, \forall x \in X$, is a necessary condition for f to be a not constant pseudolinear function.

The following lemma from Cambini and Carosi [34] points out the existence of a sort of canonical form for all the generalized linear quadratic fractional functions.

Lemma 15.1 *Suppose that the function f is defined by (15.3). Then,*

(i) *For each $x \in X, \nabla f(x) \in Q(\mathbb{R}^n)$ if and only if there exist $\bar{x}, \bar{y} \in \mathbb{R}^n$, such that*

$$Q\bar{x} = q \text{ and } Q\bar{y} = b.$$

In particular, for any given $x \in X$:

(ii) $Qu = \nabla f(x) \Leftrightarrow u = \frac{x + \bar{x} - f(x)\bar{y}}{b^T x + b_0} + k$ *with $k \in Ker(Q)$;*

(iii) $b^T x = b^T \bar{x}$ *and* $q^T x = q^T \bar{x}, \forall x \in \mathbb{R}^n$, *such that $Qx = q$;*

(iv) $b^T y = b^T \bar{y}$ *and* $q^T y = q^T \bar{y}, \forall y \in \mathbb{R}^n$, *such that $Qx = b$;*

(v) $u^T Q u = \frac{p(x)}{(b^T x + b_0)^2}$, *with*

$$p(x) = (f(x))^2 b^T \bar{y} + 2f(x)[b_0 - b^T \bar{x}] + (q^T \bar{x} - 2q_0). \tag{15.5}$$

Proof (i) Let us suppose that

$$\nabla f(x) = \frac{Qx + q - f(x)b}{b^T x + b_0} \in Q(\mathbb{R}^n), \forall x \in X.$$

Now, we prove that there exists $\bar{x}, \bar{y} \in \mathbb{R}^n$, such that

$$Q\bar{x} = q \text{ and } Q\bar{y} - b.$$

From Remark 15.1, it is clear that f is not constant, Hence, there exists $x_1, x_2 \in X$, such that $f(x_1) \neq f(x_2)$. Therefore, there exists $u_1, u_2 \in \mathbb{R}^n$, such that

$$Qu_1 = Qx_1 + q - f(x_1)b \text{ and } Qu_2 = Qx_2 + q - f(x_2)b.$$

This implies that

$$Q\left(\frac{u_1 - u_2 - x_1 + x_2}{f(x_2 - f(x_1))}\right) = b.$$

Therefore, there exists $\bar{y} \in \mathbb{R}^n$, such that $Qy = b$. It follows also that $Qu_1 = Qx_1 + q - f(x_1)Q\bar{y}$, which implies $q = Q(u_1 - x_1 + f(x_1)\bar{y})$ and hence there exists $\bar{x} \in \mathbb{R}^n$, such that $Q\bar{x} = q$.

To prove the converse, we assume that there exists $\bar{x}, \bar{y} \in \mathbb{R}^n$, such that $Q\bar{x} = q$ and $Q\bar{y} = b$. Then,

$$\nabla f(x) = Q\left(\frac{x + \bar{x} - f(x)\bar{y}}{b^T x + b_0}\right), \tag{15.6}$$

so that $\nabla f(x) \in \mathbb{R}^n, \forall x \in X$.

(ii) From (15.6) we note that

$$Qu = \nabla f(x) \Leftrightarrow Q\left(u - \frac{x + \bar{x} - f(x)\bar{y}}{b^T x + b_0}\right) = 0.$$

This happens if and only if

$$\left(u - \frac{x + \overline{x} - f(x)\overline{y}}{b^T x + b_0}\right) = k \in Ker(Q).$$

(iii) Since $Qx = Q\overline{x} = q$, therefore $Q(x - \overline{x}) = 0$ and hence, $x = \overline{x} + k$ with $k \in ker(Q)$. The result then follows being $b^T k = \overline{y}^T Qk = 0 = \overline{x}^T Qk = q^T k, \forall k \in Ker(Q)$.

(iv) The proof follows along the lines of (iii).

(v) We note that:

$$u^T Qu = \frac{1}{(b^T x + b^0)^2}[(x + \overline{x} - f(x)\overline{y})^T Q(x + \overline{x} - f(x)\overline{y})]$$

$$= \frac{1}{(b^T x + b^0)^2}[(x + \overline{x} - f(x)\overline{y})^T (Qx + q - f(x)b)]$$

$$= \frac{(f(x))^2 b^T \overline{y} + 2f(x)(b_0 - b^T \overline{x}) + (q^T \overline{x} - 2q_0)}{(b^T x + b^0)^2}.$$

This completes the proof.

The following lemma from Cambini and Carosi [34] is a key tool in characterizing the generalized linearity of f.

Lemma 15.2 *Suppose the function f is defined by (15.3). It results*

$$\nu_0(Q) = n - 1, \nabla f(x) \in Q(\mathbb{R}^n), \forall x \in X,$$

if and only if there exist $\alpha, \beta, \gamma \in \mathbb{R}, \alpha \neq 0$, such that

$$f(x) = ab^T x + \beta + \frac{\alpha\gamma}{b^T x + b_0}.$$

Moreover, $\nabla f(x) \neq 0, \ \forall x \in X \Leftrightarrow \gamma \leq 0.$

Proof It is easy to calculate that $Q = [2\alpha bb^T]$ and hence $\nu_0(Q) = n - 1$. Since, $\nabla f(x) = \alpha[1 - \gamma/(b^T x + b_0)^2]b$, it is $\nabla f(x) \in Q(\mathbb{R}^n), \forall x \in X$. From Lemma 15.1, it follows that $\nu_0(Q) = n-1$ and there exists $\overline{x}, \overline{y} \in \mathbb{R}^n$, such that $Q\overline{x} = q$ and $Q\overline{y} = b$. Since, $b \neq 0$ and $\dim Q(\mathbb{R}^n) = 1$, it is $Q\overline{x} = q$ if and only if there exists $\mu \in \mathbb{R}$, such that $q = \mu b$.

Since, $b \in Q(\mathbb{R}^n)$ and $\dim Q(\mathbb{R}^n) = 1$, vector b is an eigenvector of Q. Since, Q is symmetric, there exists $\alpha \in \mathbb{R}, \alpha \neq 0$, such that $Q = [2\alpha bb^T]$ and $\overline{y} = \frac{1}{2\alpha\|b\|^2}b, 2b^T\overline{y} = \frac{1}{\alpha}$. Consequently,

$$f(x) = \frac{\alpha(b^T x)^2 + \mu b^T x + q_0}{b^T x + b_0}$$

$$= \frac{\alpha[(b^T x + b_0) - b_0]^2 + \mu b^T x + \mu b_0 - \mu b_0 + q_0}{b^T x + b_0}$$

$$= ab^T x + (\mu - \alpha b_0) + \frac{\alpha b_0^2 - \mu b_0 + q_0}{b^T x + b_0}.$$

Defining $\beta = (\mu - \alpha b_0)$ and $\gamma = b_0^2 + \frac{1}{\alpha}(q_0 - \mu b_0)$, the result follows.

To prove the second part of the lemma, we note that

$$\nabla f(x) = \alpha \left[1 - \frac{\gamma}{(b^T x + b_0)^2} \right] b,$$

with $\alpha \neq 0, b \neq 0$. Hence it results $\nabla f(x) \neq 0 \forall x \in X$, if and only if

$$(b^T x + b_0)^2 \neq \gamma, \ \forall x \in X. \tag{15.7}$$

By definition $\{y \in \mathbb{R} : y = b^T x + b_0, x \in X\} = \mathbb{R}_{++}$, so that (15.7) holds if and only if $\gamma \leq 0$.

Using the results of Lemmas 15.1 and 15.2, Cambini and Carosi [34] provided the following characterization of quadratic fractional generalized linear functions.

Theorem 15.2 *The function f defined by (15.3) is pseudolinear (or quasi-linear) on X if and only if f is affine or there exist $\alpha, \beta, \gamma \in \mathbb{R}, \alpha \neq 0$, such that f can be rewritten in the following form:*

$$f(x) = \alpha b^T x + \beta + \frac{\alpha\gamma}{b^T x + b_0}, \ \text{with } \gamma < 0. \tag{15.8}$$

Proof (Necessary) Since f is pseudolinear, either condition (iv–a) or condition (iv–b) of Theorem 15.1 is satisfied.

If (iv–a) is satisfied the results follow from Lemma 15.2 by noting that f is affine when $\gamma = 0$. Now, we supose that condition (iv–b) holds. It follows from Lemma 15.1, that $v_-(Q) = v_+(Q) = 1$, there exists $\bar{x}, \bar{y} \in \mathbb{R}^n$, such that

$$Q\bar{x} = q \text{ and } Q\bar{y} = b, p(x) = 0, \forall x \in X,$$

with $\nabla f(x) \neq 0, \forall x \in X$. Since f is not a constant function, we have

$$p(x) = 0, \forall x \in X \ \Leftrightarrow \ b^T \bar{y} = 0, b^T \bar{x} = b_0 \text{ and } q^T \bar{x} = 2q_0.$$

Since $\nu_-(Q) = \nu_+(Q) = 1$, from the canonical form of Q, we get $Q = [uu^T - \nu\nu^T]$, where u and ν are eigenvectors of Q with $u^T\nu = 0$. From $Q\bar{y} = b, b^T \bar{y} = \bar{y}^T Q\bar{y} = 0$, it follows that

$$\bar{y}^T Q\bar{y} = (u^T \bar{y})^2 - (\nu^T \bar{y})^2 = 0,$$

so that $\nu^T \bar{y} = \pm u^T \bar{y}$. Then,

$$b = Q\bar{y} = u(u^T \bar{y}) - (\nu^T \bar{y})\nu = u^T \bar{y}(u + \delta\nu),$$

where $\delta = \pm 1$. By defining $a = \frac{1}{2u^T \bar{y}}(u - \delta v)$ and performing simple calculations, we get

$$(ab^T + ba^T) = [uu^T - \nu\nu^T] = Q.$$

We note that a and b are linearly independent. Let $\bar{x} \in \mathbb{R}^n$, such that $Q\bar{x} = q$ and define $a_0 = a^T\bar{x}$. It follows that

$$q = ab^T\bar{x} + ba^T\bar{x} = ab_0 + ba_0,$$

$$q_0 = \frac{1}{2}q^T\bar{x} = \frac{1}{2}b_0 a^T\bar{x} + a_0 b^T\bar{x} = a_0 b_0,$$

$$\frac{1}{2}x^T Q x + q^T x + q_0 = (b^T x + b_0)(a^T x + a_0).$$

Hence, $f(x) = a^T x + a_0$.

(Sufficient) If f is affine, it is trivially pseudolinear. Using Theorem 15.1 and Lemma 15.2, the result follows.

Remark 15.2 *From the proof of Theorem 15.2, it is clear that:*

(i) *If $\nu_-(Q) = \nu_+(Q) = 1$, then f is generalized linear if and only if it is affine,*

(ii) *f may be affine, when $\nu_0(Q) = n - 1$ (case $\gamma = 0$),*

(iii) *f is generalized linear but not affine only if $\nu_0(Q) = n - 1$ (case $\gamma < 0$).*

From Theorem 15.2, it is obvious that when $\gamma > 0$, it is not possible to have a generalized linear function on the whole set X (see, Example 15.1). However, for the case $\gamma > 0$, it is possible to prove that the function may be generalized linear at least on two disjoint convex sets.

Corollary 15.3 *Consider the function f defined by (15.3) and suppose that there exist $\alpha, \beta, \gamma \in R, \alpha \neq 0$, such that f can be rewritten in the following form:*

$$f(x) = \alpha b^T x + \beta + \frac{\alpha\gamma}{b^T x + b_0}.$$

(i) *If $\gamma \leq 0$, then, f is pseudolinear (quasilinear) on X.*

(ii) *If $\gamma > 0$, then f is pseudolinear (quasilinear) on*

$$X_1 := \left\{ x \in R^n : b^T x + b_0 > \sqrt{\gamma} \right\} \text{ and } X_2 := \left\{ x \in R^n : 0 < b^T x + b_0 < \sqrt{\gamma} \right\}.$$

Proof (i) The proof follows from Theorem 15.2.

(ii) We note that

$$\nabla f(x) = \frac{\alpha}{(b^T x + b_0)^2}[(b^T x + b_0)^2 - \sqrt{\gamma}\,]b.$$

Therefore, $\nabla f(x) \neq 0$ on X_1 and X_2. The result trivially follows from Theorem 15.1, being $Q = [2\alpha bb^T]$.

The following examples from Cambini and Carosi [34] justify the significance of Theorems 15.1 and 15.2.

Example 15.1 *Consider the problem (15.3) where*

$$f(x) := \frac{9x_1^2 + 24x_1x_2 + 16x_2^2 + 6x_1 - 8x_2 + 1}{3x_1 + 4x_2}.$$

We note that f is not pseudolinear, since it is not constant and $\nabla f(x)$ vanishes at $3x_1 + 4x_2 = 1$. In this case, we have:

$$Q = \begin{bmatrix} 18 & 24 \\ 24 & 32 \end{bmatrix}, q = \begin{bmatrix} 6 \\ 8 \end{bmatrix}, b = \begin{bmatrix} 3 \\ 4 \end{bmatrix}. \; b_0 = 0, q_0 = 1.$$

It is easy to calculate that

$$\nu_0(Q) = 1 = \nu_+(Q) \; and \; f(x) = 3x_1 + 4x_2 + 2 + \frac{1}{3x_1 + 4x_2}.$$

Hence $\gamma = 1 > 0$.

Example 15.2 *Consider the problem (15.3), where*

$$f(x) = \frac{8x_1^2 + 2x_2^2 + 18x_3^2 + -8x_1x_2 - 24x_1x_3 + 12x_2x_3 + 10x_1 - 5x_2 - 15x_3 - 4}{-2x_1 + x_2 + 3x_3 - 3}.$$

In this case, we have:

$$Q = \begin{bmatrix} 16 & -8 & 24 \\ -8 & 4 & 12 \\ 0 & 64 & 64 \end{bmatrix}, q = \begin{bmatrix} 10 \\ -5 \\ 15 \end{bmatrix}, b = \begin{bmatrix} -2 \\ 1 \\ 3 \end{bmatrix}, b_0 = -3, q_0 = -4.$$

Since Q is positive semidefinite with $\nu_0(Q) = n-1$, we need to verify condition (iv–a) of Theorem 15.1. Now, for all $x \in X$, we note that

$$\nabla f(x) = \left(2 + \frac{1}{(-2x_1 + x_2 + 3x_3 - 3)^2}\right), b \neq 0, \nabla f(x) \in Q(\mathbb{R}^2).$$

Hence, f is pseudolinear. Applying Theorem 15.2, we get the same result. It is easy to calculate that

$$f(x) = -4x_1 + 2x_2 + 6x_3 + 1 - \frac{1}{-2x_1 + x_2 + 3x_3 - 3} \quad so \; that \; \alpha = 2, \gamma = -\frac{1}{2} < 0.$$

Hence, f is pseudolinear.

Example 15.3 *Consider problem (15.3), where*

$$f(x) = \frac{-8x_1^2 - 24x_2^2 + 32x_3^2 + 16x_1x_2 + 64x_2x_3 + 16x_2 + 16x_3 + 2}{-8x_1 - 8x_2 - 16x_3 - 4}.$$

In this case, we have

$$Q = \begin{bmatrix} 16 & -16 & 0 \\ 16 & 48 & 64 \\ 0 & 64 & 64 \end{bmatrix}, q = \begin{bmatrix} 0 \\ 16 \\ 16 \end{bmatrix}, b = \begin{bmatrix} -8 \\ -8 \\ -16 \end{bmatrix}, b_0 = -4, q_0 = 2.$$

It is clear that the matrix Q is indefinite with $\nu_+(Q) = \nu_-(Q) = 1$. Hence, f is pseudolinear if and only if it is affine (see, also Remark 15.2). In fact, it is easy to calculate that

$$f(x) = x_1 - 3x_2 - 2x_3 - \frac{1}{2}.$$

The following theorem from Rapcsak [235] characterizes the pseudolinearity of quadratic fractional functions of (15.3).

Theorem 15.3 (Rapcsak [235]) *Assume that $\nabla f(x) \neq 0, x \in X$. Then, function f defined by (15.3) is pseudolinear if and only if*
(i) there exist real constants $\hat{\alpha} \neq 0 \in \mathbb{R}$ and $\hat{\beta} \in \mathbb{R}$, such that

$$Q = \hat{\alpha}bb^T, q = \hat{\beta}b; \tag{15.9}$$

or
(ii) there exist real constants $\hat{\beta}, \hat{\gamma} \in \mathbb{R}$ with $q = \hat{\beta}b$ and a constant vector $d \in \mathbb{R}^n$, whose direction is different from that of b, such that

$$Q = \hat{\gamma}(db^T + bd^T) \text{ and } b_0 = 0, q_0 = 0, \tag{15.10}$$

which is the affine case.

Proof (Necessary) We suppose that function f defined by (15.3) is pseudolinear. By Theorem 14.1, the function f is pseudolinear if and only if there exist continuously differentiable functions $l(x), \eta_i f(x), i = 1, ..., n, x \in X$, such that

$$\frac{Qx + q - f(x)b}{b^T x + b_0} = l(x)\eta f(x), x \in X. \tag{15.11}$$

Thus,

$$Qx = (b^T x + b_0)l(x)\eta(f(x) - q + f(x))b, x \in X \tag{15.12}$$

Differentiating (15.12), we get

$$Q(x) = l(x)\eta(f(x))b^T + (b^T x + b_0)\eta(f(x))\nabla l(x) + (b^T x + b_0)l^2(x)$$

$$\frac{d\eta(f(x))}{df}\eta(f(x))^T + l(x)b\eta(f(x))^T, x \in X.$$

That is,

$$Q(x) = l(x)\left(\eta(f(x))b^T + b\eta(f(x))^T\right) + (b^T x + b_0)l^2(x)\frac{d\eta(f(x))}{df}\eta(f(x))^T$$

$$+ (b^T x + b_0)\eta(f(x))\nabla l(x), x \in X. \tag{15.13}$$

By (15.13), the matrix function

$$(b^T x + b_0)l^2(x)\frac{d\eta(f(x))}{df}\eta(f(x))^T + (b^T x + b_0)\eta(f(x))\nabla l(x), x \in X, \tag{15.14}$$

has to be symmetric. Since $l \neq 0$ on X and $(b^T x + b_0) > 0$, (15.14) is equivalent to

$$(b^T x + b_0) l^2(x) \left(\frac{d\eta(f(x))}{df} \eta(f(x))^T + \eta(f(x)) \nabla(-\frac{1}{l(x)})^T \right), x \in \hat{A}. \quad (15.15)$$

Since the function f is pseudolinear on X, the equality level sets are hyperplanes. Let us fix a hyperplane given by

$$lev_{=f(x_0)} = \{x \in X | f(x) = f(x_0)\},$$

where $x_0 \in X$ is an arbitrary point. Therefore, a necessary condition for the matrix function (15.15) to be symmetric is that the vector function

$$\frac{1}{l^2(x)} \nabla l(x) = \nabla(-\frac{1}{l(x)}), x \in lev_{=f(x_0)},$$

is a constant vector.

Now, we show that the function l is constant on the equality level set $lev_{=f(x_0)}$. Suppose to the contrary that the statement is not true, i.e., there exist two points $x_1, x_2 \in lev_{=f(x_0)}$, such that

$$l(x_1) \neq l(x_2).$$

By (15.13), (15.14), and (15.15), we have

$$Q = l(x_i) \left(\eta f(x_0) b^T + b\eta(f(x_0))^T \right) + \left(b^T x_i + b_0 \right) l^2(x_i)$$

$$\left(\frac{d\eta(f(x_0))}{df} \eta(f(x_0))^T + \eta(f(x_0)) \nabla \left(-\frac{1}{l(x_0)} \right)^T \right), i = 1, 2. \quad (15.16)$$

From which, it follows that

$$(b^T x_2 + b_0) l^2(x_2) - (b^T x_1 + b_0) l^1(x_1) \neq 0$$

and

$$\frac{l(x_1) - l(x_2)}{(b^T x_2 + b_0) l^2(x_2) - (b^T x_1 + b_0) l^2(x_1)} \left(\eta f(x_0) b^T + b\eta(f(x_0))^T \right)$$

$$= \frac{d\eta(f(x_0))}{df} \eta(f(x_0))^T + \eta(f(x_0)) \left(\nabla - \frac{1}{l(x_0)} \right)^T. \quad (15.17)$$

Therefore, the direction of $\frac{d\eta(f(x_0))}{df}$ coincides with that of b. It follows that this is true for $\eta(f(x_0))$, thus, $b^T x + b_0$ is a constant on $lev_{=f(x_0)}$. By (15.17), we have

$$\frac{1}{(l(x_1) + l(x_2)) b^T x + b_0} (\eta f(x_0) b^T + b\eta(f(x_0)^T))$$

$$= \frac{d\eta(f(x_0))}{df}\eta(f(x_0)^T + \eta(f(x_0)\nabla(-\frac{1}{l(x_0)}),$$

for every pair of points $x_1, x_2 \in lev_{=f(x_0)}$.

Since l is a continuous function, it follows that

$$l(x_1) + l(x_2), x_1, x_2 \in lev_{=f(x_0)},$$

is not a constant on every pair of points, which is a contradiction. So, the function l is constant on the equality level set $lev_{=f(x_0)}$.

By (15.8), the following two cases may arise: case (a) and case (b), which respectively, leads to the results of (i) and (ii).

Case (a) If $b^T x + b_0$ is a constant on the level set $lev_{=f(x_0)}$, then, there exists a constant $\alpha_2 \neq 0 \in \mathbb{R}$, such that

$$\eta f(x_0) = \alpha_2 b,$$

$Q = \hat{\alpha}bb^T$, for some finite $\hat{\alpha} \in \mathbb{R}$, and by (15.12), $q = \hat{\beta}b$ for some finite $\hat{\beta} \in \mathbb{R}$. This completes the proof of case (i).

Case (b) If $b^T x + b_0$ is not a constant on the equality level set $lev_{=f(x_0)}$, then the directions of $\eta(f(x_0))$ and b are different. Moreover, $\frac{d\eta(f(x_0))}{df} = 0$ and the gradient of the function l has to be zero on X, i.e., l is a constant function on X. Thus, by (15.12), we have

$$Q = l(x_0)(\eta(f(x_0))b^T x + b\eta(f(x_0))^T).$$

$$l(x_0)(b^T x)\eta(f(x_0)) + l(x_0)(\eta(f(x_0))^T x))b$$
$$= (b^T x + b_0)l(x_0)\eta(f(x_0)) - q + f(x)b, x \in X.$$

From which, it follows that

$$l(x_0)(\eta(f(x_0)^T x - f(x))b + q = l(x_0)b_0\eta(f(x_0), x \in X. \tag{15.18}$$

If the directions of q and $\eta(f(x_0)$ coincide, then (15.18) does not hold. It follows that $q = \hat{\beta}b$ for some finite $\hat{\beta} \in \mathbb{R}, b_0 = 0$, and

$$f(x) = l(x_0)\eta(f(x_0))^T x + \hat{\beta}, x \in X, \tag{15.19}$$

which is the affine case. By (15.3), q_0 has to be zero. Let $d = \eta(f(x_0)), \hat{\gamma} = l(x_0)$. This completes the proof for case (ii).

(Sufficient) Let us assume that relation (15.9) holds. Then, we have

$$\nabla f(x) = \frac{\hat{\alpha}b^T x + \hat{\beta} - f(x)}{b^T x + b_0}b^T, \quad x \in X, \tag{15.20}$$

$$\hat{\alpha}b^T x + \hat{\beta} \neq f(x), \quad x \in X.$$

Setting

$$l(x) = \frac{\hat{a}b^T x + \hat{\beta} - f(x)^T}{b^T x + b_0}, x \in X, \eta(f(x)) = b, x \in X,$$

and applying Theorem 14.1, it follows that the function f is pseudolinear on X.

Let us assume that relation (15.10) holds. By (15.3), we get

$$f(x) = \hat{\gamma}d^T x + \hat{\beta}, x \in X, \qquad (15.21)$$

which is clearly, an affine function. This completes the proof.

Corollary 15.4 *The function f defined by (15.3) is pseudolinear and not affine if and only if there exist real constants $\hat{\alpha} \neq 0 \in \mathbb{R}$ and $\hat{\beta} \in \mathbb{R}$, such that*

$$f(x) = \frac{1}{2}\hat{a}b^T x + \hat{\beta} - \frac{1}{2}\hat{a}b_0 + \frac{q_0 - \left(\hat{\beta} - \frac{1}{2}\hat{a}b_0\right)b_0}{b^T x + b_0}, x \in X \qquad (15.22)$$

$$l(x) = \frac{1}{2}\hat{a} - \frac{q_0 - (\hat{\beta} - \frac{1}{2}\hat{a}b_0)b_0}{(b^T x + b_0)^2}^2, x \in X \qquad (15.23)$$

$$\eta(f(x)) = b, x \in X. \qquad (15.24)$$

Proof Substituting the expression (15.9) for (15.3), we get

$$f(x) = \frac{\frac{1}{2}\hat{a}(b^T x)^2 + \hat{\beta}b^T x + q_0}{b^T x + q_0}$$

$$= \frac{\frac{1}{2}\hat{a}b^T x(b^T x + b_0) - (\hat{\beta} - \frac{1}{2}\hat{a}b_0)b^T x + q_0}{b^T x + b_0}$$

$$= \frac{1}{2}\hat{a}b^T x + \frac{(\hat{\beta} - \frac{1}{2}\hat{a}b_0)b^T x + q_0}{b^T x + b_0} \qquad (15.25)$$

$$= \frac{1}{2}\hat{a}b^T x + \frac{(\hat{\beta} - \frac{1}{2}\hat{a}b_0)(b^T x + b_0) + q_0 - (\hat{\beta} - \frac{1}{2}\hat{a}b_0)b_0}{b^T x + b_0}$$

$$= \frac{1}{2}\hat{a}b^T x + \hat{\beta} - \frac{1}{2}\hat{a}b_0 + \frac{q_0 - (\hat{\beta} - \frac{1}{2}\hat{a}b_0)b_0}{b^T x + b_0}, x \in X.$$

Differentiating (15.25), the result follows.

To justify the significance of Theorem 15.3, Rapcsak [235] considered Example 15.2.

Example 15.4 *Consider problem (15.3), where*

$$f(x) = \frac{8x_1^2 + 2x_2^2 + 18x_3^2 - 8x_1x_2 - 24x_1x_3 + 12x_2x_3 + 10x_1 - 5x_2 - 15x_3 - 4}{-2x_1 + x_2 + 3x_3 - 3},$$

$$x \in X \subseteq \mathbb{R}^3.$$

Hence, we have $Q = \begin{bmatrix} -16 & -8 & -24 \\ -8 & 4 & 12 \\ -24 & 12 & 36 \end{bmatrix}, q = \begin{bmatrix} 10 \\ -5 \\ -15 \end{bmatrix}, b = \begin{bmatrix} -2 \\ 1 \\ 3 \end{bmatrix},$

$b_0 = -3, q_0 = -4.$ *In Example 15.1,* $\hat{\alpha} = 4, \hat{\beta} = -5.$ *Therefore, by Corollary 15.4, we have,*

$$f(x) = -4x_1 + 2x_2 + 6x_3 + 1 - \frac{1}{(-2x_1 + x_2 + 3x_3 - 3)^2}^2, \quad x \in X,$$

$$l(x) = (2 + \frac{1}{(-2x_1 + x_2 + 3x_3 - 3)^2}), \quad x \in X,$$

$$\eta(f(x)) = b, \quad x \in X.$$

Recently, Ujvari [276] remarked that Theorem 15.3 by Rapcsak [235] is a reformulation of Theorem 15.2 by Cambini and Carosi [34]. Therefore both these results are equivalent.

The following lemma from Ujvari [276] will be needed in the sequel.

Lemma 15.3 *Let* $g : \subseteq \mathbb{R}^n \to \mathbb{R}$ *be a differentiable function defined on an open convex set* X. *Then, between the statements:*

(a) *the function* g *is pseudolinear on* X,

(b) *for every* $x_1, x_2 \in X$,

$$\langle \nabla g(x_1), (x_2 - x_1) \rangle = 0 \Leftrightarrow g(x_1) = g(x_2);$$

(c) *for every* $\bar{x} \in X$,

$$\{x \in X : g(x) = g(\bar{x})\} = \{x \in X : \langle \nabla g(\bar{x}), (x - \bar{x}) \rangle = 0\};$$

(d) *there exist a map* $l : X \to \mathbb{R}; \eta : g(X) \to \mathbb{R}^n$, *such that*

$$\nabla g(x) = l(x).\eta(g(x)), x \in X.$$

The following implication holds:

$$(a) \Leftrightarrow (b) \Leftrightarrow (c) \, and \, (c) \Rightarrow (d).$$

Moreover, the implication $(d) \Rightarrow (a)$ *holds if the maps* l, η *are continuously differentiable functions.*

Employing the reformulation given by Rapcsak [235], Ujvari [276] has given an alternative proof of Theorem 15.3. This proof of Theorem 15.3 is shorter than the proof given by Cambini and Carosi [34] and is based on the following two propositions.

Proposition 15.4 *Suppose that the function f defined by (15.3) is pseudo-linear and $\nabla f(x) \neq 0, x \in X$. Then,*

(a) the matrix Q has rank 1 or 2;

(b) the matrix Q has rank 1 if and only if there exist constants $\hat{\alpha}, \hat{\beta} \in \mathbb{R}$, such that
$$Q = \alpha bb^T, q = \hat{\beta}b.$$

Proof (a) By part (d) of Lemma 15.3, there exist maps $l : X \to \mathbb{R}, \eta : f(X) \to \mathbb{R}^n$, such that
$$\nabla f(x) = l(x)\eta f(x), x \in X. \tag{15.26}$$
Let $\bar{x} \in X$. Then, for any $x \in X$,
$$f(x) = f(\bar{x}) \Rightarrow \eta(f(x)) = \eta(f(\bar{x})) \tag{15.27}$$
From (15.4) and (15.27), we get
$$\frac{1}{l(x)} \frac{Qx + Q - f(x)b}{b^T x + b_0} = \frac{1}{l(\bar{x})} \frac{Q\bar{x} + Q - f(\bar{x})b}{b^T x + b_0}.$$
Rewriting the equation, we get
$$Q(x - \bar{x}) = \left(\frac{l(x)(b^T x + b_0)}{l(\bar{x})(b^T \bar{x} + b_0)} - 1 \right) (Q\bar{x} + q - f(\bar{x})b). \tag{15.28}$$

For every $x \in X, f(x) = f(\bar{x})$. Moreover, by part (c) of Lemma 15.3, we have
$$\text{lin} \{x - \bar{x} : x \in X, f(x) = f(\bar{x})\} = \{\nabla f(\bar{x})\}^\perp, \tag{15.29}$$
where lin denotes a linear hull and $\perp$ denotes an orthogonal complement. From (15.28) and (15.29), it follows that
$$Q\left(\{\nabla f(\bar{x})\}^T\right) \subseteq \text{lin}\{\nabla f(\bar{x})\}. \tag{15.30}$$
Therefore,
$$Q(\mathbb{R}^n) \subseteq \text{lin}\{\nabla f(\bar{x}), Q\nabla f(\bar{x})\}, \tag{15.31}$$
where $Q(\mathbb{R}^n)$ denotes the image space of the matrix Q. Hence, $Q(\mathbb{R}^n)$ is at-most two dimensional. Hence, the rank of the matrix is at most two.

(b) If the matrix Q has rank 1, then there exists a constant $\gamma \in \mathbb{R}, \gamma \neq 0$ and a vector $c \in \mathbb{R}^n, c^T c = 1$ such that $Q = \gamma cc^T$. Therefore, from (15.30), for any $\bar{x} \in X$, it follows that
$$Q\left(\{\nabla f(x)\}^\perp\right) = \{0\} \text{or} Q\left(\{\nabla f(x)\}^\perp\right) = \text{lin}\{\nabla f(\bar{x})\}.$$

The latter case cannot hold if the matrix Q has rank 1, otherwise, there would exist a vector $z \in \mathbb{R}^n$, such that
$$\nabla f(\bar{x})^T z = 0 \text{ and } Qz = \nabla f(\bar{x}).$$

Then, $z^T Q z = 0$, that is $\gamma (c^T z)^2 = 0$, which implies that $c^T z = 0$. As $\nabla f(\bar{x}) = Q z = \gamma (c^T z) c$, it results that $\nabla f(\bar{x}) = 0$, a contradiction to our assumption that $\nabla f(x) \neq 0, \forall x \in X$. This contradiction implies that

$$Q \left(\{\nabla f(x)\}^\perp \right) = \{0\} (x \in X).$$

In other words, the vectors that are orthogonal to vector $\nabla f(x)$ are also orthogonal to vector c. Therefore,

$$\{\nabla f(x)\}^\perp = \{c\}^\perp$$

or taking the orthogonal complements

$$\text{lin}\{\nabla f(x)\} = \text{lin}\{c\}.$$

Especially, for $x \in X$, we have

$$\nabla f(x) \in Q(\mathbb{R}^n) \tag{15.32}$$

Now, we will show that (15.32) implies that $b, q \in Q(\mathbb{R}^n)$. In fact, if (15.32) holds, then by (15.4), we have

$$q - f(x)b \in Q(\mathbb{R}^n), \forall x \in X.$$

Therefore,

$$q' - f(x)b' = 0,$$

where q' and b' denote the vectors, we get projecting the vectors q and b, respectively, to the null space of the matrix Q. Since, the function f is not a constant function on X, we get

$$q' = 0 = b',$$

which imply that $b, q \in Q(\mathbb{R}^n)$.

Therefore, vectors b, q are the constant multiple of vector c. Then, there exist constants $\hat{\alpha}, \hat{\beta} \in \mathbb{R}$, such that

$$Q = \hat{\alpha} b b^T, q = \hat{\beta} b,$$

which proves the implication in part (b) of the proposition.

Example 15.5 *Consider the affine function $f : \mathbb{R}^2 \to \mathbb{R}$, given by*

$$f(x) = \frac{x_1^2 - x_2^2}{x_1 + x_2}.$$

Then by Proposition 15.4, the matrix Q can have rank 2.

Proposition 15.5 *Suppose that the function f, defined by (15.3) is pseudo-linear, and $\nabla f(x) \neq 0, x \in X$. Moreover, suppose that $b_0 = 0$. Then,*

(a) If $\bar{q} \neq 0$, then, the matrix Q has rank 1.

(b) If $\bar{q} = 0$ and the matrix Q has rank 2, then the function f is affine.

Proof (a) On the contrary, suppose that the matrix Q has rank 2. Then, there exist constants $\gamma, \delta \in \mathbb{R}, \gamma \neq 0 \neq \delta$ and vectors $c, d \in \mathbb{R}^n, c^T c = 1 = d^T d, c^T d = 0$, such that $Q = \gamma c c^T + \delta d d^T$. We assume that $p^T c \geq 0 \leq p^T d$. Since, the matrix Q has rank 2, therefore, by (15.31), we get

$$Q(\mathbb{R}^n) = \text{lin}\,\{\nabla f(x), Q \nabla f(x)\}, x \in X.$$

Then, (15.32) holds and similarly as in the proof of Proposition 15.4, we get $q, b \in Q(\mathbb{R}^n)$. We note that the vectors $\nabla f(x), Q \nabla f(x)$ are necessarily linearly independent. Therefore, for every $x \in X$, the following two vectors

$$(b^T x).\nabla f(x) = (\gamma(c^T x) + (q^T c) - f(x)(b^T c))c + (\delta(d^T x) + (q^T d) - f(x)(b^T d))d,$$

$$(b^T x).\nabla f(x) = \gamma(\gamma(c^T x) + (q^T c) - f(x)(b^T c))c + (\delta(d^T x) + (q^T d) - f(x)(b^T d))d$$

are linearly independent also.

Obviously, the linearly independency of these two vectors is obviously equivalent with the following statement:

$$\gamma \neq \delta$$

and

$$\gamma(c^T x) + (q^T c) - f(x)(b^T c) \neq 0 \neq \delta(d^T x) + (q^T d) - f(x)(b^T d), x \in X. \quad (15.33)$$

These inequalities will lead to a contradiction.

First if $b^T c = 0$, then $b^T d > 0$. Let us take

$$x := -\frac{q^T c}{\gamma} c + d.$$

Then $x \in X$ and $\gamma(c^T x) + (q^T c) = 0$, which is a contradiction. Similarly, the case $b^T d = 0$ can be dealt similarly.

Now, we suppose that $b^T c > 0 < b^T d$. The contradiction will be shown by a vector of the form $x = \lambda c, \lambda > 0$ (these vectors are all in half-space X). For these vectors, (15.32) implies that for $\lambda > 0$, we have

$$\gamma \lambda + (q^T c) - \frac{\frac{1}{2}\gamma \lambda^2 + (q^T c)\lambda + q_0}{(b^T c)\lambda}(b^T c) \neq 0, \quad (15.34)$$

$$(q^T d) - \frac{\frac{1}{2}\gamma \lambda^2 + (q^T c \lambda + q_0)}{(b^T c)\lambda}(b^T d) \neq 0. \quad (15.35)$$

Multiplying the left hand side of both inequalities by the positive constant $(b^T c)\lambda$, we get two quadratic polynomials of the variable λ, let us denote them by $\pi_1\lambda$ and $\pi_2\lambda$, respectively. A quadratic polynomial in the variable λ has a positive (and a negative) root if the coefficients of λ^2 and λ^0 have different signs. From this observation, it follows that

$$0 \in \{\pi_1(\lambda) : \lambda > 0\} \text{ if } \gamma q_0 > 0;$$

$$0 \in \pi_2(\lambda) : \lambda > 0 \text{ if } \gamma q_0 < 0.$$

This statement contradicts (15.34) or (15.35) and thus (15.33), implies that the matrix Q cannot have rank 2 if $q_0 \neq 0$.

(b) Let $a \in \{p\}^\perp$ and let us define a function $\bar{f}$, as follows:

$$\bar{f}(y) := f(y - a)(y \in \mathbb{R}^n, b^T y > 0).$$

The pseudolinearity of the function f implies the pseudolinearity of the function $\bar{f}$. Moreover,

$$\bar{f}(y) = \frac{\frac{1}{2}y^T Q y + \bar{q}y + q_0}{b^T y}(y \in \mathbb{R}^n, b^T y > 0),$$

where $\bar{q} = q - Qa, \bar{q} = \frac{1}{2}a^T Qa - q^T a$.

Since, the matrix Q has rank 2, by part (a) $\bar{q}_0 \neq 0$ cannot hold. Thus,

$$\frac{1}{2}a^T Qa - q^T a = 0(a \in \{b\}^\perp)$$

or equivalently,

$$a^T Qa = 0 = q^T a(a \in \{b\}^\perp).$$

Hence, there exist a constant $\bar{\beta} \in \mathbb{R}$ follows such that $q = \bar{\beta}b$. Furthermore,

$$x^T V Q V x = 0(x \in X),$$

where $V \in \mathbb{R}^{n \times n}$ denotes the matrix of the projection map to the subspace $b^\perp$. For notational convenience, assume that $b^T b = 1$. Then,

$$x = (b^T x)b + Vx, \forall x \in \mathbb{R}^n.$$

It is easy to calculate that

$$f(x) = \frac{\frac{1}{2}x^T Q x + q^T x}{b^T x} = \frac{\frac{1}{2}b^T Qb(b^T x)^2 + (b^T QVx)(b^T x) + \hat{\beta}b^T x}{b^T x}$$

$$= \left(\frac{1}{2}(b^T Qb)b + VQb\right)^2 x + \hat{\beta},$$

holds. Therefore, f is an affine function. This completes the proof of part (b).

Remark 15.3 *Rapcsak and Ujvari [236] have remarked that the assumption that $\nabla f(x) \neq 0, x \in X$ is superfluous in Propositions 15.4 and 15.5. Indeed, if $\nabla f(\hat{x}) = 0$ held for some $\hat{x} \in X$, then, by the pseudolinearity of the function f would be a constant function, ∇f identically zero on X and by differentiating the numerator of (15.3), Q would be the zero matrix, which is a contradiction.*

With the help of Propositions 15.4 and 15.5, Ujvari [276] has given a shorter proof of Theorem 15.3 by Rapcsak [235]. We restate the theorem and present the proof.

Theorem 15.4 *Assume that f is the function defined by (15.3) and $\nabla f(x) \neq 0, x \in X$. The function f is pseudolinear if and only if f is affine or there exist constants $\hat{\alpha}, \hat{\beta} \in \mathbb{R}$, such that*

$$Q = \hat{\alpha} b b^T, q = \hat{\beta} b.$$

Proof If the function f is pseudolinear, then the function $\hat{f}$, defined by

$$\hat{f}(y) = f\left(y - \frac{b_0}{b^T b} b\right) (y \in \mathbb{R}^n, b^T y > 0),$$

is also pseudolinear. Therefore,

$$\hat{f}(y) = \frac{\frac{1}{2} y^T Q y + q^T y + q_0}{b^T y} (y \in \mathbb{R}^n, b^T y) > 0),$$

where $q \in \mathbb{R}^n, q_0 = 0$, so Proposition 15.5 can be applied. If $q_0 \neq 0$, then, part (a) of Proposition 15.5 implies that the matrix Q has rank 1. Again part (b) of Proposition 15.5 implies that there exist constants $\bar{\alpha}, \bar{\beta} \in R$ such that $Q = \alpha b b^T, q = \beta b$. In this case, the proof is complete.

Similarly, the case when $q_0 = 0$, and the rank of Q is 1 can be dealt. Finally, we consider the case, $q_0 = 0$, and the rank of Q is 2. In this case, by part (b) of Proposition 15.5, the function $\hat{f}$ and therefore the function f is affine.

Conversely, suppose that f is not affine and there exists constants $\hat{\alpha}, \hat{\beta} \in \mathbb{R}$, such that

$$Q = \hat{\alpha} b b^T, q = \hat{\beta} b.$$

Then, for all $x \in X$, we have

$$f(x) = \alpha b^T x + \beta + \frac{\alpha \gamma}{b^T x + b_0}$$

and

$$\nabla f(x) = \left(\alpha - \frac{\alpha \gamma}{(b^T x + b_0)^2}\right) b, x \in X,$$

where

$$\alpha = \frac{1}{2} \hat{\alpha}, \beta = \hat{\beta} - \frac{1}{2} \hat{\alpha} b_0, \alpha \gamma = q_0 - \hat{\beta} - \frac{1}{2} \hat{\alpha} b_0 () b_0.$$

Due to the assumption that $\nabla f(x) \neq 0, x \in X$, we get $\gamma < 0$. Now, for $x_1, x_2 \in X$, we can show that

$$f(x_1) = f(x_2) = \left(\alpha - \frac{\alpha \gamma}{(b^T x_1 + b_0)(b^T x_2 + b_0)}\right) . b^T (x_2 - x_1).$$

As $\gamma < 0$, we get

$$\alpha - \frac{\alpha\gamma}{(p^T x_1 + p_0)(p^T x_2 + p_0)} \neq 0, x_1, x_2 \in X.$$

Therefore, for every $x_1, x_2 \in X$, we have

$$f(x_1) = f(x_2) \Leftrightarrow b^T(x_2 - x_1) = 0.$$

Moreover, we know that for every $x_1, x_2 \in X, \langle \nabla f(x_1), (x_2 - x_1) \rangle = 0$ holds if and only if $b^T(x_2 - x_1) = 0$.

Hence, by part (b) of Lemma 15.3, the pseudolinearity of the function f follows.

Ujwari [276] has shown that Theorem 15.3 due to Rapcsak [235] is a reformulation of Theorem 15.2 due to Cambini and Carosi [34]. Therefore, both the theorems are equivalent. Ujvari [276] has shown the equivalence between Theorems 15.3 and 15.2 as follows.

(Theorem 15.3 $\Rightarrow$ Theorem 15.2) Let us suppose that the function f is pseudolinear. If $\nabla f(\bar{x}) = 0$, for some $\bar{x} \in X$, then f will be a constant function. By Remark 15.1, it is clear that the function f is not a constant function. Therefore, $\nabla f(x)) \neq 0, x \in X$ and Theorem 15.3 can be applied. If f is not affine, then there exist constants $\bar{\alpha}, \bar{\beta} \in \mathbb{R}$, such that

$$Q = \hat{\alpha} b b^T, q = \hat{\beta} b.$$

In this case (similarly as in the proof of Theorem 15.4), there exists constants $\alpha, \beta, \gamma \in \mathbb{R}$, such that $\gamma < 0$ and (15.3) holds. This proves the "only if" part of Theorem 15.2.

To prove the "if" part of Theorem 15.2, let us suppose that there exist constants α, β, k with the properties described in Theorem 15.2. Then, it is easy to see that $\nabla f(x)) \neq 0, x \in X$. Therefore, Theorem 15.3 can be applied. Moreover,

$$\frac{1}{2} x^T Q x + q^T x + q_0 = f(x)(b^T x + b_0) = \alpha x^T b b^T x + (\alpha b_0 + \beta)(b^T x) + (\beta b_0 + k).$$

Now, the equality of the second, first, and zeroeth derivatives implies that

$$Q = 2\alpha p p^T, q = (\alpha p + \beta)p, q_0 = \beta p_0 + k.$$

Hence, with the constants $\hat{\alpha} := 2\alpha, \hat{\beta} = (\alpha p_0 + \beta)$, it follows that $Q = \alpha p p^T$ and $q = \beta p$. By Theorem 15.3, the function f is pseudolinear.

This completes the proof of Theorem 15.2.

(Theorem 15.2 $\Rightarrow$ Theorem 15.3) Through similar arguments, we can see that Theorem 15.2 implies Theorem 15.3.

Hence, we conclude that Theorem 15.2 and Theorem 15.3 are equivalent.

15.4   Affinity of Quadratic Fractional Functions

In this section, we present the characterization for the affinity of quadratic fractional functions.

The following theorem is from Rapcsak and Ujvari [236].

Theorem 15.5 *Let the function f be defined by (15.3). Then, function f is affine if and only if one of the following conditions hold:*

(i) *The rank of Q is equal to 1, formula (15.9) holds and*

$$\frac{1}{2}\hat{a}b_0^2 - \hat{\beta}b_0 + q_0 = 0. \tag{15.36}$$

(ii) *The rank of Q is equal to 2, $b_0 \neq 0$, and the vector $d = \frac{1}{b_0}\left(q - \frac{q_0}{b_0 b}\right)$ fulfills the equality*

$$q = bd^T + db^T. \tag{15.37}$$

(iii) *The rank of Q is equal to 2, $b_0 = 0$, there exists a vector $d \in \mathbb{R}^n$ such that $q = bd^T + db^T$, and there exists a constant $\hat{\beta} \in \mathbb{R}$, such that $q = \hat{\beta}b$, $q_0 = 0$.*

Proof (Necessary) First, we assume that the rank of the matrix Q is equal to 1. By (ii) of Proposition 15.4, formula (15.9) holds. Let a vector $v \in \mathbb{R}^n$ be such that $b^T v = b_0$. Therefore, the function

$$\hat{f}(y) = f(y - v), y \in \mathbb{R}^n, b^T y > 0, \tag{15.38}$$

is affine. Now, we have

$$\hat{f}(y) = \frac{\frac{1}{2}y^T Q y + (q - Qv)^T + \frac{1}{2}v^T Q v - q^T v + q_0}{b^T y} \tag{15.39}$$

$$= \frac{1}{2}\hat{a}b^T y + (\hat{\beta} - \hat{a}b_0) + \frac{\frac{1}{2}\hat{a}b_0^2 - \hat{\beta}b_0 + q_0}{b^T y}.$$

From which, it follows that

$$\frac{1}{2}\hat{a}b_0^2 - \hat{\beta}b_0 + q_0 = 0. \tag{15.40}$$

This completes the proof of the statement (i).

To prove statements (ii) and (iii), we proceed as follows: Assume that the rank of Q is equal to 2. Since the function f defined by (15.39) is affine, the equality

$$\frac{1}{2}v^T Q v - q^T v + q_0 = 0 \tag{15.41}$$

is valid. If formula (15.41) was not true, then, $\hat{f}$ would be the quotient of an inhomogeneous quadratic function and a homogeneous affine function, so by (i) of Proposition 15.4, the rank of Q should be 1, which is a contradiction. Thus, there exist a vector d and a real number μ, such that

$$\frac{\frac{1}{2}y^T Q y + (q - Q v)^T y}{b^T y} = d^T y + \mu, y \in \mathbb{R}^n, b^T y > 0. \tag{15.42}$$

Multiplying (15.42) by $b^T y$, and differentiating it, we have that

$$Q = bd^T + db^T, q - Qv = \mu b. \tag{15.43}$$

From (15.43) and (15.41), it follows that

$$q = (d^T v + \mu)b + b_0 d, q_0 = b_0 (d^T v + \mu).$$

Thus statements (ii) and (iii) are proved.

(Sufficient) If the conditions of (i), (ii), and (iii) hold, then by substituting them for (15.3), respectively, we have the statements.

The following corollary is an equivalent form of Theorem 15.6.

Corollary 15.5 *A quadratic fractional function is affine if and only if there exists a constant $\hat{\beta} \in \mathbb{R}$ and a vector $d \in \mathbb{R}^n$, such that*

$$Q = bd^T + db^T, q - Qv = \hat{\beta}b,$$

$$\frac{1}{2}v^T Q v - q^T v + q_0 = 0 \text{ with } v = \frac{b_0}{b^T b}b. \tag{15.44}$$

Proof It is easy to see by simple calculations that in all the three cases in Theorem 15.5, formula (15.44) holds. On the other hand, formula (15.44) implies that the function $\hat{f}$ in (15.38) (and thus, f also) is affine.

Applying Theorem 15.2 to the function $\hat{f}$ in (15.38), we have the following corollary for nonaffine case.

Corollary 15.6 *A quadratic fractional function is pseudolinear and non-affine if and only if there exist real numbers $\hat{\alpha}, \hat{\beta} \in \mathbb{R}$, such that*

$$Q = \hat{\alpha}bb^T, q - Qv = \hat{\beta}b,$$

$$\hat{\alpha}\left(\frac{1}{2}v^T Q v - q^T v + q_0\right) < 0 \text{ with } v = \frac{b_0}{b^T b}b. \tag{15.45}$$

We know that the pseudolinearity of a function implies quasilinearity in general. Cambini and Carosi [34] have established that in the case of quadratic fractional functions, also the reverse implication holds. Here, we take a different approach to this statement.

Theorem 15.6 *A quadratic fractional function is quasilinear if and only if it is pseudolinear.*

This statement is a direct consequence of the following two lemmas.

Lemma 15.4 *Let f be a real differentiable function defined on an open convex set $X \subseteq \mathbb{R}^n$. If the function f is pseudolinear, then f is quasilinear. If $\nabla f(x) \neq 0, x \in X$ and f is quasilinear, then it is pseudolinear.*

Lemma 15.5 *If the function f in (15.3) is quasilinear, then $\nabla f(x) \neq 0, x \in X$.*

Proof On the contrary, suppose that a point $x_0 \in X$ exists, such that $\nabla f(x_0) = 0$. This condition is equivalent to

$$f(x_0)b = Qx_0 + q. \tag{15.46}$$

By (15.46) and the identity,

$$f(x_0)(b^T x_0 + b_0) = \frac{1}{2}x_0^T Qx_0 + q^T x_0 + q_0,$$

it follows that

$$f(x_0)b_0 = -\frac{1}{2}x_0^T Qx_0 + q_0.$$

From which, we get

$$f(x) = f(x_0) + \frac{1}{2}\frac{(x - x_0)^T Q(x - x_0)}{b^T x_0 + b_0}, x \in X. \tag{15.47}$$

Now, we will show that Q is a positive semidefinite matrix. Again, on the contrary suppose that there exist a vector $v \neq 0$ and a real value $\lambda < 0$ exist with $Qv = \lambda v$. Let $\epsilon > 0$ be such that $x_0 \pm \epsilon v \in X$. Then,

$$x_0 = \frac{1}{2}(x_0 + \epsilon v) + \frac{1}{2}(x_0 - \epsilon v),$$

and

$$f(x_0 \pm \epsilon v) < f(x_0),$$

which is a contradiction to the quasiconvexity of f. Similarly, we can prove that Q is negative semidefinite. From which, it follows that $Q = 0$, which is a contradiction. These, complete the proof.

15.5   Gradient of Pseudolinear Quadratic Fractional Functions

In this section, we present characterizations for the gradient of pseudolinear quadratic fractional functions derived by Rapcsak and Ujvari [236]. Let us introduce the notation of the equality level set

$$lev_{=f(x_0)} := \{x \in X | f(x) = f(x_0)\}, \tag{15.48}$$

where $x_0 \in X$ is an arbitrary point.

Theorem 15.7 *The gradient of a pseudolinear quadratic fractional function f defined by (15.3) is equal to*

$$\nabla f(x) = l(x)c, c \in \mathbb{R}^n, x \in X, \tag{15.49}$$

where c is a constant vector and the function l is constant on every equality level set. Moreover, in the nonaffine case,

$$l(x) = \frac{1}{2}\hat{\alpha} - \frac{\frac{1}{2}\hat{\alpha}b_0^2 - \hat{\beta}b_0 + q_0}{(b^T x + b_0)^2}, x \in X, \tag{15.50}$$

$$\eta(f(x)) = b, x \in X.$$

In the affine cases (i), (ii), and (iii) of Theorem 15.5, we have

$$\eta(f(x)) = \frac{1}{2}\hat{\alpha}b, x \in X, \text{ for some } \hat{\alpha} \in \mathbb{R}, \tag{15.51}$$

$$\eta(f(x)) = \frac{1}{b_0}\left(q - \frac{q_0}{b_0}b\right), x \in X, \tag{15.52}$$

$$\eta(f(x)) = d, x \in X, \text{ for some } d \in \mathbb{R}^n, \tag{15.53}$$

respectively.

Proof By Theorem 14.1, the f function defined by (15.3) is pseudolinear if and only if there exist continuously differentiable functions $l(x), \eta_i(f(x)), i = 1, ..., n, x \in X$, such that

$$\nabla f(x) = l(x)\eta(f(x)), x \in X. \tag{15.54}$$

First, we show that the function l in formula (15.54) is constant on the equality level set (15.48). By (15.4) and (15.54), it follows that

$$\frac{Qx + q - f(x)b}{b^T x + b_0} = l(x)\eta(f(x)), x \in X. \tag{15.55}$$

Therefore

$$Qx = (b^T x + b_0)l(x)\eta(f(x)) - q + f(x)b, x \in X$$

and

$$Qx = (b^T x + b_0)l(x)\eta(f(x_0)) - q + f(x_0)b, \ x \in lev_{=f(x_0)}.$$

Since, $\eta(f(x_0)) \neq 0$, there exists at least one component of $\eta(f(x_0))$ different from zero. Considering the equality related to this component, it follows that l is constant on $lev_{=f(x_0)}$. By Corollary 15.4, the formula (15.50) holds true. In order to obtain (15.51), let us substitute formula (15.9) for function (15.3). We get

$$f(x) = \frac{\frac{1}{2}\hat{\alpha}(b^T x)^2 + \hat{\beta}b^T x + q_0}{b^T x + b_0} = \frac{\frac{1}{2}\hat{\alpha}(b^T x) + b_0 + \left(\hat{\beta} - \frac{1}{2}\hat{\alpha}(b_0)\right) b^T x + q_0}{b^T x + b_0}$$

$$= \frac{1}{2}\hat{\alpha}b^T x + \frac{\left(\hat{\beta} - \frac{1}{2}\hat{\alpha}b_0\right) b^T x + q_0}{b^T x + b_0}$$

$$= \frac{1}{2}\hat{\alpha}b^T x + \frac{\left(\hat{\beta} - \frac{1}{2}\hat{\alpha}b_0\right)(b^T x + b_0) + q_0 - \left(\hat{\beta} - \frac{1}{2}\hat{\alpha}b_0\right) b_0}{b^T x + b_0}$$

$$= \frac{1}{2}\hat{\alpha}b^T x + \hat{\beta} - \frac{1}{2}\hat{\alpha}b_0 + \frac{\frac{1}{2}\hat{\alpha}b_0^2 - \hat{\beta}b_0 + q_0}{b^T x + b_0}, x \subset X, \qquad (15.56)$$

from which formula (15.51) follows. Formulae (15.52) and (15.53) are direct consequences of Theorem 15.6.

Remark 15.4 *It is clear from formulae (15.50) and (15.51) that the meaning of the inequality $\alpha k < 0$, is that the gradient of the function f is different from zero, and show how to determine the constants in Theorem 15.4.*

The following example from Rapcsak and Ujvari [236] justifies the significance of the affine cases (i), (ii), and (iii) of Theorem 15.5.

Example 15.6 *Consider the following quadratic fractional functions:*

$$f^1(x) = \frac{x_1^2 + 2x_1 x_2 + x_2^2 + x_1 + x_2}{x_1 + x_2 + 1}, x \in X^1 \subseteq \mathbb{R}^2;$$

$$f^2(x) = \frac{x_1 x_2 - x_2^2 + x_1 + 1}{x_2 + 1}, x \in X^2 \subseteq \mathbb{R}^2;$$

$$f^3(x) = \frac{x_1^2 - x_2^2}{x_2 + x_2}, x \in X^3 \subseteq \mathbb{R}^2;$$

In this case, we have

$$Q^1 = \begin{bmatrix} 2 & 2 \\ 2 & 2 \end{bmatrix}, q^1 = \begin{bmatrix} 1 \\ 1 \end{bmatrix}, b^1 = \begin{bmatrix} 1 \\ 1 \end{bmatrix}, q_0^1 = 0, b_0^1 = 1.$$

$$Q^2 = \begin{bmatrix} 0 & 1 \\ 1 & -2 \end{bmatrix}, q^2 = \begin{bmatrix} 1 \\ 0 \end{bmatrix}, b^2 = \begin{bmatrix} 0 \\ 1 \end{bmatrix}, q_0^1 = 1, b_0^2 = 1.$$

$$Q^3 = \begin{bmatrix} 2 & 0 \\ 2 & -2 \end{bmatrix}, q^3 = \begin{bmatrix} 0 \\ 0 \end{bmatrix}, b^3 = \begin{bmatrix} 1 \\ 1 \end{bmatrix}, q_0^3 = 0, b_0^3 = 0.$$

It is easy to see that the functions f^1, f^2, *and* f^3 *are affine with*

$$\eta^1(f^1(x)) = \begin{bmatrix} 1 \\ 1 \end{bmatrix}, \eta^2(f^2(x)) = \begin{bmatrix} 1 \\ -1 \end{bmatrix}, \eta^3(f^3(x)) = \begin{bmatrix} 1 \\ -1 \end{bmatrix}.$$

In all the three cases

$$l^i(x) = 1, x \in X^i, i = 1, 2, 3.$$

15.6 A Larger Class of Quadratic Fractional Functions

In this section, we present characterizations for the generalized linearity of a class of quadratic fractional functions larger than the one considered in (15.3). Our aim is to characterize the generalized linearity of the following types of functions:

$$g(x) = \frac{\frac{1}{2}x^T Q x + q^T x + q_0}{b^T x + b_0} + c^T x = f(x) + c^T x, \tag{15.57}$$

where $X := \{x \in \mathbb{R}^n : b^T x + b_0 > 0\}, n \geq 2, Q$ is a $n \times n$ symmetric matrix, $q, x, b, c \in \mathbb{R}^n, b \neq 0$, and $q_0, b_0 \in \mathbb{R}$. We note that g is defined as in (15.3) when $c = 0$ and $Q \neq 0$.

The following example from Cambini and Carosi [34] points out that it may happen that $f[g]$ is pseudolinear, even if $g[f]$ is not.

Example 15.7 *We consider problem (15.58), where*

$$g(x) = \frac{x_1^2 + x_1 x_2 - x_1 + 2x_2 + 1}{x_1 + x_2} - x_1.$$

We note that

$$f(x) = \frac{x_1^2 + x_1 x_2 - x_1 + 2x_2 + 1}{x_1 + x_2} = x_1 - 1 + \frac{3x_2 + 1}{x_1 + x_2}$$

is not pseudolinear. However,

$$g(x) = \frac{-x_1 + 2x_2 + 1}{x_1 + x_2}$$

is pseudolinear since it is a linear fractional function.

Let us consider problem (15.58), where

$$g(x) = \frac{8x_1^2 + 8x_1x_2 + 2x_2^2 - 1}{2x_1 + x_2} + x_1 - x_2.$$

We note that

$$g(x) = \frac{8x_1^2 + 8x_1x_2 + 2x_2^2 - 1}{2x_1 + x_2} = 2(2x_1 + x_2) - \frac{1}{2x_1 + x_2}$$

is pseudolinear. However,

$$g(x) = 5x_1 + x_2 - \frac{1}{2x_1 + x_2}$$

is not pseudolinear. The characterization of the generalized linearity of g follows from Theorem 15.2.

Theorem 15.8 *Consider the function g defined by (15.58). Then, the following statements hold:*
(i) g is affine if and only if f is affine;
(ii) g is pseudolinear (quasilinear) but not affine if and only if either $Q + bc^T + cb^T = 0$ or there exist $\alpha, \xi \in R, \alpha \neq 0$, such that:

$$Q + bc^T + cb^T = 2\alpha bb^T, q + b_0 c = \xi b, b_0^2 < \frac{\xi b_0 - q_0}{\alpha}.$$

Proof (i) Since, g is the sum of f and an affine function, the result holds trivially.

(ii) By simple calculations, it is easy to show that the function g can be rewritten as follows:

$$g(x) = \frac{\frac{1}{2}x^T[Q + bc^T + cb^T]x + [Q + b_0c]^Tx + q_0}{b^Tx + b_0}.$$

We note that if $Q + bc^T + cb^T = [0]$, then g is a linear fractional function which is known to be pseudolinear. If otherwise $Q + bc^T + cb^T = 0$, from Theorem 15.2, g is pseudolinear but not affine if and only if there exist $\alpha, \beta, \gamma \in R, \alpha \neq 0$, such that, it can be rewritten in the following form:

$$g(x) = \alpha b^Tx + \beta + \frac{\alpha\gamma}{b^Tx + b_0} \quad \text{with } \gamma < 0.$$

Therefore,

$$g(x) = \frac{\frac{1}{2}x^T[2\alpha bb^T]x + (\beta + \alpha b_0)b^Tx + (\beta b_0 + \alpha\gamma)}{b^Tx + b_0}.$$

This means that:

$$Q + bc^T + cb^T = 2\alpha bb^T, q + b_0 c = (\beta + \alpha b_0)b, q_0 = \beta b_0 + \alpha \gamma.$$

Defining $\xi = \beta + \alpha b_0$, so that $\beta = \xi - \alpha b_0$ and $\gamma = \frac{q_0 - \beta b_0}{\alpha} = b_0^2 - \frac{\xi b_0 - q_0}{\alpha}$, the result then follows from $\gamma < 0$.

We note that Theorem 15.8 can also be applied to functions defined as in (15.3) with $c = 0$. The next examples justify the significance of Theorem 15.8.

Example 15.8 *Let us consider again function g in Example 15.6. We note that*

$$Q = \begin{bmatrix} 2 & 1 \\ 1 & 0 \end{bmatrix}, q = \begin{bmatrix} -1 \\ 2 \end{bmatrix}, q_0 = 1, b = \begin{bmatrix} 1 \\ 1 \end{bmatrix}, b_0 = 1, c = \begin{bmatrix} -1 \\ 0 \end{bmatrix},$$

and

$$Q + bc^T + cb^T = 0.$$

Hence g is pseudolinear.

Example 15.9 *Let us consider problem (15.35), where*

$$g(x) = \frac{x_2^2 - 2x_1 x_2 + 3x_2 x_3 - 2x_1 - 2x_2 + 3x_3 - 4}{-2x_1 + x_2 + 3x_3 - 3} - 4x_1 + x_2 + 6x_3.$$

We note that

$$Q = \begin{bmatrix} 0 & -2 & 0 \\ -2 & 0 & 3 \\ 0 & 3 & 0 \end{bmatrix}, q = \begin{bmatrix} -2 \\ -2 \\ 3 \end{bmatrix}, b = \begin{bmatrix} -2 \\ 1 \\ 3 \end{bmatrix}, c = \begin{bmatrix} -4 \\ 1 \\ 6 \end{bmatrix},$$

and

$$q_0 = -4, b_0 = -3.$$

Therefore, we have

$$Q + bc^T + cb^T = \begin{bmatrix} 16 & -8 & 24 \\ -8 & 2 & 12 \\ -24 & 12 & 36 \end{bmatrix} = 2\alpha bb^T, \text{ with } \alpha = 2,$$

$$q + b_0 c = \begin{bmatrix} 10 \\ -5 \\ -15 \end{bmatrix} = \xi b \text{ with } \xi = -5, b_0^2 = 9 < \frac{19}{2} = \frac{\xi b_0 - q_0}{\alpha}.$$

Hence g is pseudolinear.

15.7   Pseudolinear Quadratic Fractional Functions in Optimization

It is obvious from Theorem 15.2 that every generalized linear quadratic fractional function f is the sum of a linear and a linear fractional function. This property, together with Theorem 15.8 and Corollary 15.3, can be efficiently used in order to study the following class of problems:

$$\min / \max_{x \in S \subseteq X} g(x) = \frac{\frac{1}{2}x^T Q x + q^T x + q_0}{b^T x + b_0} + c^T x = f(x) + c^T x, \qquad (15.58)$$

where g is defined as in (15.57) and the matrix $Q + bc^T + cb^T$ has at least $n - 2$ zero eigenvalues.

Case (1) $-g$ is an affine function.
By Theorem 15.8, this case occurs, whenever f is an affine function, that is, when there exist $a_0 \in \mathbb{R}$, such that $q_0 = a_0 b_0$, there exists $a \in \mathbb{R}^n$, such that $q - a_0 b = b_0 a$ and therefore,

$$Q = ab^T + ba^T.$$

In this case, $f(x) = a^T x + a_0$ and hence $g(x) = (a + c)^T x + a_0$. Therefore, the problem (15.58) can be solved by means of a linear problem:

$$\arg\min / \arg\max_{x \in S \subseteq X} g(x) = \arg\min / \arg\max_{x \in S \subseteq X} (a + c)^T x$$

Case (2) g is a linear fractional function.
By Theorem 15.15, this case occurs when

$$Q + bc^T + cb^T = 0.$$

The objective function becomes

$$g(x) = \frac{(q + b_0 c)^T x + q_0}{b^T x + b_0}.$$

In that case, problem (15.58) can be solved with any algorithm for linear fractional functions. For details see, Cambini and Martein [37], Cambini [38], and Ellero [78].

Case (3) g is a pseudolinear but not affine function.
Suppose that g is not a linear fractional function. By Theorem 15.8, g is pseudolinear but not affine when there exist $\alpha \in \mathbb{R}, \alpha \neq 0$, such that $Q + bc^T + cb^T = 2\alpha bb^T$ and there exists $\xi \in \mathbb{R}$ such that $q + b_0 c = \xi b$.

This yields that

$$b_0^2 < (\xi b_0 - q_0)/\alpha.$$

Defining $\beta = \xi - \alpha b_0, \gamma = b_0^2 - [(\xi b_0 - q_0)/\alpha]$ and the function $\varphi(t) = \alpha t + \beta + [\alpha\gamma/(t + b_0)]$, it follows that:

$$g(x) = \alpha b^T x + \beta + \frac{\alpha\gamma}{b^T x + b_0} = \varphi(b^T x) \text{ with } \gamma < 0.$$

Since $\varphi'(t) = \alpha(1 - \gamma/(t + b_0)^2)$, we have that

$$\varphi'(t) > 0(< 0) \iff \alpha > 0(< 0).$$

Therefore,

$$\alpha > 0 \Rightarrow \arg\min_{x \in S \subseteq X} / \arg\max g(x) = \arg\min_{x \in S \subseteq X} / \arg\max b^T x;$$

$$\alpha < 0 \Rightarrow \arg\min_{x \in S \subseteq X} / \arg\max g(x) = \arg\min_{x \in S \subseteq X} / \arg\max b^T x.$$

Hence, problem (15.59) can be solved by means of a linear one.

Case (4) g is pseudolinear on subsets of X.

Suppose that there exists $\alpha \in R, \alpha \neq 0$, such that $Q + bc^T + cb^T = 2\alpha bb^T$ and there exists $\xi \in R$, such that $q + b_0 c = \xi b$, and it results $b_0^2 > (\xi b_0 - q_0)/\alpha$. Let us define $\beta = \xi - \alpha b_0, \gamma = b_0^2 - [(\xi b_0 - q_0)/\alpha]$ and the function $\varphi(t) = \alpha t + \beta + [\alpha\gamma/(t + b_0)]$. Therefore, from Theorem 15.5, it follows that

$$g(x) = \alpha b^T x + \beta + \frac{\alpha\gamma}{b^T x + b_0} = \varphi(b^T x) \text{ with } \gamma > 0.$$

From Corollary 15.3, g is not pseudolinear on X but it is pseudolinear on

$$X_1 = \{x \in R^n : b^T x + b_0 > \sqrt{\gamma}\} \text{ and } X_2 = \{x \in R^n : 0 < b^T x + b_0 < \sqrt{\gamma}.\}$$

Assume now $X_3 := \{x \in R^n : b^T x + b_0 = \sqrt{\gamma}\}$. Since,

$$g(x) = 2\alpha\sqrt{\gamma} - \alpha b_0 + \beta = 2\alpha(\sqrt{\gamma} - b_0) + \xi \ \forall x \in X_3,$$

g is constant on X_3. Consequently, problem (15.58) can be studied by determining the sets:

$$S_1^m(x) = \arg\min_{x \in S \cap X_1} \{b^T x\}, S_1^M(x) = \arg\min_{x \in S \cap X_1} \{b^T x\},$$

$$S_2^m(x) = \arg\min_{x \in S \cap X_2} \{b^T x\}, S_2^M(x) = \arg\min_{x \in S \cap X_2} \{b^T x\},$$

and by denoting (if the related sets are nonempty)

$$x_1^m \in S_1^m, x_1^M \in S_1^M, x_2^m \in S_2^m, x_2^M \in S_2^M, x_3 \in S \cap X_3.$$

If $\alpha > 0$ then $\varphi'(t) > 0$ when $t + b_0 > \sqrt{\gamma}$ while it is $\varphi'(t) < 0$ when $0 < t + b_0 < \sqrt{\gamma}$. Therefore, we get

$$\min_{x \in S \subseteq X}\{g(x)\} = \min\{g(x_1^m), g(x_2^M), g(x_3)\},$$

$$\max_{x \in S \subseteq X}\{g(x)\} = \max\{g(x_1^M), g(x_2^m), g(x_3)\}.$$

Similarly, for $\alpha < 0$, we have

$$\min_{x \in S \subseteq X}\{g(x)\} = \min\{g(x_1^M), g(x_2^m), g(x_3)\}$$

$$\max_{x \in S \subseteq X}\{g(x)\} = \max\{g(x_1^m), g(x_2^M), g(x_3)\}$$

Again, problem (15.58) can be solved by means of linear ones.

15.8 Local Analysis of Pseudolinearity through the Implicit Function Approach

In Chapter 2, we studied that the pseudolinearity can be investigated via the implicit-function approach. This approach carries out a local analysis on the given function. To this end, we have to localize the global concept of pseudolinearity. The usefulness of this concept lies in the fact that global analysis can be carried out via the local one on the basis of the following theorem. As the next theorem asserts, local pseudolinearity can be analyzed via a more simple condition.

Proposition 15.6 (Komlosi [147]) *Let $\nabla f(x_0) \neq 0$. Then $f(x)$ is locally pseudolinear at x_0, if and only if there exists a neighborhood G of x_0 such that the following condition holds:*

$$(LPLIN) \quad x \in G, f(x) = f(x_0) \Rightarrow \nabla f(x_0)(x - x_0) = 0.$$

We recall from Chapter 2 that condition (LPLIN) opens the way for applying the well-known implicit-function theorem in our local analysis. Let $f(x)$ be defined and continuously differentiable on a neighborhood of $x_0 \in R^n$ and let $\nabla f(x_0) \neq 0$. Assume that $\nabla f'_{x_n}(x_0) \neq 0$. Introduce the following notations:

$$(x_1, x_2, \ldots, x_{n-1}) = u, x_n = \nu, x = (u, \nu) \text{ and } x_0 = (u_0, \nu_0).$$

Definition 15.3 (The implicit-function theorem) *The level curve $\{x \in X : f(x) = f(x_0)\}$ can be represented locally (on a certain neighborhood G of x_0) by the help of a uniquely determined implicit function $p_{x_0}(u)$ defined on a suitable neighborhood N of u_0, as follows:*

$$\forall x \in (u, \nu) \in G, f(x) = f(x_0) \Leftrightarrow \nu = p_{x_0}(u), u \in N.$$

In Chapter 2, Theorem 2.8 establishes that any continuously differentiable function $f(x)$ satisfying condition $\nabla f(x_0) \neq 0$ at a point x_0 is locally pseudolinear at x_0, if and only if the implicit function $p_{x_0}(u)$ is linear on N. From Proposition 15.2 and Proposition 15.3, one can infer the following Rapcsak type characterization of local pseudolinearity, the global version of which was proved in Bhatia and Jain [25]. According to the following theorem, local pseudolinearity can be investigated by the help of the implicit function $p_{x_0}(u)$. The following lemma plays an important role in our further investigation:

Lemma 15.6 *Assume that the twice continuously differentiable function $f(x)$ satisfies condition $\nabla f(x_0) \neq 0$. Then $f(x)$ is locally pseudolinear at x_0 if and only if $p_{x_0}(u)$ (in short: $p(u)$) satisfies the following condition: for all $u \in N$*

$$\nabla_{uu}^2 f(u, p(u)) + 2\nabla_u f_\nu'(u, p(u))\nabla(p(u))^T + f_{\nu\nu}''(u, p(u))\nabla p(u)\nabla p(u)^T \equiv 0.$$
(15.59)

Proof Differentiating twice the implicit equation $f(u, p(u)) \equiv f(x_0)$, the Hessian $\nabla^2 p(u)$ of the implicit function $p(u)$ is given by

$$f_\nu'(u, p(u))\nabla^2 p(u) = -\nabla_{uu}^2 f(u, p(u)) - 2\nabla_u f_\nu'(u, p(u))\nabla p(u)^T - f_{\nu\nu}''(u, p(u))$$
$$\nabla p(u)\nabla p(u)^T.$$

We note that (15.59) holds if and only if

$$\nabla^2(p(u)) = 0, \ \forall u \in N,$$

which holds true if and only if $p(u)$ is linear on N.

Theorem 15.9 *Assume that the twice continuously differentiable function $f(x)$ is locally pseudolinear at $x_0 \in X$, where $\nabla f(x_0) \neq 0$. For the sake of simplicity, we assume that $f_{x_0}'(x_0) \neq 0$. Then there exist $r \in R^{n-1}$ and $\lambda, \mu \in R$ such that*

$$\nabla^2 f(x_0) = \begin{bmatrix} \lambda rr^T & \mu r \\ \mu r^T & -(\lambda + 2\mu) \end{bmatrix}.$$
(15.60)

Proof We consider the implicit function $p(u)$ which describes the level curve $f(x) = f(x_0)$ around x_0. By Theorem 2.8, $p(u)$ is linear. Now, by Lemma 15.6 and for all $u \in N$, we have

$$\nabla p(u) = -\frac{\nabla_u f(u, p(u))}{f_\nu'(u, p(u))} \equiv r = \text{ constant.}$$
(15.61)

and

$$\nabla_{uu}^2 f(u, p(u)) \equiv -2\nabla_u f_\nu'(u, p(u))^T \nabla p(u) - f_{\nu\nu}''(u, p(u))\nabla p(u)^T p(u). \quad (15.62)$$

Let us consider matrix $A = \nabla^2 f(x_0)$ in its (u, ν) decomposition:

$$\nabla^2 f(x_0) = A = \begin{bmatrix} \lambda A_{uu} & a_u \\ a_u^T & a_{\nu\nu} \end{bmatrix},$$

which is similar to the (u, ν) decomposition of x. In view of this decomposition, by a simple computation, we get

$$\nabla^2_{uu} f(u, p(u)) = A_{uu}, \nabla_u f'_\nu(u, p(u)) \nabla p(u)^T = a_u \nabla p(u)^T = a_u r^T,$$

$$f''_{\nu\nu}(u, p(u)) \nabla p(u) \nabla p(u)^T = a_{\nu\nu} \nabla p(u) \nabla p(u)^T = a_{\nu\nu} r r^T.$$

In view of (15.62), it follows that $A_{uu} = -(2a_u + a_{\nu\nu} r) r^T$. Since A_{uu} is a symmetric submatrix in A, therefore $-2a_u - a_{\nu\nu} r = \lambda r$ should hold for some $\lambda \in R$ and thus $A_{uu} = \lambda r r^T$, and $a_u = \lambda r$ with $\mu = -\frac{\lambda + a_{\nu\nu}}{2}$. Therefore, $a_{\nu\nu} = -(\lambda + 2\mu)$.

The special form of $A = \nabla^2 f(x_0)$, can be further investigated to get more useful information on it. The analysis is based on Haynsworth's inertia theorem (see, Haynsworth [101].

Definition 15.4 *The inertia of a symmetric matrix A is defined to be the triple $(\nu_-(A), \nu_0(A), \nu_+(A))$, where $\nu_-(A), \nu_0(A), \nu_+(A)$ are respectively the numbers of negative, zero, and positive eigenvalues of A counted with multiplicities.*

$$Iner \ A = (\nu_-(A), \nu_0(A), \nu_+(A)).$$

Haynsworth's inertia theorem uses the concept of the Schur complement.

Definition 15.5 *(Schur complement) Let A be a symmetric matrix and P is a nonsingular submatrix. Consider the following partitioning of the symmetric matrix A :*

$$A = \begin{bmatrix} P & Q \\ Q^T & R \end{bmatrix}$$

where P is a nonsingular submatrix. Then the matrix S, given by

$$S = R - Q^T P^{-1} Q$$

is called the Schur complement of P in A.

Definition 15.6 *(Haynsworth's inertia theorem) Haynsworth's inertia theorem states that if the nonsingular P submatrix is principal in A, then*

$$Iner(A) = Iner(P) + Iner(S), \tag{15.63}$$

where addition means componentwise addition. The principality of P ensures that P and S are symmetric matrices. Since the Schur complement plays a central role in any pivot-algorithms, therefore, Equation (15.64) may be employed to evaluate the inertia of any symmetric matrix by a sequence of pivot transformations, following a special pivoting rule. For details, we refer to Cottle [50].

Theorem 15.10 *(Komlosi [148])* *Suppose that the twice continuously differentiable function $f(x)$ is locally pseudolinear at $x_0 \in X$, where $\nabla f(x_0) \neq 0$. Assume that $A = \nabla^2 f(x_0) \neq 0$ and A admits partitioning (15.61). If $r \neq 0$ then the following statements hold true:*

(ia) If $\lambda + \mu \neq 0$, then $Iner(A) = (1, n - 2, 1)$;

(ib) If $\lambda + \mu = 0$ and $\lambda + 2\mu > 0$, then $Iner(A) = (1, n - 1, 0)$;

(ii) If $\lambda + \mu = 0$ and $\lambda + 2\mu < 0$, then $Iner(A) = (0, n - 1, 1)$.

If $r = 0$, then

$$Iner(A) = (1, n - 1, 0), \text{ if } \lambda + 2\mu > 0,$$

and

$$Iner(A) = (0, n - 1, 1) \text{ if } \lambda + 2\mu < 0.$$

Proof (ia) First, we consider the subcase that $\lambda + 2\mu \neq 0$ and choose $P = [-(\lambda + 2\mu)]$ in the Haynsworth inertia formula. By the Schur complement formula, we have

$$S = \left[\frac{(\lambda + \mu)^2}{\lambda + 2\mu} rr^T\right].$$

Since $r \neq 0$, therefore $Iner(rr^T) = (0, n - 2, 1)$ and thus

$$Iner(S) = Iner\left[\frac{(\lambda + \mu)^2}{\lambda + 2\mu}\right] + (0, n - 2, 0).$$

Since,

$$Iner(A) = Iner(P) + Iner(S) = Iner[-(\lambda + 2\mu)] + Iner\left[\frac{(\lambda + \mu)^2}{\lambda + 2\mu}\right]$$
$$+(0, n - 2, 0),$$

the result follows.

(ib) Now, we consider the subcase $\lambda + 2\mu = 0$. In this case, $\lambda \neq 0$ must hold. Choose $P_1 = [\lambda r_i^2]$ with $r_i \neq 0$. For the sake of simplicity, we may assume that $i = 1$. The Schur complement formula gives us in this case

$$S_1 = \begin{bmatrix} 0 & 0 \\ 0^T & -\lambda/4 \end{bmatrix}.$$

Let us choose $P_2 = -\lambda/4$. The Schur complement of P_2 in S_1 is the 0 matrix of order $(n - 2)$ and thus

$$Iner(A) = [\lambda r_i^2] + Iner[-\lambda/4] + (0, n - 2, 0) = (1, n - 2, 1).$$

(ii) Let us take $P = [-(\lambda + 2\mu)]$ in the Haynsworth inertia formula. Then by the Schur complement formula, we have

$$S = \left[-\frac{(\lambda + \mu)^2}{\lambda + 2\mu} rr^T\right] = 0.$$

Since,

$$Iner(A) = Iner(P) + Iner(S) = Iner[-(\lambda + 2\mu)] + (0, n - 1, 0),$$

the result follows. The proof for the case $r = 0$ is similar to that of (ii).

Theorem 15.11 *Let us assume that the twice continuously differentiable function $f(x)$ is locally pseudolinear at $x_0 \in X$, where $\nabla f(x_0) \neq 0$. Assume that $A = \nabla^2 f(x_0) \neq 0$ and admits partitioning (15.61). Let $g^T = [r^T - 1]$. Then the following statements hold true:*
(i) $rank(A)$ equals to either 1 or 2;
(ii) if $rank(A) = 1$, then $range(A) = Lin(g)$, if $rank(A) = 2$, then $range(A) = Lin(g, Ag)$ (here $range(A) = A(R^n)$ and $lin\{.\}$ denotes the linear hull of the given vectors);
(iii) $\nabla f(x_0) \in lin\{g\}$;
(iv) if $rank(A) = 2$, then

$$Ax - \nabla f(x_0) \nrightarrow x^T \nabla f(x_0) - x^T Ax - 0$$

and

$$x^T \nabla f(x_0) = 0 \Rightarrow x^T Ax = 0.$$

Proof (i) Since $rank(A) = \nu_-(A) + \nu_+(A)$, therefore, by Theorem 15.10, $rank(A) = 1$ or 2.

(ii) By Theorem 15.10, $rank(A) = 1$ is equivalent to any of the following two conditions: either one has $r \neq 0$ and $\lambda + \mu = 0$ or $r = 0$. In both of the two cases, by simple calculation one can see that

$$A = k \begin{bmatrix} rr^T & -r \\ -r^T & 1 \end{bmatrix} = k \begin{bmatrix} r \\ -1 \end{bmatrix} \begin{bmatrix} r & -1 \end{bmatrix}$$

with $k = \lambda$ in the first case and with $k = (-\lambda + 2\mu)$ in the second case. It follows that $A = kgg^T$, and $range(A) = lin\{g\}$. By Theorem 15.10, $rank(A) = 2$ is equivalent to $r \neq 0$ and $\lambda + \mu \neq 0$. First we show that $g \in range(A)$, that is the problem of $Ax = g$ is solvable in $x \in R^n$. Consider x, g and A in their (u, ν) decomposition, where $u \in R^{n-1}$ and $\nu \in R$.

$$Ax = \begin{bmatrix} \lambda rr^T & \mu r \\ r^T & -(\lambda + 2\mu) \end{bmatrix} \begin{bmatrix} u \\ \nu \end{bmatrix} = \begin{bmatrix} (\lambda(r^T u) + u\nu)r \\ \mu(r^T u) - (\lambda + 2\mu)\nu \end{bmatrix}. \quad (15.64)$$

From this decomposition, it follows that $Ax = g$ holds if and only if $x = (u, \nu)$ satisfies the following linear system:

$$\lambda(r^T u) + \mu\nu = 1, \quad (15.65)$$

$$\mu(r^T u) - (\lambda + 2\mu)\nu = -1.$$

Since, the determinant of this system equals to $\lambda + \mu$, therefore, by a simple computation there is a unique solution in $r^T u$ and ν, such that

$$r^T u = \nu = \frac{1}{\lambda + \mu}. \tag{15.66}$$

By (15.66), we note one more important fact, that is,

$$Ax = g \Rightarrow g^T x = r^T u - \nu = 0. \tag{15.67}$$

Introduce now

$$\hat{g} = \frac{1}{(\lambda + \mu)r^T r} \begin{bmatrix} r \\ r^T r \end{bmatrix}.$$

By a simple computation, we can see that $A\hat{g} = g$ and thus $g \in range(A)$.

Now we show that g and Ag are linearly independent. First, we show that $Ag \neq 0$. On the contrary, suppose that $Ag = 0$. Its (u, ν) decomposition provides us with

$$Ag = \begin{bmatrix} \lambda r r^T & \mu r \\ r^T & -(\lambda + 2\mu) \end{bmatrix} \begin{bmatrix} r \\ -1 \end{bmatrix} = \begin{bmatrix} \lambda(r^T r) - \mu r \\ \mu(r^T r) + (\lambda + 2\mu) \end{bmatrix} = \begin{bmatrix} 0 \\ 0 \end{bmatrix}.$$

This condition holds if and only if

$$\lambda(r^T r) - \mu = 0,$$

$$\mu(r^T r) + \lambda + 2\mu = 0,$$

and

$$\lambda \mu r^T r = \mu^2 = -\lambda^2 - 2\lambda\mu.$$

Hence, we get $(\lambda + \mu)^2 = 0$. Since $\lambda + \mu \neq 0$, it is impossible. This contradiction proves the result.

Now, we prove that g and Ag are linearly independent. On the contrary, suppose that $Ag = \alpha g$ with some $\alpha \neq 0$. Let $\delta = 1/\alpha \neq 0$. According to our assumption $A(\delta g) = 0$ should hold. From (15.51), it follows that $g^T(\delta g) = \delta g^T g = 0$. Since $g \neq 0$ and $\delta \neq 0$ it is impossible. It proves that g and Ag are linearly independent. Since $rank(A) = 2$, it follows that

$$range(A) = lin\{g, Ag\}.$$

(iii) From (15.62), we have

$$\nabla f(x_0)^T = -f'_\nu(x_0)[r^T \quad -1] = -f'_\nu(x_0)g^T,$$

where $f'_\nu(x_0) \neq 0$ and thus

$$\nabla f(x_0) \in lin\{g\}.$$

(iv) Finally, we consider the case when $rank(A) = 2$. Now, we have to prove that

$$Ax = \nabla f(x_0) \Rightarrow \nabla f(x_0)^T x = 0 \text{ and } \nabla f(x_0)^T x = 0 \Rightarrow x^T Ax = 0.$$

Let $x = (u, \nu)$. If x is a solution of $Ax = g$, then by (15.67), we get

$$g^T x = r^T u - \nu = 0.$$

Since $\nabla f(x_0) = \eta g$, we have

$$Ax = \nabla f(x_0) \Rightarrow \nabla f(x_0)^T x = 0.$$

Now, we suppose that $\nabla f(x_0)^T x = 0$, which is equivalent to $g^T x = r^T u - \nu = 0$. From (15.66) it follows that $Ax = (\lambda + \mu)g$ and thus $x^T Ax = (\lambda + \mu)g^T x = 0$.

The following result by Komlosi [147] presents a pure matrix algebraic result and prepares the way for further investigations.

Proposition 15.7 *Let us suppose that A is a symmetric matrix of order n, with $Incr(A) = (1, n-2, 1)$. Then, for any vector $d \in \mathbb{R}^n, d \neq 0$, such that $d = A\hat{d}$ and $d^T\hat{d} = 0$, for some $\hat{d} \in \mathbb{R}^n$, there exists a unique $p \in \mathbb{R}^n$, satisfying*

$$A = pd^T + dp^T.$$

The vector p admits the following properties:

$$p = A\hat{p} \text{ and } p^T\hat{p} = 0, \text{ for some } \hat{p} \in R^n$$

and

$$p^T\hat{d} = d^T\hat{p} = 1.$$

Proof Let us assume that $s_1, s_2, \ldots, s_n$ be a basis in $\mathbb{R}^n$ consisting of orthonormal eigenvectors of A. Suppose that $As_1 = \sigma_1 s_1$, with $\sigma_1 < 0, As_2 = \sigma_2 s_2$ with $\sigma_2 > 0$. Therefore,

$$range(A) = lin\{s_1, s_2\}.$$

Let $d \in \mathbb{R}^n$ be given with the requested property. Then $d \in range(A) = Lin\{s_1, s_2\}$. and we have a unique decomposition for d, as follows:

$$d = \delta_1 s_1 + \delta_2 s_2.$$

Since, $d = \hat{A}d$ and $d^T\hat{d} = 0$, without loss of the generality we may assume that $\hat{d} \in Lin\{s_1, s_2\}$. Therefore, we have

$$\hat{d} = \frac{\delta_1}{\sigma_1}s_1 + \frac{\delta_2}{\sigma_2}s_2 \text{ and } d^T\hat{d} = \frac{\delta_1^2}{\sigma_1} + \frac{\delta_2^2}{\sigma_2} = 0.$$

Since $d \neq 0$, it follows that $\delta_1 \neq 0$ and $\delta_2 \neq 0$. Let

$$p = \frac{\sigma_1}{2\delta_1} s_1 + \frac{\sigma_2}{2\delta_2} s_2 \text{ and } \hat{p} = \frac{1}{2\delta_1} s_1 + \frac{1}{2\delta_2} s_2.$$

By a simple calculation, we get

$$p = A\hat{p},$$

$$p^T \hat{p} = \frac{\sigma_1}{4\delta_1^2} + \frac{\sigma_2}{4\delta_2^2} = \frac{4\sigma_1\sigma_2}{\delta_1^2\delta_2^2}\left(\frac{\delta_1^2}{\sigma_1} + \frac{\delta_2^2}{\sigma_2}\right) = \frac{4\sigma_1\sigma_2}{\delta_1^2\delta_2^2} d^T \hat{d} = 0$$

and

$$d^T \hat{p} = \delta_1 \frac{1}{2\delta_1} + \delta_2 \frac{1}{2\delta_2} = 1 \text{ and } p^T \hat{d} = \frac{\sigma_1\delta_1}{2\delta_1\sigma_1} + \frac{\sigma_2\delta_2}{2\delta_2\sigma_2} = 1.$$

Now, we show that
$$A = pd^T + dp^T.$$

We consider the nonsingular matrix $S = [s_1, \ldots, s_n]$. By construction, we have $S^{-1} = S^T$ and thus $A = pd^T + dp^T$ holds if and only if

$$S^T A S = \begin{bmatrix} P & 0 \\ 0 & 0 \end{bmatrix}, \text{ where } P = \begin{bmatrix} \sigma_1 & 0 \\ 0 & \sigma_2 \end{bmatrix},$$

and

$$(S^T p)(d^T S) + (S^T d)(p^T S) = \begin{bmatrix} Q & 0 \\ 0 & 0 \end{bmatrix},$$

$$\text{where } Q = \begin{bmatrix} 2\frac{\sigma_1}{2\delta_1}\delta_1 & \frac{\sigma_1}{2\delta_1}\delta_2 + \frac{\sigma_2}{2\delta_2}\delta_1 \\ \frac{\sigma_1}{2\delta_1}\delta_2 + \frac{\sigma_2}{2\delta_2}\delta_1 & 2\frac{\sigma_2}{2\delta_2}\delta_2 \end{bmatrix}.$$

It is clear that,
$$P = Q \Leftrightarrow \frac{\sigma_1}{2\delta_1}\delta_2 + \frac{\sigma_2}{2\delta_2}\delta_1 = 0.$$

We note that

$$\frac{\sigma_1}{2\delta_1}\delta_2 + \frac{\sigma_2}{2\delta_2}\delta_1 = \frac{\sigma_1\sigma_2}{2\delta_1\delta_2}\left(\frac{\delta_1^2}{\sigma_1} + \frac{\delta_2^2}{\sigma_2}\right) = \frac{\sigma_1\sigma_2}{2\delta_1\delta_2} d^T \hat{d} = 0,$$

which proves that $P = Q$ and $A = pd^T + dp^T, A = pd^T + dp^T$.

15.9   Pseudolinearity of Generalized Quadratic Fractional Functions

We consider the case, when

$$f(x) = \frac{\frac{1}{2}x^T Bx + b^T x + \beta}{\frac{1}{2}x^T Cx + c^T x + \gamma}, \quad x \in X, \tag{15.68}$$

where $X \subseteq \mathbb{R}^n$ is an open convex set, B, C are nonzero symmetric matrices of order $n; b, c \in \mathbb{R}^n; \beta\gamma \in \mathbb{R}$, and $\frac{1}{2}C^T x + c^T x + \gamma > 0$ on X.

Now, our aim is to seek conditions on the input parameters B, C, b, c and β, γ ensuring pseudolinearity of $f(x)$.

Komlosi [148] has pointed out that the characterization for pseudolinear functions obtained in Proposition 15.2 (Cambini and Carosi [34]) and Theorem 15.3 (Rapcsak [235]) can be obtained by the help of our local analysis. Komlosi [148] has shown that both the statements can be proved under milder conditions. Since local analysis is based on Lemma 15.6, which requires the first and second order partial derivatives of function $f(x)$ defined by (15.69), therefore the following reduction scheme may help us avoid unnecessary technical difficulties.

A reduction scheme for fractional functions: We consider the fractional function

$$f(x) = \frac{g(x)}{h(x)}, \quad x \in X,$$

where X is an open convex set in R^n, functions $g(x)$ and $h(x)$ are continuously differentiable over X, and $h(x) > 0$ on X. In case of fractional functions the direct application of condition (15.60) is a difficult one, but the technical difficulties can be overcome in the following way.

We note that the solution set of the level curve equation $f(x) = f(x_0)$ is the same as the solution set to the level curve equation $\varphi(x) = \varphi(x_0)$, where

$$\varphi(x) = g(x) - \frac{g(x_0)}{h(x_0)} h(x).$$

Therefore, one can replace function $f(x)$ with $\varphi(x)$ in condition (15.60). The precise statement based on the previous reasoning may be expressed as follows.

Lemma 15.7 *The fractional function $f(x)$ is locally pseudolinear at x_0, where $\nabla f(x_0) \neq 0$, if and only if the auxiliary function $\varphi(x)$ is locally pseudolinear at x_0.*

From a technical point of view, function $\varphi(x)$ admits more simple first and second order partial derivatives than $f(x)$ does. This reduction gives us almost trivially the well-known result on linear fractional functions. The case with quadratic fractional function is not so simple. We shall demonstrate the efficiency and easy applicability of our method in the sequel for quadratic fractional function of form of (15.68). The only thing we have to take into consideration is that the auxiliary function $\varphi(x)$ is defined as follows:

$$\varphi_{x_0}(x) = \frac{1}{2}x^T A x + a^T x + \alpha - f(x_0)(b^T x + \beta). \qquad (15.69)$$

The most favorable property of this quadratic function is that its Hessian is independent from the choice of x_0, namely for all $x_0, x \in X$ one has $\nabla^2 \varphi_{x_0}(x) \equiv A$ and thus local analysis can be easily extended to global one. The next proposition will be needed in the sequel.

Proposition 15.8 *Let us assume that $f(x)$ defined by (15.69) is locally pseudolinear at least at two different places with different function values and nonvanishing gradients. Then $a, b \in range(A)$.*

Proof Suppose that $f(x)$ is locally pseudolinear at x_1, x_2 where the gradients are nonvanishing and the function has different values. By Lemma 15.7, it follows that $\varphi_{x_i}(x)$ is locally pseudolinear at $x_i, i = 1, 2$. In view of (iii) of Theorem 15.11, one has $\nabla\varphi_{x_i}(x_i) = Ax_i + a - f(x_i)b \in range(A)$ and thus $a - f(x_i)b \in range(A)$ for $i = 1, 2$. Since $f(x_1) \neq f(x_2)$, it follows that $a, b \in range(A)$.

For global pseudolinear functions, the following result has been derived by Cambini and Carosi [34]. Komlosi [148] has proved the same result by taking into consideration the local pseudolinearty.

Proposition 15.9 *Let us assume that $f(x)$ defined by (15.69) is locally pseudolinear at least at three different places with different function values and nonvanishing gradients. Assume that $rank(A) = 2$. Then there exist $\hat{a}, \hat{b} \in R^n$ such that $a = A\hat{a}, b = A\hat{b}$ and*

$$b^T\hat{b} = 0, \beta = a^T\hat{b} \text{ and } 2\alpha = a^T\hat{a}. \qquad (15.70)$$

Proof By Proposition 15.7, $a, b \in range(A)$, therefore there exist $\hat{a}, \hat{b} \in R^n$ with $a = A\hat{a}, b = A\hat{b}$. Therefore, for all we have

$$\nabla\varphi_{x_i}(x_i) = Ax_i + a - f(x_i)b = Ax_i + A\hat{a} - f(x_i)A\hat{b} = A(x_i + \hat{a} - f(x_i)\hat{b}).$$

According to (iv) of Theorem 15.11, we have

$$Ax = \nabla\varphi_{x_i}(x_i) \Rightarrow x^T\nabla\varphi_{x_i}(x_i) = 0.$$

Therefore, we have arrived at the following condition:

$$(x_i+\hat{a}-f(x_i)\hat{b})^T(Ax_i+a-f(x_i)b) = b^T\hat{b}f^2(x_i)+2(\beta-a^T\hat{b})f(x_i)+a^T\hat{a}-2\alpha = 0.$$

From which it follows that $\lambda_i = f(x_i), i = 1, 2, 3$ are three different solutions of the quadratic equation:

$$b^T \hat{b} \lambda^2 + 2(\beta - a^T \hat{b})\lambda + a^T \hat{a} - 2\alpha = 0,$$

which is possible only if when every coefficient is equal to 0.
This completes the proof.

Theorem 15.12 *Let $f(x)$ be defined by (15.69) and suppose that $rank(A) = 1$. Assume that $f(x)$ is locally pseudolinear at least at two different places, with different function values and nonvanishing gradients. Then $f(x)$ is locally pseudolinear at x_0, where $\nabla f(x_0) \neq 0$, if and only if there exist constants $\hat{\alpha} \neq 0, \hat{\beta}$ and $\hat{\gamma}$, such that*

$$f(x) = \hat{\alpha} b^T x + \hat{\beta} + \frac{\hat{\gamma}}{b^T x + \beta}, \tag{15.71}$$

and

$$\hat{\gamma} \neq \hat{\alpha}(b^T x + \hat{\beta})^2. \tag{15.72}$$

Proof (Necessity) A simple consequence of Theorem 15.11 and Proposition 15.10 is the existence of constants $\tilde{\alpha}$ and $\tilde{\beta}$, such that $A = \tilde{\alpha} b b^T$ and $a = \tilde{\beta} b$. By using this information, we get (15.72) with

$$\hat{\alpha} = \frac{\tilde{\alpha}}{2}, \hat{\beta} = \tilde{\beta} - \beta \frac{\tilde{\alpha}}{2} \text{ and } \hat{\gamma} = \alpha - \beta \tilde{\beta} - \beta^2 \frac{\tilde{\alpha}}{2}.$$

Since

$$\nabla f(x) = \left(\hat{\alpha} - \frac{\hat{\gamma}}{(b^T x + \beta)^2} \right) b, \tag{15.73}$$

$b \neq 0$ and $\nabla f(x_0) \neq 0$ therefore condition (15.58) holds.
(Sufficiency) From Equations (15.72) and (15.73), it is clear that $f(x)$ in (15.71) satisfies Rapcsak's condition (2.18) of Theorem 2.9, which is sufficient for local pseudolinearity at x_0.

Remark 15.5 *If $f(x)$ is pseudolinear on $X = \{x \in \mathbb{R}^k : b^T x + \beta > 0\}$, then, condition (15.73) holds for all $x \in X$ if and only if $\hat{\alpha}\hat{\gamma} < 0$.*

Theorem 15.13 *Let us assume that $f(x)$ is defined by (15.69) and suppose that $rank(A) = 2$. Assume that $f(x)$ is locally pseudolinear at least at three different places with different function values and nonvanishing gradients. Assume that $rank(A) = 2$. Then there exist $p \in R^n$ and $\pi \in \mathbb{R}$, such that $f(x)$ can be rewritten as*

$$f(x) = p^T x + \pi.$$

Proof From Theorem 15.10, $rank(A) = 2$ is equivalent to have $Iner(A) = (1, n - 2, 1)$. Therefore, Proposition 15.9 ensures the existence of $\hat{a}, \hat{b} \in \mathbb{R}^n$ satisfying

$$A\hat{a} = a, a^T\hat{a} = 2\alpha, A\hat{b} = b, b^T\hat{b} = 0 \text{ and } a^T\hat{b} = \beta. \qquad (15.74)$$

By Proposition 15.6 there exists $p \in \mathbb{R}^n$, such that

$$A = bp^T + pb^T. \qquad (15.75)$$

It follows from decomposition (15.75) and condition (15.74) that

$$a = A\hat{a} = (p^T\hat{a})b + (b^T\hat{a})p = \pi b + \beta p$$

and

$$\alpha = \frac{1}{2}a^T\hat{a} = \frac{1}{2}(\pi b^T\hat{a} + \beta p^T\hat{a}) = \frac{1}{2}(\pi\beta + \beta\pi) = \beta\pi$$

where $\pi = p^T\hat{a}$.

In view of the above conditions, we get

$$f(x) = \frac{\frac{1}{2}x^T Ax + a^T x + \alpha}{b^T x + \beta} = \frac{(b^T x)(p^T x) + \pi b^T x + \beta p^T x + \beta\pi}{b^T x + \beta} = p^T x + \pi.$$

This completes the proof.

15.10   The Carosi-Martein Theorem

Carosi and Martein [39] investigated the pseudoconvexity and pseudolinearity of the following class of quadratic fractional functions

$$f(x) = \frac{\frac{1}{2}x^T Bx + b^T x + \beta}{(c^T x + \gamma)^2}, \quad x \in X, \qquad (15.76)$$

where $X \subseteq \mathbb{R}^n$ is an open convex set, $B \neq 0$ is a quadratic and symmetric matrix of order n; $b, c \in R^n, c \neq 0$; $\beta, \gamma \in R$ and $(c^T x + \gamma)^2 > 0$ on X.

Komlosi [148] has noted that the application of the reduction scheme is efficient only for investigating fractional functions in the form of (15.3). But, it could only be carried out with serious technical difficulties for functions in the form of (15.77). Fortunately, there exists another possibility of reducing the analysis of function (15.77) to the analysis of an auxiliary function with a much more simple structure. This reduction scheme applies a special transformation of the independent variable, called the Charnes-Cooper transformation, which is defined as

$$y(x) = \frac{x}{c^T x + \gamma}, \qquad (15.77)$$

whose inverse is

$$x(y) = \frac{\gamma y}{1 - c^T y}.$$ (15.78)

This transformation shares a remarkable property, it preserves pseudo-convexity/concavity and thus pseudolinearity. It follows that this transformation preserves local pseudolinearity, as well.

Now, apply the Charnes-Cooper transformation on $f(x)$ defined by (15.76). We get a pure quadratic function

$$\varphi(y) = \frac{1}{2} y^T A y + a^T y + \alpha,$$ (15.79)

where

$$A = B \mid \frac{2\beta}{\gamma^2} cc^T - \frac{1}{\gamma}(cb^T + bc^T), a - \frac{1}{\gamma}b - \frac{2\beta}{\gamma^2}c \text{ and } \alpha = \frac{\beta}{\gamma^2}.$$ (15.80)

Since the Charnes-Cooper transformation preserves pseudolinearity, therefore the local pseudolinearity of $f(x)$ in (15.76) can be tested by testing the local pseudolinearity of $\varphi(y)$ in (15.79).

Proposition 15.10 *Let us suppose that $\varphi(y)$ defined by (15.79) is locally pseudolinear at $y_0 \in y$, where $\nabla \varphi(y_0) \neq 0$. Then the following statements hold true:*

(i) $rank(A)$ equals to either 0, 1, or 2.

(ii) If $rank(A) = 0$, then,
$$\varphi(y) = a^T y + \alpha.$$

(iii) If $rank(A) = 1$, there exists $k \in \mathbb{R}$, such that

$$\varphi(y) = k(a^T y)^2 + a^T y + \alpha.$$

(iv) If $rank(A) = 2$ function $\varphi(y)$ cannot be locally pseudolinear at two different points with different function values.

Proof (i) It follows from Theorem 15.11.
   (ii) holds if and only if $A = 0$ and it proves the statement.
   (iii) Now, we consider the case when $rank(A) = 1$. Then by Theorem 15.11

$$A = \lambda g g^T, \nabla \varphi(y_0) = \tau g \text{ and } a = \psi g \text{ with some } \lambda, \tau, \psi \in \mathbb{R}.$$

It follows that $A = 2k a a^T$ and

$$\varphi(y) = k(a^T y)^2 + a^T y + \alpha.$$

(iv) Now, we consider the case when $rank(A) = 2$. By Theorem 15.11, we get $\nabla\varphi(y_0) \in range(A)$. Therefore, it follows that $a = \nabla\varphi(y_0) \in range(A)$ and thus $A\hat{a} = a$ holds with some $\hat{a} \in \mathbb{R}^n$. Since

$$\nabla\varphi(y_0) = Ay_0 + a = A(y_0 + \hat{a}).$$

Therefore from (iv) of Theorem 15.11, it follows that

$$(Ay_0 + a)^T(y_0 + \hat{a}) = y_0^T Ay_0 + 2a^T y_0 + a^T\hat{a} = 2\varphi(y_0) - 2\alpha + a^T\hat{a} = 0,$$

which is equivalent to

$$2\varphi(y_0) = 2\alpha - a^T\hat{a}. \tag{15.81}$$

An important consequence of condition (15.81) is that the function $\varphi(y)$ cannot be locally pseudolinear at two different points with different function values.

On the contrary, suppose that there exist y_0 and y_1, with $\nabla\varphi(y_0) \neq 0$, $\nabla\varphi(y_1) \neq 0$ and $\varphi(y_0) \neq \varphi(y_1)$. Then, taking into account necessary condition (15.81) the following equations hold:

$$2\varphi(y_0) = 2\alpha - a^T\hat{a} \text{ and } 2\varphi(y_1) = 2\alpha - a^T\hat{a}.$$

Since these equations hold true if and only if $\varphi(y_0) = \varphi(y_1)$ the result follows.

Theorem 15.14 *Let us suppose that $f(x)$ defined by (15.76) is locally pseudolinear at two different points with nonvanishing gradients and different function values. Then $f(x)$ can be rewritten in one of the following two forms:*

$$f(x) = \frac{a^T x}{c^T x + \gamma} + \frac{\beta}{\gamma^2},$$

$$f(x) = k\frac{(a^T x)^2}{(c^T x + \gamma)^2} + \frac{a^T x}{c^T x + \gamma} + \frac{\beta}{\gamma^2},$$

where $a = \frac{1}{\gamma}b - \frac{2\beta}{\gamma^2}c$.

Proof In view of Proposition 15.9, we have to consider only the following two cases: $rank(A) = 0$ or $rank(A) = 1$. From Proposition 15.9, the results show that either

$$\varphi(y) = a^T y + \alpha$$

or

$$\varphi(y) = k(a^T y)^2 + a^T y + \alpha.$$

Applying (15.79) and taking into account (15.81), it follows that either

$$f(x) = \varphi(y(x)) = \frac{a^T x}{c^T x + \gamma} + \alpha$$

or

$$f(x) = \varphi(y(x)) = k\frac{(a^T x)^2}{(c^T x + \gamma)^2} + \frac{a^T x}{c^T x + \gamma} + \alpha,$$

where $a = \frac{1}{\gamma}b - \frac{2\beta}{\gamma^2}c$ and $\alpha = \frac{\beta}{\gamma^2}$.

Chapter 16

Pseudolinear Fuzzy Mapping

16.1 Introduction

The concepts of fuzzy mappings and fuzzy models play an important role in several areas of modern research such as finance (Verdegay [278]; Buckley [30]; Fedrizzi *et al.* [79]), networking and transportation (Jimnez and Verdegay [120] and Moreno *et al.* [208]), interest rate forecasting (Maciel *et al.* [174]), and ecological models (Mouton *et al.* [209]). The concept of fuzzy mapping has been introduced by Chang and Zadeh [42]. Nanda and Kar [212] introduced the concept of convex fuzzy mapping and established that fuzzy mapping is convex if and only if its epigraph is convex. Since then, generalized convexity of fuzzy mapping has been extensively studied. Syau [265] introduced the concepts of pseudoconvexity, invexity, and pseudoinvexity for fuzzy mappings of one variable by using the notion of differentiability and the results proposed by Goestschel and Voxman [93]. Panigrahi [218] employed Buckley-Feuring's [31] approach for fuzzy differentiations to extend and generalize these notions to fuzzy mappings of several variables. Wu and Xu [286] introduced the concept of fuzzy pseudoconvex, fuzzy invex, fuzzy pseudoinvex, and fuzzy preinvex mappings from $\mathbb{R}^n$ to the set of fuzzy numbers by employing the concept of differentiability of fuzzy mappings defined by Wang and Wu [279]. Moreover, Mishra *et al.* [204] introduced the concept of pseudolinear fuzzy mappings by relaxing the definition of pseudoconvex fuzzy mappings. The concept of η-pseudolinear fuzzy mappings is also introduced by relaxing the definition of pseudoinvex fuzzy mappings. By means of the basic properties of pseudolinear fuzzy mappings, the solution set of a pseudolinear fuzzy program is characterized. Then, characterizations of the solution set of an η-pseudolinear program are also derived.

In this chapter, we will study the characterizations and properties of fuzzy pseudolinear and fuzzy η-pseudolinear functions. We present several characterizations for the solution set of programming problems involving pseudolinear fuzzy functions and η-pseudolinear functions.

16.2   Definition and Preliminaries

Let $u : \mathbb{R} \to [0,1]$ be a fuzzy set. For each such fuzzy set u, we denote by $[u]_\alpha = \{x \in \mathbb{R}^n : u(x) \geq \alpha\}$ for any $\alpha \in]0,1]$ its $\alpha-$cut set. By supp u, we denote the support of u, i.e. $\{x \in \mathbb{R}^n : u(x) > 0\}$. Let $[u]_\alpha$ denote the closure of supp u.

Definition 16.1 *A fuzzy number is a fuzzy set u with the following properties:*

(i) u is normal, i.e., there exists $\bar{x} \in \mathbb{R}^n$ such that $u(\bar{x}) = 0$;

(ii) u is fuzzy convex set, i.e., $u(\lambda x + (1 - \lambda)y) \geq \min(u(x), u(y))$, $\forall x, y \in \mathbb{R}^n$, $\lambda \in [0,1]$;

(iii) u is an upper semicontinuous function;

(iv) $[u]_0$ is compact.

Let $\mathfrak{I}_0$ denote the family of fuzzy numbers. Since, each $r \in \mathbb{R}$ can be considered as a fuzzy number $\tilde{r}$, defined by

$$\tilde{r}(t) = \begin{cases} 1, & \text{if} \quad t = r, \\ 0, & \text{if} \quad t \neq r. \end{cases}$$

Thus, a precise number $r \in \mathbb{R}$ is a special case of fuzzy number and $\mathbb{R}$ can be embedded in $\mathfrak{I}_0$. It is known from Diamond and Kloeden [64] that the α-level set of a fuzzy number $u \in \mathfrak{I}_0$ is a closed and bounded interval, given by

$$[u^-(\alpha), u^+(\alpha)] = [u]_\alpha = \begin{cases} \{x \in \mathbb{R} : u(\alpha) \geq \alpha\}, & \text{if} \ \ 0 < \alpha < 1, \\ (\text{cl supp } u), & \text{if} \ \ \alpha = 0. \end{cases}$$

It is easy to verify that a fuzzy set $u : \mathbb{R} \to [0,1]$ is a fuzzy number if and only if

(i) $[u]_\alpha$ is a closed and bounded interval for each $\alpha \in [0,1]$ and

(ii) $[u]_1 \neq \phi$.

Thus, we can identify a fuzzy number u with the parametrized triples $\{(u^-(\alpha), u^+(\alpha), \alpha) : 0 \leq \alpha \leq 1\}$, where $u^-(\alpha)$ and $u^+(\alpha)$ denote the left and right hand end points of $[u]_\alpha$, respectively.

For fuzzy numbers $u, \nu \in \mathfrak{I}_0$, represented by

$$\{(u^-(\alpha), u^+(\alpha), \alpha) : 0 \leq \alpha \leq 1\} \ and \ \{(\nu^-(\alpha), \nu^+(\alpha), \alpha) : 0 \leq \alpha \leq 1\},$$

respectively, and each real number r, we define the addition $u + \nu$ and scalar multiplication ru, as follows:

$$(u + \nu)(x) = \sup_{y+z=x} \min [u(y), \nu(z)]$$

and

$$(\lambda u)(x) = \begin{cases} u(\lambda^{-1}x), & \text{if} \quad \lambda \neq 0, \\ 0, & \text{if} \quad \lambda = 0. \end{cases}$$

It is well-known that for any $u, \nu \in \mathfrak{I}_0$ and $\lambda u \in \mathfrak{I}_0$, we have

$$[u + \nu]^\alpha = [u]^\alpha + [\nu]^\alpha$$

and

$$[\lambda u]^\alpha = \lambda [u]^\alpha,$$

that is,

$$u + \nu = \{(u^-(\alpha), u^+(\alpha), \alpha) : 0 \leq \alpha \leq 1\} + \{(\nu^-(\alpha), \nu^+(\alpha), \alpha) : 0 \leq \alpha \leq 1\}$$
$$= \{(u^-(\alpha) + \nu^-(\alpha), (u^+(\alpha) + \nu^+(\alpha), \alpha) : 0 \leq \alpha \leq 1\},$$

$$ru = \{(ru^-(\alpha), ru^+(\alpha), \alpha) : 0 \leq \alpha \leq 1\}.$$

We note that for $u \in \mathfrak{I}_0$, ru is not a fuzzy number for $r < 0$.

The family of parametric representations of members of $\mathfrak{I}_0$ and the parametric representations of their negative scalar multiplications form subsets of the vector space

$$\aleph =: \{\{(u^-(\alpha), u^+(\alpha), \alpha) : 0 \leq \alpha \leq 1\}; \ u^- : [0,1] \to \mathbb{R} \text{ are bounded functions}.$$

We metricize $\aleph$, by the metric

$$d\left(\{(u^-(\alpha), u^+(\alpha), \alpha) : 0 \leq \alpha \leq 1\}, \{(\nu^-(\alpha), \nu^+(\alpha), \alpha) : 0 \leq \alpha \leq 1\}\right)$$

$$\sup\{\max\{|u^-(\alpha) - \nu^-(\alpha)|, |u^+(\alpha) - \nu^+(\alpha)|\} : 0 \leq \alpha \leq 1\}.$$

Let

$$H^* = \{(u^-(\alpha), u^+(\alpha), \alpha) : 0 \leq \alpha \leq 1\};$$

where u^- and $u^+ : [0,1] \to [0, \infty)$ are bounded functions.

It is obvious that $\mathfrak{I}_0^* \subset H^* \subset H$ where $\mathfrak{I}_0^*$ denotes the set of all nonnegative fuzzy numbers of $\mathfrak{I}_0$. Syau [265] has shown that H^* is a closed convex cone in the topological vector space (H, d). Suppose that $u, \nu \in \mathfrak{I}_0$ are fuzzy numbers represented by $\{(u^-(\alpha), u^+(\alpha), \alpha) : 0 \leq \alpha \leq 1\}$ and $\{(\nu^-(\alpha), \nu^+(\alpha), \alpha) : 0 \leq \alpha \leq 1\}$, respectively.

Definition 16.2 (Partial ordering $\preceq$ in $\mathfrak{I}_0$) *For $u, \nu \in \mathfrak{I}_0$ we say $u \preceq \nu \Leftrightarrow u^-(\alpha) \leq \nu^-(\alpha)$ and $u^+(\alpha) \leq \nu^+(\alpha), \forall \alpha \in [0,1]$;*

$$u = \nu \Leftrightarrow u \preceq \nu \text{ and } \nu \preceq u.$$

We say $u \prec \nu$ if $u \preceq \nu$ and there exists $\alpha_0 \in [0,1]$, such that $u^-(\alpha_0) < \nu^-(\alpha_0)$ and $u^+(\alpha_0) < \nu^+(\alpha_0)$.

For $u, \nu \in \mathfrak{I}_0$ if either $u \preceq \nu$ or $\nu \preceq u$, then we say that u and ν are comparable; otherwise, they are noncomparable. It is clear that the relation $\preceq$ is a partial order relation on $\mathfrak{I}_0$. It is often convenient to write $\nu \succeq u$ (respectively $\nu \succ u$) in place of $u \preceq \nu$ ($u \prec \nu$).

For $u, \nu \in \mathfrak{I}_0$ it is clear that

$$u - \nu \in H^* \Leftrightarrow u \succeq \nu;$$

$$u - \nu \in H^*_{\{\tilde{0}\}} \Leftrightarrow u \succ \nu.$$

It is also obvious, that addition and nonnegative scalar multiplication preserve the order on $\mathfrak{I}_0$. A fuzzy number $u : \mathbb{R} \to [0, 1]$ is called nonnegative if $u(t) = 0$, for all $t < 0$. It can be seen easily that for $u \in \mathfrak{I}_0, u$ is nonnegative if and only if $u \succeq \tilde{0}$.

Definition 16.3 *A subset S^* of $\mathfrak{I}_0$ is said to be bounded above, if there exists a fuzzy number $u \in \mathfrak{I}_0$ called an upper bound of S^* such that $\nu \preceq u$, for every $\nu \in S^*$.*

Definition 16.4 *Further, a fuzzy number $u_0 \in \mathfrak{I}_0$ is called the least upper bound (sup in short) for S^*, if*

(i) u_0 is an upper bound of S^* and

(ii) $u_0 \preceq u$, for every upper bound u of S^*.

A lower bound and the greatest lower bound (inf in short) is defined similarly.

Syau [264] has shown that every nonnegative set $S^* \subseteq \mathfrak{I}_0$, which is bounded above (resp. bounded below) has a least upper (resp. greatest lower) bound. In particular, $\sup\{u, \nu\}$ and $\inf\{u, \nu\}$ exist in $\mathfrak{I}_0$, for every pair $\{u, \nu\} \subset \mathfrak{I}_0$.

Furthermore, according to Syau [264], we have

$$\inf\{u, \nu\} \preceq \lambda u + (1 - \lambda)\nu \preceq \sup\{u, \nu\},$$

$\forall u, \nu \in \mathfrak{I}_0$ *and* $\lambda \in [0, 1]$.

Definition 16.5 (Invex set) *A set $K \subseteq \mathbb{R}^n$ is said to be invex with respect to $\eta : K \times K \to \mathbb{R}^n$, if for $x, y \in K$, we have*

$$y + \lambda\eta(x, y) \in K.$$

Definition 16.6 (Preinvex fuzzy mapping) (Wu and Xu [286]) *A fuzzy mapping $F : T \to \mathfrak{I}_0$ is said to be preinvex on invex set K with respect to $\eta : K \times K \to \mathbb{R}^n$, if for any $x, y \in K$, $\lambda \in [0, 1]$, we have*

$$F(y + \lambda\eta(x, y)) \preceq \lambda F(x) + (1 - \lambda)F(y)$$

Example 16.1 *(Wu and Xu [286]) Let the following fuzzy mapping F : $T \to \mathfrak{I}_0$ represent the variation in this month's output compared with the last month's output for some factory.*

$$F(x)(t) = \begin{cases} 1 + \frac{t}{x}, & \text{if} \quad t \in [-x, 0], \\ 1 - \frac{t}{x}, & \text{if} \quad t \in [0, x], \\ 0, & \text{if} \quad t \in [-x, x]. \end{cases}$$

Then, $[F(x)]^{\alpha} = [-(1-\alpha)x, (1-\alpha)x], \alpha \in [0,1]$. If $\eta(x,y) = x-y$ then $F(x)$ is a preinvex fuzzy mapping, where $K = [0, \infty[, x \in K$, represents the monthly quantity of the previous month and t represents the variation in monthly output.

Finally, we give preliminary definitions of differentiable fuzzy mappings of several variables and fuzzy mappings from the standpoint of convex analysis. Let K be a nonempty convex subset of $\mathbb{R}^n$. Let T be a nonempty open and convex subset of $\mathbb{R}^n$.

Definition 16.7 (Differentiable fuzzy mapping) (Syau [266]) *A fuzzy mapping $F : T \to \mathfrak{I}_0$ is differentiable at $x_0 \in T$, if there is a unique linear transformation $\lambda : \mathbb{R}^n \to H$ such that*

$$\lim_{||\triangle x|| \to 0} \frac{1}{||\triangle x||} d(F(x_0 + \triangle x), (x_0 + \lambda \triangle x)) = 0,$$

where $||.||$ denotes the usual Euclidean norm in $\mathbb{R}^n$.

Remark 16.1 *The linear transformation λ is denoted by $D_{x_0}^{(F)}$ and called the differential of F at x_0.*

Definition 16.8 (Fuzzy convex mapping) (Wu and Xu [286]) *A differentiable fuzzy mapping $F : T \to \mathfrak{I}_0$ is called fuzzy convex, if for all $x, y \in T$, we have*

$$F(x) - F(y) \succeq D_y^{(F)}(x, y).$$

Definition 16.9 (Fuzzy pseudoconvex and fuzzy pseudoconcave mappings) (Syau [266]) *A differentiable fuzzy mapping $F : T \to \mathfrak{I}_0$ is called (i) pseudoconvex, if for all $x, y \in T$,*

$$D_y^{(F)}(x - y) \in H^* \Rightarrow F(x) - F(y) \in H^*$$

(ii) pseudoconcave, if for all $x, y \in T$,

$$-D_y^{(F)}(x - y) \in H^* \Rightarrow F(y) - F(x) \in H^*.$$

Mishra *et al.* [204] introduced the notion of fuzzy pseudolinear functions, as follows:

Definition 16.10 (*Pseudolinear fuzzy mapping*) *A differentiable fuzzy mapping* $F : T \to \mathfrak{I}_0$ *is called pseudolinear, if for all* $x, y \in T$ *both* F *and* $-F$ *are pseudoconvex, that is,*

$$D_y^{(F)}(x - y) \in H^* \Rightarrow F(x) - F(y) \in H^*$$

and

$$-D_y^{(F)}(x - y) \in H^* \Rightarrow F(y) - F(x) \in H^*.$$

Definition 16.11 (*Fuzzy invex mapping*) (*Wu and Xu [286]*) *A differentiable fuzzy mapping* $F : T \to \mathfrak{I}_0$ *is called a fuzzy invex with respect to a function* $\eta : T \times T \to \mathbb{R}^n$, *if*

$$F(x) - F(y) \succeq D_y^{(F)}(\eta(x, y)).$$

Example 16.2 (*Wu and Xu [286]*) *Let* $F : T \to \mathfrak{I}_0$ *represent the reproduction rate of some germ:*

$$F(x)(t) = \begin{cases} \frac{t}{x^2}, & \text{if } t \in \left[0, x^2\right], \\ 1 - \frac{t - x^2}{x^2}, & \text{if } t \in \left]x^2, 2x^2\right], \\ 0 & \text{if } t \notin \left[0, 2x^2\right]. \end{cases}$$

Then, $[F(x)]^\alpha = [\alpha x^2, (2 - \alpha)x^2], \alpha \in [0, 1]$. *Let* $\eta(x, y) = x - y$, *then,* $F(x)$ *is a fuzzy invex mapping, where* $T =]0, \infty[, x \in T$ *represented the predicted quantity and* t *represents the actual reproduction quantity.*

Definition 16.12 (*Fuzzy pseudoinvex and fuzzy pseudoincave mappings*) *A differentiable fuzzy mapping* $F : T \to \mathfrak{I}_0$ *is called pseudoinvex with respect to a function* $\eta : T \times T \to \mathbb{R}^n$ *if for all* $x, y \in T$,

$$D_y^{(F)}(\eta(x, y)) \in H^* \Rightarrow F(x) - F(y) \in H^*.$$

Moreover, $F : T \to \mathfrak{I}_0$ *is called pseudoincave with respect to a function* $\eta : T \times T \to \mathbb{R}^n$ *if*

$$\forall x, y \in T, -D_y^{(F)}(\eta(x, y)) \in H^* \Rightarrow F(y) - F(x) \in H^*.$$

Definition 16.13 (*η-Pseudolinear fuzzy mapping*) *A differentiable fuzzy mapping* $F : T \to \mathfrak{I}_0$ *is called η-pseudolinear with respect to a function* $\eta : T \times T \to \mathbb{R}^n$, *if for all* x, $y \in T$, F *and* $-F$ *are pseudoinvex, that is,*

$$D_y^{(F)}(\eta(x, y)) \in H^* \Rightarrow F(x) - F(y) \in H^*$$

and

$$-D_y^{(F)}(\eta(x, y)) \in H^* \Rightarrow F(y) - F(x) \in H^*.$$

Remark 16.2 *Note that every pseudolinear fuzzy mapping is η-pseudolinear with $\eta(x, y) = x - y$, but the converse is not true. For the converse part, consider a fuzzy mapping $F : T \to \mathfrak{I}_0$ in two variables, defined by*

$$F(x) = x_1 + \sin x_2, \forall x = (x_1, x_2) \in D,$$

where $T = \left\{ (x_1, x_2) \in \mathbb{R} \times \mathbb{R} : x_1 \succ -1, \frac{\pi}{2} \prec x_2 \prec \frac{\pi}{2} \right\}$ and $\eta(x, y) = \left(x_1 - y_1, \frac{\sin x_2 - \sin y_2}{\cos y_2} \right)^T$. Then, F is η-pseudolinear, but not pseudolinear; as at $x = \left(\frac{\pi}{3}, 0 \right)$ and $y = \left(0, \frac{\pi}{3} \right)$, then $D_y^{(F)}(x - y) = 0$ but $F(x) \prec F(y)$.

16.3 Characterizations of Pseudolinear Fuzzy Mappings

Now, we present some characterizations of η-pseudolinear fuzzy mappings and pseudolinear fuzzy mappings, derived by Mishra *et al.* [204].

Proposition 16.1 *Let F be a differentiable fuzzy mapping defined on an open set T in $\mathbb{R}^n$ and K be an invex subset of T, such that $\eta : K \times K \to \mathbb{R}^n$ satisfies*

$$\eta(x, y + t\eta(x, y)) = (1 - t)\eta(x, y)$$

and

$$\eta(y, y + t\eta(x, y)) = -t\eta(x, y), \forall t \in [0, 1].$$

Suppose that F is η-pseudolinear on K. Then,

$$\forall x, y \in K, D_y^{(F)}(\eta(x, y)) = 0 \Leftrightarrow F(x) = F(y).$$

Proof Suppose that F is η-pseudolinear on K. Then, for all $x, y \in K$, we have

$$D_y^{(F)}(\eta(x, y)) \in H^* \Rightarrow F(x) - F(y) \in H^*$$

and

$$-D_y^{(F)}(\eta(x, y)) \in H^* \Rightarrow F(y) - F(x) \in H^*.$$

Combining these two, it follows that

$$D_y^{(F)}(\eta(x, y)) = 0 \Rightarrow F(x) = F(y), \forall x, y \in K.$$

Now, we prove that $F(x) = F(y) \Rightarrow D_y^{(F)}(\eta(x, y)) = 0, \forall x, y \in K$. For that, we show that for any $x, y \in K$, such that

$$F(x) = F(y) \Rightarrow F(y + t\eta(x, y)) = F(y), \forall t \in]0, 1[.$$

If $F(y + t\eta(x, y)) \succ F(y)$, then, by the definition of pseudoinvexity of F, we have

$$D_z^{(F)}(\eta(x, z)) \prec 0, \tag{16.1}$$

where $z = y + t\eta(x, y)$.

From the assumption of the proposition, we have

$$\eta(y, z) = \eta(y, y + t\eta(x, y)) = -t\eta(x, y) = \frac{-t}{1-t}\eta(x, z).$$

Therefore, from (16.1), we get

$$D_z^{(F)}\left(\frac{-t}{1-t}\eta(x, z)\right) \prec 0.$$

Hence

$$D_z^{(F)}(\eta(x, z)) \succ 0.$$

By η-pseudolinearity of F, we have

$$F(x) \succeq F(z).$$

This contradicts the assumption that $F(z) \succ F(y) = F(x)$.

Similarly, we can also show that $F(y + t\eta(x, y)) \prec F(y)$ leads to a contradiction, using the pseudoinvexity of $-F$. Thus,

$$F(y + t\eta(x, y)) = F(y), \forall t \in \,]0,1[.$$

Hence,

$$D_y^{(F)}(\eta(x, y)) = 0.$$

Corollary 16.1 *Let F be a differentiable fuzzy mapping defined on an open set T in $\mathbb{R}^n$ and K be a convex subset of T. Suppose that F is pseudolinear on K. Then, for all $x, y \in K$,*

$$D_y^{(F)}(x - y) = 0 \Leftrightarrow F(x) = F(y).$$

Remark 16.3 *Proposition 16.1 and Corollary 16.1 are extensions of Theorem 13.1 and Theorem 2.1, respectively, to the case of fuzzy mappings.*

Proposition 16.2 *Let F be a differentiable fuzzy mapping defined on an open set T in $\mathbb{R}^n$ and K be an invex subset of T. Then F is η-pseudolinear if and only if there exists a function p defined on $K \times K$ such that $p(x, y) \succ 0$ and*

$$F(x) = F(y) + p(x, y)D_y^{(F)}(\eta(x, y)) \; \forall x, y \in K.$$

Proof Let F be an η-pseudolinear function. We have to construct a function p on $K \times K$, such that

$$p(x, y) \succ 0 \text{ and } F(x) = F(y) + p(x, y)D_y^{(F)}(\eta(x, y)) \; \forall x \, y \in K.$$

If $D_y^{(F)}(\eta(x, y)) = 0$, for all $x, y \in K$, then we define $p(x, y) = 1$. In this case, we have $F(x) = F(y)$, due to Proposition 16.1. On the other hand, if $D_y^{(F)}(\eta(x, y)) \neq \tilde{0}$, then, we define

$$p(x, y) = \frac{F(x) - F(y)}{D_y^{(F)}(\eta(x, y))}.$$

Now, we have to show that $p(x, y) \succ 0$. Suppose that $F(x) \succ F(y)$. Then, by pseudoinvexity of $-F$, we have $D_y^{(F)}(\eta(x,y)) \succ 0$. Hence, $p(x, y) \succ \tilde{0}$.

For the converse part, we first show that F is pseudoinvex, i.e., for any $x, y \in K$, $D_y^{(F)}(\eta(x, y)) \in H^*$. Then, we have

$$F(x) - F(y) = p(x,y) D_y^{(F)}(\eta(x,y)) \succeq 0.$$

Thus, $F(x) \succeq F(y)$. Likewise, we can prove that $-F$ is pseudoinvex. Hence, F is η-pseudolinear.

Corollary 16.2 *Let F be a differentiable fuzzy mapping defined on an open set T in $\mathbb{R}^n$ and K be a convex subset of T. Then F is pseudolinear if and only if there exists a function p defined on $K \times K$, such that*

$$p(x,y) \succ 0 \text{ and } F(x) = F(y) + p(x,y) D_y^{(F)}(x-y), \forall x, y \in K.$$

Proof Replace $\eta(x, y) = x - y$ in the proof of Proposition 16.2.

Remark 16.4 *Proposition 16.2 and Corollary 16.2 are extensions of Theorem 13.2 and Theorem 2.2, respectively, to the case of fuzzy mappings.*

Proposition 16.3 *Let $F : T \to \mathbb{R}^n$ be an η-pseudolinear fuzzy mapping defined on an open set T of $\mathbb{R}^n$ and $G : \mathbb{R} \to \mathbb{R}$ be differentiable fuzzy mapping with $D^{(G)}(t) \succ \tilde{0}$ or $D^{(G)}(t) \prec \tilde{0}$, for all $t \in \mathbb{R}$. Then, the composition mapping $G \circ F$ is also η-pseudolinear fuzzy mapping.*

Proof Let $H(x) = G(F(x))$, for all T. It suffices to prove the result for $D^{(G)}(t) \succ 0$, since the negative of an η-pseudolinear function is η-pseudolinear. We have,

$$D_y^{(H)}(\eta(x,y)) = D^{(G)}(F(x)) D_y^{(F)}(\eta(x,y)).$$

Then,

$$D_y^{(H)}(\eta(x,y)) \succeq \tilde{0}(\tilde{0}) \Rightarrow D_y^{(H)}(\eta(x,y)) \succeq \tilde{0}(\tilde{0})$$

Since G is strictly increasing, this yields $F(x) \succeq F(y)(F(x) \preceq F(y))$, due to η-pseudolinearity of F. Thus, $H(x) \succeq H(y)(H(x) \preceq H(y))$ since G is strictly increasing. Hence, H is η-pseudolinear.

Corollary 16.3 *Let $F : T \to \mathbb{R}^n$ be a pseudolinear fuzzy mapping defined on an open set T of $\mathbb{R}^n$ and $G : \mathbb{R} \to \mathbb{R}$ be differentiable fuzzy mapping with $D^{(F)}(t) \succ \tilde{0}$ or $D^{(F)}(t) \prec \tilde{0}$ for all $t \in \mathbb{R}$. Then, the composition mapping $F \circ G$ is also pseudolinear fuzzy mapping.*

Remark 16.5 *Proposition 16.3 and Corollary 16.3 are extensions of Theorem 13.3 and Theorem 2.5, respectively, to the case of fuzzy mappings.*

16.4 Characterization of Solution Sets

Let F be a fuzzy mapping defined on a nonempty open subset T of $\mathbb{R}^n$ and K is an invex set of T, such that $\inf \{F(x) : x \in K\}$ exists in $\Im_0$. Let $\mu = \inf \{F(x) : x \in K\}$. Syau [266] has established that: $\Gamma := \{x \in K : F(x) = \mu\}$ is an invex set if F is preinvex fuzzy mapping.

Mishra *et al.* [204] have established the following results.

Theorem 16.1 *Let $F : T \to \Im_0$ be a differentiable fuzzy mapping on an open set T of $\mathbb{R}^n$ and F be η-pseudolinear on an invex subset K of T, where η satisfies*

$$\eta(x, \ y) + \eta(y, \ x) = 0,$$

$$\eta(y, \ y + t\eta(x, y)) = -t\eta(x, y)$$

and

$$\eta(x, y + t\eta(x, y)) = (1 - t)\eta(x, y), \forall x, y \in K \ and \forall t \in [0, 1].$$

Let $\bar{x} \in \Gamma$, then, $\Gamma = \tilde{\Gamma} = \hat{\Gamma}$, where

$$\tilde{\Gamma} := \left\{ x \in K : \ D_x^{(F)}(\eta(x, \ \tilde{x})) = \tilde{0} \right\}$$

and

$$\hat{\Gamma} := \left\{ x \in K : \ D_x^{(F)}(\eta(x, \ \bar{x})) = \tilde{0} \right\}.$$

Proof The point $x \in \Gamma$ if and only if $F(x) = F(\bar{x})$. By Proposition 16.1, we have $F(x) = F(\bar{x})$ if and only if $D_x^{(F)}(\eta(x, y)) = \tilde{0}$. Also $F(\bar{x}) = F(x)$ if and only if $D_{\bar{x}}^{(F)}(\eta(x, \bar{x})) = \tilde{0}$. The latter is equivalent to $D_{\bar{x}}^{(F)}(\eta(\bar{x}, \ x)) = \tilde{0}$, since $\eta(\tilde{x}, \ x) = -\eta(x, \ \tilde{x})$.

Corollary 16.4 *Let F and η be the same as in Theorem 16.1. Then $\Gamma = \tilde{\Gamma}_1 = \hat{\Gamma}_1$, where*

$$\tilde{\Gamma}_1 := \left\{ x \in K : \ D_x^{(F)}(\eta(x, \ \tilde{x})) \succeq \tilde{0} \right\}$$

and

$$\hat{\Gamma}_1 := \left\{ x \in K : \ D_x^{(F)}(\eta(x, \ \bar{x})) \succeq \tilde{0} \right\}.$$

Proof It is clear from Theorem 16.1 that $\Gamma \subset \tilde{\Gamma}_1$. We prove that $\tilde{\Gamma}_1 \subset \Gamma$. Assume that $x \in \tilde{\Gamma}_1$, i.e., $x \in K$, such that $D_x^{(F)}(\eta(x, \ \bar{x})) \succeq \tilde{0}$. In view of Proposition 16.2, there exists a function p defined on $K \times K$, such that $p(x, \ \tilde{x}) \succ 0$ and $F(x) = F(\bar{x}) + p(x, \bar{x}) D_x^{(F)}(\eta(x, \ \bar{x})) \succeq F(\bar{x})$. This implies that $x \in \Gamma$, and hence $\tilde{\Gamma}_1 \subset \Gamma$. Similarly, we can prove that $\Gamma = \hat{\Gamma}_1$, using the identity $\eta(x, \ \bar{x}) = -\eta(\bar{x}, \ x)$.

Corollary 16.5 *Let $F : T \to \Im_0$ be a differentiable fuzzy mapping on an open set T of R^n, and F be pseudolinear on a convex subset K of T. Let $\bar{x} \in \bar{\Gamma}$. Then, $\bar{\Gamma} = \tilde{\bar{\Gamma}} = \hat{\bar{\Gamma}}$, where $\bar{\Gamma}$ = solution set of pseudolinear program,*

$$\tilde{\bar{\Gamma}} = \left\{ x \in K : D_x^{(F)} (x - \tilde{x}) = \tilde{0} \right\}$$

and

$$\hat{\bar{\Gamma}} = \left\{ x \in K : D_x^{(F)} (x - \bar{x}) = \tilde{0} \right\}.$$

Proof The proof is similar to the proof of Theorem 16.1 with $\eta(x, \bar{x}) = x - \bar{x}$.

Corollary 16.6 *Let F be as in Corollary 16.5 and $\bar{x} \in \bar{\Gamma}$. Then,*

$$\bar{\Gamma} = \tilde{\bar{\Gamma}}_1 = \hat{\bar{\Gamma}}_1,$$

where

$$\tilde{\bar{\Gamma}}_1 = \left\{ x \in K : D_x^{(F)} (x - \tilde{x}) \succeq \tilde{0} \right\}$$

and

$$\hat{\bar{\Gamma}}_1 = \left\{ x \in K : D_x^{(F)} (x - \bar{x}) \succeq \tilde{0} \right\}.$$

Theorem 16.2 *Let F and η be the same as in Theorem 16.1. If $\bar{x} \in \Gamma$, then $\Gamma = \Gamma^* = \Gamma_1^*$, where*

$$\Gamma^* = \left\{ x \in K : D_{\bar{x}}^{(F)} (\eta(x, \bar{x})) = D_x^{(F)} (\eta(\bar{x}, x)) \right\},$$

and

$$\Gamma_1^* = \left\{ x \in K : D_{\bar{x}}^{(F)} (\eta(x, \bar{x})) \succeq D_x^{(F)} (\eta(\bar{x}, x)) \right\}.$$

Proof (i) $\Gamma \subset \Gamma^*$. Let $x \in \Gamma$. It follows from Theorem 16.1 that

$$D_x^{(F)} (\eta(x, \bar{x})) = \tilde{0} = D_{\bar{x}}^{(F)} (\eta(\bar{x}, x)).$$

Since $\eta(\bar{x}, x) = -\eta(x, \bar{x})$, we have

$$D_x^{(F)} (\eta(\bar{x}, x)) = \tilde{0} = D_{\bar{x}}^{(F)} (\eta(x, \bar{x})).$$

Thus, $x \in \Gamma^*$, and hence $\Gamma \subset \Gamma^*$.
(ii) $\Gamma^* \subset \Gamma_1^*$ is obvious.
(iii) $\Gamma_1^* \subset \Gamma$. Assume that $x \in \Gamma_1^*$. Then $x \in K$ satisfies

$$D_{\bar{x}}^{(F)} (\eta(x, \bar{x})) \succeq D_x^{(F)} (\eta(\bar{x}, x)). \tag{16.2}$$

Suppose that $x \in \Gamma$. Then $F(x) \succ F(y)$. By pseudoinvexity of $-F$, we have $-D_{\bar{x}}^{(F)} (\eta(\bar{x}, x)) \in H^*$. Since, $\eta(x, \bar{x}) = -\eta(\bar{x}, x)$, we have $D_{\bar{x}}^{(F)} (\eta(x, \bar{x})) \in H^* \backslash \{\tilde{0}\}$. Using (16.2), we have, $D_x^{(F)} (\eta(x, \bar{x})) \prec 0$ or $D_x^{(F)} (\eta(\bar{x}, x)) \succ 0$. In view of Proposition 16.2, there exists a function p defined on $K \times K$, such that $p(x, \bar{x}) \succ 0$, and $F(x) = F(\bar{x}) + p(x, \bar{x}) D_x^{(F)} (\eta(x, \bar{x})) \succeq F(\bar{x})$, a contradiction. Hence, $x \in \Gamma$.

Corollary 16.7 *Let F be as in Corollary 16.5. If $\bar{x} \in \bar{\Gamma}$, then $\bar{\Gamma} = \bar{\Gamma}^* = \bar{\Gamma}_1^*$,*
where

$$\bar{\Gamma}^* := \left\{ x \in K : D_{\bar{x}}^{(F)} (x - \bar{x}) = D_x^{(F)} (\bar{x} - x) \right\}$$

and

$$\bar{\Gamma}_1^* := \left\{ x \in K : D_{\bar{x}}^{(F)} (x - \bar{x}) \succeq D_x^{(F)} (\bar{x} - x) \right\} .$$

Proof The proof of this corollary will follow from the proof of Theorem 16.2, with $\eta (x, \bar{x}) = x - \bar{x}$.

Chapter 17

Pseudolinear Functions and Their Applications

17.1 Introduction

In this chapter, we shall discuss applications of pseudolinear functions and their properties, especially in hospital management, economics, and in developing simplex-type algorithms for quadratic fractional programming problems. In the first section, we present the studies of Kruk and Wolkowicz [150]. Kruk and Wolkowicz [150] have revisited and reformulated the hospital management problem studied by Mathis and Mathis [180] to provide the mathematical background and convergence proof for the algorithm. Kruk and Wolkowicz have slightly modified the algorithm of Mathis and Mathis [180] and explained that the cause of simplicity of the algorithm is that its objective function is a pseudolinear function. They have shown that the hospital management problem described by Mathis and Mathis [180] falls into the class of linear constrained pseudolinear optimization problems. The study of Kruk and Wolkowicz justifies the significance of the class of pseudolinear functions and related programming problems. Kruk and Wolkowicz have remarked that this could only be possible by using the properties of pseudolinear functions. Moreover, in the case of a stationary point, the behavior of a pseudolinear function is as good as a linear function. The properties of this class of functions have helped to conclude that: The simplex method is much more than its tableau representation, and the class of problems to which it applies is much larger than that of linear programs.

In the second section of the chapter, we present a simplex-like algorithm developed by Ujvari [276] for pseudolinear quadratic fractional programming. Ujvari [276] has presented a simplex-type algorithm for optimizing a pseudolinear quadratic fractional function over a polytope. This algorithm works in a more general setting than convex simplex algorithms adopted to the above-mentioned problem. The final section of the chapter is about the application of pseudolinear functions in the Le Chatelier principle.

17.2   A Hospital Management Problem and Algorithm

Mathis and Mathis [180] developed a simple algorithm to solve a nonlinear hospital management problem in the United States. This algorithm depends on the theory of nonlinear programming problems and the classical Karush-Kuhn-Tucker condition. The algorithm selects a hospital's charge structure to maximize reimbursement in Texas (United States). Mathis and Mathis [180] have considered the following problem of hospital management:

Hospital Management Problem Suppose that the manager of a hospital in the United States wishes to increase charges a fixed percentage over the charges of the past year. However, the strategy to assign an across-the-board increase in each department might not increase in the most favorable increase in the revenue for the hospital due to Medicare support to a large number of patients. Hospital services are divided into several departments, where each department administers a number of procedures. The hospital sets a specific charge for each procedure and return for these charges depends on whether or not a patient is a Medicare beneficiary. Reimbursement for Medicare patients are of two kinds — whether the service is inpatient or outpatient. If the charge is for a Medicare inpatient service, the hospital receives a fixed amount that does not depend on the charges. Since, this source of revenue is constant with respect to increases in hospital charges, it is ignored in the study. If the charge is to a Medicare patient for outpatient service, the hospital is reimbursed through a cost accounting procedure. This cost is the sum of the fixed Medicare cost for each department according to the government guidelines and Medicare outpatient reimbursement for each department which is the fixed cost multiplied by the percentage of Medicare outpatient charges for that particular department. Therefore, the problem is to find an optimal increase in the charges of procedures of departments, which results in the most favorable increase in the revenue of the hospital.

Some important points of the model formulation of Mathis and Mathis [180] may be presented as follows:

Suppose that

(i) Number of departments in the hospital $= d$;

(ii) Procedures performed by the department $i = p_i$;

(iii) Charges are assigned to procedure j in department i by:

(a) $m_{ij} \geq 0$ represents Medicare/Medicaid charges;

(b) $o_{ij} \geq 0$ represents other charges;

(c) $c_{ij} \geq 0$ represents total charges $(c_{ij} = m_{ij} + o_{ij})$;

(iv) The government-fixed outpatient cost for department i is $C_i > 0$;

(v) The decision variable r_{ij} represents the fraction of increase in the charge for procedure j, department i.

(vi) The upper and lower bound for increase in the charges is given by

$$l_{ij} \leq r_{ij} \leq u_{ij}.$$

Suppose that the overall charge increase is a constant $q \times 100$ percent. Then the model problem can be formulated as the following maximization problem:

$$(P) \quad \text{Max} \, F(r) := \sum_{i=1}^{d} \left[\sum_{j=1}^{p_i} o_{ij} (1 + r_{ij}) + C_i \frac{\sum_{j=1}^{p_i} m_{ij} (1 + r_{ij})}{\sum_{j=1}^{p_i} c_{ij} (1 + r_{ij})} \right]$$

$$\text{subject to} \quad \sum_{i=1}^{d} \sum_{j=1}^{p_i} c_{ij} r_{ij} = q \sum_{i=1}^{d} \sum_{j=1}^{p_i} c_{ij}, \quad l_{ij} \le r_{ij} \le u_{ij}.$$

Mathis and Mathis [180] have given the following, Algorithm 1, for solving the problem (P):

Algorithm 1 Set all r_{ij} to q and sort the procedure in ascending order of $Q_{ij} : \left(\frac{\partial F}{\partial r_{ij}} \right) \left(\frac{1}{c_{ij}} \right)$.

while the sort produces a different ordering **do**

Set all r_{ij} to u_{ij}.

Beginning with the smallest Q_{ij}, assign l_{ij} to r_{ij} until the solution is nearly feasible.

Then adjust the last r_{ij} to make it so.

Compute the new Q_{ij} and sort.

end while

17.2.1 Mathematical Background of the Algorithm and Pseudolinear Functions

Kruk and Wolkowicz [150] have remarked that Algorithm 1 is unexpectedly simple, in contrast to the fact that it solves a difficult nonlinear programming problem. Moreover, a convergent proof for the algorithm is not given, which is essential.

One of the characteristics of Algorithm 1 is that it uses a feasible region as the polytope, that is, iterates proceeds from vertex to vertex. To observe this behavior, let n be the number of variables $\left(n := \sum_{i=1}^{i=d} p_i \right)$ and notice that the program has 1 equality constraint and $2n$ inequalities. We know that a vertex is a basic feasible solution and that the algorithm, forcing all variables except one to either their upper or lower bound, will satisfy with equality exactly $n-1$ of the inequalities. Adding the single equality constraint, we obtain n constraints in n-space, a vertex.

Moreover, according to Mathis and Mathis [180], the algorithm seems never to remain stuck at a local optimum; it finds, the global optimum. Motivated by this surprsing fact, Kruk and Wolkowicz [150] revisited the problem and algorithm studied by Mathis and Mathis [180]. To analyze the simplicity of the problem structure, they have reformulated the problem to highlight the characteristics that explain the algorithm's behavior. They have slightly modified the algorithm and have given it a strong mathematical background. They viewed the simplex algorithm as an active set algorithm and showed that the simplex algorithm is the driving force behind a successful algorithm for an almost linear problem.

However, the algorithm presented by Kruk and Wolkowicz is not very different from that of Mathis and Mathis. It differs in the choice of constraints to drop and pick up. In other words, it differs in the choice of entering and leaving variables. Although this different pivoting rule does not produce a different algorithm, it drastically alters performance. The main differences in the description of the algorithms of Kruk and Wolkowicz [150] and Mathis and Mathis [180] lie in the approach used to derive it. Moreover, Kruk and Wolkowicz [150] have explained the behavior and proved its correctness under certain stated assumptions on the problem data.

Since, the objective function in the problem (P) is nonlinear and not necessarily convex, therefore, the well-known simplex method (see Dantzig [61]) and interior point method (see Nesterov and Nemirovski [215]) may not help to recognize the special characteristics of the problem. Moreover, a full-featured nonlinear solver, possibly based on sequential quadratic programming (see Powel and Yuan [221]) will not help. To simplify the analysis and recognize the special characteristics of the problem (P), Kruk and Wolkowicz have reformulated the modeling problem (P). With the help of certain substitutions, they have shown that the nonlinear programming problem (P) can be reformulated as a less nonlinear programming problem.

17.2.2   A Less Nonlinear Program

Let the solution space be $\mathbb{R}^n := \mathbb{R}^{p_1} \oplus ... \oplus \mathbb{R}^{pd}$, where the p_i, indicate the number of procedures per department, to obtain

$$\tilde{x}_i \in \mathbb{R}^{p_i}, \ \tilde{x}_i := \left[1 + r_{i1}^T, 1 + r_{i2}^T, ..., 1 + r_{ip_i}^T \right]^T,$$

$$\tilde{x} \in \mathbb{R}^n, \ \tilde{x} := \left[\tilde{x}_1^T, \tilde{x}_2^T ..., \tilde{x}_d^T \right]^T.$$

We observe that the decision variable is now nonnegative. Let $\tilde{x}$ refers to the whole vector, $\tilde{x}_i$ refers to the subvector corresponding to department i, and x_i refers to a component of x. Corresponding substitutions to the other parameters are given by

$$\tilde{a}_i \in \mathbb{R}^{p_i}, \ \tilde{a}_i := \left[o_{i1}^T, o_{i2}^T ..., o_{ip_i}^T \right]^T,$$

$$\tilde{b}_i \in \mathbb{R}^{p_i}, \ \tilde{b}_i := C_i \left[m_{i1}^T, m_{i2}^T ..., m_{ip_i}^T \right]^T,$$

$$\tilde{c}_i \in \mathbb{R}^{p_i}, \ \tilde{c}_i := \left[c_{i1}^T, c_{i2}^T ..., c_{ip_i}^T \right]^T,$$

$$\tilde{l}_i \in \mathbb{R}^{p_i}, \ \tilde{l}_i := \left[1 + l_{i1}^T, 1 + l_{i2}^T ..., 1 + l_{ip_i}^T \right]^T,$$

$$\tilde{u}_i \in \mathbb{R}^{p_i}, \ \tilde{u}_i := \left[1 + u_{i1}^T, 1 + u_{i2}^T ..., 1 + u_{ip_i}^T \right]^T,$$

$$t := (1 + q) \sum_{i=1}^{d} \sum_{j=1}^{j=p_i} c_{ij},$$

where the coefficients now satisfy $\tilde{c}_i = \tilde{a}_i + C_i^{-1} \tilde{b}_i$.

After these substitutions the problem is reformulated as

$$(P) \quad \max \left\{ \sum_{i=1}^{d} \left[\tilde{a}_i^t + \frac{\tilde{b}_i^t}{\tilde{c}_i^T \tilde{x}_i} \right] \mid \sum_{i-1}^{d} \tilde{c}_i^t \tilde{x}_i = t, \ l \le x \le u \right\},$$

which is a fractional program. These problems have been widely studied due to their applications in finance (see Schaibel [247]) and recent attempts have been made to develop the interior point algorithms (see Freund *et al.* [84], Nemirovski [214]).

With this formulation, it is clear that
(i) The objective function is separable by department and it is a sum of functions, each concerned with different vectors.
(ii) The product $\tilde{c}_i^t \tilde{x}_i$ must be strictly positive.
(iii) The feasible region is the intersection of a hyperplane with a box.
(iv) Each term is either a linear $\left(\tilde{a}_i^T \tilde{x}_i \right)$ or a linear fractional $\left(\frac{\tilde{b}_i^T \tilde{x}_i}{\tilde{c}_i^T \tilde{x}_i} \right)$ transformation (also known as a projective transformation).

Now, we note the linearity of the feasible region and each term of the objective function is pseudolinear. The cause of interest in pseudolinearity conditions is that for a linear constrained pseudolinear optimization problem, we can find an optimal solution as easily as for a linear program. Moreover, in case of a stationary point the behavior of pseudolinear function is as good as a linear function. The following lemma illustrates this fact.

Lemma 17.1 *If the directional derivative of a pseudolinear function vanishes in a direction d, the function is constant on the line containing d.*

Proof Suppose that at some point z, we have

$$\langle \nabla f(z), d \rangle = 0.$$

Then both inequalities describing pseudolinearity apply, and we get

$$f(z + \alpha d) = f(z), \forall \alpha.$$

The function f is constant along the line containing d.

This implies that a pseudolinear function has no stationary point, unless it is constant. This property justifies the behavior of Algorithm 1, that is, a pseudolinear function, optimized over a polytope, only has global optima and attains its extrema at vertices. However, the crucial unstated assumption the algorithm makes is that the sum $\tilde{a}_i^t \tilde{x}_i + \frac{\tilde{b}_i^t \tilde{x}_i}{\tilde{c}_i^t \tilde{x}_i}$ maintains its (pseudo) convexity properties over the feasible region.

Schaible [247] has studied in detail the conditions under which such sums remain either quasiconvex or quasiconcave. Here, we note that characterization will depend on the values of $\tilde{c}_i, \tilde{b}_i$, and $\tilde{a}_i$ and that, in the absence of formal restrictions on these values, we cannot claim that the sum of linear and linear-fractional transformations remains pseudolinear. Kruk and Wolkowicz have mentioned that since this condition is ignored by Mathis and Mathis, while some convexity assumption is essential to their algorithm, one can easily construct examples of failure.

Moreover, Kruk and Wolkowicz [150] have noted that at the cost of losing the simplicity of the algorithm one can study a larger class of problems. However, to give a solid mathematical foundation to the algorithm, suppose that the assumption holds.

17.2.3   Characterization of a Global Solution

It is well-known that in the constrained optimization algorithm, the usual goal is to find the Karush-Kuhn-Tucker (KKT) point due to the fact that under a constraint qualification, a Karush-Kuhn-Tucker (KKT) point characterizes the necessary optimality conditions for a constrained optimization problem. This search is easily justified in our case.

Lemma 17.2 *An optimal solution of* (P) *is a Karush-Kuhn-Tucker point.*

Proof Since, the feasible region is defined by affine functions, therefore, the Karush-Kuhn-Tucker constraint qualification is satisfied.

We know that to obtain the constraint qualification, the recognition of the shape of the feasible region is required only. In particular, we need not insist on the linear independence of the gradients of the active constraints. We are therefore justified in looking for a Karush-Kuhn-Tucker point.

Now, the following lemma from Kruk and Wolkowicz [150] justifies an important characterisitc of the probelem (P).

Lemma 17.3 *Under the pseudolinearity assumption, any point satisfying the Karush-Kuhn-Tucker condition is a solution of* (P).

Proof Since the constraints are affine, they are convex. Under the assumption of pseudoconcavity of the objective function, therefore, by Theorem 1.45 (Mangasarian [175], Theorem 10.1.1]), the Karush-Kuhn-Tucker conditions are sufficient for optimality.

The following lemma from Kruk and Wolkowicz [150] restricts our search for Karush-Kuhn-Tucker points to the vertices of the polytope.

Lemma 17.4 *Under the pseudolinearity assumption, if* (P) *is feasible, it has an optimal solution at a vertex.*

Proof Let $\bar{x}$ be an optimal solution for (P) and for some nonzero direction d and $\alpha_1, \alpha_2 > 0$, $\bar{x}$ lies on an edge described by the interval $[\bar{x} - \alpha_1 d, \bar{x} + \alpha_2 d]$. By optimality, f must be nondecreasing from $\bar{x}$, that is,

$$\langle \nabla f(\bar{x}), (\bar{x} + \alpha_2 d - \bar{x}) \rangle \geq 0 \, and \, \langle \nabla f(\bar{x}), (\bar{x} - \alpha_1 d - \bar{x}) \rangle \geq 0.$$

From which, it follows that

$$-\left\langle \nabla f(\bar{x})^T, d \right\rangle \geq 0 \, and \, \left\langle \nabla f(\bar{x})^T, d \right\rangle \geq 0.$$

This implies that $\nabla f(\bar{x})^T = 0$. By Lemma 17.1, f is constant along d, and therefore the vertices adjacent to $\bar{x}$ are also optimal.

We note that, now, we have a complete characterization of optimal solutions and a finite subset of points of the feasible region that we need to investigate. Thus, we observe that all the conditions required for a successful application of the simplex method for linear programming are satisfied. Now, using an active set approach, the simplex method is easy to transpose to our problem, which we now proceed to do. Kruk and Wolkowicz [150] have described the implementation of the simplex method for the pseudolinear program (P) as follows:

17.2.4 A Pseudolinear Simplex Algorithm

Kruk and Wolkowicz [150] provided a MATLAB ® implementation for the problem (P) as follows:
1. Variable Declarations The algorithm is described in consecutive sections, starting with the parameter declarations in Algorithm 2.
Algorithm 2 (Pseudolinear simplex method, parameters)
$$\text{plSimplex}\,(d, p, c, m, a, lb, ub, t)$$

integer d {Number of departments}
integer $p\,[1..d]$ {Number of procedures per department}
integer $nb := \sum_{i=1}^{d} p\,[i]$ {Total number of procedures}
real $c\,[1..nb]$ {Total charges}
real $b\,[1..nb]$ {Transformed Medicare charges}
real $a\,[1..nb]$ {Other charges}
real $l_b\,[1..nb]$ {Lower bound on x}
real $u_b\,[1..nb]$ {Upper bound on x}
real t {Transformed total charge increase}
2. Finding an Initial Feasible Vertex The first problem faced in any simplex-type approach is the initial vertex. In our case, with only a box constraint intersecting a hyperplane, a basic feasible solution is within easy reach. The first step described in Algorithm 1 will work: set all variables to their upper (or lower) bound and decrease (or increase) them one by one until equality is satisfied.

Algorithm 3 (Pseudolinear simplex method, phase I.)
real x [1..nb] {Decision variable}
integer k {Index of inactive constraint}
if $(c^t l_b > t \lor c^t u_b < t)$ **then**
return 0; {Program is infeasible}
else
$x = l_b$; {Try variable at lower bound}
for $i = 1...nb$ **do**
$x[i] = u_b[i] - \left((c^T x - T)/c[i]\right)$; {Move component to upper bound}
if $(c^T x > t)$ **then**
$x[i] = u_b[i] - \left((c^t x - t)/c[i]\right)$; {Adjust to feasibility}
$k = i$; {Record inactive constraint}
end if
end for
end if

Kruk and Wolkowicz [150] have remarked that Algorithm 3 will detect infeasibility or provide a feasible vertex and record which constraint (there is exactly one) is not an element of the active set, the set of constraints forced to equality. This so-called inactive constraint might not actually be slack. Moreover, it might also be satisfied with equality. We will say that it is saturated. The case of such a degenerate vertex will be studied in Section 17.1.4. Some heuristic for choosing the ordering of the variables based on the vectors m and c might prove effective, but is not required.

3. Iterating to the Optimal Vertex Now, we compute the Lagrange multiplier estimates by trying to solve for dual feasibility. In fact, we will solve for everything but nonnegativity of the multipliers.

Let the Lagrange multipliers be denoted by:

$\tilde{\lambda}_i \in \mathbb{R}_+^{p_i}$, Upper bound multipliers of department p_i,

$$\lambda := \left[\tilde{\lambda}_1^T, \tilde{\lambda}_2^T, ..., \tilde{\lambda}_d^T\right]^T ;$$

$\tilde{\gamma}_i \in \mathbb{R}_+^{p_i}$, Lower bound multipliers of department p_i,

$$\gamma := \left[\tilde{\gamma}_1^T, \tilde{\gamma}_2^T, ..., \tilde{\gamma}_d^T\right]^T ;$$

$\mu \in \mathbb{R}$, Multiplier of hyperplane constraint;

and the Lagrangian be

$$L := \sum_{i=1}^d \left[\tilde{a}_i^T \tilde{x}_i + \frac{\tilde{b}_i^T \tilde{x}_i}{\tilde{c}_i^T \tilde{x}_i}\right] + \sum_{i=1}^d \tilde{\lambda}_i^T \left(\tilde{u}_i - \tilde{x}_i\right) + \sum_{i=1}^d \tilde{\gamma}_i^T \left(\tilde{x}_i - \tilde{l}_i\right)$$

$$+\mu \left(\sum_{i=1}^d \left(\tilde{c}_i^T \tilde{x}_i - t\right)\right).$$

The algorithm has to solve the following system resulted from the stationarity of the Lagrangian, together with complementarity:

$$\tilde{a}_i + \frac{\tilde{b}_i\left(\tilde{c}_i^t \tilde{x}_i\right) - \tilde{c}_i\left(\tilde{b}_i^t \tilde{x}_i\right)}{\left(\tilde{c}_i^t \tilde{x}_i\right)^2} - \tilde{\lambda}_i + \tilde{\gamma}_i + \mu \tilde{c}_i = 0, \ \ 1 \leq i \leq d, \tag{17.1}$$

$$\tilde{\lambda}_i^T\left(\tilde{x}_i - \tilde{u}_i\right) = 0, \ \ 1 \leq i \leq d, \tag{17.2}$$

$$\tilde{\gamma}_i^T\left(-\tilde{x}_i + \tilde{l}_i\right) = 0, \ \ 1 \leq i \leq d. \tag{17.3}$$

We note that the objective is separable, and constraints are the box constraints, so that a variable cannot be both at its upper bound and at its lower bound, otherwise we will take the variable out of the problem altogether. Therefore, it is easy to solve the system (17.1)–(17.3). Therefore, at least half of the multipliers λ and γ must be zero.

Kruk and Wolkowicz [150] have noted that the active set consists of one for every pair of box constraints except exactly one, usually corresponding to a slack primal variable, a component x_k that is at neither its lower nor its upper bound. Although this constraint could be saturated (what we will call the degenerate case), we can force its two corresponding multipliers to zero. If the constraint is truly slack, then we have no choice. In either case, this leads to one pair $\lambda_k = \gamma_k = 0$, and we can solve first for μ and then for every other multiplier by simple substitution.

We then consider the sign of the multipliers. If all of them are nonnegative, we have a KKT point and, therefore, an optimal solution. If not, we need to move to another vertex. The classical way to do this, somewhat different from the Mathis and Mathis approach, is to choose the largest multiplier of the wrong sign (either λ_l or γ_l), drop the corresponding constraint ($x_l \leq u_l$ or $x_l \geq l_l$), and move to the adjacent vertex. In the context of linear programming, this choice is usually known as Dantzig's rule but it makes sense in much more general settings.

Motivated by this, Kruk and Wolkowicz [150] considered the following program

$$\min\left\{f\left(x\right) | g\left(x\right) \leq b\right\}.$$

Under appropriate conditions, the optimal solution and the multipliers in terms of the right hand side, can be rewritten as

$$x\left(b\right) \ and \ \lambda\left(b\right).$$

Let us consider a small perturbation of the right-hand side of active constraint k, namely, b_k. A well-known sensitivity result from Gill *et al.* [87] relates the change in the objective function to the change in the right hand side, under appropriate conditions, as

$$\nabla f\left(x\left(b\right)\right) = -\lambda b.$$

The interpretation of this result is that if $\lambda_k(b) < 0$, then a change of b_k to $b_k - \varepsilon$ or equivalently, a move of distance ε away from the constraint k, within the feasible region, will yield a first-order change in the objective function of

$$f\left(x\left(b + \varepsilon e_k\right)\right) - f\left(x\left(b\right)\right) \approx -\varepsilon \nabla f\left(x\left(b\right)\right)^T e_k = \varepsilon \lambda_k < 0,$$

where e_k has one in the k th position and zeros elsewhere. We therefore have a decrease in the objective function proportional to the magnitude of the multiplier.

To find the direction in which to move or, equivalently, which constraint to pick up, since we are moving from vertex to vertex, we need to solve an even simpler system. We need a direction d, satisfying $d_j = 0$, for all active constraints (all the currently active ones except the dropped constraint). Yet we need to remain feasible, which translates into $\sum_{i=1}^d \bar{c}_i^t d_i = 0$. Since we have one component corresponding to the constraint not in the active set (x_k) and one component corresponding to the dropped constraint (x_l), the condition reduces to

$$c_k x_k + c_l x_l = 0,$$

and the step length is just enough to get to the next vertex.

Algorithm 4 (Pseudolinear simplex method, phase II.)
real $\gamma\,[1..nb]$ {Multiplier upper bound constraints}
real $\lambda\,[1..nb]$ {Multiplier lower bound constraints}
while 1 do
Given x and $\lambda_k = \gamma_k = 0$, solve system (17.1)–(17.3) for λ,γ,μ
Say $\lambda\,[l_\lambda] = \min\{\lambda\}$; $\gamma\,[l_\lambda] = \min\{\gamma\}$; {Find most negative multipliers}
if $\lambda\,[l_\lambda] < 0 \wedge \lambda\,[l_\lambda] < \gamma\,[l_\gamma]$ {We should drop an upper bound} **then**
if $((x\,[k] - u_b\,[k])\,c\,[k]\,/c\,[l_\lambda] > l_b\,[l_\lambda] - x\,[l_\lambda])$ **then**

$$x\,[l_\lambda] = x\,[l_\lambda] - c\,[k]\,(u_b\,[k] - x\,[k])\,/c\,[l_\lambda]\,; x\,[k] = u_b\,[k]\,; k = l_\lambda;$$

else

$$x\,[k] = x\,[k] - c\,[l_\lambda]\,(l_b\,[l_\lambda] - x\,[l_\lambda])\,/c\,[k]\,; x\,[l_\lambda] = l_b\,[l_\lambda]\,;$$

end if
else if $(\gamma\,[l_\lambda] < 0)$ {We should drop a lower bound} **then**
if $((l_b\,[k] - x\,[k])\,c\,[k]\,/c\,[l_\gamma] > x\,[l_\gamma] - u_b\,[l_\gamma])$ then

$$x\,[k] = x\,[k] - c\,[l_\gamma]\,(u_b\,[l_\gamma] - x\,[l_\gamma])\,/c\,[k]\,; x\,[l_\gamma] = u_b\,[l_\gamma]\,;$$

else

$$x\,[l_\gamma] = x\,[l_\gamma] - c\,[k]\,(l_b\,[k] - x\,[k])\,/c\,[l_\gamma]\,; x\,[k] = l_b\,[k]\,; k = l_\gamma;$$

end if
else
return x; {We are optimal}
end if

end while

The following lemma from Kruk and Wolkowicz [150] justifies the fact this algorithm will increase the objective function at each step where we take a nonzero step.

Lemma 17.5 *The objective function f increases at each nondegenerate step of the algorithm.*

Proof Suppose that k is the index of x corresponding to the constraints not in the active set and that 1 is the index of x corresponding to the dropped constraint (because either $x_k = l_k$ or $x_k = u_k$). The algorithm solves the system (17.1)–(17.3), from which we have that

$$\nabla f(x) - \lambda + \gamma + \mu c = 0, \quad \lambda_k = \gamma_k = 0,$$

and either of

$$\lambda_l = 0 \quad \text{or} \quad \gamma_l = 0,$$

whether we are dropping an upper or a lower bound constraint. In order to compute a direction to move, the algorithm ensures that

$$c_k d_k + c_l d_l = 0.$$

Moving in the direction d (for a small step), the change in the objective function may be estimated by

$$\nabla f = f(x+d) - f(x)$$

$$= \nabla f(x)^t d + o(\|d\|)$$

$$= (\lambda - \gamma - \mu c)^t d + o(\|d\|)$$

$$= (\lambda_l - \gamma_l) d_l + (\lambda_k - \gamma_k) d_k - \mu (c_k d_k - c_l d_l) + o(\|d\|)$$

$$= (\lambda_l - \gamma_l) d_l + o(\|d\|).$$

Now, the following two cases may arise:
1. If we dropped an upper bound. Then $\lambda_l < 0, \gamma_l = 0$, and x_l decreased so that d_l is negative. Then the last line above reads $\Delta f = \lambda_l d_l > 0$.
2. If we dropped a lower bound. Then $\lambda_l = 0, \gamma_l < 0$, and x_l increased so that d_l is positive. Then the last line above reads $\Delta f = -\lambda_l d_l > 0$.

In both cases, the objective function increases in the direction away from the dropped constraint, and since the directional derivative cannot vanish on the edge we are following (by Lemma 17.1), the objective function must increase monotonically from the current vertex to the next one along the edge.

Finally, Kruk and Wolkowicz [150] have discuussed the possiblity of degeneracy and its solution.

TABLE 17.1:   Iterations in Case of Degeneracy

Iter	μ	x^t	k	l	λ^t	γ^t
0	$-2/2.4$	$[3,1,1]$	2	3	$[-1,0,0]$	$[0,0,-1,]$
1	$-3/2.4$	$[3,1,1]$	3	1	$[-2,0,0]$	$[0,1,0,]$
2	$-4/2.4$	$[1,1,3]$	3		$[0,0,0]$	$[2,1,0]$

17.2.5   Degeneracy

We note that a degenerate vertex is a vertex where a constraint not in the active set is nevertheless saturated. As an example, consider the simple three-dimensional case where all vertices are degenerate and the objective function is linear. This is clearly a special case of (P) :

$$\max \{x_1 + 2x_2 + 3x_3 | 2.4x_1 + 2.4x_2 + 2.4x_3 = 12, \ 1 \le x_i \le 3\}.$$

Phase I of the algorithm (Algorithm 3) will, as coded, produce $[3,1,1]^T$ as the initial vertex and consider the constraint $1 \le x_2$ as outside of the active set $(k = 2)$ even if $x_2 = 1$. The iterations will be as shown in Table 17.1, and the vector $[3,1,1]^T$ is optimal since all multipliers are nonnegative. The first iteration is usually known as a degenerate pivot. We did not move in the primal space. The algorithm tried to increase both x_2, corresponding to the inactive constraint, and x_3, corresponding to the dropped constraint $(1 \le x_3)$, but that is not possible while ensuring feasibility. The consequence was that we picked up a new constraint $(1 \le x_1)$. The next iteration works.

An alternative to this would have been to recognize a degenerate vertex and react accordingly. In general, the simplex method for linear programming can be made to handle degenerate vertices by rules governing the choice of inclusion of components into the basis. Such rules can be shown to work but generally degrade performance. Because of the special structure of our feasible region, we can simplify the degeneracy handling by ensuring that if our inactive component x_k is actually at its upper (respectively, lower) bound, we choose to drop a lower (respectively, upper) bound constraint (corresponding, possibly, to the next most negative multiplier). In this way, we guarantee improvement of the objective function. One way to implement this is to replace the third line in Algorithm 4 by if

$$\text{if } (\lambda [l_\lambda] < 0 \wedge x [k] < u [k]).$$

This chooses to drop a lower bound constraint if our slack variable is at its upper bound and therefore allows us to move. Like most degeneracy-avoiding routines, this slows down the algorithm (see Table 17.1).

Kruk and Wolkowicz have given the following example to illustrate the entire algorithm after transforming the variables.

TABLE 17.2: Iterations in Phase I and Phase II of the Algorithm

Charges			Solutions ($\bar{x}$)	
$\tilde{a}$	$\tilde{b} \times 10$	$\tilde{c}$	Phase I	Optimal
16.35	901.2907	64.89	1.15	1.15
74.12	187.7225	84.23	1.15	1.15
39.26	43.4491	41.60	1.15	1.15
48.20	105.6519	53.89	1.15	1.15
12.63	133.6896	19.83	1.15	1.15
13.22	576.0705	61.66	1.15	1.0473
12.67	248.9766	33.61	1.15	1.15
22.71	772.2555	87.66	1.15	1.15
5.44	272.0432	28.32	1.15	0.95
7.24	207.9561	24.73	1.15	1.15
12.66	182.5256	26.26	1.15	1.15
6.84	93.5444	13.81	1.15	1.15
3.72	300.4962	26.11	1.15	0.95
6.66	740.5708	61.84	1.15	0.95
15.89	851.4282	79.33	1.0845	0.95
33.03	233.8502	47.07	0.95	1.15
0.18	117.7579	7.25	0.95	1.15
5.23	22.1525	6.56	0.95	1.15
7.60	691.7237	49.13	0.95	1.15
67.54	395.4134	91.28	0.95	1.15

17.2.6 Example

The overall charge increase is 10% ($q = 0.10$). We have four departments ($d = 4$), each with five procedures ($p_i = 5, 1 \le i \le d$). The fraction of increase must be between -5% and 15%, which translates to the box constraints, for all ($1 \le i \le d$),

$$\tilde{l}_i = 0.95 \le \tilde{x}_i \le \tilde{u}_i = 1.15e,$$

where e is the all-ones vector. The right-hand side of the equality constraint is $t = 999.966$.

The charges, suitably transformed, are given in Table 17.2, along with the feasible solution produced by phase I and the optimal solution produced by phase II. This solution, transformed back to the problem space, is the same as the one provided by Mathis and Mathis [180], modulo some rounding of values in their original table.

17.3   Pseudolinear Quadratic Fractional Functions and Simplex Algorithm

In Chapter 15, we studied in detail about the pseudolinear quadratic fractional functions. Now, we shall study about a simplex-type algorithm developed for pseudolinear quadratic fractional programming problems described by Ujvari [276].

17.3.1   Description of the Algorithms

Ujvari [276] has described a simpex type algorithm for the optimization of pseudolinear quadratic fractional functions over a polytope.

Let Q be any $n \times n$ symmetric matrix, $n \geq 2$ and $Q \neq 0$. Suppose that $f : H \to \mathbb{R}$ be given by

$$f(x) =: \frac{\frac{1}{2} x^T Q x + q T x + \bar{q}}{p^T x + \bar{p}},$$

where $q, p \in \mathbb{R}^n, p \neq 0$ and $q_0, p_0 \in \mathbb{R}$. Let $H =: \{x \in \mathbb{R}^n : p^T x + \bar{p} > 0\}$.

Suppose that f is a pseudolinear function and it is not affine. Then by Theorem 15.2, there exists constants $\alpha, \beta, \kappa \in \mathbb{R}$, such that $\alpha \kappa < 0$ and for $x \in H$, we have

$$f(x) = \alpha p^T x + \beta + \frac{\kappa}{p^T x + \bar{p}}.$$

Consider the problem

$$(QFP) \qquad \min f(x)$$

$$\text{subject to } x \in P := \{x \in \mathbb{R}^n : Ax = b, x \geq 0\},$$

where $P \subseteq H$ is a polytope and $A \in \mathbb{R}^{m \times n}, b \in \mathbb{R}^m$.

This algorithm works under the following assumption:

(1) Suppose that the rank of the matrix is m. Suppose an initial feasible basis and corresponding simplex tableau is known, that is, we have an invertible submatrix $B \in \mathbb{R}^{m \times n}$, such that $B^{-1} b \geq 0$ and, the corresponding simplex tableau is

$$\begin{pmatrix} B & 0 \\ p_B^T & 1 \end{pmatrix}^{-1} \cdot \begin{pmatrix} A & b \\ p^T & -\bar{p} \end{pmatrix}^{-1} = \begin{pmatrix} B^{-1} A & B^{-1} b \\ p^T - p_B^T B^{-1} A & -p_B^T B^{-1} b - \bar{p} \end{pmatrix},$$

where $p_B \in \mathbb{R}^m$ is the subvector of $p \in \mathbb{R}^n$ such that the matrix (B^T, p_B) is a submatrix of the matrix (A^T, p). Using the first phase of a finite version of a simplex method, such a basis and tableau can be calculated. For details, we refer to Murty [211], Prekopa [227], and Schrijver [251].

(2) Let $P \neq \phi$ and it is a polytope. In other words, there is no nonzero vector $z \in \mathbb{R}^n$, such that $Az = 0$, $z \geq 0$. By Gordan's theorem, this is equivalent to a vector $y \in \mathbb{R}^m$, such that $Ay > 0$.

(3) The problem (QFP) is nondegenerate, that is, $B^{-1}b > 0$ hold for every feasible basis B. This assumption simplifies the calculation of all the one-dimensional faces of the polytope P.

17.3.2 Basic Definitions and Preliminaries

Let $T(B)$ denote the simplex table corresponding to a feasible basis $B \in \mathbb{R}^{m \times n}$. Let K denote the following set of positions in the matrix $B^{-1}.(A \backslash B)$:

$$K =: \left\{ (i,j) : 0 < \frac{t_{i,m+1}}{t_{i,j}} = \min_{1 \leq i \leq m, t_{i,j} \geq 0} \frac{t_{i,m+1}}{t_{i,j}} \right\}, \qquad (17.4)$$

where $t_{i,j}$ $(1 \leq i \leq m+1, 1 \leq j \leq n+1)$ denotes the (i,j)th element of simplex tableau $T(B)$. We know that if $(i,j) \in K$ and $\bar{B}$ denote the $m \times m$ matrix obtained after exchanging the ith column vector of B for the jth column vector of A, then $\bar{B}$ is a feasible basis. Moreover, the corresponding simplex tableau $T(\bar{B})$ can be obtained by pivoting the (i,j)th position of the tableau $T(B)$.

Definition 17.1 (Basic feasible solution) *A vector $x \in \mathbb{R}^n$ is called a basic feasible solution or a BFS corresponding to a feasible basis $B \in \mathbb{R}^{m \times n}$ if*

$$x_B = B^{-1}b \text{ and } x_{A/B} = 0.$$

Corresponding to a feasible basis, basic feasible solutions are exactly the extremal point of the polytope P.

Definition 17.2 (Extremal subset or face of a polytope) *A convex set $C \subseteq \mathbb{R}^n$ is called an extremal subset or a face of a polytope P, denoted by $C \triangleleft P$, if*

$$\forall x, y \in P, \lambda \in \,]0,1[\,, \lambda x + (1 - \lambda) y \in P \Rightarrow x, y \in X.$$

The vector $x \in \mathbb{R}^n$ is an extremal point of P, if $\{x\} \triangleleft P$.

Definition 17.3 (Adjacent extremal point) *Two extremal points x and y of a polytope P are adjacent, if*

$$[x,y] \triangleleft P,$$

where $[x, y]$ denotes the set of all convex combinations of the points x and y.

Definition 17.4 (Neighbor of an extremal point) *The vector $\bar{x} \in \mathbb{R}^n$ is a neighbor of a vector $x \in \mathbb{R}^n$ if x and $\bar{x}$ are adjacent extremal points of the polytope. The set of neighbors of x is denoted by $N(x)$.*

Remark 17.1 *(i) Assumption (2) implies that there is no nonpositive column in the matrix $B^{-1}A$. In fact, if $B^{-1}Ae_j \leq 0$, for some j, where e_j denotes the jth column vector of identity matrix, then let $z \in \mathbb{R}^n$ be the vector, such that*

$$z_B = -B^{-1}Ae_j \text{ and } z_{A/B} = e_j.$$

Then $z \neq 0$ and $Az = 0, z \geq 0$, a contradiction.
Therefore, in every column of $B^{-1}A$, we must have at leat one positive element.

(ii) By Theorem 1.7.7 of Prekopa [227], it is known that the basic feasible solutions x and $\bar{x}$ corresponding to the feasible bases B and $\bar{B}$, respectively, are adjacent. However, due to the nondegeneracy assumption (3), the converse also holds. The following proposition illustrates this fact.

Proposition 17.1 *Suppose $B \in \mathbb{R}^{m \times n}$ is a feasible basis with corresponding basic feasible solution x. Let $\bar{x}$ be an extremal point of (QFP) such that x and $\bar{x}$ are adjacent. Then there exists a sequence of feasible bases $B_1, B_2, .., B_i$, such that the following statement holds:*

1. *the first element of the sequence is $B_1 = B$;*

2. *the basic feasible solution corresponding to $B_1, B_2, .., B_{i-1}$ is $\bar{x}$;*

3. *each two consecutive bases from the sequences $B_1, B_2, .., B_i$, have $m - 1$ common column vectors.*

Moreover, if $x_B > 0$, then $i = 2$.

Proof The simplex method is used to prove the proposition. Let S denote the finite set of extremal points of P. The extremal point $\bar{x}$ is also an exposed point of P, that is there exists a vector $a_1 \in \mathbb{R}^n$, such that

$$a_1^T \bar{x} < a_1^T \tilde{x}, \text{ for } \tilde{x} \in P \backslash \{\bar{x}\}.$$

Similarly, there exists a vector $a_2 \in \mathbb{R}^n$ and a constant $\beta \in \mathbb{R}$, such that

$$\beta = a_2^T \tilde{x}_1 < a_2^T \tilde{x}_2, \text{ for } \tilde{x}_1 \in [x, \bar{x}] \, ; \tilde{x}_2 \in P \backslash [x, \bar{x}] \, .$$

Then for every $\varepsilon > 0$, we get

$$(a_2 + \varepsilon a_1)^T (x - \bar{x}) > 0. \tag{17.5}$$

For an approximately chosen $\varepsilon > 0$,

$$(a_2 + \varepsilon a_1)^T (x - \bar{x}) > 0, \text{ for } \tilde{x} \in S; \tilde{x} \neq x, \bar{x}. \tag{17.6}$$

Let $a =: a_2 + \varepsilon a_1$, and let us minimize the corresponding linear (cost) function over P, using the simplex method, with the initial feasible basis B. For a while the value of the cost function does not decrease, the BFS generated by the algorithm remains the same extremal point x, only the feasible bases change. Let these feasible bases be denoted by $B_1, B_2, ..., B_{i-1}$. It is clear that the cost function decreases after a finite number of steps. Therefore, from Equations (17.5) and (17.6), the current basic feasible solution will be $\bar{x}$. Let B_i denote the current feasible basis. It is obvious that the feasible bases $B_1, B_2, ..., B_i$ satisfy the statement (i)–(iv).

Now, we assume that $x_B > 0$, then B is the only feasible basis such that the corresponding basic feasible solution is x. Hence, the cost function decreases in the first step. This proves the last statement.

Remark 17.2 *(i) We note that if there is a tie in (17.4), then while pivoting on the positions $(i_1, j) \in K$ and $(i_2, j) \in K$, respectively, we get the same basic feasible solution $\bar{x}$, corresponding to different feasible bases. Therefore, in case, $x_B > 0$, then while calculating all the basic feasible solutions $\bar{x}$ adjacent to x, it suffices to choose only one pivot position from each column.*

Proposition 17.2 *Let x be an extremal point of P, then*

$$\text{cone}\left\{\bar{x} - x : \bar{x} \in N\left(x\right)\right\} = \text{cone}\left\{\bar{x} - x : \bar{x} \in P\right\}.$$

Proof For the convenience in notation, we take $x = 0$. Now, we have to prove only the inclusion $\text{cone}\left(P\right) \subseteq \text{cone}\left(N\left(x\right)\right)$, as the other inclusion is trivial. By the Minkowski theorem (Corollary 1.3) it follows that the polytope P is the convex hull of its extreme points. Hence, it is sufficient to prove that for every extremal point $\bar{x} \in P$ imply that

$$\bar{x} \in \text{cone}\left(N\left(x\right)\right).$$

Let S be the set (finite) of extremal points of P and let $\bar{x} \in S$. It is known that extremal point x is also an exposed point of P. Therefore, there exists a vector $a \subset \mathbb{R}^n$ such that

$$a^T x < a^T \bar{x}, \text{ for } \bar{x} \in P \backslash \{x\}.$$

Let $\alpha \in \mathbb{R}$ be a constant between the values

$$a^T x \text{ and } \min\left\{a^T \bar{x} : \bar{x} \in S \backslash \{x\}\right\}.$$

Let

$$M =: \left\{\bar{x} : a^T \bar{x} = \alpha\right\}.$$

Then the set $P \bigcap M$ is a polytope, $\bar{P}$ with extremal points in $\bar{S}$. Therefore, by Corollary 1.3, $\bar{P}$ is the convex hull of $\bar{S}$. There exists a constant $\varepsilon \in \mathbb{R}, 0 < \varepsilon < 1$, such that $\varepsilon\bar{x} \in \bar{P}$. Then the vector $\varepsilon\bar{x}$ is the convex combination of points from $\bar{S}$. Hence, $\bar{x} \in \text{cone}\left(\bar{S}\right)$.

Now, we show that $\text{cone}\left(\bar{S}\right) \subseteq \text{cone}\left(N\left(x\right)\right)$. Suppose that $\hat{x} \in \bar{S}$. Let Λ denote the minimal face of P, that contains $\hat{x}$. Then, $\hat{x} \in \text{ri}\left(\Lambda\right)$, therefore, $\hat{x} \in M$ is in the relative interior of $\Lambda \bigcap M$. Hence, $\Lambda \bigcap M$ is a face of $P \bigcap M$.

Since, $\hat{x} \in \bar{S}$, so $\hat{x} \in \text{ri}(\Lambda \bigcap M)$ if $\Lambda \bigcap M = \{\hat{x}\}$. Let $\bar{\Lambda}$ denote the affine hull of Λ. Since, $\dim(\bar{\Lambda} \bigcap M) = 0$ and $\dim(\bar{\Lambda} \bigcup M) \leq n$, therefore,

$$\dim\left(\bar{\Lambda} \bigcap M\right) + \dim\left(\bar{\Lambda} \bigcup M\right) = \dim\left(\bar{\Lambda}\right) + \dim\left(M\right).$$

Since, the $\dim(M) = n - 1$, hence, $\dim(\bar{\Lambda}) \leq 1$.

On the other hand, $\Lambda = \{\hat{x}\}$ is impossible as otherwise, $\hat{x} \in S$ and $\hat{x} \in M$, but $S \bigcap M = \emptyset$. Hence, Λ is the convex hull of two extreme points of P, taken from the two sides of the hyperplane M. Now, x will be one of the extremal points and the other extremal point will be the element of $N(x)$. Thus, $\hat{x} \in \text{cone}(N(x))$. Hence, we have shown that

$$\text{cone}(\bar{S}) \subseteq \text{cone}(N(x)).$$

Now, since, the inclusions $\text{cone}(S) \subseteq \text{cone}(\bar{S})$ and $\text{cone}(P) \subseteq \text{cone}(S)$ holds, it follows that $\text{cone}(P) \subseteq \text{cone}(N(x))$. This completes the proof.

Remark 17.3 *The geometrical interpretation of Proposition 17.2 is that if a convex cone with apex at an extremal point of P contains the neighbors of this extremal point, then cone contains the whole polytope P.*

The following proposition states that if an extremal point is optimal over the set of its neighbors then it is optimal over the whole polytope P. Here, we use the fact that every pseudolinear function is quasiconcave and strictly quasiconvex.

Proposition 17.3 *Let x be an extremal point of the polytope P. If $f(x) \leq f(\bar{x})$ holds for every $\bar{x} \in N(x)$, then $f(x) \leq f(\bar{x})$, for every $\bar{x} \in P$.*

Proof Similarly as in Proposition 17.2, we assume that $x = 0$. On the contrary, suppose that

$$f(x) > f(\bar{x}), \text{ for some } \bar{x} \in P.$$

By Proposition 17.2, we have $\bar{x} \in \text{cone}(N(x))$. For some $\varepsilon \in \mathbb{R}$, $0 < \varepsilon < 1$, it follows that

$$\varepsilon\bar{x} \in \text{conv}\left(N(x) \bigcup \{x\}\right).$$

By the strict quasiconvexity and quasiconcavity of the function f, we have

$$f(x) > f(\varepsilon\bar{x}) \geq \min\left\{f(x) : \bar{x} \in N(x) \bigcup \{x\}\right\}.$$

But by our assumption, at the right hand side the minimum is at least $f(x)$, which is a contradiction. This completes the proof.

Now, we present the algorithm and its correctness given by Ujvari [276]. Suppose that the assumptions (i)–(iii) hold.

Algorithm 5 Assume that a feasible basis $B \in \mathbb{R}^{m \times m}$ is given. The existence of the basis is guaranteed by the assumption (i). Let $T(B)$ and x be the corresponding simplex tableau and basic feasible solution. Let λ be the element of $T(B)$ in the lower right corner. Suppose after testing one by one the element of K, we get a position $(i,j) \in K$, such that on pivoting on position (i,j), and denoting by $\bar{\lambda}$ the lower right element of the new simplex table, the following holds:

1. If $\left(\lambda - \bar{\lambda}\right)\left(\alpha - \frac{\kappa}{\lambda\bar{\lambda}}\right) < 0$, and B be the new basis obtained after pivoting on the position (i,j), start again the algorithm.

2. If $\left(\lambda - \bar{\lambda}\right)\left(\alpha - \frac{\kappa}{\lambda\bar{\lambda}}\right) \geq 0$, holds for every $(i,j) \in K$, then the basic feasible solution x is optimal.

Theorem 17.1 *The algorithm finds an optimal solution of the problem (QFP) after a finite number of steps.*

Proof From Propositions 17.1 and 17.3 and the fact that

$$f(x) - f(\bar{x}) = \left(\lambda - \bar{\lambda}\right)\left(\alpha - \frac{\kappa}{\lambda\bar{\lambda}}\right),$$

the correctness of the algorithm follows, where $\bar{x}$ denotes the basic feasible solution corresponding to the feasible basis obtained from B after pivoting on the (i,j)th position. We note that during algorithm the value $f(x)$ decreases, so there can be no repetition in the sequence of feasible bases B. Therefore, the algorithm is finite.

Ujvari [276] has shown that the convex simplex algorithm may be adopted to solve the problem (QFP).

17.3.3 The Convex Simplex Algorithm

If x is a Karush-Kuhn-Tucker point, that is (as $x_B > 0$ and $x_{A\setminus B} = 0$) if

$$\nabla f(x)^T - (\nabla f(x))_B^T B^{-1} A \geq 0 \tag{17.7}$$

holds. By a simple calculation, we can show that (17.7) is equivalent to

$$\left(p^T - p_B^T B^{-1} A\right)\left(\alpha - \frac{\kappa}{\lambda^2}\right) \geq 0. \tag{17.8}$$

If condition (17.8) holds, then x is a Karush-Kuhn-Tucker point. Therefore, due to the pseudolinearity of f, x is an optimal solution.

If condition (17.8) does not hold, then for some index j, we have

$$\left(p^T - p_B^T B^{-1} A\right) e_j \left(\alpha - \frac{\kappa}{\lambda^2}\right) \geq 0.$$

Since, $x_B > 0$, select an index i such that $(i,j) \in K$ and make a pivot on the position (i,j). For the basic feasible solution $\bar{x}$ corresponding to the new basis $\bar{B}$, we have $f(\bar{x}) < f(x)$, which holds. This can be seen as follows: by pseudoconvexity of f we get

$$\langle \nabla f(x), \bar{x} - x \rangle < 0.$$

Since, $\nabla f(\tilde{x}) \neq 0$, for every $\tilde{x} \in P$, therefore for $0 < \varepsilon < 1$, we have

$$\langle \nabla f(x + \varepsilon(\bar{x} - x)), \bar{x} - x \rangle < 0.$$

Hence, the function

$$\varphi : z \to f(x + \varepsilon(\bar{x} - x)), \ 0 \leq \varepsilon \leq 1$$

decreases strictly on the interval $[0, 1]$. Consequently, $f(\bar{x}) < f(x)$. Let $x =: \hat{x}$, $B = \hat{B}$ and repeat the above step until the optimality is reached.

Theorem 17.2 *The convex simplex algorithm adopted to problem (QFP) calculates an optimal solution in a finite number of steps.*

Remark 17.4 *We note that the optimality criteria of algorithm for (QFP) and convex simplex algorithm are equivalent. In fact an extremal point x is a Karush-Kuhn-Tucker point if and only if x is optimal and by Proposition 17.3, x is optimal if and only if x is optimal over its neighbors.*

However, the difference between the two algorithms is that while convex simplex algorithm relies on the nondegeneracy assumption (iii), algorithm V for (QFP) can be generalized easily to the cases, when only the assumption (i) and (ii) hold. In the generalized algorithm, we consider all the feasible bases B, corresponding to the current extremal point x instead of one feasible basis B. These bases can be determined by pivoting on positions (i, j) such that $x_j = 0$ in a finite number of steps. The resulting algorithm finds an optimal solution for the problem (QFP) in a finite number of steps. The proof runs along the lines of Theorem 17.1.

17.4   The Le Chatelier Principle in Pseudolinear Programming

In 1884, the French chemist formulated a very nice principle regarding the interaction of parameters and variables. We can do no better than cite this principle the way it has been stated in the Eichhorn and Oettli [74] paper. We quote (page 711):

If a system is in stable equilibrium and one of the conditions is changed, then the equilibrium will shift in such a way as to tend to annul the applied change in the conditions.

The above statement is not lucid – in fact quite vague. In spite of this vagueness the principle has been used in at least three important areas: economics, physics, and chemistry.

Nobel Laureate Samuelson [243] was the first economist to use the Le Chatelier principle in economics and he also extended in [244]. The simplest example of this principle is provided by the short and long run response of a firm's supply behavior. Consider a parametric change in the price of a commodity. The firm's response in the long run will be greater than in the short run as in the long run it has more flexibility regarding the use of its factors. This response although very intuitive can be proved rigorously by using the Le Chatelier principle. In another important paper, Leblanc and Moeseke [158] applied this principle to the marginal value of a resource. However, in all such applications convexity is invoked.

In this chapter, we present some results of Mishra and Hazari [190], which extends the Le Chatelier principle to the pseudolinear and η-pseudolinear programs. This work extends both Eichhorn and Oettli [74] and Leblanc and Moeseke [158] to a more general class of problems.

17.4.1 Karush-Kuhn-Tucker Optimality Conditions

We now recall the Kuhn-Tucker necessary and sufficient optimality conditions for a maximization problem under pseudolinearity assumption, though for a minimization problem this has been done in the literature.

Consider a maximization problem:

(P)

$$\text{Max} f(x)$$

subject to

$$X \equiv \left\{ x \in X^0 : h(x) \leqq b \right\}, \tag{17.9}$$

where $X^0 \subset \mathbb{R}^n$ is an open convex set; $f : \mathbb{R}^n \to \mathbb{R}$ is a pseudolinear scalar function; $h : \mathbb{R}^n \to \mathbb{R}^m$ is a pseudolinear vector function; $b \in \mathbb{R}^m$.

In economics, the above concepts can be translated into a problem in production theory. Let x be a vector of production levels, $f(x)$ the firm's objective function and b a vector of available resources and the function $h(x)$ the vector of resource use. In this set up the variations in b are analyzed by the Le Chatelier principle.

Theorem 17.3 (Kuhn-Tucker necessary optimality conditions) *If x^* is an optimal solution for (P) then there exists a $v^* \in R^m$, $v_i^* \geqq 0$, such that*

$$\nabla f(x^*) - \sum_{i=1}^{m} v_i^* \nabla h_i(x^*) = 0,$$

$$v_i^* (b_i - h_i(x^*)) = 0, i = 1, \dots, m,$$

$$h_i(x^*) \leqq b_i, \quad i = 1, \dots, m.$$

Theorem 17.4 (Kuhn-Tucker sufficient optimality conditions) *Let x^* be a feasible solution for the maximization problem. Suppose that f is pseudolinear and each h_i for $i = 1, \ldots, m$, is pseudolinear and there exists $v^* \in R^m$, $v_i^* \geq 0$, such that*

$$\nabla f(x^*) - \sum_{i=1}^{m} v_i^* \nabla h_i(x^*) = 0, \qquad (17.10)$$

$$v_i^*(b_i - h_i(x^*)) = 0, i = 1, \ldots, m, \qquad (17.11)$$

$$h_i(x^*) \leq b_i, \quad i = 1, \ldots, m. \qquad (17.12)$$

Then x^* is an optimal solution for (P).

Proof The proof is an easy exercise and left to the reader.

Now define the Lagrangian function for (P) as follows:

$$L(x, v) \equiv f(x) + \sum_{i=1}^{m} v_i [b_i - h_i(x)].$$

Using Theorem 17.3, one can establish that: x^* solves (P)-assuming the Slater regularity condition $h(x^0) < b$, for some $x^0 \in X^0$-if and only if there exists $v^* \in R^m$, $v^* \geq 0$, such that (x^*, v^*) is a saddle point of the Lagrangian

$$L(x, v) \equiv f(x) + \sum_{i=1}^{m} v_i [b_i - h_i(x)].$$

Note that (x^*, v^*) is a saddle point of the Lagrangian $L(x, v) \equiv f(x) + \sum_{i=1}^{m} v_i [b_i - h_i(x)]$, i.e.,

$$f(x) + v^*(b - h(x)) \leq f(x^*) + v^*(b - h(x^*)) \leq f(x^*) + v(b - h(x^*)), \qquad (17.13)$$

for all $x \in X^0$, $v \geq 0$.

Clearly, the second inequality holds if and only if

$$v^*(b - h(x^*)) = 0. \qquad (17.14)$$

17.4.2   The Le Chatelier Principle

For the first principle, we consider a problem $(\bar{P})$, where $\bar{b}$ has been substituted for b, so that

$$(\bar{P}) \quad \max f(x)$$

$$\text{subject to } x \in \bar{X} \equiv \{x \in X^0 : h(x) \leq \bar{b}\}. \qquad (17.15)$$

Let $(\bar{x}, \bar{v})$ be a saddle point of $(\bar{P})$ so that

$$f(x) + \bar{v}(\bar{b} - h(x)) \leq f(\bar{x}) + \bar{v}(\bar{b} - h(\bar{x})) \leq f(\bar{x}) + v(\bar{b} - h(\bar{x})), \qquad (17.16)$$

for all $x \in X^0$, $v \geq 0$.

Again, the second inequality holds if and only if

$$\bar{v}\left(\bar{b} - h\left(\bar{x}\right)\right) = 0, \tag{17.17}$$

denote $\Delta b = \bar{b} - b$, $\Delta v = \bar{v} - v^*$.

Proposition 17.4 *Let* f *and* h *be pseudolinear. If there is a saddle point* $(x^*,\ v^*)$ *for (P) and a saddle point* $(\bar{x},\ \bar{v})$ *for (P̄), then*

$$\Delta v \Delta b \leq 0 . \tag{17.18}$$

Proof The inequalities (17.13) and (17.16) will be verified for $\bar{x} \in X^0$, $\bar{v} \geq 0$ and $x^* \in X^0$, $v^* \geq 0$, respectively. Adding the corresponding inequalities, and canceling, we get (17.18).

Proposition 17.5 *Let* f *and* h *be* η*-pseudolinear. If there is a saddle point* $(x^*,\ v^*)$ *for (P) and a saddle point* $(\bar{x},\ \bar{v})$ *for (P̄), then*

$$\Delta v \Delta b \leq 0 .$$

Bibliography

[1] Aggarwal, S., Bhatia, D.: Pseudolinearity and efficiency via Dini derivatives. *Ind. J. Pure Appl. Math.* 20, 1173-1183 (1989)

[2] Ansari, Q.H., Lalitha, C.S., Mehta, M.: *Generalized Concavity, Nonsmooth Variational Inequalities and Nonsmooth Optimization.* CRC Press, Taylor & Francis Group, NY (2014)

[3] Ansari, Q.H., Rezaei, M.: Existence result for Stampacchia and Mity type vector variational inequalities. *Optimization* 59, 1053-1065 (2010)

[4] Ansari, Q.H., Rezaei, M.: Generalized pseudolinearity, *Optim. Lett.* 6, 241-251 (2012)

[5] Ansari, Q.H., Schaible, S., Yao, J.C.: η-Pseudolinearity. *Riv. Mat. Sci. Econ. Soc.* 22, 31-39 (1999)

[6] Antczak, T.: Nonsmooth minimax programming under locally Lipschitz (ϕ, ρ)-invexity. *Appl. Math. Comput.* 217, 9606-9624 (2011)

[7] Arrow, K.J., Hurwicz, L., Uzawa, L.: *Studies in Linear and Nonlinear Programming.* Stanford University Press, Stanford, CA (1958)

[8] Aussel, D.: Subdifferential properties of quasiconvex and pseudoconvex functions. *J. Optim. Theory Appl.* 97, 229-245 (1998)

[9] Avriel, M., Diewert, W. E., Schaible, S., Zang, I.: *Generalized Concavity.* Plenum Publishing Corporation, NY (1988)

[10] Avriel, M., Schaible. M.: Second order characterizations of pseudoconvex functions. *Math. Program.* 14, 170-185 (1978)

[11] Baer, R.: *Linear Algebra and Projective Geometry.* Academic Press, NY (1952)

[12] Bajona-Xandri, C., Martinez Legaz, J.E.: Lower subdifferentiability in minimax fractional programming. *Optimization* 45, 1-12 (1998)

[13] Barani, A., Pouryayevali, M.R.: Invex sets and preinvex functions on Riemannian manifolds. *J. Math. Anal. Appl.* 328, 767-779 (2007)

[14] Barani, A., Pouryayevali, M.R.: Vector optimization problems under *d*-invexity on Riemannian manifolds. *Diff. Geom. Dyn. Syst.* 13, 34-44 (2011)

[15] Barrodale, I.: Best rational approximation and strict-quasiconvexity, *SIAM J. Numer. Anal.* 10, 8-12 (1973)

[16] Barros, A.I.: *Discrete Fractional Programming Techniques for Location Models.* Kluwer Academic Publishers, Dordrecht (1998)

[17] Bazaraa, M.S., Sherali, H.D., Shetty, C.M.: *Nonlinear Programming: Theory and Algorithms.* John Wiley & Sons, NJ (2006)

[18] Bector, C.R.: Duality in nonlinear fractional programming. *Z. Fur Oper. Res.* 17, 183-193 (1973)

[19] Bector, C.R., Bhatia, B.L.: Sufficient optimality and duality for a minimax problem. *Util. Math.* 27, 229-247 (1985)

[20] Bector, C.R., Chandra, S., Durga Prasad, M.V.: Duality in pseudolinear multiobjective programming. *Asia Pac. J. Oper. Res.* 5, 150-159 (1988)

[21] Ben-Israel, A., Mond, B.: What is invexity? *J. Austral. Math. Soc.* Ser. B, 28, 1-9 (1986)

[22] Berge, C.: *Topological Spaces, Including a Treatment of Multi-Valued Functions, Vector Spaces and Convexity.* Oliver & Boyd, Edinburgh and London (1963)

[23] Bertsekas, D.P., Nedic, A., Ozdagler, A.E.: *Convex Analysis and Nonlinear Optimization.* Athena Scientific, Belmont, MA (2003)

[24] Bhatia, D., Jain, P.: Generalized (F, ρ)-convexity and duality for nonsmooth multiobjective programs. *Optimization* 31, 153-164 (1994)

[25] Bhatia, D., Jain, P.: Non-differentiable pseudo-convex functions and duality for minimax programming problems. *Optimization* 35, 207-214 (1995)

[26] Bianchi, M., Hadjisavvas, N., Schaible, S.: On Pseudomonotone Maps T for which $-T$ is also Pseudomonotone. *J. Convex Anal.* 10, 149-168 (2003)

[27] Bianchi, M., Schaible, S.: An extension of pseudolinear functions and variational inequality problems. *J. Optim. Theory Appl.* 104, 59-71 (2000)

[28] Borwein, J.M.: Direct theorem in semi-infinite convex programming. *Math. Program.* 21, 301-318 (1981)

[29] Boyd, S., Vandenberghe, L.: *Convex Optimization*, Cambridge University Press (2004)

[30] Buckley, J.J.: The fuzzy mathematics of finance. *Fuzzy Sets Syst.* 21, 57-63 (1987)

[31] Buckley, J.J., Feuring, T.: Fuzzy differential equations. *Fuzzy Sets Syst.* 110, 43-54 (2000)

[32] Burke, J.V., Ferris, M.C.: Characterization of solution sets of convex programs. *Oper. Res. Lett.* 10, 57-60 (1991)

[33] Cambini, R., Carosi, L.: On generalized convexity of quadratic fractional functions, Proceedings of the IV International Conference in Stochastic Geometry, Convex Bodies and Empirical Measures, Tropea (Italy), September 24-28, 2001, Supp. Rendiconti del Circolo Matematico di Palermo, 70, Series II, 155-176, (2002)

[34] Cambini, R., Carosi, L.: On generalized linearity of quadratic fractional functions. *J. Glob. Optim.* 30, 235-251, (2004)

[35] Cambini, A., Castagnoli, E., Martein, L., Mazzoleni, P. and Schaible, S. (eds.): Generalized convexity and fractional programming with economic applications. Lecture Notes in Economics and Mathematical Systems. Springer-Verlag, Berlin, NY (1990)

[36] Cambini, A., Martein, L.: *Generalized Convexity and Optimization*, Springer-Verlag, Berlin, Heidelberg (2009)

[37] Cambini, A., Martein, L.: A modified version of Martos algorithm, *Methods Oper. Res.* 53, 33-44 (1986)

[38] Cambini, R.: A class of non-linear programs: theoretical and algorithmical results. In: S. Komlosi, T. Rapcsak and S. Schaible eds., Generalized Convexity, Lecture Notes in Economics and Mathematical Systems. Springer, Berlin, 405, 294-310 (1994)

[39] Carosi, L., Martein, L.: On the pseudoconvexity and pseudolinearity of some classes of fractional functions. *Optimization*, 56(3), 385-398 (2007)

[40] Chandra, S., Craven, B.D., Mond, B.: Vector-valued Lagrangian and multiobjective fractional programming duality. *Numer. Funct. Anal. Optim.* 11, 239-254 (1990)

[41] Chandra, S., Kumar, V.: Duality in fractional minimax programming. *J. Austral. Math. Soc.* (Series A), 58, 376-386 (1995)

[42] Chang, S.S.L., Zadeh, L.A.: On fuzzy mappings and control. *IEEE Trans. Syst. Man Cybern.* 2(1), 30-34 (1972)

[43] Chankong, V., Haimes, Y.Y.: *Multiobjective Decision Making: Theory and Methodology.* North-Holland, NY (1983)

[44] Chen, W., Sahai, A., Messac, A., Sundararaj, G.: Exploration of the effectiveness of the physical programming in robust design, *J. Mech. Des.* 122, 155-163 (2000)

[45] Chen, G.-Y., Cheng, G.M.: Vector variational inequalities and vector optimization. In: *Lecture Notes in Economics and Mathematical Systems.* 285, pp. 408-416. Springer, Berlin (1987)

[46] Chew, K.L.: Pseudolinear minimax programming. *Pac. J. Oper. Res.* 1, 53-64 (1984)

[47] Chew, K.L., Choo, E.U.: Pseudolinearity and efficiency, *Math. Program.* 28, 226-239 (1984)

[48] Choo, E.U.: Proper efficiency and the linear fractional fractional vector maximization. *J. Math. Anal. Appl.* 22, 618-630 (1968)

[49] Clarke, F.H. *Optimization and Nonsmooth Analysis.* Wiley-Interscience, NY (1983)

[50] Cottle, R.W.: Manifestations of the Schur complement. *Linear Algebra Appl.* 8, 189-211 (1974)

[51] Craven, B.D.: Nonsmooth multiobjective programming. *Numer. Funct. Anal. Optim.* 10, 49-64 (1989)

[52] Craven, B. D.: On duality with generalized convexity. In: *Generalized Concavity in Optimization and Economics*, S. Schaible, W.T. Ziemba (eds.), Academic Press, NY, pp. 291-305 (1981a)

[53] Craven, B.D.: Invex functions and constrained local minima. *Bull. Austral. Math. Soc.* 24, 357-366 (1981b)

[54] Craven, B,D.: Duality for generalized convex fractional programs. In: *Generalized Concavity in Optimization and Economics.* S., Schaible, W.T., Ziemba, Academic Press, NY, 473-489 (1981c)

[55] Craven, B.D., Glover, B.M.: Invex functions and duality. *J. Austral. Math. Soc.*, Ser. A, 39, 1-20 (1985)

[56] Crouzeix, J.P.: Characterizations of generalized convexity and monotonicity, a survey. In: *Generalized Convexity, Generalized Monotonicity.* J.P. Crouzeix, J.E. Martinez-Legaz and M. Volle (eds.), Kluwer Academic Publisher, Dordrecht, pp. 237-256 (1998)

[57] Crouzeix, J.P., Ferland, J.A.: Criteria for differentiable generalized monotone maps. *Math. Program.* 75, 399-406 (1996)

[58] Dafermos, S.: Exchange price equilibrium and variational inequalities. *Math. Program.* 46, 391-402 (1990)

[59] Da Cunha, N.O., Polak, E.: Constrained minimization under vector-valued criteria in finite dimensional spaces. *J. Math. Anal. Appl.* 19 (1), 103-124 (1967)

[60] Daniilidis, A., Hadjisavvas, N.: On the subdifferential of quasiconvex and pseudoconvex functions and cyclic monotonicity. *J. Math. Anal. Appl.* 237, 30-42 (1999)

[61] Dantzig, G.: *Linear Programming and Extensions*, Princeton University Press, Princeton, NJ (1963)

[62] De Finetti, B.: Sulle stratification converse. *Ann. Mat. Pura Appl.* 30, 173-183 (1949)

[63] Dhara, A., Dutta, J.: *Optimality Conditions in Convex Optimization-A Finite Dimensional View*. CRC Press, Taylor & Francis Group, NY (2012)

[64] Diamond, P., Kloeden, P.: *Metric Spaces of Fuzzy Sets: Theory and Applications.* World Scientific, Singapore (1994)

[65] Diewert W. E.: Generalized concavity and economics. In: Generalized Concavity in Optimization and Economics, S., Schaible, W.T., Ziemba (eds.), Academic Press, NY (1981)

[66] Diewert, W.E., Avriel, M., Zang, I.: Nine kinds of quasiconcavity and concavity. *J. Econom. Theory* 25, 397-420 (1981)

[67] Dinh, N., Jeyakumar, V., Lee, G.M.: Lagrange multiplier characterizations of solution sets of constrained pseudolinear optimization problems. *Optimization* 55, 241-250 (2006)

[68] Do Carmo, M.P.: *Riemannian Geometry*, Birkhauser, Boston (1992)

[69] Dolezal, J.: Necessary conditions for Pareto optimality. *Prob. Control Inform. Theory* 14(2), 131-140 (1985)

[70] Dugundji, J., Granas, A.: *Fixed Point Theory*, Vol. 1, Polish Scientific Publishers, Warsaw (1982)

[71] Edgeworth, F.Y.: *Mathematical Physics.* C. Kegan Paul, London (1881)

[72] Egudo, R.R.: Multiobjective fractional duality. *Bull. Austral. Math. Soc.* 37, 367-388 (1988)

[73] Egudo, R.R., Mond, B.: Duality with generalized convexity. *J. Austral. Math. Soc.*, Ser. B, 28, 10-21 (1986)

[74] Eichhorn, W., Oettli, W.: A General Formulation of the Le Chatelier-Samuelson Principle. *Econometrica* 40, 711-717 (1972)

[75] Elster, K.H., Nehse, R: *Optimality Conditions for Some Nonconvex Problems*. Springer-Verlag, NY (1980)

[76] Evans, L.C., Gariepy, R.F.: *Measure Theory and Fine Properties of Functions*. CRC Press, Boca Raton, FL (1992)

[77] Ewing, G. M.: Sufficient conditions for global minima of suitable convex functionals from variational and control theory. *SIAM Rev.* 19, 202-220 (1977)

[78] Ellero, A.: The optimal level solutions method, *J. Inform. Optim. Sci.* 17, 355-372 (1996)

[79] Fedrizzi, M., Fedrizzi, M., Ostasiewicz, W.: Towards fuzzy modelling in economics. *Fuzzy Sets Syst.* 54, 259-268 (1993)

[80] Fenchel, W.: On conjugate convex functions. *Canad. J. Math.* 1, 73-77 (1949)

[81] Fenchel, W.: Convex Cones, Sets and Functions. Mimeographed Lecture Notes. Princeton University, Princeton, NJ (1953)

[82] Fichera, G.: Sul problema elastostatico di Signorini con ambigue condizioni al contorno (On the elastostatic problem of Signorini with ambiguous boundary conditions), Rendiconti della Accademia Nazionale dei Lincei Classe di Scienze Fisiche. *Mat. Natur. Series VIII*, (in Italian) 34 (2), 138-142 (1963)

[83] Flaudas, C.A., Pardalos, P.M. (eds.): *Encyclopedia of Optimization*, 2nd edn. Springer, NY (2009)

[84] Freund, R.W., Jarre, F., Schaible, S.: On self-concordant barrier functions for conic hulls and fractional programming. *Math. Program.* Ser. A, 74, 237-246 (1996)

[85] Fritz, J.: Extremum problems with inequalities as subsidiary conditions. In: *Fritz John Collected Papers*, J. Moser (ed.), Berkhauser Verlag, pp. 543-560 (1985), First published in (1948)

[86] Geoffrion, A.M.: Proper efficiency and the theory of vector maximization. *J. Math. Anal. Appl.* 22, 618-630 (1968)

[87] Gill, P.E., Murray, W., Wright, M.H.: *Practical Optimization*. Academic Press, NY, London, Toronto, Sydney, San Francisco (1981)

[88] Giorgi, G., Komlosi, S.: Dini derivatives in optimization—Part I. *Riv. Mat. Sci. Econ. Soc.* 15 (1), 3-30 (1993a)

[89] Giorgi, G., Komlosi, S.: Dini derivatives in optimization—Part II. *Riv. Mat. Sci. Econ. Soc.* 15, 3-24 (1993b)

[90] Giorgi, G., Komlosi, S.: Dini derivatives in optimization—Part III. *Riv. Mat. Sci. Econ. Soc.* 18, 47-63 (1995)

[91] Giorgi, G., Rueda, N.G.: η-Pseudolinearity and efficiency. *Int. J. Optim. Theory Methods Appl.* 1, 155-159 (2009)

[92] Goberna, M.A., Lopez, M.A., Pastor, J.: Farkas-Minkowsky system in semi-infinite programming. *Appl. Math. Optim.* 7, 295-308 (1981)

[93] Goestschel, R., Voxman, W.: Elementary fuzzy calculus. *Fuzzy. Sets. Syst.* 18, 31-43 (1986)

[94] Giannessi, F.: Theorems of the alternative, quadratic programming and complementarity problems. In: *Variational Inequalities and Complementarity Problems*, R.W. Cottle (ed.), John Wiley and Sons, NY, pp. 151-186 (1980)

[95] Giannessi, F.: On Minty variational principle. In: *New Trends in Mathematical Programming*, F. Giannessi, S. Komlosi, T. Rapcsak (eds.), Kluwer Academic Publishers, Dordrecht, Netherlands, pp. 93-99 (1997)

[96] Giannessi, F., Maugeri, A., Pardalos, P.M.: *Equilibrium Problems: Nonsmooth Optimization and Variational Inequality Models*. Kluwer Academic Publishers, Dordrecht, Holland (2001)

[97] Hachimi, M., Aghezzaf, B.: Sufficiency and duality in differentiable multiobjective programming involving generalized type I functions. *J. Math. Anal. Appl.* 296, 382-392 (2004)

[98] Hadjisavvas, N.: The use of subdifferentials for studying generalized convex functions. *J. Stat. Manag. Syst.* 5, 125-139 (2002)

[99] Hanson, M.A.: On sufficiency of Kuhn-Tucker conditions. *J. Math. Anal. Appl.* 30, 545-550 (1981)

[100] Hartman, P., Stampacchia, G.: On some nonlinear elliptic differential functional equations. *Acta Math.* 115, 153-188 (1966)

[101] Haynsworth, E.V.: Determination of the inertia of a partitioned hermitian matrix. *Linear Algebra Appl.* 1, 73-81 (1968)

[102] Hettich, R., Kortanek, K.O.: Semi-infinite programming: Theory, methods and applications. *SIAM Rev.* 35, 380-429 (1993)

[103] Hettich, R., Still, G.: Semi-infinite programming models in robotics. In: *Parametric Optimization and Related Topics II*. Guddat et al. (eds.), Academic Verlag, Berlin, pp. 112-118 (1991)

[104] Hicks, N.J.: Notes on Differential Geometry. D. van Nostrand Company, Princeton, NJ (1965)

[105] Hiriart-Urruty, J.B., Lemarechal, C.: *Convex Analysis and Minimization Algorithms I and II : Fundamental Principles of Mathematical Sciences.* Springer-Verlag, Berlin (1993)

[106] Hiriart-Urruty, J.B., Strodiot, J.J., Nguyen, V.H.: Generalized Hessian matrix and second order optimality conditions for problems with data. *Applied Math. Optim.* 11, 43-56 (1984)

[107] Ho, S.C., Lai, H.C.: Nonsmooth minimax fractional programming problem with exponential (p, r)-invex functions. *J. Nonlinear Convex Anal.* 13(3), 433-447 (2012)

[108] Isermann, H.: On some relation between dual pair of multiple objective linear programs. *Zeit. fur Oper. Res.* 22, 33-41 (1978)

[109] Ishizuka, Y., Shimijhu, K.: Necessary and sufficient conditions for the efficient solutions of nondifferentiable mutiobjective problems. *IEEE Trans. Syst. Man Cyber.* 14, 625-629 (1984)

[110] Ivanov, E.H., Nehse, R.: Some results on dual vector optimization problems. *Optimization* 16, 505-517 (1985)

[111] Jaiswal, M., Mishra, S.K., Al Shamary, B.: Optimality conditions and duality for semi-infinite programming involving semilocally Type I-preinvex and related functions. *Commun. Korean Math. Soc.* 27, 411-423 (2012)

[112] Jensen, J.L.W.V.: Sur les fonctions convexes et les inegalities entre les valeurs moyennes. *Acta Math.* 30, 175-193 (1906)

[113] Jeroslow, R.G.: Uniform duality in semi-infinite convex optimization. *Math. Program.* 27, 144-154 (1983)

[114] Jeyakumar, V. Infinite dimensional convex programming with applications to constrained approximation. *J. Optim. Theory Appl.* 75 (3), 469-486 (1992)

[115] Jeyakumar, V., Lee, G.M., Dinh, N.: Lagrange multiplier conditions characterizing optimal solution sets of cone-constrained convex programs. *J. Optim. Theory Appl.* 123, 83-103 (2004)

[116] Jeyakumar, V., Lee, G.M., Dinh, N.: Characterization of solution sets of convex vector minimization problems. *Eur. J. Oper. Res.* 174, 1380-1395 (2006)

[117] Jeyakumar, V., Wolkowicz, H.: Generalizations of Slater's constraint qualification for infinite convex programs. *Math. Program.* 57, 551-571 (1992)

[118] Jeyakumar, V., Yang, X.Q.: Convex composite multiobjective non-smooth programming. *Math. Program,* 59, 325-343 (1993)

[119] Jeyakumar, V., Yang, X.Q.: On characterizing the solution sets of pseudolinear programs. *J. Optim. Theory Appl.* 87, 747-755 (1995)

[120] Jimenez, F., Verdegay, J.L.: Solving fuzzy solid transportation problems by an evolutionary algorithm based parametric approach. *Eur. J. Oper. Res.* 117, 485-510 (1999)

[121] John, F.: Extremum problems with inequalities as subsidiary conditions. In: *Studies and Essays Presented to R. Courant on His 60th Birthday,* Interscience Publishers, Inc., NY, pp. 187-204 (1948)

[122] Jahn, J.: *Introduction to the Theory of Nonlinear Optimization.* Springer-Verlag, Berlin, Heidelberg (2007)

[123] John, R.: A First-Order Characterization of Generalized Monotonicity, Discussion Paper A-490, University of Bonn, Bonn, Germany (1995)

[124] Kanniappan, P.: Necessary conditions for optimality of nondifferentiable convex multiobjective programming. *J. Optim. Theory Appl.* 40 (2), 167-174 (1983)

[125] Kanzi, N.: Necessary optimality conditions for nonsmooth semi-infinite programming problems. *J. Global Optim.* 49, 713-725 (2011)

[126] Kanzi, N.: Constraint qualifications for semi-infinite systems and their applications in nonsmooth semi-infinite problems with mixed constraints. *SIAM J. Optim.* 24, 559-572 (2014)

[127] Kanzi, N., Nobakhtian, S.: Nonsmooth semi-infinite programming with mixed constraints. *J. Math. Anal. Appl.* 351, 170-181 (2009)

[128] Kanzi, N., Nobakhtian, S.: Optimality condition for non-smooth semi-infinite programming problems. *Optimization* 59, 717-727 (2010)

[129] Kanzi, N., Nobakhtian, S.: Optimality conditions for nonsmooth semi-infinite multiobjective programming. *Optim. Lett.* DOI 10.1007/s11590-013-0683-9 (2013)

[130] Kaplan, A., Tichatschke, R.: On the numerical treatment of a class of terminal problems. *Optimization* 41, 1-36 (1997)

[131] Karamardian, S.: Generalized complementarity problems. *J. Optim. Theory Appl.* 8, 161-168 (1971)

[132] Karamardian, S.: Complementarity over cones with monotone and pseudomonotone maps. *J. Optim. Theory Appl.* 18, 445-454 (1976)

[133] Karamardian, S., Schaible, S., Crouzeix, J.P.: Characterizations of generalized monotone maps. *J. Optim. Theory Appl.* 76, 399-413 (1993)

[134] Karney, D.F., Morley, T.D.: Limiting Lagrangian: A primal approach. *J. Optim. Theory Appl.* 48, 163-174 (1986)

[135] Karush, W.: *Minima of Functions of Several Variables with Inequalities as Side Conditions*. M.S. Thesis, Department of Mathematics, University of Chicago (1939)

[136] Kaul, R.N., Kaur, S.: Generalization of convex and related functions. *Eur. J. Oper. Res.* 9, 369-377 (1982a)

[137] Kaul, R.N., Kaur, S.: Optimality criteria in nonlinear programming involving nonconvex functions. *J. Math. Anal. Appl.* 105, 104-112 (1985)

[138] Kaul, R.N., Lyall, V., Kaur, S.: Semilocal pseudolinearity and efficiency. *Eur. J. Oper. Res.* 36, 402-409 (1988)

[139] Kaul, R.N., Suneja, S.K., Lalitha, C.S.: Duality in pseudolinear multiobjective fractional programming. *Indian J. Pure Appl. Math.* 24, 279-290 (1993)

[140] Kim, D.S.: Nonsmooth multiobjective fractional programming with generalized invexity. *Taiwanese J. Math.* 10, 467-478 (2006)

[141] Kim, D.S., Bae, K.D.: Optimality conditions and duality for a class of nondifferentiable multiobjective programming problems. *Taiwanese J. Math.* 13 (2B), 789-804 (2009)

[142] Kinderlehrer, D., Stampacchia, G.: *An Introduction to Variational Inequality and Their Applications*. Academic Press, London, (1980)

[143] Kiwiel, K.C.: *Methods of Descent for Nondifferentiable Optimization Optimization*. Lecture Notes in Mathematics 11133, Springer Verlag, Berlin, Heidelberg (1985)

[144] Komlosi, S.: Second order characterization of pseudoconvex and strictly pseudoconvex functions in terms of quasi-Hessian. In: *Contribution to the Theory of Optimization*, F. Forgo (ed.), University of Economics, Budapest, pp. 19-46 (1983)

[145] Komlosi, S.: Second order characterization of generalized convexity and local optimality in nonlinear programming: The quasi Hessian approach, Studia Oeconomica Auctoritate Universitetis Pecs Publicata, pp. 20-40 (1985)

[146] Komlosi, S.: Generalized monotonicity of generalized derivatives. In: *Proceedings of the Workshop on Generalized Concavity for Economic Applications*, P. Mazzoleni, (ed.), Technoprint Bologna, Pisa, Italy, pp. 1-7 (1992)

[147] Komlosi, S.: First and second order characterizations of pseudolinear functions. *Eur. J. Oper. Res.* 67, 278-286 (1993)

[148] Komlosi, S.: On pseudolinear fractional functions - the implicit function approach. *Math. Pan.* 20, 257-274 (2009)

[149] Kortanek, K.O., Evans, J.P.: Pseudoconcave programming and Lagrange regularity. *Oper. Res.* 15, 882-892 (1967)

[150] Kruk, S., Wolkowicz, H.: Pseudolinear programming. *SIAM Rev.* 41, 795-805 (1999)

[151] Kuhn, H.W., Tucker, A.W.: Nonlinear programming. In: *Proceedings of the Second Berkeley Symposium on Mathematical Statistics and Probability*, University of California Press, Berkeley and Los Angeles, pp. 481-492 (1951)

[152] Kuk, H.: Duality for nonsmooth multiobjective fractional programming with (V, ρ)-invexity. *JKSIAM.* 4, 1-10 (2000)

[153] Kuk, H., Lee, G.M., Kim, D.S.: Nonsmooth multiobjective programs with V-invexity. *Indian J. Pure Appl. Math.* 29, 405-412 (1998)

[154] Kuk, H., Tanino, T.: On nonsmooth minimax programming problems containing (V, ρ)-invexity. *J. Inform. Optim. Sci.* 21, 437-444 (2000)

[155] Lalitha, C.S., Mehta, M.: A note on pseudolinearity in terms of bifunctions. *Asia Pac. J. Oper. Res.* 24, 1-9 (2007a)

[156] Lalitha, C.S., Mehta, M.: Characterizations of solution sets of mathematical programs in terms of Lagrange multipliers. *Optimization* 58, 995-1007 (2009)

[157] Lalitha, C.S., Mehta, M.: Characterizations of solution sets of pseudolinear program and pseudoaffine variational inequality problems. *J. Nonlinear Convex Anal.* 8 (1), 87-98 (2007)

[158] Leblanc, G., Moeseke, P.V.: The Le Chatelier principle in convex programming. *Rev. Economic Studies* 43, 143-147 (1976)

[159] Lee, G.M.: On relations between vector variational inequality and vector optimization problem. In: *Progress in Optimization, II: Contributions from Australasia*, X.Q. Yang, A.I. Mees, M.E. Fisher, L.S. Jennings (eds.) Kluwer Academic, Dordrecht, pp. 167-179 (2000)

[160] Lee, G.M., Kim, D.S., Lee, B.S., Yen, N.D.: Vector variational inequality as a tool for studying vector optimization problems. *Nonlinear Anal.* 34, 745-765 (1998)

[161] Lee, G.M., Lee, K.B.: Vector variational inequalities and nondifferentiable convex vector optimization problems. *J. Glob. Optim.* 32, 597-612 (2005)

[162] Liang, Z.A., Huang, H.X., Pardalos, P.M.: Optimality conditions and duality for a class of nonlinear fractional programming problems. *J. Optim. Theory Appl.* 110, 611-619 (2001)

[163] Liang, Z.A., Huang, H.X., Pardalos, P.M.: Efficiency conditions and duality for a class of multiobjective fractional programming problems. *J. Global Optim.* 27, 447-471 (2003)

[164] Liang, Z.A., Shi, Z.W.: Optimality conditions and duality for a minimax fractional programming problem with generalized convexity. *J. Math. Anal. Appl.* 277, 474-488 (2003)

[165] Liu, J.C.: Optimality and duality for multiobjective fractional programming involving nonsmooth (F, ρ)-convex functions. *Optimization* 36, 333-346 (1996)

[166] Liu, J.C., Wu, C.S., Shew, R.L.: Duality for fractional minmax programming. *Optimization* 41, 117-123 (1997)

[167] Liu, J.C., Wu, C.S.: On minimax fractional optimality conditions with invexity. *J. Math. Anal. Appl.* 219, 21-35 (1998a)

[168] Liu, J.C., Wu, C.S.: On minimax fractional optimality conditions with (F, ρ)-convexity. *J. Math. Anal. Appl.* 219, 36-51 (1998b)

[169] Lopez, M., Still, G.: Semi-infinite programming. *Eur. J. Oper. Res.* 180, 491-518 (2007)

[170] Lopez, M., Vercher, E.: Optimality conditions for nondifferentiable convex semi-infinite programming. *Math. Prog.* 27, 307-319 (1983)

[171] Lu, Q.H., Zhu, D.L.: Some characterizations of locally Lipschitz pseudolinear functions. *Math. Appl.* 18, 272-278 (2005)

[172] Luc, D.T.: On generalized convex nonsmooth functions. *Bull. Austral. Math. Soc.* 49, 139-149 (1994)

[173] Luenberger, D.G.: Linear and nonlinear programming. Addison-Wesley, Reading, MA (1984)

[174] Maciel, L., Gomide, F., Ballini, R.: MIMO evolving functional fuzzy models for interest rate forecasting. In: *2012 IEEE Conference on Computational Intelligence for Financial Engineering and Economics (CIFER)*, pp. 120-127 (2012)

[175] Mangasarian, O.L.: Pseudoconvex functions. *SIAM J. Optim. Control.* 3, 281-290 (1965)

[176] Mangasarian, O.L.: Nonlinear Programming. McGraw-Hill, NY (1969)

[177] Mangasarian, O.L.: A simple characterization of solution sets of convex programs. *Oper. Res. Lett.* 7, 21-26 (1988)

[178] Martinez-Legaz, J.E.: Quasiconvex duality theory by generalized conjugation methods. *Optimization* 19, 603-652 (1988)

[179] Martos, B.: *Nonlinear Programming Theory and Methods*. North Holland, Amsterdam (1975)

[180] Mathis, F.H., Mathis, L.J.: A nonlinear programming algorithm for hospital management. *SIAM Rev.* 37, 230-234 (1995)

[181] Mc Shane, E.J.: *Integration*. Princeton University Press, Princeton, NJ (1944)

[182] Mehra, A., Bhatia, D.: Optimality and duality for minmax problems involving arcwise connected and generalized arcwise connected functions. *J. Math. Anal. Appl.* 231, 425-445 (1999)

[183] Meister, B., Oettli, W.: On the capacity of a discrete constant channel. *Inform. Control.* 11, 341-351 (1998)

[184] Miettinen, K.: *Nonlinear Multiobjective Optimization*. Kluwer Academic Publishers (1998)

[185] Miglierina, E.: Invex functions on differentiable manifolds. In: *Generalized Convexity and Optimization for Economics and Financial Decisions,* G. Giorgi, F. Rossi (eds.), Pitagora, Bologna, pp. 299-311 (1999)

[186] Minty, G.J.: On the generalization of a direct method of the calculus of variations. *Bull. Amer. Math. Soc.* 73, 314-321 (1967)

[187] Mishra, S.K.: Pseudolinear minimax fractional programming. *Indian J. Pure Appl. Math.* 26, 763-772 (1995)

[188] Mishra, S.K., Giorgi, G.: Vector optimization involving type I and related functions on differentiable manifolds. *Anal. Univ. Bucu. Mat.* Anul LVI, 39-52 (2007)

[189] Mishra, S.K., Giorgi, G.: *Invexity and Optimization*. Springer-Verlag, Berlin, Heidelberg (2008)

[190] Mishra, S.K., Hazari, B.R.: The Le Chatelier principle in invex programming. In: *Econometric and Forecasting Models*, C. Putcha, B. Sloboda, K. Colulibaly (eds.), Lewiston, NY, pp. 125-138 (2013)

[191] Mishra, S.K., Jaiswal, M., Le Thi, H.A.: Nonsmooth semi-infinite programming problem using limiting subdifferentials. *J. Glob. Optim.* 53, 285-296 (2012)

[192] Mishra, S.K., Jaiswal, M.: Optimality conditions and duality for nondifferentiable multiobjective semi-infinite programming. *Vietnam J. Math.* 40, 331-343 (2012)

[193] Mishra, S.K. Jaiswal, M., Hoai An, L.T.: Duality for nonsmooth semi-infinite programming problems. *Optim Lett.* 6, 261-271 (2012)

[194] Mishra, S.K., Laha, V.: On approximately star-shaped functions and approximate vector variational inequalities. *J. Optim. Theory Appl.* 156 (2), 278-293 (2013)

[195] Mishra, S.K., Mukherjee, R.N.: Generalized convex composite multiobjective non-smooth programming and conditional proper efficiency. *Optimization* 34, 53-66 (1995)

[196] Mishra, S.K., Shukla, K.: Nonsmooth minimax programming problems with $V - r$-invex functions. *Optimization* 59 (1), 95-103 (2010)

[197] Mishra, S.K., Upadhyay, B.B.: Duality in nonsmooth multiobjective programming involving η-pseudolinear functions. *Ind. J. Ind. Appl. Math.* 3, 152-161 (2012)

[198] Mishra, S.K., Upadhyay, B.B.: Efficiency and duality in nonsmooth multiobjective fractional programming involving η-pseudolinear functions. *Yugoslav J. Oper. Res.* 22 (1), 3-18 (2012)

[199] Mishra, S.K., Upadhyay, B.B.: Some relations between vector variational inequality problems and nonsmooth vector optimization problems using quasi efficiency. *Positivity* 17, 1071-1083 (2013)

[200] Mishra, S.K., Upadhyay, B.B., Hoai An, L.T.: Lagrange multiplier characterization of solution sets of constrained nonsmooth pseudolinear optimization problems. *J. Optim. Theory Appl.* 160, 763-777 (2014)

[201] Mishra, S.K. Upadhyay, B.B.: Nonsmooth minmax fractional programming involving η-pseudolinear functions. *Optimization* 63 (5), 775-788 (2014)

[202] Mishra, S.K., Wang, S.Y., Lai, K.K.: *V-Invex Functions and Vector Optimization.* Springer Science + Business Media, LLC, New York (2008)

[203] Mishra, S.K., Wang, S.Y., Lai, K.K.: *Generalized Convexity and Vector Optimization.* Springer-Verlag, Berlin, Heidelberg (2009)

[204] Mishra, S.K., Wang, S.Y., Lai, K.K.: Pseudolinear fuzzy mappings. *Eur. J. Oper. Res.* 182 (2), 965-970 (2007)

[205] Mohan, S.R., Neogy, S.K.: On invex sets and preinvex functions. *J. Math. Anal. Appl.* 189, 901-908 (1994)

[206] Mond, B.: Mond-Weir duality. In: *Optimization-Structure and Applications,* Pearce, C., Hunt, E. (eds.) Springer Optimization and Its Applications, pp. 157-165 (2009)

[207] Mond, B., Weir, T.: Generalized concavity and duality. In: *Generalized Concavity in Optimization and Economics,* Schaible, S., Ziemba, W.T. (eds.), Academic Press, NY, pp. 263-280 (1981)

[208] Moreno, J.A., Moreno, J.M, Verdegay, J.L.: Fuzzy location problems on networks. *Fuzzy Sets Syst.* 142, 393-405 (2004)

[209] Mouton, A.M., Jowett, I., Goethals, P., De Baets, B.: Prevalence-adjusted optimization of fuzzy habitat suitability models for aquatic invertebrate and fish species in New Zealand. *Ecol Inform.* 4, 215-255 (2012)

[210] Mukherjee, R.N., Rao, C.P.: Multiobjective fractional programming under generalized invexity. *Indian J. Pure Appl. Math.* 27, 1175-1183 (1996)

[211] Murty, K.G.: *Linear and Combinatorial Programming.* John Wiley & Sons, NY (1976).

[212] Nanda, S., Kar, K.: Convex fuzzy mappings. *Fuzzy Sets Syst.* 48, 129-132 (1992)

[213] Nagurney, A.: *Network Economics: A Variational Inequality Approach.* Kluwer Academic Publishers (1993)

[214] Nemirovski, A.S.: The large step method of analytic centres for fractional problems. *Math. Program.* Ser. B, 77, 191-224 (1997)

[215] Nesterov, Y.E., Nemirovski, A.S.: *Interior Point Polynomial Algorithms in Convex Programming.* SIAM, Philadelphia (1994)

[216] Nobakhtian, S.: Optimality and duality for nonsmooth multiobjective fractional programming with mixed constraints. *J. Glob. Optim.* 41, 103-115 (2008)

[217] Ortega, J.M., Rheinboldt, W.C.: *Iterative Solution of Nonlinear Equations in Several Variables.* Academic Press, NY (1970)

[218] Panigrahi, M.: Convex fuzzy mapping with differentiability and its application in fuzzy optimization. *Eur. J. Oper. Res.* 185, 47-62 (2007)

[219] Pareto, V.: *Manuale di economia politica.* Translated into English by Schwier, A.S. (1971) Manual of Political Economy, The MacMillan Company, NY (1906)

[220] Penot, J.P.: Characterization of solution sets of quasiconvex programs. *J. Optim. Theory Appl.* 117 (3), 627-636 (2003)

[221] Powell, M.J.D., Yuan, Y.: A recursive quadratic programming algorithm that uses differentiable penalty functions. *Math. Program.* 7, 265-278 (1986)

[222] Pini, R.: Convexity along curves and invexity. *Optimization* 29, 301-309 (1994)

[223] Preda, V.: On duality with generalized convexity. *Bull. U.M.I.* 291-305 (1991)

[224] Preda, V.: On sufficiency and duality for multiobjective programs. *J. Math. Anal. Appl.* 166, 365-377 (1992)

[225] Preda, V.: Optimality conditions and duality in multiple objective programming involving semilocally convex and related functions. *Optimization* 36, 219-230 (1996)

[226] Preda, V., Stancu-Minasian, I.M.: Duality in multiple objective programming involving semilocally preinvex and related functions. *Glas. Mat.* 32, 153-165 (1997)

[227] Preda, V.: Optimality and duality in fractional multiple objective programming involving semilocally preinvex and related functions. *J. Math. Anal. Appl.* 288, 365-382 (2003)

[228] Prekopa, A.: *Linear Programming I.* Bolyai Janos Matematikai Tarsulat, Budapest (1968) (in Hungarian)

[229] Pripoae, C.L., Pripoae, G.T.: Invexity vs. generalized convexity. BSG Proceedings 18, The International Conference of Differential Geometry and Dynamical Systems (DGDS-2010), Oct. 8-11, Bucharest, Romania, 86-93 (2010)

[230] Pripoae, C.L., Pripoae, G.T., Preda, V.: η-Pseudolinearity on differentiable manifolds (2013)

[231] Rapcsak, T.: On second order optimality conditions. *Alkalmazot Matematikai Lapok* 4, 109-116 (1978)

[232] Rapcsak, T.: On the second order sufficiency conditions. *J. Inform. Optim. Sci.* 4, 183-191 (1983)

[233] Rapcsak, T.: On pseudolinear functions. *Eur. J. Oper. Res.* 50, 353-360 (1991)

[234] Rapcsak, T.: *Smooth Nonlinear Optimization in Nonconvex Optimization and Its Applications.* Kluwer Academic Publishers, Dordrecht, Boston, London (1997)

[235] Rapcsak T.: On the pseudolinearity of quadratic fractional functions. *Optim. Lett.* 1, 193-200 (2007)

[236] Rapcsak, T., Ujvari, M.: Some results on pseudolinear quadratic fractional functions. *CEJOR* 16, 415-424 (2008)

[237] Roberts, A.W., Varberg, D.E.: Another proof that convex functions are locally Lipschitz. *Am. Math. Mon.* 81, 1014-1016 (1974)

[238] Rockafellar, R.T.: *Convex Analysis*. Princeton University Press, Princeton, NJ (1970)

[239] Rockafellar, R.T., Wets, R.J.B.: *Variational Analysis*. Springer-Verlag, Berlin, Heidelberg (2009)

[240] Rueda, N.G.: Generalized convexity in nonlinear programming. *J. Inform. Optim. Sci.* 10, 395-400 (1989)

[241] Ruckmann, J.J., Shapiro, A.: First-order optimality conditions in generalized semi-infinite programming. *J. Optim. Theory Appl.* 101, 677-691 (1999)

[242] Sach, P.H., Penot, J.P.: Characterizations of generalized convexities via generalized directional derivative. *Numer. Funct. Anal. Optim.* 19, 615-634 (1998)

[243] Samuelson, P.A.: An extension of the Le Chatelier principle, *Econometrica* 28, 368-379 (1960)

[244] Samuelson, P.A.: *Foundations of Economic Analysis*. 8th Printing. Cambridge, Harvard University Press (1966)

[245] Schirotzek, W.: *Nonsmooth Analysis*. Springer-Verlag, Berlin, Heidelberg (2007)

[246] Schaible, S., Ibaraki, T.: Fractional programming. *Eur. J. Oper. Res.* 12, 325-338 (1983)

[247] Schaible, S.: Fractional programming, In: *Handbook of Global Optimization, Nonconvex Optimization and Its Applications,* Kluwer, Dordrecht, Netherlands, pp. 495-608 (1995)

[248] Schonfeld, P.: Some duality theorems for the nonlinear vector maximum problem. *Unternehmensforschung* 14, 51-63 (1970)

[249] Schroeder, R.G.: Linear programming solutions to ratio games. *Oper. Res.* 18, 300-305 (1970)

[250] Schmitendorf, W.E.: Necessary conditions and sufficient conditions for static minimax problems. *J. Math. Anal. Appl.* 57, 683-693 (1977)

[251] Schrijver, A.: *Theory of Linear Integer Programming*. John Wiley & Sons, NY 1986

[252] Shapiro, A.: On duality theory of convex semi-infinite programming. *Optimization* 54, 535-543 (2005)

[253] Shapiro, A.: Semi-infinite programming, duality, discretization and optimality condition. *Optimization* 58, 133-161 (2009)

[254] Shimizu, K., Ishizuka, Y., Bard, J.F.: *Nondifferentiable and Two-Level Mathematical Programming.* Kluwer Academic, Boston (1997)

[255] Singh, C.: Optimality conditions in multiobjective differentiable programming. *J. Optim. Theory Appl.* 53 (1), 411-415 (1987)

[256] Slater, M.L.: A note on Motzkin's transposition theorem. *Econometrica* 19, 185-186 (1951)

[257] Smietanski, M.J.: A note on characterization of solution sets to pseudolinear programming problems. *Applicable Anal.* 91, 2095-2104 (2012)

[258] Stampacchia, G.: Formes bilineaires coercitives sur les ensembles convexes. *C.R. Acad. Sci.* Paris, 9, 4413-4416 (1960)

[259] Stancu-Minasian, I.M.: Optimality conditions and duality in fractional programming involving semilocally preinvex and related functions. *J. Inf. Optim. Sci.* 23, 185-201 (2002)

[260] Stancu-Minasian, I.M., Dogaru, G., Stancu, A.M.: Duality for multiobjective fractional programming problems involving d-type-I n-set-functions. *Yugoslav J. Oper. Res.* 19, 63-73 (2009)

[261] Still, G.: Generalized semi-infinite programming: theory and methods. *Eur. J. Oper. Res.* 119, 301-313 (1999)

[262] Studniarski, M., Taha, A.W.A.: Characterizations of strict local minimizers of order one for non-smooth static minimax problems. *J. Math. Anal. Appl.* 259, 368-376 (2001)

[263] Suneja, S.K., Gupta, S.: Duality in multiobjective nonlinear programming involving semilocally convex and related functions. *Eur. J. Oper. Res.* 107, 675-685 (1998)

[264] Syau, Y.R.: On convex and concave fuzzy mappings. *Fuzzy Sets Syst.* 103, 163-168 (1999)

[265] Syau, Y.R.: Invex and generalized convex fuzzy mappings. *Fuzzy Sets Syst.* 115, 455-461 (2000)

[266] Syau, Y.R.: Generalization of preinvex and B-vex fuzzy mappings. *Fuzzy Sets Syst.* 120, 533-542 (2001)

[267] Szilagyi, P.: A class of differentiable generalized convex functions. In: *Generalized Convexity*, S. Komlosi, T. Rapcsak, S. Schaible (eds.) Springer-Verlag, Berlin, pp. 104-115 (1994)

[268] Tanimoto, S.: Duality for a class of nondifferentiable mathematical programming problems. *J. Math. Anal. Appl.* 79, 286-294 (1981)

[269] Tanino, T., Sawaragi, Y.: Duality theory in multiobjective programming. *J. Optim. Theory Appl.* 27, 509-529 (1979)

[270] Thompson, A.W., Parke, D.W.: Some properties of generalized concave functions. *Oper. Res.* 21, 305-313 (1973)

[271] Tigan, S.: Sur le probleme de la programmation vectoriclle fractionaire. *Mathematica-Revue d' Analyse Numerique et de Theorie de l' Approximation* 4, 99-103 (1975)

[272] Tsai, J.F.: Global optimization of nonlinear fractional programming in engineering design. *Eng. Optim.* 37, 399-409 (2005)

[273] Tuy, H.: *Convex Analysis and Global Optimization*. Kluwer Academic Publishers, Dordrecht, Netherlands (1998)

[274] Udriste, C.: *Convex Functions and Optimization Methods on Riemannian Manifolds*. Kluwer Academic Publisher (1994)

[275] Udriste, C., Ferrara, M., Opris, D.: *Economic Geometric Dynamics*. Geometry. Balkan Press, Bucharest, Romania, 2004.

[276] Ujvari, M.: Simplex-type algorithm for optimizing a pseudolinear quadratic fractional function over a polytope. *P.U.M.A.* 18, 189-202 (2007)

[277] Von Neumann, J., Morgenstern, O.: *Theory of Games and Economic Behavior*. Princeton University Press, Princeton, NJ (1944)

[278] Verdegay J.L.: Applications of fuzzy optimization in operational research. *Control Cybern.* 13, 229-239 (1984)

[279] Wang, G.X., Wu, C.X.: Directional derivatives and subdifferential of convex fuzzy mappings and applications in convex fuzzy programming. *Fuzzy Sets Syst.* 138, 559-591 (2003)

[280] Ward, D.E., Lee, G.M.: On relations between vector optimization problems and vector variational inequalities. *J. Optim. Theory Appl.* 113, 583-596 (2002)

[281] Weir, T.: A duality theorem for a multiobjective fractional optimization problem. *Bull. Austral. Math. Soc.* 34, 415-425 (1986)

[282] Weir, T.: Pseudoconvex minmax programming. *Util. Math.* 42, 234-240 (1992)

[283] Weir, T., Mond, B.: Preinvex functions in multiple objective optimization. *J. Math. Anal. Appl.* 137, 29-38 (1988)

[284] Weir, T., Mond, B.: Generalized convexity and duality in multiple objective programming. *Bull. Austral. Math. Soc.* 39, 287-299 (1989)

[285] Wolfe, P.: A duality theorem for nonlinear programming. *Quart. Appl. Math.* 19, 239-244 (1961)

[286] Wu, Z., Xu, J.: Nonconvex fuzzy mappings and the fuzzy pre-variational inequality. *Fuzzy Sets. Syst.* 159, 2090-2103 (2008)

[287] Yang, X.Q.: Vector variational inequality and vector pseudolinear optimization. *J. Optim. Theory Appl.* 95, 729-734 (1997)

[288] Yang, X.Q., Goh, C.J.: On vector variational inequality: application to vector equilibria. *J. Optim. Theory Appl.* 95 431-443 (1997)

[289] Yang, X.M., Hou, S.H.: On minimax fractional optimality and duality with generalized convexity. *J. Glob. Optim.* 31, 235-252 (2005)

[290] Yang, X.M., Yang, X.Q., Teo, K.L.: Criteria for generalized invex monotonicities. *Eur. J. Oper. Res.* 164, 115-119 (2005)

[291] Yang, X.M., Yang, X.Q., Teo, K.L.: Generalized invexity and generalized invariant monotonicity. *J. Optim. Theory Appl.* 117, 607-625 (2003)

[292] Yang, X.M., Yang, X.Q., Teo, K.L.: Some remarks on the Minty vector variational inequality. *J. Optim. Theory Appl.* 121 (1), 193-201 (2004)

[293] Yuan, D.H., Liu, X.L., Chinchuluun, A., Pardalos, P.M.: Nondifferentiable minimax fractional programming problems with (C, α, ρ, d)-convexity. *J. Optim. Theory Appl.* 120, 185-199 (2006)

[294] Yuan, D.H., Liu, X.L.: Generalized minimax programming with nondifferentiable ρ-invexity. *J. Appl. Math.* DOI: 10.1155/2013/854125 (2013)

[295] Zamora, J.M., Grossmann, I.E.: MINLP model for heat exchanger networks. *Comput. Chem. Eng.* 22, 367-384 (1998)

[296] Zhang, Q.X.: Optimality conditions and duality for arcwise semi-infinite programming with parametric inequality constraints. *J. Math. Anal. Appl.* 196, 998-1007 (1995)

[297] Zhang, Q.X.: Optimality conditions and duality for semi-infinite programming involving B-arcwise connected functions. *J. Glob. Optim.* 45, 615-629 (2009)

[298] Zhao, K.Q., Tang, L.P.: On characterizing solution set of non-differentiable η-pseudolinear extremum problem. *Optimization* 61(3), 239-249 (2012)

[299] Zhao, K.Q., Yang, X.M.: Efficiency for a class of multiobjective optimization problems. *Oper. Res. Trans.* 15, 1-9 (2011)

[300] Zheng, X.J., Cheng, L.: Minimax fractional programming under nonsmooth generalized (F, ρ, θ)-d-univexity. *J. Math. Anal. Appl.* 328, 676-689 (2007)

Index